T0181971

Lecture Notes in Computer Science   10360

Commenced Publication in 1973
Founding and Former Series Editors:
Gerhard Goos, Juris Hartmanis, and Jan van Leeuwen

More information about this series at http://www.springer.com/series/7409

Jordi Cabot · Roberto De Virgilio
Riccardo Torlone (Eds.)

Web Engineering

17th International Conference, ICWE 2017
Rome, Italy, June 5–8, 2017
Proceedings

 Springer

Editors
Jordi Cabot
ICREA
Barcelona
Spain

Riccardo Torlone
Roma Tre University
Rome
Italy

Roberto De Virgilio
Roma Tre University
Rome
Italy

ISSN 0302-9743   ISSN 1611-3349   (electronic)
Lecture Notes in Computer Science
ISBN 978-3-319-60130-4   ISBN 978-3-319-60131-1   (eBook)
DOI 10.1007/978-3-319-60131-1

Library of Congress Control Number: 2017943001

LNCS Sublibrary: SL3 – Information Systems and Applications, incl. Internet/Web, and HCI

Printed on acid-free paper

This Springer imprint is published by Springer Nature
The registered company is Springer International Publishing AG
The registered company address is: Gewerbestrasse 11, 6330 Cham, Switzerland

Preface

The International Conference on Web Engineering (ICWE) is the prime yearly international conference on different aspects of designing, building, maintaining, and using Web applications. ICWE is supported by the International Society for Web Engineering (ISWE), and brings together researchers and practitioners from various disciplines in academia and industry to tackle the emerging challenges in the engineering of Web applications and in the problems of its associated technologies, as well as the impact of those technologies on society, media, and culture.

This volume collects the full research papers (technical and vision), short research papers, application papers, demonstrations, and PhD symposium papers presented at the 17th International Conference on Web Engineering (ICWE 2017), held in Rome, Italy, during June 5–8, 2017. Previous editions of ICWE took place in Lugano, Switzerland (2016); Rotterdam, The Netherlands (2015); Toulouse, France (2014); Aalborg, Denmark (2013); Berlin, Germany (2012); Paphos, Cyprus (2011); Vienna, Austria (2010); San Sebastián, Spain (2009); Yorktown Heights, NY, USA (2008); Como, Italy (2007); Palo Alto, CA, USA (2006); Sydney, Australia (2005); Munich, Germany (2004); Oviedo, Spain (2003); Santa Fe, Argentina (2002); and Cáceres, Spain (2001).

The 17th edition of ICWE accepted contributions related to different dimensions of Web applications: Web application modeling and engineering, human computation and crowdsourcing applications, Web applications composition and mashup, social Web applications, Semantic Web applications, Web of Things applications, and big data. The different publication types mentioned above, each one with its own separate scientific chairs, allowed us to offer the community the opportunity to submit their work at different stages of maturity and target multiple, staggered paper submission deadlines.

The main research track (covering technical and vision papers) continued with the tiered Program Committee (PC) organization introduced last year, with a senior PC composed of well-known experts from the field in charge of monitoring the work and animating the discussions of the broader regular PC. This made it easier to run the virtual PC meeting of the full research papers track and the discussion about each paper.

This year's call for papers attracted 139 submissions from around the world, out of which two PCs selected 20 full papers (18 technical and two vision), for a 28% total acceptance rate, and 12 short papers, for a 37% acceptance rate. Additionally, six application papers, six demonstrations, and six contributions to the PhD symposium were accepted by their respective chairs. ICWE 2017 also accepted three tutorials on cutting-edge topics in the field of Web engineering, entitled: "Model-Based Development of JavaScript Web Applications," "Liquid Web Applications," and "Big Web Data: Warehousing and Analytics – Recent Trends and Future Challenges." Five workshops were co-located with ICWE 2017 as well.

The excellent program would not have been possible without the support of all the people who contributed to the organization of this event. We would like to thank all the different chairs for their hard work: In-young Ko, Piero Fraternali, Irene Garrigos,

Manuel Wimmer, Erik Wilde, Santiago Melia, Oscar Diaz, Boualem Benatallah, Silvia Abrahao, and Luis Olsina. Our thanks also go to Carlo Batini, Barry Smith, and Francesco Bonchi, who accepted to be our keynote speakers.

Special thanks are extended to Marco Brambilla and Flavius Frasincar for their advice and encouragement in their role of SC liaisons for the conference. We are also grateful to our local organizers Consulta Umbria for their logistical support, to Florian Daniel for leading the compilation of this volume, and to Springer for publishing it. In addition, we thank all PC members and additional reviewers for their meticulous work in selecting the best papers to be presented at ICWE 2017. Last, but not least, we would like to thank the authors who submitted their work to this conference and all the participants who contributed to the success of this event.

April 2017   Jordi Cabot
                                                              Roberto De Virgilio
                                                               Riccardo Torlone

Organization

Program Committee

Senior Program Committee

Alessandro Bozzon	Delft University of Technology, The Netherlands
Gianluca Demartini	University of Sheffield, UK
Schahram Dustdar	TU Wien, Austria
Geert-Jan Houben	TU Delft, The Netherlands
Oscar Pastor Lopez	Universitat Politecnica de Valencia, Spain
Gustavo Rossi	UNLP, Argentina
Daniel Schwabe	PUC-Rio, Brazil
Steffen Staab	Institut WeST, University Koblenz-Landau, Germany and WAIS, University of Southampton, UK
Stefan Tai	TU Berlin, Germany

Program Committee

Benjamin Adams	University of Canterbury, UK
Sören Auer	University of Bonn and Fraunhofer IAIS, Germany
Wolf-Tilo Balke	Institut für Informationssysteme, TU Braunschweig, Germany
Olivier Barais	IRISA/Inria/University of Rennes 1, France
Devis Bianchini	University of Brescia, Italy
Maria Bielikova	Slovak University of Technology in Bratislava, Slovakia
Javier Luis Canovas Izquierdo	IN3 - UOC
Cinzia Cappiello	Politecnico di Milano, Italy
Rubby Casallas	University of los Andes, Colombia
Fabio Casati	University of Trento, Italy
Sven Casteleyn	Universitat Jaume I, Spain
Michele Catasta	EPFL, Switzerland
Soon Ae Chun	CUNY
Mirel Cosulschi	University of Craiova, Romania
Valter Crescenzi	Università degli Studi Roma Tre, Italy
Alexandra Cristea	University of Warwick, UK
Alfredo Cuzzocrea	ICAR-CNR and University of Calabria, Italy
Olga De Troyer	Vrije Universiteit Brussel, Belgium
Emanuele Della Valle	DEIB, Politecnico di Milano, Italy
Angelo Di Iorio	University of Bologna, Italy
Vania Dimitrova	University of Leeds, UK
Peter Dolog	Aalborg University, Denmark

Francesco M. Donini	Università della Tuscia, Italy
Laurence Duchien	University of Lille, France
Enrico Francesconi	ITTIG-CNR
James Geller	New Jersey Institute of Technology, USA
Cristina Gómez	UPC
Sebastian Hellmann	AKSW/KILT, Universität Leipzig, Germany
Gerti Kappel	Vienna University of Technology, Austria
Chin-Laung Lei	National Taiwan University, Taiwan
Lei Liu	Huawei
Zakaria Maamar	Zayed University, UAE
Cristian Mateos	CONICET-ISISTAN, UNICEN
Maristella Matera	Politecnico di Milano, Italy
Paolo Merialdo	Università degli Studi Roma Tre, Italy
Philippe Merle	Inria, France
Amedeo Napoli	LORIA Nancy, France
Wolfgang Nejdl	L3S and University of Hannover, Germany
Moira Norrie	ETH Zurich, Switzerland
Gennady Pekhimenko	Carnegie Mellon University, USA
Vicente Pelechano	Universidad Politecnica de Valencia, Spain
Alfonso Pierantonio	University of L'Aquila, Italy
Sherif Sakr	The University of New South Wales, Australia
Jocelyn Simmonds	Universidad de Chile, Chile
Olga Streibel	National Institute of Informatics
Eleni Stroulia	University of Alberta, Canada
Fernando Sánchez	Universidad de Extremadura, Spain
Massimo Tisi	AtlanMod team (Inria, Mines Nantes, LINA), France
Manolis Tzagarakis	Patras University, Greece
Stratis Viglas	University of Edinburgh, UK
Marco Winckler	Paul Sabatier University, France
Erik Wittern	IBM Research
Makoto Yamada	Yahoo! Labs
Nicola Zannone	Eindhoven University of Technology, The Netherlands

Additional Reviewers

Marwan Al-Tawil	Linda Di Geronimo	Alexandra Mazak
Ciro Baron Neto	Tommaso Di Noia	Najmeh Mousavi Nejad
Seyed-Mehdi-Reza Beheshti	John Duprey	Alfonso Murolo
	Joan Fons	Gustavo Publio
Robert Bill	Johannes Frey	Diego Serrano
Gabriela Bosetti	Garces	Ivan Srba
Erwan Bousse	Kelly Gómez	William Van Woensel
Jonathan Carlton	Abel Evgeny Krivosheev	Sabine Wolny
Sandro Coelho	Dorian Leroy	Jakub Ševcech
Diego Collarana	Xiaomo Liu	
Kevin Corre	Ioanna Lytra	

Short Paper Committee

Alessandro Bozzon Adams	Delft University of Technology, The Netherlands
Florian Daniel	Politecnico di Milano, Italy
Davide Di Ruscio	Università degli Studi dell'Aquila, Italy
Angel Jimenez-Molina	University of Chile, Chile
Davide Martinenghi	Politecnico di Milano, Italy
Fernando Sánchez	Universidad de Extremadura, Spain
Massimo Tisi	AtlanMod team (Inria, Mines Nantes, LINA), France
Mario Matias Urbieta	LIFIA, Universidad Nacional de La Plata, Argentina
Manuel Wimmer	Vienna University of Technology, Austria

Application Track Committee

Benjamin Adams	The University of Auckland, New Zealand
Claudio Bartolini	Cloud4Wi
Joseph Busch	Taxonomy Strategies
Elena Cabrio	Université Cote d'Azur, CNRS, Inria, I3S, France
Mauro Dragoni	Fondazione Bruno Kessler - FBK-IRST, Italy
Danh Le Phuoc	Technische Universität Berlin, Germany
Maria Maleshkova	AIFB, Karlsruhe Institute of Technology, Germany
Jindrich Mynarz	University of Economics Prague, Czech Republic
Phuong Nguyen	Duy Tan University, Vietnam
Sergio Oramas	Universidad Politécnica Madrid, Spain
Vito Claudio Ostuni	Pandora Media Inc.
Carlos R. Rivero	Rochester Institute of Technology, USA
Domenico Rotondi	Fincons Group S.p.A
Dezhao Song	Thomson Reuters
Dhavalkumar Thakker	University of Bradford, UK
Jürgen Umbrich	Vienna University of Economy and Business (WU), Austria
Ruben Verborgh	Ghent University/imec

Demonstration Track Committee

Marco Brambilla	Politecnico di Milano, Italy
Hugo Bruneliere	AtlanMod Team (IMT Atlantique, Inria, LS2N), France
Sven Casteleyn	Universitat Jaume I, Spain
Florian Daniel	Politecnico di Milano, Italy
Ivano Malavolta	Vrije Universiteit Amsterdam, The Netherlands
Simon Mayer	ETH Zurich, Switzerland
Carmen Santoro	ISTI-CNR, Italy
Jean-Sébastien Sottet	Luxembourg Institute for Sciences and Technology
Juan Carlos Preciado	Universidad de Extremadura, Spain
Manuel Wimmer	Vienna University of Technology, Austria
Marco Winckler	Paul Sabatier University, France
Erik Wittern	IBM Research

PhD Symposium Committee

Carlos Canal	University of Malaga, Spain
Florian Daniel	Politecnico di Milano, Italy
Flavius Frasincar	Erasmus University Rotterdam, The Netherlands
Martin Gaedke	Chemnitz University of Technology, Germany
In-Young Ko	Korea Advanced Institute of Science and Technology, South Korea
Cesare Pautasso	University of Lugano, Switzerland
Roberto Rodriguez-Echeverria	Universidad de Extremadura, Spain
Gustavo Rossi	UNLP, Argentina
Marco Winckler	Paul Sabatier University, France

Contents

Vision Papers

Short Papers

Application Papers

PhD Symposium Papers

Technical Research Papers

Evaluating Knowledge Anchors in Data Graphs Against Basic Level Objects

Marwan Al-Tawil[1](✉), Vania Dimitrova[1], Dhavalkumar Thakker[2],
and Alexandra Poulovassilis[3]

[1] School of Computing, University of Leeds, Leeds, UK
scmata@leeds.ac.uk
[2] School of Electrical Engineering and Computer Science,
University of Bradford, Bradford, UK
[3] Knowledge Lab, Birkbeck, University of London, London, UK

Abstract. The growing number of available data graphs in the form of RDF Linked Data enables the development of semantic exploration applications in many domains. Often, the users are not domain experts and are therefore unaware of the complex knowledge structures represented in the data graphs they interact with. This hinders users' experience and effectiveness. Our research concerns intelligent support to facilitate the exploration of data graphs by users who are not domain experts. We propose a new navigation support approach underpinned by the subsumption theory of meaningful learning, which postulates that new concepts are grasped by starting from familiar concepts which serve as knowledge anchors from where links to new knowledge are made. Our earlier work has developed several metrics and the corresponding algorithms for identifying knowledge anchors in data graphs. In this paper, we assess the performance of these algorithms by considering the user perspective and application context. The paper address the challenge of aligning basic level objects that represent familiar concepts in human cognitive structures with automatically derived knowledge anchors in data graphs. We present a systematic approach that adapts experimental methods from Cognitive Science to derive basic level objects underpinned by a data graph. This is used to evaluate knowledge anchors in data graphs in two application domains - semantic browsing (Music) and semantic search (Careers). The evaluation validates the algorithms, which enables their adoption over different domains and application contexts.

Keywords: Data graphs · Basic level objects · Knowledge anchors · Usable semantic data exploration

1 Introduction

With the recent growth of linked data graphs, a plethora of interlinked domain entities is available for users' exploratory search tasks, such as learning and topic investigation [1]. Gradually, data graphs are also being exposed to users in different Semantic Web applications, taking advantage of the exploration of the rich knowledge encoded in the graphs. Among the applications for supporting user exploration, the two closest to the

© Springer International Publishing AG 2017
J. Cabot et al. (Eds.): ICWE 2017, LNCS 10360, pp. 3–22, 2017.
DOI: 10.1007/978-3-319-60131-1_1

context of this paper are semantic data browsers [2–4] and semantic search systems [5, 6]. A broad range of users interact with such applications. Often, the users are not domain experts and struggle to formulate queries that represent their needs. Furthermore, the users are usually exposed to an overwhelming amount of unfamiliar options for exploration of the data graph, which can lead to confusion, high cognitive load, frustration and a feeling of being lost. This hinders the users' exploration experience and effectiveness. A way to overcome these challenges is to suggest 'good' trajectories through the graph which can bring some utility to the users (e.g. increase effectiveness, improve motivation, or expand knowledge). Our work focuses on *knowledge utility* – expanding one's domain knowledge while exploring the graph.

Lay users, who are not experts in the corresponding domain, are unaware of the underlying complex knowledge structures encoded in a data graph [1, 7]. In other words, the *users' cognitive structures* about the domain may not match the *semantic structure of the data graph*. To address this challenge, we propose a novel approach to support graph exploration that can expand a users' domain knowledge. Our approach is underpinned by the subsumption theory for meaningful learning [8]. It postulates that a human cognitive structure is hierarchically organized in terms of highly inclusive concepts which can be used as anchors to introduce new knowledge [8]. A core algorithmic component for adopting subsumption theory for generating 'good' trajectories is the automatic identification of knowledge anchors in a data graph (KA_{DG}), i.e. entities that refer to anchoring concepts in human cognitive structures.

Our earlier research has developed several metrics and corresponding algorithms for identifying KA_{DG}, which are presented in detail in [9]. To utilize the KA_{DG} metrics in applications for data graph exploration, a systematic evaluation approach that examines the performance of the metrics is needed. Such an approach is presented in this paper. As the KA_{DG} should align with anchoring concepts in human cognitive structures, we develop an original way to derive such familiar concepts in a domain that corresponds to a data graph and considers the domain coverage of the graph. We adapt Cognitive Science experimental approaches of free-naming tasks to identify basic level objects (BLO) in human cognitive structures, i.e. domain concepts that are highly familiar and inclusive, so that people are able to recognize them quickly [10].

The evaluation approach presented in this paper contributes to developing usable semantic data graph exploration applications by providing:

- formal description of an algorithm for identifying basic level objects which correspond to human cognitive structures over a data graph;
- implementation of the BLO algorithm and utilization to evaluate KA_{DG} metrics over two application contexts for data graph exploration - semantic browsing (in musical instrument domain) and semantic search (in Career domain); and
- analysis of the performance of KA_{DG} metrics, including hybridization heuristics, using the benchmarking sets of BLO identified by humans.

The rest of the paper is structured as follows. Section 2 positions the work in the relevant literature and points at the main contribution. Section 3 briefly outlines the KA_{DG} metrics, summarizing [9]. An algorithm for identifying a benchmarking set of BLO is presented in Sect. 4. Sections 5 and 6 describe experimental studies where we apply the algorithm for identifying BLO using data graphs of two semantic exploration applications – music browser (MusicPinta) and career guidance (L4All). The BLO are used to evaluate the derived KA_{DG}. Section 7 discusses the evaluation findings, points at generality and applicability of the algorithms, and concludes the paper.

2 Related Work

Recent research on data exploration over the semantic Web examines different approaches to reduce users' cognitive load, especially when the users are exposed to complex domains which they are not familiar with. This has brought together research from Semantic Web, personalization, and HCI to shape user-oriented application for data exploration [1, 3, 6]. Personalized exploration based on user interests has been presented in [11]. A web-based graph visualization approach was used in [12] to help domain experts with analysis tasks. A co-clustering approach that organizes semantic links and entity classes was presented in [13] to support iterative navigation of entities over RDF data. The notion of relevance based on the relative cardinality and the in/out degree centrality of a graph node has been used to produce graph summaries [14]. Our work brings a new dimension to this research effort by looking at the *knowledge utility of the exploration*, i.e. providing ways to expand the user's awareness of the domain. This is crucial for the usability of semantic exploration applications, especially when the users are not domain experts.

Our approach is based on identifying knowledge anchors in data graphs. Relevant work on finding key concepts in a data graph was developed by research on ontology summarization [15] and formal concept analysis [16]. Ontology summarization aims at helping ontology engineers to make sense of an ontology in order to reuse and build new ontologies [17]. The closest ontology summarization approach to this paper's context is [18], which highlighted the value of cognitive natural categories for identifying key concepts. The work in [19] has formalized the main psychological approaches for identifying basic level concepts in formal concept analysis. In [9] we have operationalized these approaches, allowing automatic identification of KA_{DG}.

According to [17], there are two main approaches for evaluating a user-driven ontology summary: gold standard evaluation, where the quality of the summary is expressed by its similarity to a manually built ontology by domain experts, or corpus coverage evaluation, in which the quality of the ontology is represented by its appropriateness to cover the topic of a corpus. The evaluation approach used in [18] included identifying a gold standard by asking ontology engineers to select a number of concepts they considered the most representative for summarizing an ontology. To the best of our knowledge, there are no approaches that consider key concepts in data graphs which correspond to cognitive structures of lay users who are not domain experts. We identify such concepts in data graphs including both an automatic method to derive KA_{DG} and an experimental method to derive BLO that correspond to human cognitive structures. We evaluate KA_{DG} against benchmarking sets of BLO over the data graphs of two semantic exploration applications – browsing (Music) and search (Careers). By providing a systematic evaluation approach, the paper facilitates the adoption of the KA_{DG} metrics, and the corresponding hybridization methods, to enhance the usability of semantic web applications that offer user exploration of data graphs.

3 Identifying Knowledge Anchors in Data Graphs

A Data Graph DG describes entities (vertices) and attributes (edges), represented as *Resource Description Framework (RDF)* statements. Each statement is a triple of the form *<Subject, Predicate, Object>* [20]. Formally, a data graph is as a labeled directed graph $DG = \langle V, E, T \rangle$, depicting a set of RDF triples where:

- $V = \{v_1, v_2, \ldots, v_n\}$ is a finite set of entities;
- $E = \{e_1, e_2, \ldots, e_m\}$ is a finite set of edge labels;
- $T = \{t_1, t_2, \ldots, t_k\}$ is a finite set of triples where each t_i is a proposition in the form of a triple $\langle v_s, e_i, v_o \rangle$ with $v_s, v_o \in V$, where v_s is the *Subject* (source entity) and v_o is the *Object* (target entity); and $e_i \in E$ is the *Predicate* (relationship type).

The set of entities V is divided further by using the subsumption relationship `rdfs:subClassOf` (denoted as $\subseteq$) and following its transitivity inference. This includes **category entities** ($C \subseteq V$ which is the set of all entities that have at least one subclass, at least one superclass, and at least one instance) and **leaf entities** ($L \subseteq V$ which is the set of entities that have no subclasses).

The set of edge types E is divided further considering two relationship categories: **hierarchical relationships** (H: is a set of subsumption relationships between the *Subject* and *Object* entities in the corresponding triples) and **domain-specific relationships** (D: represent relevant links in the domain, other than hierarchical links, e.g. in a Music domain, instruments used in the same *performance* are related).

Our work in [9] has formally adopted the Cognitive science notion of basic level objects [10], to describe two groups of metrics and their corresponding algorithms for identifying knowledge anchors in data graphs (KA_{DG}).

Distinctiveness metrics. These are adapted from the formal definition of cue validity, to identify the most differentiated categories whose attributes are associated exclusively with the category members but are not associated to the members of other categories. For example, in Fig. 1, the AV value for entity v_2 is the aggregation of the AV values of entities (e_3, e_4, e_5) linked to members of v_2 (v_{21}, v_{22}, v_{23}, v_{24}) using the domain-specific relationship D. The AV value for e_3 equals the number triples between e_3 (*Source* vertex) and the members of v_2 (*Target* vertices v_{21}, v_{22}) via relationship D (2 triples), divided by the number of triples between e_3 (*Source* vertex) and all entities in the graph (*Target* vertices v_{12}, v_{21}, v_{22}) via relationship D (3 triples).

Distinctiveness metrics include:

- **Attribute Validity (AV)** – represents the proportion of relationships involving the category's members.

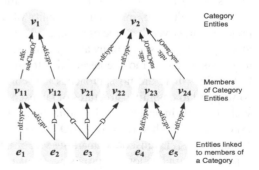

Fig. 1. A data graph showing entities and relationship types between entities.

- **Category Attribute Collocation (CAC)** – uses frequency of an attribute within the category's members; gives preference to categories with many attributes shared by members.
- **Category Utility (CU)** - considers whether a category has many attributes shared by its members, and at the same time has attributes not related to many other categories.

Homogeneity Metrics. These metrics aim to identify categories whose members share many entities among each other. In this work, we have utilized three set-based similarity metrics [9]: **Common Neighbors (CN), Jaccard (Jac), and Cosine (Cos)**. For example (see Fig. 1), consider the entity v_2 and the hierarchical relationship rdf:-type and the domain-specific relationship D. Entity v_2 has three entities (e_3, e_4, e_5) linked to its members (v_{21}, v_{22}, v_{23}, v_{24}), with *two* entities (e_3, e_5) shared among the four members through the hierarchical relationship rdf:type and relationship D, whereas the entity v_1 has no entities shared by similar relationship types with its members (v_{11}, v_{12}). This indicates that entity v_2 is more homogenous than v_1.

4 Identifying Basic Level Objects Over Data Graphs

The notion of basic level objects was introduced in Cognitive Science research, illustrating that domains of concrete objects include familiar categories that exist at a highly inclusive level of abstraction in humans' cognitive structures, more than categories at the superordinate level (i.e. above the basic level) or the subordinate level (i.e. below the basic level) [10, 21]. An example from [10] of a BLO is Guitar - most people are likely to recognize objects that belong to the category Guitar (*basic level*). However, users who are not experts in the music domain are unlikely to be able to recognize the category Folk Guitar (*subordinate level*) and name it with its exact name; instead, users may consider such objects equivalent to Guitar (closest basic level) rather than Musical Instrument (*superordinate level*).

4.1 Cognitive Science Experimental Approaches for Deriving BLO

While studying the notion of basic level objects, Rosch et al. [10] conducted several experiments comprising free-naming tasks testing the hypothesis that object names at the basic level should be the names by which objects are most generally designated by adults. In a free-naming task, objects in a taxonomy are shown to a participant as a series of images in fixed portions of times, and the participant is asked to identify the names of the objects shown in the images as quickly as possible. Three types of packets of images were shown to the participants: those in which one picture from each superordinate category appeared; one in which one image from each basic level category appeared; and one in which all images appeared. The participants overwhelmingly used names at the basic level while naming objects in the images [10].

To identify BLO, accuracy and frequency were considered. Accuracy considers whether a participant provides an accurate name for the object in the taxonomy, while frequency indicates how many times an object was named correctly by different participants. In the example of Guitar, when participants were shown members of Guitar (e.g. Folk Guitar, Classical Guitar) in a packet, they named them with their parent Guitar at the basic level more frequently than with names at the superordinate level (e.g. Musical instrument) or with their exact names (e.g. Folk Guitar, Classical Guitar) at the subordinate level.

The selection of object names used in the free-naming tasks in [10] was based on the population of categories of concrete nouns in common use in English. Every noun with a word frequency of 10 or greater from a sample of written English [22] was selected as a basic level object. A superordinate category was considered in common use if at least four of its members met this criterion.

However, the Cognitive Science approach for selecting BLO cannot be applied directly in the context of a data graph. The principal difference is that we need to constrain the human cognitive structures upon the data graph, as opposed to using a bag of words from popular dictionaries. This is because a data graph presents a lesser number of concepts from a domain, which belong to the graph scope, and there can be concepts that have been omitted. Moreover, the Cognitive Science studies included concrete domains where images of the objects could be shown to participants. Many semantic web applications utilize data graphs which include more abstract concepts for which images cannot be reliably shown to users (e.g. medical illnesses, environmental concepts, professions). Therefore, we adapt the Cognitive science experimental approach for deriving BLO to take into account the domain coverage of a data graph, which is applicable to any domain presented with a data graph.

4.2 Algorithm for Identifying BLO Over Data Graphs

Following Cognitive Science experimental studies outlined above, we present two strategies with the corresponding algorithm for identifying BLO in a data graph.

Strategy 1. Takes into account whether a leaf entity $v \in L$ that has no subclasses is presented to a user and named with its parents (i.e. superclasses).
Strategy 2. Takes into account whether a category entity $v \in C$ that has one or more subclasses is presented and named with its exact name, or with the name of a parent that is a superclass or a category member (i.e. subclass that is not a leaf entity).

Algorithm 1 describes the two strategies for identifying BLO using accuracy and frequency. *Accuracy* refers to naming an entity correctly. It considers whether a user names an entity with its exact name, or with a parent (superclass) or with a category member (subclass) of the entity. *Frequency* indicates how many times a particular category was accurately identified by different participants.

The algorithm takes a data graph as input and returns two sets of BLO. For any class entity $v \subseteq V$, we identify the number of users to be asked to name the entity (line 2). For *Strategy 1* (lines 3–7), we consider accurate naming of a category entity (a parent) when a leaf entity $v \in L$ that is a member of this category is seen. For *Strategy 2* (lines 8–14), we consider naming a category entity $v \in C$ with its exact name (lines 10, 11) or a name of its superclasses (parents) or subclasses (members) (lines 12–13).

In each strategy, we use a representation function $show(r, v)$ to create a representation of an entity v to be shown to the user. The representation of a leaf entity $v \in L$ (in Strategy 1) will consider the leaf itself (e.g. show a single label or a single image for the leaf entity), while the representation of a category entity $v \in C$ (in Strategy 2) will consider all (or some) of the category leaves (e.g. showing a random listing of a set of labels of entity leaves or showing a group of images of leaves as a collage).

Algorithm 1: Identifying Basic Level Objects in Data Graphs

Input $DG = \langle V, E, P \rangle$
Output two sets of entities: **Set1** and **Set2**
1. **for a set of entities** $v \subseteq V$ **do**
2.   **for all** $(i := 1; i \leq n; i + +)$ //show the entity v to n users
3.   **if** $v \in L$ **then**   //Strategy1
4. $\quad$ $show(r, v)$ **and ask a user to name** v
5. $\quad$ **if** $answer(a, v) \in parent(p, v)$ **then**   //check accuracy
6. $\quad\quad$ $count_a + +$ //count frequency
7. $\quad$ **end if;**
8.   **else if** $v \in C$ **then**   //Strategy2
9. $\quad$ $show(r, v)$ **and ask a user to name** v
10. $\quad$ **if** $answer(a, v) = label(b, v)$ **then**   //check accuracy
11. $\quad\quad$ $count_a + +$ //count frequency
12. $\quad$ **else if** $answer(a, v) \in \{parent(p, v) \cup member(m, v)\}$ **then** //check accuracy
13. $\quad\quad$ $count_a + +$ //count frequency
14. $\quad$ **end if;**
15.   **end if;**
16.   **end for;**
17. **end for;**
18. **Set1** $= \{answer(a, v) : v \in L \wedge count_a \geq k\}$ //K is number of different users
19. **Set2** $= \{answer(a, v) : v \in C \wedge count_a \geq k\}$ //K is number of different users

---

For an entity v, the following SPARQL query is used to get the set of entity leaves:

```
SELECT  ?leaf ?leaf_label
WHERE   {?leaf rdfs:subClassOf v.
         ?leaf rdfs:label ?leaf_label.
FILTER NOT EXISTS
         {?member rdfs:subClassOf ?leaf.}}
```

The two strategies in Algorithm 1 for obtaining BLO are applied as follows:

Strategy 1, when a user is shown a representation of a leaf entity $v \in L$ (line 4), the following steps are conducted:

– The function $answer(a, v)$ assigns a user's *answer* a to the leaf entity v.

- The function *parent*(*p*, *v*) returns a set of labels (i.e. names) of the *parent*(s) *p* of the leaf entity *v* via the following SPARQL query:

```
SELECT  ?parent_label ?label
 WHERE  {v rdfs:subClassOf ?parent.
         ?parent rdfs:label ?parent_label.}
```

- The algorithm in (line 5) checks if the user named the leaf entity *v* with one of its parents. If an accurate name of a parent was provided, then the frequency of the parent entity will be increased by one (line 6).

Strategy 2, when a user is shown a representation of a category entity $v \in C$ (line 9), the following steps are conducted:

- The function *answer*(*a*, *v*) assigns a user's *answer a* to the category entity *v*.
- The function *parent*(*p*, *v*) returns a set of labels of *parent*(s) *p* of the category entity *v* via SPARQL queries similar to Strategy 1 above.
- The function *member*(*m*, *v*) returns a set of labels (i.e. names) of *member*(s)*m* of the category entity *v* via the following SPARQL query:

```
SELECT  ?member_label
 WHERE  {?member rdfs:subClassOf v.
         ?member rdfs:label ?member_label.}
```

- The function *label*(*b*, *v*) returns the *label* (i.e. name) of the category entity *v* via the following SPARQL query:

```
SELECT ?label
 WHERE  {v rdfs:label ?label.}
```

- The algorithm in (lines 10, 12) checks if the user named the category entity *v* with its exact name, or a name of its parents or its members. If there was accurate naming of the category, a parent or a member, the frequency of the category name (line 11), the parent name or the member name (line 13) will be increased by one.

4.3 Application Contexts Used for Experimental Evaluation

Linked Data graphs represented as a set of RDF triples can be ideal structures for Semantic exploration applications [23]. One class of applications is semantic data browsers which operate on semantically tagged content and present browsing trajectories using relationships in the underpinning ontologies [1, 2], supporting uncertain or complex information needs [3]. They enable the users to initiate a data exploration session from a single entry point in the graph and move through entities by following RDF links [2]. Another class of widely used semantic Web applications are semantic data search engines [24]. Such applications allow the users to enter search queries though keyword-based search interfaces and provide the users with a list of search results obtained by using semantic queries automatically generated by the system [6].

In this paper, we present experimental studies over two different application domains for evaluating KA_{DG} metrics against BLO. The first study is in the context of a **semantic data browser in the Music domain**, called **MusicPinta** [2]. MusicPinta enables users to navigate through musical instruments extracted from DBpedia, and get information about these instruments together with musical performances and artists using these instruments. MusicPinta provides context for studying BLO in a concrete domain, as users can see images of musical instruments (as in [10, 25]). The second study is in the context of a **semantic search engine in Career guidance**, called **L4All** [26]. L4All is a proprietary semantic search application which enables learners to explore various career options to plan their career progression [26]. L4All provides context for studying basic level objects in an abstract domain, where the users cannot be shown concrete representations of the graph entities.

The data graphs of the two applications are used for the evaluation studies.

MusicPinta. The dataset includes several open sources. DBpedia[1] for musical instruments and artists - this dataset is extracted from dbpedia.org/sparql using CONSTRUCT and made available as open source at the sourceforge[2]. DBTune[3] for music-related structured data - this dataset is made available by the DBTune.org in linked data fashion. Among the datasets on DBTune.org we utilize: (i) Jamendo - a large repository of Creative Commons licensed music; (ii) Megatune - an independent music label; and (iii) MusicBrainz - a community-maintained open source encyclopaedia of music information. All datasets are available as RDF datasets and the Music ontology[4] is used as a schema to interlink them. For the experimental study, we use the top level class Music Instrument and all its entities (classes and instances).

L4All. The dataset is drawn from the "LifeLong Learning in London for All" (L4All) project [26], bringing together experts from lifelong learning and careers guidance, content providers, and groups of students and tutors. It provided lifelong learners with access to information and resources that would support them in exploring learning and career opportunities and in planning and reflecting on their learning. The L4All dataset uses the ontology developed by the L4All project, and users' data collected during the project (anonymised for privacy). Among five class hierarchies in the L4All ontology, the *Occupation* and *Subject* class hierarchies have the richest class representation and depth (see Table 1).

---

[1] http://dbpedia.org/About.

[2] http://sourceforge.net/p/pinta/code/38/tree/.

[3] http://dbtune.org/.

[4] http://musicontology.com/.

Table 1. Main characteristics of the MusicPinta and L4All data sets

Dataset	Hierarchy root class	Depth	No. of classes	No. of instances/leaves
MusicPinta	Instrument	7	364	256
L4All	Occupation	4	464	3737
	Subject	2	160	2194

5   MusicPinta: Evaluating KA_{DG} Against BLO

As a use case in a representative domain for evaluating knowledge anchors over a data graph, we used a typical semantic data browser, MusicPinta, which was developed in our earlier research [2]. Knowledge anchors would lead to extending MusicPinta to suggest exploration paths that can improve the user's domain knowledge.

5.1   Obtaining BLO

To enable impartial comparison of the outputs of the KA_{DG} algorithms and BLO, we conducted a user study in the Musical Instrument domain following Algorithm 1.

Participants. 40 participants, university students and professionals, age 18–55, recruited on a voluntary basis. None of them had expertise in Music.

Method. The participants were asked to freely name objects that were shown in image stimuli, under limited response time (10 s). Overall, 364 taxonomical musical instruments were extracted from the MusicPinta dataset by running SPARQL queries over the MusicPinta triple store to get all musical instrument concepts linked via the `rdfs:subClassOf` relationship. The entities included: *leaf entities* (total 256) and *category entities* (total 108). Applying the two strategies in Algorithm 1, for each leaf entity, a representative image was collected from the Musical Instrument Museums Online (MIMO)[5] to ensure that pictures of high quality were shown[6]. For a category entity, all leaves from that category entity were shown as a group in a single image (similarly to a packet of images in [10]). Ten online surveys[7] were run: (i) leaf entities: eight surveys presented 256 leaf entities, each showed 32 leaves; (ii) category entities: two surveys presented 108 category entities, each showed 54 categories.

Free-naming task. Each image was shown for 10 s on the participant's screen. She was asked to type the name of the given object (for leaf entities) or the category of objects (for category entities). The image allocation in the surveys was random. Every survey had four respondents from the study participants (corresponds to line 2 in Algorithm 1). Each participant was allocated only to one survey (either leaf entities or

---

[5] http://www.mimo-international.com/MIMO/.

[6] MIMO provided pictures for most musical instruments. In the rare occasions when an image did not exist in MIMO, Wikipedia images were used instead.

[7] The study was conducted with Qualtrics (www.qualtrics.com). Examples from the surveys are available at: https://drive.google.com/drive/folders/0B5ShywKndSLXaVhrSWpiYVZ3WjA.

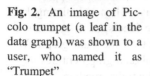

Fig. 2. An image of Piccolo trumpet (a leaf in the data graph) was shown to a user, who named it as "Trumpet"

Fig. 3. An image of Trumpet (a Category concept in the data graph with two subclasses) was shown to a user, who named it as "Trumpet".

Fig. 4. An image of Brass (Category concept in the data graph) shown to a user, who named it as "Trumpet".

category entities). Figures 2, 3 and 4 show example instrument images and participant answers (Fig. 2 from Strategy 1, and Figs. 3, 4 from Strategy 2).

Applying Algorithm 1 over the MusicPinta dataset, two sets of BLO were identified. *Set1* (Strategy 1) was derived from presenting leaf entities. We consider accurate naming of a category entity (parent) when a leaf entity that belongs to this category is seen. For example (see Fig. 2), a participant was shown the image of `Piccolo trumpet`, a leaf entity in the data graph, and named it with its parent category `Trumpet`. This will be counted as an accurate naming and will increase the count for `Trumpet`. The overall count for `Trumpet` will include all cases when participants named `Trumpet` while seeing any of its leaf members. *Set2* (Strategy 2) was derived from presenting category entities. We consider naming a category entity with its exact name or a name of its parent or subclass member. For example (see Fig. 3), a participant was shown the image of category `Trumpet` and named it with its exact name. This will increase the count for `Trumpet`. In Fig. 4, a participant saw the category `Brass` and named it as its member category `Trumpet`.

In each of the two sets, entities with frequency equal or above two (i.e. named by at least two different users) were identified as potential BLO. The union of *Set1* and *Set2* gives BLO. It includes musical instruments such as: `Bouzouki`, `Guitar` and `Saxophone`. The BLO obtained from MusicPinta are available here[8].

5.2   Evaluating KA_{DG} Against BLO

Quantitative Analysis. We used the BLO identified to examine the performance of the KA_{DG} metrics. For each metric, we aggregated (using union) the KA_{DG} entities identified using the hierarchical relationships (H). We noticed that the three homogeneity metrics have the same values; therefore, we choose one metric when reporting the results, namely Jaccard similarity[9]. A cut-off threshold point for the result lists with

---

[8] https://drive.google.com/drive/folders/0B5ShywKndSLXaVhrSWpiYVZ3WjA.

[9] The Jaccard similarity metric is widely used, and was used in identifying basic formal concepts in the context of formal concept analysis [28].

potential KA_{DG} entities was identified by normalizing the output values from each metric and taking the mean value for the *60th percentile* of the normalized lists. The KA_{DG} metrics evaluated included the three distinctiveness metrics plus the Jaccard homogeneity metric; each metric was applied over both families of relationships – hierarchical (*H*) and domain-specific (*D*). As in ontology summarization approaches [18], a name simplicity strategy was applied to reduce noise when calculating key concepts (usually, basic level objects have relatively simple labels, such as chair or dog). The name simplicity approach we use is solely based on the data graph. We identify the *weighted median* for the length of the labels of all data graph entities $v \subseteq V$ and filter out all entities whose name length is higher than the median. For the MusicPinta data graph, the weighted median is 1.2, and hence we only included entities which consist of one word. Table 2 illustrates precision and recall values comparing BLO and KA_{DG} derived using hierarchical and domain specific relationships.

Table 2. MusicPinta: performance of the KA_{DG} algorithms compared to BLO.

Relationship types	Precision				Recall			
	AV	CAC	CU	Jac	AV	CAC	CU	Jac
Hierarchical	0.58	0.55	0.59	0.6	0.64	0.73	0.73	0.55
Domain-specific	0.62	0.58	0.59	0.62	0.36	0.5	0.59	0.36

Hybridization. Further analysis of the False Positive (FP) and False Negative (FN) entities indicated that the algorithms had different performance on the different taxonomical levels in the data graph. This led to the following heuristics for hybridization.

Heuristic 1: Use Jaccard metric with hierarchical relationships for the most specific categories in the graph (i.e. the categories at the bottom quartile of the taxonomical level). There were FP entities (e.g. Shawm and Oboe) returned by distinctiveness metrics using the domain-specific relationship MusicOntology:Performance because these entities are highly associated with musical performances (e.g. Shawm is linked to 99 performances and Oboe is linked to 27 performance). Such entities may not be good knowledge anchors for exploration, as their hierarchical structure is flat. The best performing metric at the specific level was Jaccard for hierarchical attributes - it excluded entities which had no (or a very small number of) hierarchical attributes.

Heuristic 2: Take the majority voting for all other taxonomical levels. Most of the entities at the middle and top taxonomical level will be well represented in the graph hierarchy and may include domain-specific relationships. Hence, combining the values of all algorithms is sensible. Each algorithm represents a voter and provides two lists of votes, each list corresponding to hierarchical or domain-specific associated attributes (*H, D*). At least half of the voters should vote for an entity for it to be identified in KA_{DG}. Examples from the list of KA_{DG} identified by applying the above hybridization heuristics included `Accordion`, `Guitar` and `Xylophone`. The full KA_{DG} list is available here[10]. Hybridization improved Precision to 0.65 and Recall to 0.63.

6   L4All: Evaluating KA_{DG} Against BLO

The Career domain is a suitable domain for studying basic level objects due to the richness of its ontological structures and the fact that the identification of knowledge anchors can facilitate users' exploration of such structures, as discussed in [27]. We followed Algorithm 1, conducting a study with human participants to identify BLO.

6.1   Obtaining BLO

Participants. 28 participants, university students and professionals, age 25–64, recruited on a voluntary basis. Most of them were experienced mainly in Computing.

Method. The experimental study for evaluating knowledge anchors in the L4All dataset included categories from the *Occupation* and *Subject* class hierarchies, for the reasons discussed above. Categories were represented to participants (corresponding to the $show(r, v)$ function in Algorithm 1) using names (i.e. labels) of the category's leaves. Overall, 624 class entities were extracted from the two class hierarchies (464 for *Occupation* and 160 for *Subject*) by running SPARQL queries to get all class entities linked via the rdfs:subClassOf relationship. The entities included: *leaves* (349 for *Occupation* and 141 for *Subject*) and *categories* (115 for *Occupation* and 19 for *Subject*). Seven online surveys[7] were developed (six surveys presented the 114 category entities of the *Occupation* class hierarchy, with each survey showing 19 categories; and one survey presented the 19 categories of the *Subject* class hierarchy). The category allocation in each survey was random. Every survey had four respondents from the study participants. Each participant was allocated *only to one survey*.

Category identification task. A representation of each category was shown on the participant's screen and he/she was asked to identify the category name. The representation included a list of leaves' names of that category (at most four leaf names were shown on the participant's screen). The participant was provided with four different categories as candidate answers (including the category which the leaves belong to) and the participant was asked to select one category that he/she thinks the leaf entities belong to. The three additional candidate categories covered three levels of abstraction, namely: a parent from the superordinate level, a member from the subordinate level, and a sibling at the same category level. In cases where no parents or members could be added to the candidate answers, siblings were used instead.

Applying Strategy 2 in Algorithm 1 over the *Occupation* and *Subject* class hierarchies in the L4All dataset, we considered naming a category entity with its exact name or a

---

[10] https://drive.google.com/drive/folders/0B5ShywKndSLXaVhrSWpiYVZ3WjA.

Fig. 5. A representation of Housekeeping Occupation (a Category concept in the *Occupation* hierarchy with two subclasses) was shown to a user, who identified it as "Personal Service Occupation".

Fig. 6. A representation of Biological Sciences (a Category concept in the *Subject* hierarchy with four random subclasses) was shown to a user, who identified it as "Biological Sciences".

name of its parents or its non-leaf subclass members shown to the participants. Figures 5 and 6 show examples of the category identification task from the *Occupation* and *Subject* class hierarchies respectively. For instance, the participant in Fig. 5 saw two leaves (the category has two leaves only) of the category Housekeeping Occupation and the participant identified the category's parent Personal Service Occupation, which he/she thinks that the leaves belong to. This will increase the frequency for the category Personal Service Occupation. In Fig. 6, a participant was shown the leaf names of the category Biological Sciences (four random leaves where selected among 9) and selected its exact name. This will increase the count for the category Biological Sciences.

Category entities in the *Occupation* and *Subject* class hierarchies with frequency equal or above two (i.e. categories named by at least two different users) were identified as potential BLO. Examples of BLO from *Occupation* were Administrative, IT Service Delivery, Functional Managers and from *Subject* were Biological Sciences, Law, Medicine and Dentistry. The full KA_{DG} and BLO lists obtained from the L4All data set are available here[11].

6.2   Evaluating KA_{DG} Against BLO

Quantitative Analysis. The KA_{DG} metrics developed in [9] were run over the *Occupation* and *Subject* class hierarchies and the metrics outputs of KA_{DG} were tested against the BLO identified. For each KA_{DG} metric, we aggregated (using union) the entities identified using the hierarchical relationships (rdfs:subClassOf and rdf:type). One domain-specific relationship was used by the metrics (Job for *Occupation* and Qualification for *Subject*). We normalized the metrics output values and took the *60th percentile* of the normalized lists as a cut-off threshold point. Name simplification was applied using the *weighted medians* for the length of the

---

[11] https://drive.google.com/drive/folders/0B5ShywKndSLXaVhrSWpiYVZ3WjA.

Table 3. KA_{DG} metrics performance using the two varieties of attribute types for the Occupation and Subject hierarchies in the L4All dataset

Class hierarchy	Relationship type	Precision				Recall			
		AV	CAC	CU	Jac	AV	CAC	CU	Jac
Occupation	Hierarchical	0.72	0.76	0.79	0.79	0.52	0.88	0.44	0.44
	Domain-specific	0.73	0.75	0	0	0.76	0.36	0	0
Subject	Hierarchical	1	1	0	0	0.53	0.53	0	0
	Domain-specific	1	1	0	0	0.31	0.08	0	0

labels of class entities in the *Occupation* and *Subject* class hierarchies (for *Occupation* = 3.2 and for *Subject* = 2.8) to filter out entities whose name length is higher than the median. Entities with name length greater than 3 were excluded (the names of the two class hierarchies - *Occupation* and *Subject* - and conjunctions, e.g. "and", were not taken into account in counting the name length of entities).

Precision and *Recall* values for the metrics were identified (see Table 3). The three homogeneity metrics from [9] had the same values; therefore, we choose the Jaccard similarity metric in reporting the results (similarly to the MusicPinta analysis). Using the hierarchical relationships (rdfs:subClassOf and rdf:type), precision and recall values were good for *Occupation* (precision ranging from 0.72 to 0.79 and recall from 0.44 to 0.88) and very mixed for *Subject* (precision ranging from 0 to 1 and recall from 0 to 0.53). For the domain-specific relationships, the precision and recall were mixed for *Occupation* (precision ranging from 0 to 0.75 and recall from 0 to 0.76) and *Subject* (precision ranging from 0 to 1 and recall from 0 to 0.31).

By inspecting what caused the zero precision and recall values for the Category Utility (CU) distinctiveness metric and Jaccard (Jac) similarity metric, we noticed that none of these two metrics picked False Negative (FN) entities (i.e. potential KA_{DG}) using the domain-specific relationships (for *Occupation* and *Subject*) and using the hierarchical relationships (for *Subject*). The CU metric did not pick any FN entities since it multiplies the ratio [number of instances of a category divided by number of all entities, classes and instances in *Occupation*] with the total CU values for members of a category. Hence, the CU value will be decreased especially when there are 1000s of entities (i.e. classes and instances) in the graph. For instance, in the *Occupation* class hierarchy, the CU ratio for the FN category Sales Related Occupation is: *87* instances divided by *4201* (464 classes + 3737 instances in the *Occupation* hierarchy), reducing the CU value for Sales Related Occupation to become less than the *60th* percentile cut-off point (0.01). The Jaccard similarity metric did not pick FN entities since each entity has instances linked with one instance only via a domain-specific relationship (e.g. *Job*). Hence, the categories will have no intersections among their instances, producing zero values in the Jaccard metric.

Hybridization. Analysis of the False Positive (FP) and False Negative (FN) entities indicated that the algorithms had different performance on the different taxonomical levels in the L4All data graph, which is formulated in the two heuristics below.

Heuristic 1: Use the AV and CAC distinctiveness metrics with hierarchical relationships for the categories at the bottom quartile of the class taxonomy. There were FN entities (e.g. Sales Related and Science and Engineering Technicians) returned by the AV and CAC homogeneity metrics using the domain-specific relationship Job, because these entities have a low number of instances (e.g. Sales Related has 87 instances and Science and Engineering Technicians has 50 instances; the median of instances per category is 144).

Heuristic 2: Take the majority voting for all other taxonomical levels. Most of the entities at middle and top taxonomical level are well represented in the graph hierarchy. Each metric represents a voter and provides two lists of votes, each list corresponding to hierarchical or domain-specific relationships. At least half of the voters should vote for an entity for it to be identified as KA_{DG}.

Examples of KA_{DG} identified by applying the above hybridization heuristics for Occupation and Subject class hierarchies are: for *Occupation* (Engineering Professionals, Process Operatives, Science and Engineering Technicians), and for *Subject* (Business and Administrative Studies, Education). The full lists of KA_{DG} identified are available here[12].

Hybridization increased performance, as follows: for *Occupation*, Precision = 0.77 and Recall = 0.92; for *Subject*, Precision = 1 and Recall = 0.53.

7 Discussion

This paper presents a systematic evaluation approach to validate KA_{DG} metrics against basic level objects derived by humans.

Algorithm for identifying BLO. The BLO algorithm presented in Sect. 4 is generic and can be applied over different application domains represented as data graphs. In this paper, the algorithm is applied in two application domains for data exploration, Music and Careers, using the data graphs from two semantic exploration applications. Applying the BLO algorithm over two domains allows us to *illustrate two ways of instantiating the algorithm for obtaining BLO*. MusicPinta describes concrete objects - musical instruments - that can have digital representations (e.g. image, audio, video). An image stimulus was used to represent musical instruments, and free-naming tasks included showing image representations of graph entities and asking the users to quickly name the entities they see. In contrast, L4All comprises of abstract career categories, such as *Occupation* and *Subject*, which have text representations (i.e. labels of entities) but no clearly distinguishable images. In this case, a category verification task was used to obtain BLO by showing text representations of graph entities and asking the user to identify the matching entity given some answers.

An important component for applying the BLO is to identify appropriate stimuli to be used for representing graph entities and showing them to humans in either a free-naming task or in a category verification task. One of the main factors that affects

---

[12] https://drive.google.com/drive/folders/0B5ShywKndSLXaVhrSWpiYVZ3WjA.

choosing appropriate stimuli is how well the stimuli cover the entities in the data graph. In other words, the chosen stimuli should have representations for all entities in the graph hierarchies. For instance, the stimuli for MusicPinta were images - taken from an established source (MIMO[5]). The chosen stimuli have to be close enough to users' cognitive structures, so the users can understand the representation of entities.

The BLO algorithm over shallow graph hierarchies has some limitations. For instance, most categories (15 categories out of 19) in the *Subject* class hierarchy of the L4All ontology were identified as BLO. In a category verification task over a shallow hierarchy, finding candidate answers to be presented to users is challenging, especially when the shallow hierarchy does not contain the three levels of abstraction (basic, subordinate and superordinate). Furthermore, the identified BLO in data graphs can have confusing category labelling which reflect insufficiently articulated scope; for instance, vague names (e.g. 'European Language, Literature and related subject') or combining two categories in one (e.g. 'Mathematical and Computer Sciences'). Hence, the BLO algorithm is sensitive to the quality of the ontology. This points at another possible application of BLO – peculiarities in the output can indicate deficiencies of the ontology which can provide insights for re-engineering the ontology. An area of future work is to improve the L4All ontology by modifying the class labels and better articulating their scope.

Performance of KA_{DG} metrics. The identified BLO were used to examine the performance of the KA_{DG} metrics. Our analysis found that hybridization of the metrics notably improved performance. The hybridization heuristics for the upper level of the graph hierarchies tend to be the same – combine the KA_{DB} metrics using majority voting. However, the hybridization heuristics for the bottom level of the hierarchy differed depending on how instances at the bottom of the graph were associated through domain-specific relationships. The performance is sensitive to the appropriateness of the domain-specific relationships captured in the data graph. Examining the FP and FN entities for the hybridization algorithms for KA_{DG} led to the following observations:

Missing basic level entities due to unpopulated areas in the data graph. We noticed that none of the metrics picked FN entities belonging to the bottom quartile of the taxonomies and having a small number of members (such as Cello in MusicPinta and Construction Operatives in the *Occupation* class hierarchy in L4All - Cello has only one subclass and Construction Operatives has 10 instances – mean number of instances in *Occupation* is 184). While these entities belong to the cognitive structures of humans and were therefore added to the BLO sets, one could question whether such entities would be useful knowledge anchors because of their relatively small number of members. These entities could lead the user to 'dead ends' within unpopulated areas of the data graph which may be confusing. We therefore see such FN cases as 'good misses' by the KA_{DG} metrics.

Selecting entities that are superordinates of entities in BLO. The FP included entities (such as Reeds in MusicPinta and Secretarial and Related Occupation in the *Occupation* class hierarchy in L4All) which are well represented in the graph (Reeds has 36 subclasses linked to 60 DBpedia categories; Secretarial and

Related Occupation has 8 subclasses and 800 instances). Although these entities are not close to human cognitive structures, they provide direct links to entities in BLO (Reeds links to Accordion; Secretarial and Related Occupation links to Administrative and Secretarial Occupation). We therefore see such FP as 'good picks', as they provide *bridges* to BLO entities.

8   Conclusion and Future Work

Data graph exploration underpins semantic Web applications, such as browsing and search. Lay users who are not domain experts can face high cognitive load and usability challenges when exploring an unfamiliar domain because the users are unaware of the knowledge structure of the graphs. This brings forth the challenge of building systematic approaches for supporting users' exploration taking into account the knowledge utility of the exploration paths. To address this challenge, we adopt the subsumption theory for meaningful learning [8] where new knowledge is subsumed under familiar and highly inclusive entities. A core algorithmic component for adopting this theory is the automatic identification of knowledge anchors in a data graph.

The work in this paper adapts Cognitive Science experimental approaches for deriving the BLO, and presents an algorithm to capture the BLO that correspond to human cognitive structures over a data graph. Our work contributes to improving the usability of data graph exploration by presenting a methodology for aligning BLO in human cognitive structures and the corresponding knowledge anchors in a data graph. The obtained sets of BLO and KA_{DG} can have two broad implications: (i) to improve users' exploration of large data graphs; and (ii) to reengineer the ontology to better align with human cognitive structures. We are focusing on the former, and are devising navigation strategies to expand users' knowledge while exploring a data graph.

Acknowledgements. This research uses outputs from the EU/FP7 project Dicode and the UK/JISC project L4All. We are grateful to Riccardo Frosini and Mirko Dimartino in helping us prepare the L4All dataset used for the experiments in this paper. We thank all the participants in the experimental studies.

References

1. Marie, N., Gandon, F.: Survey of linked data based exploration systems. In: IESD@ISWC (2014)
2. Thakker, D., Dimitrova, V., Lau, L., Yang-Turner, F., Despotakis, D.: Assisting user browsing over linked data: requirements elicitation with a user study. In: ICWE 2013 International conference on Web Engineering, pp. 376–383 (2013)
3. Cheng, G., Zhang, Y., Qu, Y.: Explass: exploring associations between entities via top-k ontological patterns and facets. In: ISWC 2013, pp. 422–437 (2014)
4. Thellmann, K., Galkin, M., Orlandi, F., Auer, S.: LinkDaViz – automatic binding of linked data to visualizations. In: ISWC 2013, pp. 147–162 (2015)

5. Lopez, V., Fernández, M., Motta, E., Stieler, N.: PowerAqua: supporting users in querying and exploring the semantic web. Semant. Web. **3**, 249–265 (2012)
6. Cheng, G., Qu, Y.: Searching linked objects with falcons: approach, implementation and evaluation. Int. J. Semant. Web Inf. Syst. **5**, 49–70 (2009)
7. Al-Tawil, M., Thakker, D., Dimitrova, V.: Nudging to expand user's domain knowledge while exploring linked data. In: IESD@ISWC (2015)
8. Ausubel, D.P.: A subsumption theory of meaningful verbal learning and retention. J. Gen. Psychol. **66**, 213–224 (1962)
9. Al-Tawil, M., Dimitrova, V., Thakker, D., Bennett, B.: Identifying knowledge anchors in a data graph. In: HT 2016 - 27th ACM Conference on Hypertext and Social Media (2016)
10. Rosch, E., Mervis, C.B., Gray, W.D., Johnson, D.M., Boyes-Braem, P.: Basic objects in neutral categories. Cogn. Psychol. **8**, 382–439 (1976)
11. Sah, M., Wade, V.: Personalized concept-based search on the linked open data. J. Web Semant. **36**, 32–57 (2016)
12. Zimmer, B., Kerren, A.: Harnessing WebGL and WebSockets for a web-based collaborative graph exploration tool. In: Cimiano, P., Frasincar, F., Houben, G.-J., Schwabe, D. (eds.) ICWE 2015. LNCS, vol. 9114, pp. 583–598. Springer, Cham (2015). doi:10.1007/978-3-319-19890-3_37
13. Zheng, L., Xu, J., Jiang, J., Qu, Y., Cheng, G.: Iterative entity navigation via co-clustering semantic links and entity classes. In: Sack, H., Blomqvist, E., d'Aquin, M., Ghidini, C., Ponzetto, S.P., Lange, C. (eds.) ESWC 2016. LNCS, vol. 9678, pp. 168–181. Springer, Cham (2016). doi:10.1007/978-3-319-34129-3_11
14. Troullinou, G., Kondylakis, H., Daskalaki, E., Plexousakis, D.: RDF digest: efficient summarization of RDF/S KBs. In: Gandon, F., Sabou, M., Sack, H., d'Amato, C., Cudré-Mauroux, P., Zimmermann, A. (eds.) ESWC 2015. LNCS, vol. 9088, pp. 119–134. Springer, Cham (2015). doi:10.1007/978-3-319-18818-8_8
15. Zhang, X., Cheng, G., Qu, Y.: Ontology summarization based on RDF sentence graph. In: Proceedings of the 16th International World Wide Web Conference (2007)
16. Wille, R.: Formal concept analysis as mathematical theory of concepts and concept hierarchies. In: Formal Concept Analysis, pp. 1–33 (2005)
17. Li, N., Motta, E.: Evaluations of user-driven ontology summarization. In: Cimiano, P., Pinto, H.S. (eds.) EKAW 2010. LNCS, vol. 6317, pp. 544–553. Springer, Heidelberg (2010). doi:10.1007/978-3-642-16438-5_44
18. Peroni, S., Motta, E., Aquin, M.: Identifying key concepts in an ontology through the integration of cognitive principles with statistical and topological measures. In: ASWC 2008 (2008)
19. Belohlavek, R., Trnecka, M.: Basic level in formal concept analysis: Interesting concepts and psychological ramifications. In: International Joint Conference on Artificial Intelligence, pp. 1233–1239 (2013)
20. Bizer, C., Heath, T., Berners-Lee, T.: Linked data-the story so far. Int. J. Semant. Web Inf. Syst. (2009)
21. Rosch, E., Lloyd, B.B.: Cognition and Categorization, pp. 27–48. Lloydia Cincinnati (1978)
22. Henry Kucera, W.N.F.: Computational Analysis of Present-Day American English. Am. Doc. **19**, 419 (1968)
23. Cappiello, C., Noia, T., Marcu, B.A., Matera, M.: A quality model for linked data exploration. In: Bozzon, A., Cudre-Maroux, P., Pautasso, C. (eds.) ICWE 2016. LNCS, vol. 9671, pp. 397–404. Springer, Cham (2016). doi:10.1007/978-3-319-38791-8_25
24. Heath, T., Bizer, C.: Linked data: Evolving the Web into a Global Data Space, 1st edn. (2011)

25. Palmer, C.F., Jones, R.K., Hennessy, B.L., Unze, M.G., Pick, A.D.: How Is a Trumpet known? The "Basic Object Level" concept and perception of musical instruments. Am. J. Psychol. **102**, 17–37 (1989)
26. de Freitas, S., Harrison, I., Magoulas, G., Papamarkos, G., Poulovassilis, A., Van Labeke, N., Mee, A., Oliver, M.: L4All, a web-service based system for lifelong learners. In: Learning Grid Handbook: Concepts, Technologies and Applications, vol. 2, pp. 143–155 (2008)
27. Poulovassilis, A., Al-Tawil, M., Frosini, R., Dimartino, M., Dimitrova, V.: Combining flexible queries and knowledge anchors to facilitate the exploration of knowledge graphs. In: IESD@ISWC (2016)
28. Belohlavek, R., Trnecka, M.: Basic level of concepts in formal concept analysis. In: Domenach, F., Ignatov, D.I., Poelmans, J. (eds.) ICFCA 2012. LNCS, vol. 7278, pp. 28–44. Springer, Heidelberg (2012). doi:10.1007/978-3-642-29892-9_9

Decentralized Evolution and Consolidation
of RDF Graphs

Natanael Arndt[(⊠)] and Michael Martin

Agile Knowledge Engineering and Semantic Web (AKSW), Institute
of Computer Science, Leipzig University, Augustusplatz 10, 04109 Leipzig, Germany
{arndt,martin}@informatik.uni-leipzig.de

Abstract. The World Wide Web and the Semantic Web are designed
as a network of distributed services and datasets. In this network and
its genesis, collaboration played and still plays a crucial role. But currently we only have central collaboration solutions for RDF data, such as
SPARQL endpoints and wiki systems, while decentralized solutions can
enable applications for many more use-cases. Inspired by a successful distributed source code management methodology in software engineering
a framework to support distributed evolution is proposed. The system is
based on Git and provides distributed collaboration on RDF graphs. This
paper covers the formal expression of the evolution and consolidation of
distributed datasets, the synchronization, as well as other supporting
operations.

1 Introduction

On the World Wide Web, besides documents, also datasets are getting more and
more into the focus. An increasing part of the datasets on the Web is available as
Linked Data, also called the *Linked Open Data Cloud*[1] or *Giant Global Graph*[2].
These datasets need to be curated and new data is added in a collaborative way.

Projects from various domains are striving for distributed models to collaborate on common knowledge bases. Examples are in the e-humanities, the *Pfar-
rerbuch*[3], the *Catalogus Professorum*[4] [25] and the *Heloise – European Network
on Digital Academic History*[5] [24]. In libraries meta-data of more and more electronic resources is gathered and shared among stakeholders. The *AMSL*[6] project
is looking for collaborative curation and management of electronic resources as
Linked Data [2,23]. But even businesses have a need for managing data in distributed setups. In the *LUCID – Linked Value Chain Data*[7] project [13] the

---

[1] http://lod-cloud.net/.
[2] http://dig.csail.mit.edu/breadcrumbs/node/215.
[3] http://aksw.org/Projects/Pfarrerbuch.
[4] http://aksw.org/Projects/CatalogusProfessorum.
[5] http://heloisenetwork.eu/.
[6] http://amsl.technology/.
[7] http://www.lucid-project.org/.

© Springer International Publishing AG 2017
J. Cabot et al. (Eds.): ICWE 2017, LNCS 10360, pp. 23–40, 2017.
DOI: 10.1007/978-3-319-60131-1_2

communication of data along supply chains is subject of research, and in the *LEDS – Linked Enterprise Data Services*[8] project there is a need for distributed collaboration on datasets to organize business procedures.

Currently available possibilities for collaborating on Linked Data are central SPARQL endpoints and wiki systems [11,12,20]. In both cases a common version of the dataset is kept in a central infrastructure and collaborators are editing on the same computer system simultaneously. This central approach for a synchronized state has drawbacks in different scenarios, where multiple different versions of the dataset are preferable. Multiple different versions of a dataset can exist if a simultaneous access to the central dataset is not possible for all participants for instance due to limited network accessibility. Also it might be, that different levels of access rights are to be implemented, or different releases of a dataset should be available. Even on a common repository it can be intended to have multiple versions of a dataset, for instance in an ongoing discussion where a consensus on a certain topic is not yet reached. Thus it is desirable to manage multiple branches of the dataset evolution.

In the early days of computers and software the term *software crisis* was coined to describe the immaturity of the software engineering process and software engineering domain. Dijkstra described the situation as follows: "[...] *as long as there were no machines, programming was no problem at all; when we had a few weak computers, programming became a mild problem, and now we have gigantic computers, programming had become an equally gigantic problem.*"[9] The process of creating software could be made more reliable and controllable by introducing software engineering methods. In the 1970s software configuration management enabled structured collaborative processes, where version control is an important aspect to organize the evolution of software. Early source code management (SCM), such as *CVS* and *Subversion* allowed to create central repositories. The latest version on the repository represents the current state of development and the linear versioning history draws the evolution process of the software. Decentralized SCM, such as *Darcs*, *Mercurial* and *Git* were developed to allow every member of distributed teams to fork the programs source code and individually contribute new features or bug-fixes as pull-requests which then can be merged into a master branch.

Drawing a parallel from the software engineering to (Linked) Data Engineering, *central* SCM systems would correspond to *central* SPARQL endpoints and wiki systems. To support decentralized and distributed collaboration processes it is necessary to adapt DSCM to Linked Data. This would allow us to support decentralized work, ease the synchronization and manage the independent evolution of datasets. The subject of collaboration in the context of Linked Data are datasets resp. graphs instead of source code files. Similar to source code development with DSCM individual local versions of a dataset are curated by data scientists and domain experts. For consolidating those diverged datasets, a synchronization resp. merging strategy needs to be applied. Berners-Lee and

---

[8] http://www.leds-projekt.de/.

[9] https://www.cs.utexas.edu/users/EWD/transcriptions/EWD03xx/EWD340.html.

Connolly also deal with *the synchronization problem* [6] and plead for not mixing the individual parts of the problem, to perhaps get a more robust and extensible system. On the informal *Semantic Web Layer Cake* model[10] we can locate different layers, where a certain level of consistency has to be considered: *data interchange*; *ontology, rule* and *RDFS (ontological)*; and *applications*. Considering the *data interchange* layer it is important, that valid RDF data is always provided. This RDF data can then serve as input for checking the compliance to *ontological* rules. Additionally, requirements to the data might be assumed on the *application* layer.

In the ecosystem around DSCM continuous integration systems are responsible for verifying each operation on the source code, e.g. commits of a pull-request or the result of a merge process. These systems execute among others unit and integration tests before they send a positive or negative result telling the maintainers whether to accept or reject changes to the repository. For managing Semantic Web data continuous integration tools can be responsible for verifying the resulting graph on the *ontological* and *application* layer, as presented in [22] and [19].

In this paper we concentrate on the syntactic *data interchange* layer of the problem and present the following contributions. We propose a formalized model for expressing change and evolution with support for tracking, reverting, branching and merging changes. This model is generic, quad aware and supports blank nodes. To actually pursue merging operations, identifying and resolving conflicts we propose various merge strategies, which can be utilized in different scenarios. Moreover, we evaluate our model regarding correctness and performance using a respective *Git* based implementation. The paper is structured as follows. The state of the art and related work are presented and discussed in Sect. 2. An introduction to the preliminaries for the formal model with basic definitions is given in Sect. 3. The operations on a version graph of distributed evolving datasets are defined in more detail in Sect. 4. Different merge strategies are presented in Sect. 5. The presented concepts are evaluated regarding correctness and performance, using our prototypical implementation, in Sect. 6. Finally a conclusion is given together with a prospect to future work in Sect. 7.

2 Related Work

First we consider abstract models for expressing changes and evolution, and second, we examine implementations dealing with versioning of RDF data. Berners-Lee and Connolly [6] give a general overview on the problem of synchronization and on how to calculate delta on RDF graphs. This work concentrates on possibilities to transfer changes to datasets by applying patches. They introduce an ontology that describes patches in *"a way to uniquely identify what is changing"* and *"to distinguish between the pieces added and those subtracted"*. Haase and Stojanovic [16] introduce their concept of ontology evolution as follows: *"Ontology evolution can be defined as the timely adaptation of an ontology*

---

[10] https://www.w3.org/2007/03/layerCake.svg.

*to the arisen changes and the consistent management of these changes. [...]
An important aspect in the evolution process is to guarantee the consistency
of the ontology when changes occur, considering the semantics of the ontology
change.".* This paper concentrates on the linear evolution process of an individual dataset, rather than the distributed evolution process. For dealing with
inconsistency resp. consistency they define three levels: structural, logical, and
user-defined consistency. In the remainder of the paper they mainly concentrate
on the implications of the evolution with respect to OWL (DL) rather than a
generic approach. Auer and Herre [5] propose a generic framework to support
the versioning and evolution of RDF graphs. The main concept introduced in
this work are *atomic graphs*, which provides a practical approach for dealing
with blank nodes in change sets. Additionally they introduce a formal hierarchical system for structuring a set of changes and evolution patterns leading to
the changes of a knowledge base. Cassidy and Ballantine [8] present a version
control system for RDF graphs based on the model of the DSCM *Darcs*. Their
approach covers the versioning operations *commute, revert* and *merge*. In contrast to other DSCM systems the merge operation is implemented using patch
commutation, which requires history rewriting and thus loses the context of the
original changes.

TailR as presented in [21] is a system for preserving the history of arbitrary
linked data sets. It follows a combined delta and snapshot storage approach. The
system is comparable to the approach presented by Frommhold et al. [14], as both
system are linear change tracking systems. None of the systems provides support
for branches to allow independent evolution of RDF graphs. Graube et al. [15]
propose the R43ples approach, which uses named graphs for storing revisions as
deltas. For querying and updating the triple store an extended SPARQL protocol
language is introduced. R&Wbase [26] is a tool for versioning an RDF graph.
It is tracking changes which are stored in individual named graphs which are
combined on query time, this makes it impossible to use the system to manage
RDF datasets with multiple named graphs. The system also implements an
understanding of coexisting branches within a versioning graph, which is very
close to the concept of Git. In the *dat*[11] project a tool for distributing and
synchronizing data is developed. The aim is mainly on synchronizing any file
type peer to peer. It has no support for managing branches and merging diverged
versions and is not focusing on RDF data.

Table 1 provides an overview of the related work and compares the presented
approaches with regard to their used storage system, quad and blank node support, as well as the possibilities to create multiple branches, merge branches and
create distributed setups for collaboration using push and pull mechanisms.

Git4Voc, as proposed by Halilaj et al. [17] is a collection of best practices and
tools to support authors in the collaborative creation process of RDF and OWL
vocabularies. The methodology is based on Git and implements pre- and post-
commit hooks, which call various tools for automatically checking the vocabulary specification. In addition to the methodology it has also formulated very

---

[11] http://dat-data.com/.

Table 1. Comparison of the different (D)SCM systems for RDF data. All of these systems are coming with custom implementations and are not reusing existing SCMs. At the level of abstraction all of these systems can be located on the *data interchange* layer.

Approach	Storage	Quad support	Bnodes	Branches	Merge	Push/Pull
Meinhardt et al. [21]	hybrid	no[a]	yes	no[f]	no	(yes)[h]
Frommhold et al. [14]	delta	yes	yes	no[f]	no	no
Graube et al. [15]	delta	no[b,c]	(yes)[d]	yes	no	no
Vander Sande et al. [26]	delta	no[c]	(yes)[e]	yes	(yes)[g]	no
dat	chunks	n/a	n/a	no	no	yes

[a] The granularity of versioning are repositories; [b] Only single graphs are put under version control; [c] The context is used to encode revisions; [d] Blank nodes are skolemized; [e] Blank nodes are addressed by internal identifiers; [f] Only linear change tracking is supported; [g] Naive merge implementation; [h] No pull requests but history replication via memento API

important requirements for the collaboration on RDF data. Fernández et al. [10] put their focus on the creation of a benchmark for RDF Archives – BEAR. The authors are focusing on *"exploiting the blueprints to build a customizable generator of evolving synthetic RDF data"*. We hope that this can be used as generic benchmark in the future.

3 Preliminaries

In the following we are speaking of *RDF graph* resp. just *graph* and *RDF dataset* resp. just *dataset* as commonly used in RDF and also defined in [9].

According to [5] an *Atomic Graph* is defined as follows:

Definition 1 (Atomic Graph). *A graph is atomic if it may not be split into two nonempty graphs whose blank nodes are disjoint.*

This means that all graphs containing exactly one statement are atomic. Furthermore the graph is still atomic if it contains a statement with a blank node and all other statements in the same graph also contain this blank node. If a statement as well contains a second blank node the same takes effect for this blank node recursively. Basically an *Atomic Graph* is comparable to a *Minimum Self-Contained Graph* (MSG) as introduced by [29].

We define $\mathbb{G}_A$ as the set of all *Atomic Graphs*. We also define the equivalence relation on two graphs $G, H \in \mathbb{G}_A$ as $G \approx H \Leftrightarrow G$ and H are *isomorphic* as defined for RDF in [9]. Further the quotient set of $\mathbb{G}_A$ by $\approx$ is $\mathbb{A} := \mathbb{G}_A/\approx$, the quotient set of all equivalence classes of atomic graphs. Based on this we now define an *Atomic Partition* of a graph as follows:

Definition 2 (Atomic Partition). *Let* $\mathcal{P}_G \subseteq \mathbb{G}_A$ *denote the partition of* G *containing only atomic graphs.*

Then the Atomic Partition *is defined as*

$$P(G) := \mathcal{P}_G / \approx$$

This means that a graph is split into a set of sets, called partition, containing only atomic graphs. Further we can also say, that $P(G) \subseteq \mathbb{A}$. Each of the containing sets consists of exactly one statement for all statements without blank nodes. For statements with blank nodes, it consists of the complete subgraph connected to a blank node and all neighboring blank nodes. This especially means, that all sets in the *Atomic Partition* are disjoint regarding the contained blank nodes. Further they are disjoint regarding the contained triples (because its a partition).

We now also let $\check{P}(X)$ be a subset of X consisting of exactly one element of each equivalence class, where X is an *Atomic Partition*. This means that G and $\check{P}(P(G))$ are isomorphic, we just make sure that no multiple isomorphic atomic subgraphs exist.

Definition 3 (Difference). *Let* G *and* G' *be two graphs, and* $P(G)$ *resp.* $P(G')$ *the* Atomic Partitions.[12]

$$C^+ := \dot{\bigcup}\left(\check{P}\left(P(G') \setminus P(G)\right)\right)$$
$$C^- := \dot{\bigcup}\left(\check{P}\left(P(G) \setminus P(G')\right)\right)$$
$$\Delta(G, G') := (C^+, C^-)$$

Looking at the resulting tuple (C^+, C^-) we can also say that the inverse of $\Delta(G, G')$ is $\Delta^{-1}(G, G') = \Delta(G', G)$ by swapping the positive and negative sets.

Definition 4 (Change). *A change is a tuple of two graphs* (C_G^+, C_G^-) *on a graph* G, *with*

$$P(C_G^+) \cap P(G) = \emptyset$$
$$P(C_G^-) \subseteq P(G)$$
$$P(C_G^+) \cap P(C_G^-) = \emptyset$$
$$P(C_G^+) \cup P(C_G^-) \neq \emptyset$$

Since blank nodes can't be identified across graphs, a new blank node has to be completely enclosed in the set of additions. Thus $P(C_G^+)$ and $P(G)$ have to be disjoint. This means an addition can't introduce just new statements to an existing blank node. Parallel to the addition a blank node can only be removed if it is completely removed with all its statements. This is made sure by $P(C_G^-)$ being a subset of $P(G)$. Simple statements without blank nodes can be simply

---

[12] The set-minus operator $\setminus$ in this case is defined to also remove elements equivalent with respect to the $\approx$ relation.

added and removed. Further since C_G^+ and C_G^- are disjoint we avoid the removal of atomic graphs, which are added in the same change and vice versa. And since at least one of C_G^+ or C_G^- can't be empty we avoid changes with no effect.

Definition 5 (Application of a Change). *Let $C_G = (C_G^+, C_G^-)$ be a change on G. The function Apl is defined for the arguments G, C_G resp. $G, (C_G^+, C_G^-)$ and is determined by*

$$Apl(G, (C_G^+, C_G^-)) := \bigcup \left(\breve{P} \left((P(G) \setminus P(C_G^-)) \cup P(C_G^+) \right) \right)$$

We say that C_G is applied to G with the result G'.

4   Operations

Now that we have defined the calculation with additions and removals on a graph we can define basic version tracking and distributed evolution operations. Figure 1 depicts an initial commit $\mathcal{A}$ without any predecessor resp. parent commit and a commit $\mathcal{B}$ referring to its parent $\mathcal{A}$.

Fig. 1. Two commits with an ancestor reference

Let G^0 be a graph under version control. $\mathcal{A}(\{G^0\})$ is a commit containing the graph G^0. G will be the new version of G^0 after a change $(C_{G^0}^+, C_{G^0}^-)$ was applied on G^0; $Apl(G^0, (C_{G^0}^+, C_{G^0}^-)) = G$. Now we create a new commit containing G which refers to its predecessor commit, from which it was derived: $\mathcal{B}_{\{\mathcal{A}\}}(\{G\})$. Another change applied on G would result in G' and thus a new commit $\mathcal{C}_{\{\mathcal{B}_{\{\mathcal{A}\}}\}}(\{G'\})$ is created. (In the further writing, the indices and arguments of commits are sometimes omitted for better readability, while clarity should still be maintained by using distinct letters).

The evolution of a *graph* is the process of subsequently applying changes to the graph using the *Apl* function as defined in Definition 5. Each commit is expressing the complete evolution process of a set of graphs, since it refers to its predecessor, which in turn refers to its predecessor as well. Initial commits holding the initial version of the graph are not referring to any predecessor.

Since a commit is referring to its predecessor and not vice versa, nothing hinders us from creating another commit $\mathcal{D}_{\{\mathcal{B}_{\{\mathcal{A}\}}\}}(\{G''\})$. Taking the commits $\mathcal{A}$, $\mathcal{B}_{\{\mathcal{A}\}}$, $\mathcal{C}_{\{\mathcal{B}\}}$, and $\mathcal{D}_{\{\mathcal{B}\}}$ results in a directed rooted in-tree, as depicted in Fig. 2. The commit $\mathcal{D}$ is now a new *branch* or *fork* based on $\mathcal{B}$, which is diverged from $\mathcal{C}$. We know that $G \neq G'$ and $G \neq G''$, while we don't know about the relation between G' and G''. From now on the graph G is *independently* evolving in two branches, while *independent* means possibly independent, i.e. two

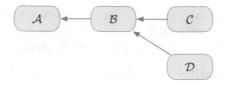

Fig. 2. Two branches evolved from a common commit

actors performing a change do not have to know of each other or do not need a direct communication channel. The actors could actually communicate, but communication is not required for those actions.

Definition 6 (Distributed Evolution). *Distributed Evolution is the (independent) evolution of a graph G with two graphs G_1 and G_2 as result, where $Apl(G, C_1) = G_1$ and $Apl(G, C_2) = G_2$. The changes C_1 and C_2 might be unequal, but can be the same. The same applies for G_1 and G_2, they can be different after the independent evolution, but can be similar.*

4.1   Merge Different Branches

After creating a second branch, the tree of commits is diverged, as shown in the example of Fig. 2. We now want to merge the branches again, in order to get a version of the graph, containing changes made in those different branches or at least take all of these changes into account. The notation of the merge is defined as follows:

Definition 7 (Merge of two Evolved Graphs). *Given are two commits $\mathbb{B}_{\{X\}}(\{G\})$ and $\mathbb{D}_{\{C\}}(\{H\})$ Merging the two graphs G and H with respect to the change history expressed by the commits $\mathbb{B}$ and $\mathbb{D}$ is a function*

$$Merge(\mathbb{B}(\{G\}), \mathbb{D}(\{H\})) = \mathbb{E}_{\{\mathbb{B},\mathbb{D}\}}(\{I\})$$

With $\mathbb{E}_{\{\mathbb{B},\mathbb{D}\}}(\{I\})$ being the merge commit and I the merged graph resulting from G and H.

The *Merge* function is taking two commits as arguments and creates a new commit dependent on the input commits. If we take our running example this is done by creating a commit $\mathcal{E}_{\{\mathcal{C}_{\{B\}}, \mathcal{D}_{\{B\}}\}}(\{G'''\})$, which has two predecessor commits it is referring to. Taking the commits $\mathcal{A}$, $\mathcal{B}_{\{A\}}$, $\mathcal{C}_{\{B\}}$, $\mathcal{D}_{\{B\}}$, and $\mathcal{E}_{\{C,D\}}$, we get an acyclic directed graph, as it is depicted in Fig. 3.

Note that the definition doesn't make any assumptions about the predecessors of the two input commits. It depends on the actual implementation of the *Merge* function whether it is required that both commits have any common ancestors. Furthermore different merge strategies can produce different results, thus it is possible to have multiple merge commits with different resulting graphs but with the same ancestors. Possible merge strategies are presented in Sect. 5.

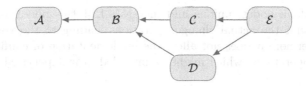

Fig. 3. Merging commits from two branches into a common version of the graph

4.2 Revert a Commit

Reverting the commit $\mathcal{B}_{\{\mathcal{A}\}}(\{G\})$ is done by creating an inverse commit $\mathcal{B}_{\{\mathcal{B}\}}^{-1}(\{\tilde{G}^0\})$ (while the commit $\mathcal{A}$ is specified as $\mathcal{A}(\{G^0\})$). This inverse commit is then directly applied to $\mathcal{B}$. The resulting graph $\tilde{G}^0$ is calculated by taking the inverse difference $\Delta^{-1}(G^0, G) = \Delta(G, G^0)$ and applying the resulting change to G. After this operation $\tilde{G}^0 = G^0$.

Fig. 4. A commit reverting the previous commit

A versioning log containing three commits is shown in Fig. 4. The latest commit reverts its parent and thus the state of the working directory is again equal to the one expressed by the first commit. While it is obvious how to revert the previous commits, it might be a problem if other commits exist between the commit to be reverted and the current top of the versioning log. In this case a three-way-merge is applied (cf. Section 5). For this merge, the merge *base* is the commit to be reverted, branch $\mathcal{A}$ is the parent commit of the commit which is to be reverted, and branch $\mathcal{B}$ the currently latest commit.

5 Merge Strategies

Since we are not only interested in the abstract branching and merging model of the commits, we want to know what a merge operation means for the created graph in the commit. In the following we present some possible implementations of the merge operations. Note, that *merging* in this context is not–generally–to be understood as in the *RDF 1.1 Semantics Recommendation* [18] as the union of two graphs.

Union Merge. Merging two *graphs* G' and G'' could be considered trivially as the union operations of the two graphs: $G' \cup G'' = G'''$. This merge–as mentioned above–is well defined in the RDF 1.1 Semantics Recommendation [18] in section

"4.1 Shared blank nodes, unions and merges". But this operation would not take into account the actual change operations leading to the versions of the graphs. Furthermore it does not allow the implementation of conflict detection or resolution operations, which might be intended in unsupervised situations.

All Ours/All Theirs. Two other merge strategies, which wouldn't produce merge conflicts are *ours* and *theirs*, which just take the complete graph $G' = G'''$ or $G'' = G'''$, while ignoring the other graph respectively. This strategy might be chosen to completely discard changes from a certain branch.

Three-Way Merge. A methodology used in distributed version control systems for software source code, such as Git and Mercurial is the *three-way-merge*[13]. The merge consists of three phases, (1) finding a common *merge base* for the two commits to merge, (2) inferring which lines where added and removed between the *merge base* and the individual branches, and (3) creating a merged version by combining the changes made in the two branches.

Taking into account the versions of the graphs in the two commits to be merged $A_{\{D\}}(\{G'\})$, $B_{\{E\}}(\{G''\})$ we find the most recent common ancestor $F(\{G\})$. This strategy relies on the existence of a common ancestor F, such that for A and B there must exist an ancestor path $A \left\{ \begin{smallmatrix} \cdot \cdot \\ \cdot_{\{F,\ldots\}} \end{smallmatrix} \right\}$ resp. $B \left\{ \begin{smallmatrix} \cdot \cdot \\ \cdot_{\{F,\ldots\}} \end{smallmatrix} \right\}$ to F.

$$(C_A^+, C_A^-) = \Delta(G, G')$$
$$(C_B^+, C_B^-) = \Delta(G, G'')$$
$$G''' = \bigcup \left((P(G') \cap P(G'')) \cup P(C_A^+) \cup P(C_B^+) \right)$$

With this merge strategy in Git changes from two different branches on the same line or two neighboring lines are marked as conflicts. In this case Git can't decide how to combine the changes or which line to take. For transferring this principle from source code files to graphs, we first see, that this definition of conflict is related to its *local* context. In graphs the local context for a statement can be other outgoing and incoming edges of its subject and object resources.

Usually in source code management with this merge strategy a merge conflict occurs, when two close by lines were added or removed in different branches. Thus it can't be decided, whether they are resulting from the same original line or if they are even contradicting. This means for two close by inserted lines one can't automatically decide on in which order the lines have to be inserted into the merge result. Since there is no order relevant in the RDF data model, we don't have to produce merge conflicts in those situations. Producing merge conflicts in a graph's local context is subject to future work.

---

[13] How does Git merge work: https://www.quora.com/How-does-Git-merge-work, 2016-05-10.

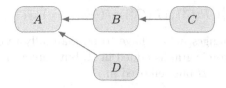

Fig. 5. A situation of one commit revising a priorly introduced change

Touch Merge. This merge strategy is an extension of the *Three-Way Merge*, which should introduce an additional manual revision of the merge operation. In a situation, where we have the commits $A(\{G\})$, $B_{\{A\}}(\{G'\})$, $C_{\{B\}}(\{G''\})$, $D_{\{A\}}(\{G'''\})$ as depicted in Fig. 5. We can have an atomic subgraph g[14], which was not in the merge base (common ancestor, A), but was added in both commits B and D, while it was again removed in C. (Following $g \not\subseteq G \vee G''$, $g \subseteq G' \wedge G'''$). When merging C and D the *Three-Way Merge* would decide on including g in the result, since it would assume it as added in D but doesn't see it in C. The intention of the editors might have been, that g was true at the point in time, when B and D were created, but the creator of C was already updated, that g doesn't hold anymore. The *Touch Merge* (as presented in Algorithm 1) in addition to dealing with adds and removals, also marks atomic graphs as *touched*, once they were changed in a commit and presents them as conflict in case of doubt.

Data: A, F
Result: $+$, $-$
Let X be a sequence of all commits $x_i \in X; 0 \leq i \leq n, n > 0$ on the path from A
 to F, ordered from $x_0 := A$ to $x_n := F$;
$+, - = \{\}$ // initialize $+$ and $-$ as two empty sets;
$i = n$;
while $i > 0$ **do**
 $G :=$ the graph in x_{i-1};
 $G' :=$ the graph in x_i;
 $(C^+, C^-) := \Delta(G, G')$;
 $+ = (+ \setminus P(C^-)) \cup P(C^+)$;
 $- = (- \setminus P(C^+)) \cup P(C^-)$;
 $i = i - 1$;
end
Algorithm 1. The touch algorithm for a path from commit A to commit F

Taking into account the two commits to be merged $A_{\{D\}}(\{G'\})$, $B_{\{E\}}(\{G''\})$ we find the most recent common ancestor $F(\{G\})$. Now we run Algorithm 1 with $A_{\{D\}}(\{G'\})$ and $F(\{G\})$ and store the results as follows $+_A := +, -_A := -$; we do the same for $B_{\{E\}}(\{G''\})$ and $F(\{G\})$ and store the results as $+_B := +, -_B := -$. The result of the merge will be the graph G'''.

---

[14] The subgraph is denoted by a lowercase letter to underline, that it represents an atomic unit of change.

$$G''' = \bigcup \left((P(G') \cap P(G'')) \cup (+_A \setminus -_B) \cup (+_B \setminus -_A) \right)$$

The conflicting changes, which have to be manually resolved are in a tuple (Y_1, Y_2). Y_1 are the atomic graphs added in A, but removed in B and Y_2 are the atomic graphs added in B but removed in A.

$$(Y_1, Y_2) = (+_A \cap -_B, +_B \cap -_A)$$

6   Evaluation

For evaluating the proposed framework we consider the correctness of the framework regarding the recorded changes and the performance, memory and storage footprint. In order to pursue this task we have taken our implementation of the *Quit Store* [3]. This currently is a prototypical implementation to prove the concept of our framework, thus we are not aiming at competitive performance results. The implementation is written in Python[15]. It is based on the RDFlib[16] for handling RDF, performing the query processing and also uses its in-memory quad store implementation. The HTTP interface to provide a *SPARQL 1.1 Endpoint* is using the Flask API[17].

The hardware setup of the machine running the benchmarks is a virtual machine on a Hyper-V cluster with Intel(R) Xeon(R) CPU E5-2650 v3 with a maximum frequency of 2.30 GHz and 62.9 GiB of main memory. As operating system Ubuntu 16.10 (yakkety) 64-Bit is used.

```
$ ./generate -pc 4000 -ud -tc 4000 -ppt 1
```
Listing 1.1. The BSBM generate command with its argument

As benchmarking framework we have decided to use the Berlin SPARQL benchmark (BSBM) [7], since it is made for executing SPARQL Query and SPARQL Update operations. The initial dataset as it is generated using the BSBM, shown in Listing 1.1, contains 46370 statements and 1379201 statements to be added and removed during the benchmark. To also execute update operations we are using the *Explore and Update Use Case*. We have executed 40 warm-up and 1500 query mix runs which resulted in 4592 commits on the underlying git repository using the testdriver as shown in Listing 1.2.

```
$ ./testdriver http://localhost:5000/sparql \
    -runs 1500 -w 40 -dg"urn:bsbm" -o run.xml \
    -ucf usecases/exploreAndUpdate/sparql.txt \
    -udataset dataset_update.nt \
    -u http://localhost:5000/sparql
```
Listing 1.2. The BSBM testdriver command with its argument

The setup for reproducing the evaluation is also available at the following link: https://github.com/AKSW/QuitEval.

---

[15] https://www.python.org/.
[16] https://rdflib.readthedocs.io/en/stable/.
[17] http://flask.pocoo.org/.

6.1   Correctness of Version Tracking

For checking the correctness of the recorded changes in the underlying git repository we have created a verification setup. The verification setup takes the git repository, the initial data set and the query execution log (`run.log`) produced by the BSBM setup. The repository is set to its initial commit, while the reference store is initialized with the initial dataset. Each update query in the execution log is applied to the reference store. When an effect, change in number of statements, is detected on the store, the git repository is forwarded to the next commit. Now the content of the reference store is serialized and compared to the content of the git repository at this point in time. This scenario is implemented in the `verify.py` script in the evaluation tool collection. We have executed this verification scenario and could verify, that the recorded repository has the same data as the store after executing the same queries.

6.2   Performance

The performance of the reference implementation was analyzed in order to identify obstacles in the conception of our approach. In Fig. 6 the queries per second for the different categories of queries in the BSBM are given and compared to the baseline. We compare the execution of our store with version tracking enabled to only the in memory store of python rdflib without version tracking. As expected the versioning has a big impact on the update queries (INSERT DATA and DELETE WHERE), while the explore queries (SELECT, CONSTRUCT and DESCRIBE) are not further impacted. We could reach 83 QMpH[18] resp.

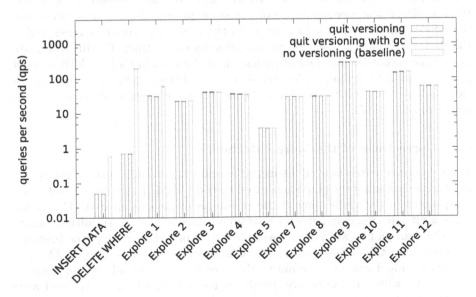

Fig. 6. Execution of the different BSBM queries.

---

[18] *QMpH* Query Mixes per Hour, Query Mixes are defined by the BSBM.

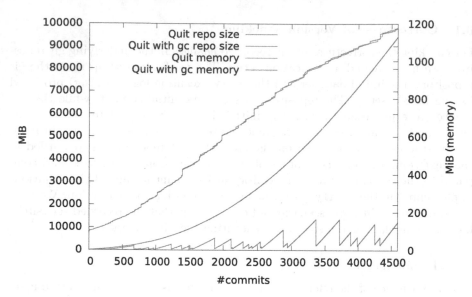

Fig. 7. Usage of system resources and number of commits in the repository during the execution of the BSBM. (gc = garbage collection)

82 QMpH with quit versioning (without resp. with garbage collection) and 787 QMpH for the baseline.

Additionally we have monitored the impact of the execution of the store on the system. The results are visualized in Fig. 7. We have measured the size of the repository, the memory usage of the store process during the execution of the BSBM and have put it in relation to the number of commits generated at that point in time. The memory consumption increased from 100 MiB to 1.1 GiB during the execution. The repository has increased from initially 15 MiB to a size of 92 GiB without garbage collection at the end of the benchmark. Enabling the garbage collection on the quit store during the benchmark could compress the repository to 12 GiB.

6.3   Correctness of the Merge Method

The functional correctness of the merge method was verified using a repository filled with data using the graph generated by the BSBM. To create the evaluation setup and run the verification of the results, a script was created. This script takes a git repository and creates a setup of three commits. An initial commit contains a graph file, which serves as base of two branches. Each of the branches is forked from this initial commit and contains an altered graph file. The files in the individual commits contain random combinations of added and removed statements, while the script also produces the graph, which is expected after merging the branches.

After creating the two branches with different graphs they are merged using `git merge`. The result is then compared to the expected graph and the result is presented to the user. We have executed the verification 1000 times and no merge conflict or failure in the merge result occurred.

7 Conclusion and Future Work

In this paper we have presented a formal framework for the distributed evolution of RDF knowledge bases. It provides a definition of atomic operations for applying on RDF graphs as well as formalized definitions of the versioning operations *commit, branch, merge* and *revert*. In contrast to the related work in Sect. 2 our approach is quad aware, can handle blank nodes, supports branches, supports merging with conflict resolution and allows distributed collaboration with push and pull operations. Also the naive merge strategies *Union, All Ours* and *All Theirs* are described. The *Three-Way Merge*, as it is used in Git is then transferred to the application on atomic graphs to be used on RDF datasets. For allowing more manual control over the merge process also the *Touch Merge* was introduced. The provided formal model and methodology is implemented in the Quit Store tool, which provides a SPARQL 1.1 read/write interface to query and update an RDF dataset in a quad store. The contents of the store are tracked for version control in Git in the local filesystem in parallel. Based on this prototypical implementation we have pursued an evaluation regarding the correctness of the model and monitored the performance of our implementation. The theoretical foundations could be confirmed, while the results regarding the performance reveal potential for improvement. Using the garbage collection we could show that the snap shot based approach doesn't put high load on storage requirements.

Based on the presented approach the application in distributed and collaborative data curation scenarios is now possible. It enables the setup of platforms similar to GitHub specialized on the needs of data scientists and data engineers for creating datasets using local working copies and sending pull requests, while the versioning history and provenance of the results is automatically tracked. Also this system can support the application of RDF in enterprise scenarios such as supply chain management as described in [13]. An integration with the distributed evolution model of distributed semantic social network [4,27], as well as the use-case of synchronization in mobile scenarios [28] is possible. Further we are planning to lift the collaborative curation and annotation in distributed scenarios such as presented in the Structured Feedback protocol [1] to the next level by directly recording the user feedback as commits, which can enable powerfull co-evolution strategies.

Acknowledgement. We want to thank Sören Auer for his valuable important remarks and Norman Radtke for his precious and tireless implementation work. This work was partly supported by a grant from the German Federal Ministry of Education and Research (BMBF) for the LEDS Project under grant agreement No 03WKCG11C.

References

1. Arndt, N., Junghanns, K., Meissner, R., Frischmuth, P., Radtke, N., Frommhold, M., Martin, M.: Structured feedback: a distributed protocol for feedback and patches on the web of data. In: Proceedings of the Workshop on Linked Data on the Web Co-located with the 25th International World Wide Web Conference (WWW 2016), CEUR Workshop Proceedings, vol. 1593, Montréal, Canada, April 2016. http://events.linkeddata.org/ldow2016/papers/LDOW2016_paper_02.pdf
2. Arndt, N., Nuck, S., Nareike, A., Radtke, N., Seige, L., Riechert, T.: AMSL: creating a linked data infrastructure for managing electronic resources in libraries. In: Horridge, M., Rospocher, M., van Ossenbruggen, J. (eds.) Proceedings of the ISWC 2014 Posters and Demonstrations Track, CEUR Workshop Proceedings, vol. 1272, pp. 309–312, Riva del Garda, Italy, October 2014. http://ceur-ws.org/Vol-1272/paper_66.pdf
3. Arndt, N., Radtke, N., Martin, M.: Distributed collaboration on RDF datasets using GIT: towards the quit store. In: 12th International Conference on Semantic Systems Proceedings (SEMANTiCS 2016), SEMANTiCS 2016, Leipzig, Germany, September 2016. https://dx.doi.org/10.1145/2993318.2993328
4. Arndt, N., Tramp, S.: Xodx: a node for the distributed semantic social network. In: Horridge, M., Rospocher, M., van Ossenbruggen, J. (eds.) Proceedings of the ISWC 2014 Posters and Demonstrations Track, CEUR Workshop Proceedings, vol. 1272, pp. 465–468, Riva del Garda, Italy, October 2014. http://ceur-ws.org/Vol-1272/paper_154.pdf
5. Auer, S., Herre, H.: A versioning and evolution framework for RDF knowledge bases. In: Virbitskaite, I., Voronkov, A. (eds.) PSI 2006. LNCS, vol. 4378, pp. 55–69. Springer, Heidelberg (2007). doi:10.1007/978-3-540-70881-0_8
6. Berners-Lee, T., Connolly, D.: Delta: an ontology for the distribution of differences between RDF graphs. Technical report, W3C (2001). http://www.w3.org/DesignIssues/Diff
7. Bizer, C., Schultz, A.: The Berlin SPARQL benchmark. Int. J. Semant. Web Inf. Syst. **5**, 1–24 (2009)
8. Cassidy, S., Ballantine, J.: Version control for RDF triple stores. In: Filipe, J., Shishkov, B., Helfert, M. (eds.) ICSOFT 2007, Proceedings of the Second International Conference on Software and Data Technologies, Volume ISDM/EHST/DC, Barcelona, Spain, 22–25 July 2007, pp. 5–12. INSTICC Press (2007)
9. Cyganiak, R., Wood, D., Lanthaler, M., Klyne, G., Carroll, J.J., McBride, B.: RDF 1.1 concepts and abstract syntax, February 2014. https://www.w3.org/TR/2014/REC-rdf11-concepts-20140225/,   https://www.w3.org/TR/2014/REC-rdf11-concepts-20140225/
10. Fernández, J.D., Umbrich, J., Polleres, A., Knuth, M.: Evaluating query and storage strategies for RDF archives. In: Proceedings of the 12th International Conference on Semantic Systems, SEMANTiCS 2016, NY, USA, pp. 41–48 (2016). https://dx.doi.org/10.1145/2993318.2993333
11. Frischmuth, P., Arndt, N., Martin, M.: OntoWiki 1.0: 10 years of development - what's new in OntoWiki. In: Joint Proceedings of the Posters and Demos Track of the 12th International Conference on Semantic Systems - SEMANTiCS2016 and the 1st International Workshop on Semantic Change and Evolving Semantics (SuCCESS 2016), CEUR Workshop Proceedings, Leipzig, Germany, September 2016. http://ceur-ws.org/Vol-1695/paper11.pdf

12. Frischmuth, P., Martin, M., Tramp, S., Riechert, T., Auer, S.: OntoWiki-an authoring, publication and visualization interface for the data web. Semant. Web J. **6**(3), 215–240 (2015). http://www.semantic-web-journal.net/system/files/swj490_0.pdf

13. Frommhold, M., Arndt, N., Tramp, S., Petersen, N.: Publish and subscribe for RDF in enterprise value networks. In: Proceedings of the Workshop on Linked Data on the Web Co-located with the 25th International World Wide Web Conference (WWW 2016) (2016). http://events.linkeddata.org/ldow2016/papers/LDOW2016_paper_05.pdf

14. Frommhold, M., Piris, R.N., Arndt, N., Tramp, S., Petersen, N., Martin, M.: Towards versioning of arbitrary RDF data. In: 12th International Conference on Semantic Systems Proceedings (SEMANTiCS 2016), SEMANTiCS 2016, Leipzig, Germany, September 2016

15. Graube, M., Hensel, S., Urbas, L.: Open semantic revision control with R43ples: Extending SPARQL to access revisions of named graphs. In: Proceedings of the 12th International Conference on Semantic Systems, SEMANTiCS 2016, NY, USA, pp. 49–56 (2016). https://dx.doi.org/10.1145/2993318.2993336

16. Haase, P., Stojanovic, L.: Consistent evolution of OWL ontologies. In: Gómez-Pérez, A., Euzenat, J. (eds.) ESWC 2005. LNCS, vol. 3532, pp. 182–197. Springer, Heidelberg (2005). doi:10.1007/11431053_13

17. Halilaj, L., Grangel-González, I., Coskun, G., Auer, S.: Git4Voc: Git-based versioning for collaborative vocabulary development. In: 10th International Conference on Semantic Computing, Laguna Hills, California, February 2016

18. Hayes, P.J., Patel-Schneider, P.F.: RDF 1.1 semantics, February 2014. https://www.w3.org/TR/2014/REC-rdf11-mt-20140225/, https://www.w3.org/TR/2014/REC-rdf11-mt-20140225/

19. Kontokostas, D., Westphal, P., Auer, S., Hellmann, S., Lehmann, J., Cornelissen, R., Zaveri, A.: Test-driven evaluation of linked data quality. In: Proceedings of the 23rd International Conference on World Wide Web, WWW 2014, International World Wide Web Conferences Steering Committee, Republic and Canton of Geneva, Switzerland, pp. 747–758 (2014)

20. Krötzsch, M., Vrandečić, D., Völkel, M.: Semantic MediaWiki. In: Cruz, I., Decker, S., Allemang, D., Preist, C., Schwabe, D., Mika, P., Uschold, M., Aroyo, L.M. (eds.) ISWC 2006. LNCS, vol. 4273, pp. 935–942. Springer, Heidelberg (2006). doi:10.1007/11926078_68

21. Meinhardt, P., Knuth, M., Sack, H.: TailR: a platform for preserving history on the web of data. In: Proceedings of the 11th International Conference on Semantic Systems, SEMANTICS 2015, NY, USA, pp. 57–64 (2015). https://dx.doi.org/10.1145/2814864.2814875

22. Meissner, R., Junghanns, K.: Using devOps principles to continuously monitor RDF data quality. In: 12th International Conference on Semantic Systems Proceedings (SEMANTiCS 2016), CEUR Workshop Proceedings, Leipzig, Germany, September 2016. http://ceur-ws.org/Vol-1695/paper34.pdf

23. Nareike, A., Arndt, N., Radtke, N., Nuck, S., Seige, L., Riechert, T.: AMSL: managing electronic resources for libraries based on semantic web. In: Plödereder, E., Grunske, L., Schneider, E., Ull, D. (eds.) Proceedings of the INFORMATIK 2014: Big Data - Komplexität meistern. GI-Edition-Lecture Notes in Informatics, vol. P-232, pp. 1017–1026. Gesellschaft für Informatik e.V. September 2014. © 2014 Gesellschaft für Informatik

24. Riechert, T., Beretta, F.: Collaborative research on academic history using linked open data: a proposal for the heloise common research model. CIAN-Revista de Historia de las Universidades 19 (2016). http://e-revistas.uc3m.es/index.php/CIAN/article/view/3147

25. Riechert, T., Morgenstern, U., Auer, S., Tramp, S., Martin, M.: Knowledge engineering for historians on the example of the *Catalogus Professorum Lipsiensis*. In: Patel-Schneider, P.F., Pan, Y., Hitzler, P., Mika, P., Zhang, L., Pan, J.Z., Horrocks, I., Glimm, B. (eds.) ISWC 2010. LNCS, vol. 6497, pp. 225–240. Springer, Heidelberg (2010). doi:10.1007/978-3-642-17749-1_15

26. Sande, M.V., Colpaert, P., Verborgh, R., Coppens, S., Mannens, E., de Walle, R.V.: R&Wbase: git for triples. In: Bizer, C., Heath, T., Berners-Lee, T., Hausenblas, M., Auer, S. (eds.) LDOW, CEUR Workshop Proceedings, vol. 996, CEUR-WS.org (2013). http://dblp.uni-trier.de/db/conf/www/ldow2013.html#SandeCVCMW13

27. Tramp, S., Ermilov, T., Frischmuth, P., Auer, S.: Architecture of a distributed semantic social network. In: Federated Social Web Europe 2011, Berlin, June 3rd–5th 2011 (2011)

28. Tramp, S., Frischmuth, P., Arndt, N., Ermilov, T., Auer, S.: Weaving a distributed, semantic social network for mobile users. In: Antoniou, G., Grobelnik, M., Simperl, E., Parsia, B., Plexousakis, D., Leenheer, P., Pan, J. (eds.) ESWC 2011. LNCS, vol. 6643, pp. 200–214. Springer, Heidelberg (2011). doi:10.1007/978-3-642-21034-1_14

29. Tummarello, G., Morbidoni, C., Bachmann-Gmür, R., Erling, O.: RDFSync: efficient remote synchronization of RDF models. In: Aberer, K., Choi, K.-S., Noy, N., Allemang, D., Lee, K.-I., Nixon, L., Golbeck, J., Mika, P., Maynard, D., Mizoguchi, R., Schreiber, G., Cudré-Mauroux, P. (eds.) ASWC/ISWC -2007. LNCS, vol. 4825, pp. 537–551. Springer, Heidelberg (2007). doi:10.1007/978-3-540-76298-0_39

The BigDataEurope Platform – Supporting the Variety Dimension of Big Data

Sören Auer, Simon Scerri, Aad Versteden, Erika Pauwels,
Angelos Charalambidis, Stasinos Konstantopoulos, Jens Lehmann,
Hajira Jabeen, Ivan Ermilov, Gezim Sejdiu, Andreas Ikonomopoulos,
Spyros Andronopoulos, Mandy Vlachogiannis, Charalambos Pappas,
Athanasios Davettas, Iraklis A. Klampanos, Efstathios Grigoropoulos,
Vangelis Karkaletsis, Victor de Boer, Ronald Siebes,
Mohamed Nadjib Mami[✉], Sergio Albani, Michele Lazzarini, Paulo Nunes,
Emanuele Angiuli, Nikiforos Pittaras, George Giannakopoulos,
Giorgos Argyriou, George Stamoulis, George Papadakis, Manolis Koubarakis,
Pythagoras Karampiperis, Axel-Cyrille Ngonga Ngomo,
and Maria-Esther Vidal

The H2020 BigDataEurope Project Consortium, c/o Fraunhofer IAIS, Sankt
Augustin, Germany
{auer,scerri,jabeen,mami}@cs.uni-bonn.de

Abstract. The management and analysis of large-scale datasets –
described with the term Big Data – involves the three classic dimensions
volume, velocity and variety. While the former two are well supported by
a plethora of software components, the variety dimension is still rather
neglected. We present the BDE platform – an easy-to-deploy, easy-to-use
and adaptable (cluster-based and standalone) platform for the execution
of big data components and tools like Hadoop, Spark, Flink, Flume and
Cassandra. The BDE platform was designed based upon the requirements gathered from seven of the societal challenges put forward by
the European Commission in the Horizon 2020 programme and targeted
by the BigDataEurope pilots. As a result, the BDE platform allows to
perform a variety of Big Data flow tasks like message passing, storage,
analysis or publishing. To facilitate the processing of heterogeneous data,
a particular innovation of the platform is the Semantic Layer, which
allows to directly process RDF data and to map and transform arbitrary
data into RDF. The advantages of the BDE platform are demonstrated
through seven pilots, each focusing on a major societal challenge.

1 Introduction

The management and analysis of large-scale datasets – described with the term
Big Data – involves the three classic dimensions *volume, velocity* and *variety*.
While the former two are well supported by a plethora of software components,
the variety dimension is still rather neglected. We present the *BigDataEurope*[1]

---

[1] BigDataEurope (https://www.big-data-europe.eu/) is a Coordination and Support
Action funded by the Horizon 2020 programme of the European Commission.

© Springer International Publishing AG 2017
J. Cabot et al. (Eds.): ICWE 2017, LNCS 10360, pp. 41–59, 2017.
DOI: 10.1007/978-3-319-60131-1_3

(BDE) platform – an easy-to-deploy (cluster-based and standalone), easy-to-use and adaptable platform where the variety dimension of Big Data is taken into account right from its inception. The BDE platform is currently being applied to the seven societal challenges put forward by the European Commission in its Horizon 2020 research programme (Health, Food and Agriculture, Energy, Transport, Climate, Social Sciences and Security)[2].

While working with stakcholder communities of the seven societal challenges over the last two years, we have identified the following requirements to be crucial for the success of Big Data technologies:

R1 Simplifying use. The use of Big Data components and the development of analytical algorithms and applications is still cumbersome. Many analytical applications are 'hard-wired', requiring lavish data 'massaging' and complex development in various languages and data models.

R2 Easing deployment. Various deployment schemes are required for different Big Data applications or even during the lifecycle of one particular Big Data application. Prototyping and development, for example, should be possible on a single machine or small cluster, testing and staging possibly on a public or private cloud infrastructure, while production systems might have to be deployed on a dedicated cluster.

R3 Managing heterogeneity. Our survey of societal challenge stakeholders [3] clearly showed that in most cases, Big Data applications originate with a large number of heterogeneous, distributed and often relatively small datasets. Only after their aggregation and integration true Big Data emerges. Hence, a major challenge is managing the heterogeneity of data in terms including data models, schemas, formats, governance schemes and modalities.

R4 Improving scalability. Finally, especially if a number of different storage and processing tools are employed, scalability is still an issue.

The BDE platform addresses these requirements and facilitates the execution of Big Data frameworks and tools like Hadoop, Spark, Flink and many others. We have selected and integrated these components based upon the requirements gathered from the seven different societal challenges. Thus, the platform allows to perform a variety of Big data flow tasks such as *message passing* (Kafka, Flume), *storage* (Hive, Cassandra), *analysis* (Spark, Flink) or *publishing* (GeoTriples).

The remainder of the article is structured as follows: Sect. 2 presents an overview of the BDE platform, while Sect. 3 introduces its Semantic Layer. We present two out of the overall seven pilots demonstrating the advantages of the BDE platform in detail in Sect. 4. We discuss related work and conclude in Sect. 5 along with directions for future work.

---

[2] https://ec.europa.eu/programmes/horizon2020/en/h2020-section/societal-challenges.

2   Platform Overview

The requirements gathered from the seven societal challenge stakeholders revealed that the BDE platform should be generic and must be able to support a variety of Big Data tools and frameworks running together. Installing and managing such a system on the native environments without running into resource or software conflicts is rather hard to achieve. Figure 1 gives a high-level overview on the BDE platform architecture, which is described in the sequel.

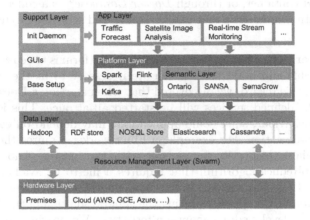

Fig. 1. High-level BDE platform architecture

Hardware Layer. We decided to use *Docker*[3] as packaging and deployment methodology so as to manage the variety of underlying hardware resources efficiently alongside the varying software requirements for different stakeholders. Docker allows to package an application into a standardized unit for development. Docker containers wrap a software in a complete file system that contains everything needed to run, making sure that the software always runs as intended, regardless of the server environment. Therewith, dockers offers a lightweight virtual environment by sharing the same operating system kernel. The images of containers are constructed from a layered file system, sharing common files, thus making the disk usage and image downloads efficient. It is based on open standards and able to run on major operating systems. We have dockerized a large number of Big Data components[4]. None of them posed major problems to be run in a Docker container. Based on our experience and the popularity of Docker, we are confident that all components can be dockerized in reasonable time.

---

[3] https://www.docker.com.
[4] https://github.com/big-data-europe/README/wiki/Components.

Resource Management. *Docker Swarm*[5] is an orchestration tool that allows to deploy Docker containers on a cluster in a transparent way. With its built-in scheduler, it offers most of the features required by the platform, i.e., scalability, interlinking the containers, networking between different containers, resource management between containers, load balancing, fault tolerance, failure recovery, log-based monitoring etc. Docker Swarm operates as a resource manager directly on top of the hardware layer. This hardware layer can vary from small set of machines in the premises of an organization to the infrastructure of some big cloud provider. On top of Docker Swarm, applications can be deployed easily as a single Docker container, or through *Docker Compose*[6] as a collection (cluster) of communicating containers that can be scaled and scheduled dynamically.

Support Layer. The technical aim of the BDE Platform is to reuse components wherever possible and build tools which are necessary to fit the societal challenge needs. In this regard, we discovered that when starting a Docker Compose application, the defined services will be started all at once. This is not always the intended behaviour, since some applications may depend on each other, or on a human intervention. For example, a Flink worker requires the Flink master to be available before it can register with the master. Another example is a Flink MapReduce algorithm that requires a file to be available on HDFS before starting the computation. At the moment, these dependencies cannot be expressed in Docker Compose. Awaiting the general solution of the Docker community, we have developed a semantic alternative. We provide an *init daemon* service that, given an application-specific workflow, orchestrates the initialization process of the components. The init daemon service is a microservice built on the mu.semte.ch platform and provides requests through which the components can report their initialization progress. The init daemon is aware of the startup workflow and, thus, it can validate whether a specific component can start based on the initialization status reported by the other components. The workflow needs to be described per application. It specifies the dependencies between services and indicates where human interaction is required. The UIs described in the next section together with the base Docker images and the init daemon service provide additional support to the user. It facilitates the tasks of building, deploying and monitoring Big Data pipelines on the BDE platform. This support is illustrated in Fig. 1 as an additional layer in the platform architecture.

User Interfaces. In order to lower the usage barrier of Big Data technologies, we have implemented several UIs, including a pipeline builder, a pipeline monitor, an integrator UI and a Swarm UI (see Fig. 2). The UIs serve different purposes, but in general make it easier for the user to build, deploy and monitor applications on the BDE platform. Most of the UI applications are built

---

[5] https://www.docker.com/products/docker-swarm.
[6] https://docs.docker.com/compose.

on mu.semte.ch, a microservices framework backed by Linked Data. Each of the applications is described in more detail in the next sections.

Workflow Builder. This interface allows users to create and edit workflows. The workflow steps reflect the dependencies between the container images or manual actions in the application. Example dependencies are:

- The Spark master needs to be started before the Spark worker such that the worker can register itself at the master.
- The input data needs to be loaded in HDFS before the MapReduce algorithm starts computing.

The order of the steps can be rearranged by dragging-and-dropping the step panels. Once finished the workflow can be exported and fed into the init daemon. *Workflow monitor.* Once an application is running on the BDE platform, this interface allows a user to follow-up the initialization process. It displays the workflow as defined in the pipeline builder application. For each step in the workflow, the corresponding status (not started, running or finished) is shown as retrieved from the init daemon service. The interface automatically updates when a status changes, due to an update through the init daemon service by one of the pipeline components. The interface also offers the option to the user to manually abort a step in the pipeline if necessary.

Swarm UI. The Swarm UI allows to clone a Git repository containing a pipeline (i.e., containing a `docker-compose.yml`) and to deploy this pipeline on a Swarm cluster. Once the pipeline is running, the user can inspect the status and logs of the several services in the pipeline. Users can also scale up/down one or more services or start/stop/restart them.

Integrator UI. Most components (e.g. Spark, Flink, HDFS) provide dashboards to monitor their status. Each of the components runs in a separate Docker container with its own IP address, ports and varying access paths. The Integrator UI displays all component dashboards in a unified interface.

3   Semantic Layer

The ability to cross-link large-scale data with each other and with structured Semantic Web data, and the ability to uniformly process Semantic Web and other data adds value both to the Semantic Web and the Big Data communities; it extends the scope of the former to include a vast data domain and increases the opportunities for the latter to process data in novel ways and combinations.

3.1   Semantic Data Lake

The term Data Lake, in the context of Big Data, appeared in recent years[7] to describe the repository of datasets that are provided for processing and analysis in their very original formats. It is often regarded as the opposing concept of

---

[7] https://jamesdixon.wordpress.com/2010/10/14/pentaho-hadoop-and-data-lakes.

Fig. 2. BDE Platform UI screenshots. Top: Integrator UI, focusing on Spark Master Dashboard. Bottom left: Swarm UI showing the services and the status of a pipeline named *webCAT*. Bottom right: Workflow Monitor showing a case of a pipeline named *Sensor demo* consisting of a set of stages – installing HDFS, installing Spark, populating HDFS with data, etc.

Data Warehouse in the sense that, in the later, data is ready for analysis only *a posteriori* of a mandatory data reorganization phase. This difference is captured by two opposing data access strategies: data *on-read* and data *on-write*, respectively. Data on-read corresponds to the Data Lake where the schema of the data is looked at only when the data is actually used. Data on-write, on the other hand, corresponds to the Data Warehouse where data is organized according to a rigid schema before processing. Another substantial difference between Data Lakes and Data Warehouses is the type of processing the data is undergoing in each case. A Data Lake, as per its definition, contains data that is open to any kind of processing, be it natural language processing, machine learning, ad-hoc (semi-)structured querying, etc. In a Data Warehouse, on the other side, data is accessible only using a specific suite of tools, namely ad-hoc OLAP queries and BI standards. In the BDE platform, we put emphasis on easing and broadening and, whenever possible, harmonizing the use of Big Data technologies. The adoption of the Data Lake concept is, therefore, less of a choice, it is rather a logical consequence. This brings us back to our main goal: addressing the problem of variety and heterogeneity of the data.

While a Data Lake allows heterogeneous data to be stored and accessed, dealing with this data is cumbersome, time-consuming and inefficient, due to the different data models, instance structures and file formats. To address this, the idea of semantifying Data Lakes was recently proposed. The idea is to equip datasets in the lake with mappings to vocabularies, ontologies or a knowledge graph, which can then be used as a semantic access layer to the underlying data.

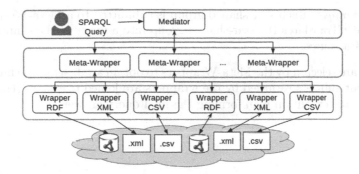

Fig. 3. Ontario multi-layer architecture for ontology-based data access to the Data Lake.

Ontario. The BDE platform comprises a Semantic Data Lake implementation, named *Ontario*. Ontario builds a Semantic Layer on top of the Data Lake, which is responsible for mapping data into existing Semantic vocabularies/ontologies. A successful mapping process, termed Semantic Lifting, provides a 'uniform' view over the whole data. As a result, the user can deal with the heterogeneous data in the lake as if it was in one unique format. Data can then be extracted, queried or analysed using a unique high-level declarative query language. The primary challenge is then to trigger a process with three main steps:

1. *Query analysis and decomposition:* The query is broken down into sub-queries. An execution plan is generated.
2. *Selection of relevant data sources:* Relevant datasets are selected starting from the generated sub-queries and using the mappings we have predefined.
3. *Extraction of the results:* The sub-queries are translated into the syntax of the selected datasets, executed and their results are put together in a certain way (following the plan generated in 1: merge, join, etc.) that the original query is accurately answered.

The main characteristics of Ontario are that the query execution process does not require any data materialization or shipping, and is fully transparent to the user. Data extraction rather happens fully on-the-fly upon a reception of a query.

Data Model. Data in Ontario is conceptually modelled around RDF classes. Each class can be seen as a star-shaped graph centred on the RDF subjects of the same type. The predicates of a class C consist of all the predicates found connected to the class C, even if they were dispersed across different data sources. Thus, the data in the Semantic Data Lake is conceptually represented as a set of class instances. Thanks to the Semantic Mappings associated to each dataset in the lake, every data instance has a class, even if it is not RDF.

Architecture. Ontario adopts a Wrapper-Mediator architecture (cf. Fig. 3) with one extra middle-layer:

– *Mediator:* Decomposes the SPARQL query into a set of star-shaped groups. Star-shaped groups are planned into a bushy tree execution plan.

- *Meta-Wrapper:* Each star-shaped group is submitted to a Meta-Wrapper.
- *Wrapper:* Translates the star-shaped group into a query of the syntax of the final data source.

Wrappers are selected by the Meta-Wrappers to obtain a sub-set of the final data. Those result sub-sets returned by the wrappers are joined together according to the bushy tree execution plan forming the final query answer.

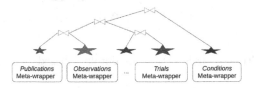

Fig. 4. Bushy execution plan and Meta-Wrapper invocation.

Example. We are interested in getting the *number of distinct publications and number of distinct deaths due to the disease Tuberculosis in India.* To obtain the answer, three datasets of different formats need to be queried: PubMed in XML, GHO in CSV and LinkedCT in RDF.

```
SELECT count(distinct(?publication))
AS ?no_of_publications
count(?deaths) AS ?no_of_deaths
WHERE {
    ?item a qb:Observation .
    ?item gho:Country ?country .
    ?item gho:Disease ?disease .
    ?item att:unitMeasure gho:Measure .
    ?item eg:incidence ?deaths .
    ?country rdfs:label "India" .
    ?disease rdfs:label "Tuberculosis".
    ?trial a ct:trials .
    ?trial ct:condition ?condition .
    ?trial ct:location ?location .
    ?trial ct:reference ?publication.
    ?condition owl:sameAs ?disease .
    ?location redd:locatedIn ?country .
    ?publication ct:citation ?citation.
}
```

```
?item a qb:Observation .
?item gho:Country ?country .
?item gho:Disease ?disease .
?item att:unitMeasure gho:Measure .
?item eg:incidence ?deaths .

?country rdfs:label "India" .

?disease rdfs:label "Tuberculosis".

?trial a ct:trials .
?trial ct:condition ?condition .
?trial ct:location ?location .
?trial ct:reference ?publication.

?condition owl:sameAs ?disease .

?location redd:locatedIn ?country .

?publication ct:citation ?citation.
```

Fig. 5. An example of a SPARQL query on the left, and the corresponding star-shaped groups the Mediator generates on the right.

Listing 1.1. Example of RML Mapping Rules

```
<#ObservationMappings>
  rml:logicalSource [
    rml:source              "hdfs://.../GHO/observations.csv";
    rml:referenceFormulation  ql:CSV ];
  rr:subjectMap [ rr:template"http://ex.com/{@OID}"];
  rr:predicateObjectMap [
    rr:predicate  gho:Country;
    rr:objectMap  [ rml:reference"country"; ];
  rr:predicateObjectMap [
    rr:predicate  gho:Disease;
    rr:objectMap  [ rml:reference"disease"; ];
  ...
```

The respective query used is shown in Fig. 5 left. The Mediator decomposes the SPARQL query into star-shaped groups and generates a corresponding bushy tree execution plan, as shown in Fig. 4. Each star-shaped group is submitted to a Meta-Wrapper. Each Meta-Wrapper checks the mapping rules, which are expressed in RML[8] in our implementation, and selects the relevant Wrappers, as depicted in Fig. 6. A snippet from the mapping rules used is shown in Listing 1.1. We can read that observations exist in the CSV file located in HDFS. The CSV columns are mapped to Ontology terms, e.g. country to `gho:Country`, disease to `gho:disease`, etc. Using those Mappings, the Wrapper is able to convert the star-shaped group, sent from the Meta-Wrapper, into an executable query. For example, in Fig. 6, the Wrapper converts the star-shaped group into an SQL query, as the CSV file is queried using *Apache Spark SQL*[9]. Wrappers' individual results are joined according to the execution plan generated earlier.

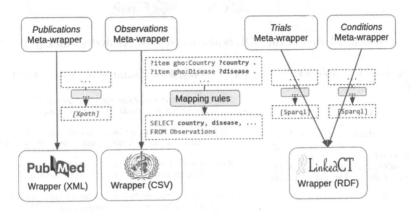

Fig. 6. Examples of Data Lake wrappers for PubMed, LinkedCT and GHO datasets.

3.2 Big Data Analytics for RDF

One of the key features of Big Data is its complexity and heterogeneity. While most of the big data applications can deal with concurrent computations of different kinds of data, there is still the need to combine these data from different simultaneous resources in a meaningful manner. RDF provides a model for encoding semantic relationships between items of data so that these relationships can be interpreted computationally. This section presents the *Semantic ANalytics StAck* (SANSA)[10] which is a Big Data platform that provides tools for implementing machine learning algorithms directly on RDF data. SANSA is divided into different layers described below (Fig. 7).

---

[8] http://rml.io.
[9] http://spark.apache.org/sql.
[10] http://sansa-stack.net/.

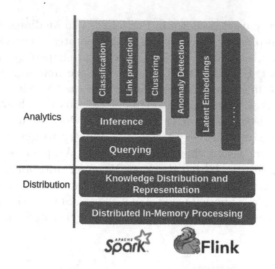

Fig. 7. Conceptual view of the BDE Scalable Semantic Analytics Stack (SANSA).

Listing 1.2. Example of reading an RDF file in SANSA

```
import
net.sansa_stack.rdf.spark.io.NtripleReader

val input ="hdfs://.../file.nt"

val triplesRDD =
NTripleReader.load(spark, new File(input))
triplesRDD.take(5).foreach(println(_))
```

Listing 1.3. Example to load an OWL file

```
// In RDD
FunctionalSyntaxOWLAxiomsRDDBuilder.build(sc,
  "path/to/functional/syntax/file.owl")
// Using Dataset
FunctionalSyntaxOWLAxiomsDatasetBuilder.build(
  spark,"path/to/functional/syntax/file.owl")
// Manchester syntax RDD
ManchesterSyntaxOWLAxiomsRDDBuilder.build(sc,
  "path/to/manchester/syntax/file.owl")
// Manchester syntax Dataset
ManchesterSyntaxOWLAxiomsDatasetBuilder.build(sc,
  "path/to/manchester/syntax/file.owl")
```

Read/write Layer. This layer provides the facility to read and write RDF data from HDFS or local drive and represent it in the distributed data structures of the frameworks.

Querying Layer. Querying an RDF graph is a major source of information extraction and searching facts from the underlying linked data. In order to efficiently answer runtime SPARQL queries for large RDF data, we are exploring different representation formats, namely graphs, tables and tensors.

Listing 1.4. Example to query an RDF file using SPARQL

```
val graphRdd = NTripleReader.load(spark, new File(input))
val partitions = RdfPartitionUtilsSpark.partitionGraph(graphRdd)
val rewriter = SparqlifyUtils3.createSparqlSqlRewriter(spark, partitions)
val qef = new QueryExecutionFactorySparqlifySpark(spark, rewriter)
```

Our aim is to have cross representational transformations for efficient query answering. Spark's GraphX is not very efficient, due to complex querying related to graph structure. On the other hand, an RDD based representation is efficient for queries like filters or applying user defined functions on specific resources, data frames have been found efficient for calculating the support of rules.

Inference Layer. The core of the inference process is to continuously apply schema related rules on the input data to infer new facts. This process can derive new facts from the knowledge base, detect inconsistencies from the KBs, and extract new rules to help in reasoning. Rules describing general regularity can help to understand the data better.

Listing 1.5. Example of inferencing an RDF Graph in SANSA

```
val graph = RDFGraphLoader.loadFromFile(new File(input).getAbsolutePath, spark, 4)
val reasoner = profile match {                    // create reasoner
   case TRANSITIVE => new TransitiveReasoner(spark, properties, parallelism)
   case RDFS => new ForwardRuleReasonerRDFS(spark, parallelism)
   case RDFS_SIMPLE =>
     val r = new ForwardRuleReasonerRDFS(spark, parallelism)
       r.level = RDFSLevel.SIMPLE
       r
   case OWL_HORST => new ForwardRuleReasonerOWLHorst(spark)
}
val inferredGraph = reasoner.apply(graph)        // compute inferred graph
```

We use an adaptive rule engine that will be able to optimize itself based on the rules available in the KB. This helps in developing an execution plan from a set of inference rules enabling applications to fine tune the rules for scalability.

Machine Learning Layer. In addition to above mentioned tasks, one of the very important tasks is to perform machine learning or analytics to gain insights of the data for relevant trends, predictions or detection of anomalies. There exists a wide range of machine learning (supervised and unsupervised) algorithms for the structured data. However, the challenging task is to distribute the data and to devise distributed versions of these algorithms to fully exploit the underlying frameworks. This distribution effort can be further divided into two separate categories; One is parallelizing the algorithms, and the other is ensemble learning or parallel modeling techniques. We are exploring different algorithms namely, tensor factorization, association rule mining, decision trees and clustering. The aim is to provide a set of out-of-the-box algorithms to work with the structured data. As an example consider the implementation of a partitioning algorithm for RDF graphs given as NTriples. The algorithm uses the structure of the underlying undirected graph to partition the nodes into different clusters.

Listing 1.6. Example for Clustering RDF

```
import net.sansa_stack.ml.spark.clustering.RDFByModularityClustering
val numIterations = 100
val input ="path_to_your_RDFgraph"
val output ="path_name_for_clusters"
RDFByModularityClustering(spark, numIterations, input, output)
```

SANSA's clustering procedure follows a standard algorithm for partitioning undirected graphs aimed to maximize a modularity function. Usage examples and further information can be found at http://sansa-stack.net/faq/.

Listing 1.7. SPARQL query that retrieves uniform event summaries.

```
PREFIX geo:<http://www.opengis.net/ont/geosparql#>
PREFIX ev:<http://bde.eu/man-made-changes/ont#>
PREFIX nev:<http://cassandra.semagrow.eu/events#>
SELECT ?id ?title ?loc ?name ?desc {
  ?e rdf:type           ev:NewsEvent ;
     ev:hasId           ?id ;
     ev:hasTitle        ?title ;
     ev:hasArea/geo:hasGeometry/geo:asWKT ?loc .
  ?s nev:event_id       ?id ;
     nev:description    ?desc ;
     nev:event_date     ?date .
  OPTIONAL{?s nev:tweet_post_ids ?tweet_post_ids}
  FILTER regex(?title, 'zaatari','i') .
}
```

Listing 1.8. Federated query retrieving events from Strabon & Cassandra.

```
SELECT ?id ?title ?loc ?name ?desc
{ ?e rdf:type          ev:NewsEvent ;
     ev:hasId          ?id ;
     ev:hasTitle       ?title ;
     ev:hasArea/geo:hasGeometry/geo:asWKT
                       ?loc .
  FILTER regex(?title, 'zaatari','i') .
} @ Strabon
{ ?s nev:event_id      ?id ;
     nev:description   ?desc ;
     nev:event_date    ?date .
  OPTIONAL { ?s nev:tweet_post_ids
                ?tweet_post_ids }
} @ Cassandra
```

3.3 Semagrow Query Federation

Another component of the BDE Platform relevant to the semantics of the data being processed, is the *Semagrow* federation engine. Semagrow is a SPARQL query processing system that federates multiple remote endpoints. Semagrow hides schema heterogeneity by applying the appropriate vocabulary transformations and also uses metadata about the contents of the remote data sources to optimize querying plans [2]. Client applications are presented with a single SPARQL endpoint, and Semagrow transparently optimizes queries, executes sub-queries to the remote endpoints, dynamically integrating results in heterogeneous data models, and joins the partial results into the response to the original query and into the original query's schema.

In the context of integrating Semagrow in the BDE Platform, we have redesigned both the query planner and the execution engine so that it can be extended to support different querying languages. In cases where the target language is less expressive than SPARQL, Semagrow itself undertakes the required additional computations. For instance, when Apache Cassandra endpoints are included in the federation, the Semagrow query planner is aware of the fact that CQL, the Apache Cassandra query language, does not support joining across tables. The resulting plan is built in such a way that the tuples to be joined are fetched by different CQL queries to the same endpoint and joined at the Semagrow side. But this does not mean that only individual query patterns are fetched: "star" queries that fetch multiple properties of the same entity are expressible in CQL and the Semagrow planner takes into account that such patterns may be bundled together in one query [6]. Consider, for example, the SPARQL query in Listing 1.7 that involves geo-location from events stored in Strabon, a geospatial triple store that supports stSPARQL, and event summaries harvested from Twitter stored in Apache Cassandra.

Semagrow provides a transparent way to access and cross-join data from both sources. This is achieved by decomposing the initial SPARQL query into valid subqueries taking the capabilities and expressivity of each store into account, deciding the order of execution by estimating the cost of each alternative execution, and lastly translating each subquery into an appropriate query that the underlying system can understand. In our example the query in Listing 1.7 will be split into two subqueries, depicted in Listing 1.8 as blocks annotated with the name of the store to be executed. Note also that the subquery to be executed in Apache Cassandra is further translated into a valid CQL query. In order to scale out the query execution across different blades of a distributed infrastructure, we are re-implementing the execution engine over Apache Flink. (possibly on different machines) rather than different threads on the same machine.

4 BDE Platform Showcases

The main goal of the BigDataEurope project is to produce an easy-to-develop, -use and -adapt platform for wildly varying Big Data challenges. To validate this, we develop pilot implementations in seven different domains, corresponding to the seven Societal Challenges (SC) of Horizon2020. These pilots are defined and developed by user partners in each of these challenges in collaboration with the technical team. A particular pilot comprises key data assets and domain-specific enabling technology in this domain and a BDE pilot implementation supports domain-specific workflows, exploration and visualization technologies. This has resulted in a versatile, but coherent set of demonstrators, which illustrate how relevant large-scale datasets or data-streams for the respective seven SC communities can be processed by the BDE infrastructure and provide novel insights that are promised by the Big Data community. We present two of these pilots, the BDE components used as well as an evaluation of the specific pilot and refer the reader to BDE deliverables[11] for details on all pilots.

Note that for the functional and non-functional requirements of the generic infrastructure part the *FURPS model* [5] is followed, classifying the software quality attributes with respect to Functionality, Usability, Reliability, Performance and Scalability. The details of each of these requirements are different for each challenge and were evaluated separately. At the same time, the generic BDE infrastructure was evaluated independently of these challenges according to the FURPS model. For each pilot, challenge-specific key evaluation questions were answered by the challenge partners. These were specified from generic evaluation questions corresponding to FURPS items. An example of such a question for SC1 is *'Are there currently vulnerabilities in the BDE infrastructure that might reveal any sort of communication to a 3rd party (e.g. queries and results, or IP addresses)?'*, relating to the functionality item.

4.1 SC1 Health, Demographic Change and Wellbeing

The first pilot[12] in SC1 "Health, demographic change and wellbeing" implements the Open PHACTS Discovery Platform [10] for drug discovery on the BDE infrastructure.

---

[11] https://www.big-data-europe.eu/results/.
[12] https://www.big-data-europe.eu/pilot-health/.

The Open PHACTS Discovery Platform has been developed to reduce barriers to drug discovery in industry, academia and for small businesses. Researchers in drug discovery use multiple different data sources; Open PHACTS integrates and links these together so that researchers can easily see the relationships between compounds, targets, pathways, diseases and tissues. The Open PHACTS platform is a good example of a Big Data solution for efficient querying over a wide variety of large data sources that are integrated via an elaborate and mostly human curated process. The platform is founded on semantic web and linked data principles and uses industrial strength tools such as *Virtuoso*[13] to provide fast and robust access to the integrated chemistry and biological data sources. This integration effort resulted in a set RDF link sets, which map the large numbers of identifiers from these various sources and are stored in Virtuoso to answer queries from users. To simplify and scale access, an abstraction layer using the *Puelia*[14] implementation of the *Linked Data API Specification*[15] is added to translate REST-full requests to instantiated SPARQL queries. The REST API is documented via the *OpenAPI specification*[16] (formerly Swagger) and available on the Open PHACTS Github repository[17].

The Open PHACTS pilot. The advantage of making the Open PHACTS functionality available as an instance on the BDE infrastructure is threefold:

- To improve security, an organisation might prefer to have the Open PHACTS functionality available on their own secure local cluster. The BDE infrastructure allows an open, almost 'one-click install' which makes this an affordable option for smaller companies and other organisations with a limited budget.
- The BDE approach, and in particular the Docker stack, provides a modular architecture where components can easily be replaced. For example, the current Open PHACTS platform uses the commercial cluster version of Virtuoso. The BDE stack makes it easy to replace this with another RDF store, for example, the open source 4Store or the Ontario and SANSA stack.
- The modularisation also makes it relatively simple to adapt for a different domain. One only has to have data available as RDF, create the link sets between the sources and describe the SPARQL templates which, via Puelia, provide the REST interface for the desired functionality.

In order to realise the Open PHACTS platform on the BDE infrastructure, all third party software and the functionally independent components from the Open PHACTS platform are 'dockerized' (cf. Sect. 2). Two third party tools used by Open PHACTS are already available as Docker containers, namely *MemCached* and *MySQL*. For data storage and management, as an alternative to

---

[13] https://virtuoso.openlinksw.com/.

[14] https://code.google.com/archive/p/puelia-php/.

[15] https://code.google.com/archive/p/linked-data-api/wikis/Specification.wiki.

[16] https://www.openapis.org/.

[17] https://github.com/openphacts/OPS_LinkedDataApi/blob/develop/api-config-files/swagger-2.0.json.

the commercial version of Virtuoso, it is possible to use the open source version (which is also available as a Docker container[18]) or 4Store integrated with the SANSA stack (cf. Subsect. 3.2). The remaining internally developed components[19] are:

- *OPS LinkedDataApi*, a Docker component containing the Puelia related code and the Swagger documentation generator.
- *Open PHACTS Explorer*, an HTML5 & CSS3 application for chemical information discovery and browsing. It is used to search for chemical compound and target information using a web search interface.
- *IdentityMappingService*, a Docker container of the native Open PHACTS service *queryExpander*, which includes the *IdentityMappingService* (IMS) and the *VoID Validator*.

The current pilot can be deployed on Linux and Windows, and the instructions can be found on the BDE Github repository[20]. BDE project Deliverable 6.3 [4] provides an evaluation of all the pilots and shows that this pilot adheres to the specified requirements, that the software can be easily deployed, and that the code is well documented. Community feedback has guided the next pilot cycle to (a) broaden the community beyond drug research to include other data and (b) add functionality to support extended domain requirements. In the next pilot cycle the existing datasets will be updated with new data from their sources, and functionality and data-sources to address the domain of food safety will be investigated.

4.2   SC7: Secure Societies

The SC7 pilot[21] combines the process of detecting changes in land cover or land use in satellite images (e.g., monitoring of critical infrastructures) with the display of geo-located events in news sites and social media. Integrating *remote sensing* with *social sensing* sources is crucial in the Space and Security domain, where useful information can be derived not only from Earth Observation products, but also from the combination of news articles with the user-generated messages of social media. The high-level architecture of the pilot is presented in Fig. 8. In total, it comprises 11 BDE platform components, which can be grouped into the following three workflows:

- The *change detection workflow* consists of the three components at the bottom of Fig. 8. The Image Aggregator receives the area and the time interval of interest from the UI and retrieves corresponding satellite images from ESA's *Sentinel Scientific Data Hub* (SciHub)[22]. The images are ingested into HDFS

---

[18] https://hub.docker.com/r/stain/virtuoso/.
[19] Available from: https://hub.docker.com/r/openphacts/.
[20] https://github.com/big-data-europe/pilot-sc1-cycle1.
[21] https://www.big-data-europe.eu/pilot-secure-societies.
[22] https://scihub.copernicus.eu.

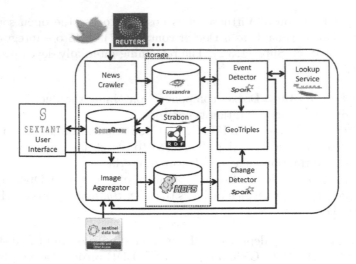

Fig. 8. High-level architecture of the SC7 pilot.

to be processed by the Change Detector, which relies on Spark in order to apply a set of established operators efficiently and in parallel for comparing satellite images.

– The *event detection workflow* comprises the four components at the top of Fig. 8. The News Crawler periodically checks various public news streams (Twitter, specific Twitter accounts, RSS feeds of Reuters[23]. These are stored in Cassandra in a way that abides by the corresponding privacy regulations. The Event Detector is periodically executed in order to cluster the news items into events using Spark for efficiency. In this process, special care is taken to associate every event with one or more geo-locations. The location names that are extracted from the text are associated with their geo-coordinates through a query in the Lookup Service, which indexes 180,000 location names from the *GADM dataset*[24] using Lucene.

– The *activation workflow* consists of the four components in the middle of Fig. 8. *GeoTriples* [8] receives the detected areas with changes in land cover or use and summaries of detected events in order to convert their descriptions into RDF. These are stored in the spatio-temporal triplestore *Strabon* [7], which efficiently executes *GeoSPARQL* and *stSPARQL* queries. Semagrow federates Cassandra and Strabon, offering a unified access interface to *Sextant* [9], the user interface of the pilot. To meet all user requirements, Sextant has been significantly extended, allowing users to call both the change and the event detection workflows and to visualize their outcomes.

Figure 9 shows an example of refugee camps located in Zaatari, Jordan. The results from the change detection and the event detection workflows are displayed

---

[23] http://www.reuters.com/tools/rss.
[24] http://www.gadm.org.

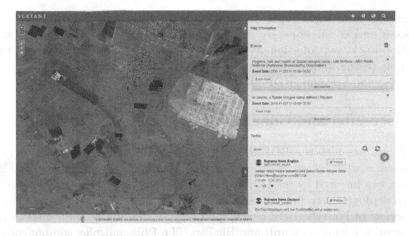

Fig. 9. Visualization of the change detection (displayed in orange on the map) and event detection results (clustered events and tweets in the right panel).

Table 1. Comparison of the BDE Stack with other Big Data distributions (SFR = Single failure recovery; MFR = Multiple failure recovery, SF = Self healing).

	Hortonworks	Cloudera	MapR	Bigtop	BDE platform
File system	HDFS	HDFS	NFS	HDFS	HDFS
Installation	Native	Native	Native	Native	Lightweight virtualization
Plug & play components	☐	☐	☐	☐	■
High availability	SFR (yarn)	SFR (yarn)	MFR, SF	SFR (yarn)	MFR
Cost	Commercial	Commercial	Commercial	Open-source	Open-source
Scaling	Freemium	Freemium	Freemium	Free	Free
Extensibility	Difficult	☐	☐	☐	■
Integration testing	■	■	■	■	☐
Operating systems	Linux	Linux	Linux	Linux	All
Management tool	Ambari	Cloudera manager	MapR control	☐ system	Docker swarm UI+ Custom

on Sextant using SemaGrow to retrieve information from Cassandra and Strabon in a uniform way (i.e., through the queries in Listings 1.7 and 1.8).

In general, variety is manifested in this pilot in the form of the different types of satellite images (e.g., SAR and optical ones) as well as the textual content from news agencies and social media. To address variety, useful information is extracted from these types of data by the change and the event detection workflows and is subsequently converted into searchable RDF triples by the activation workflow. Additionally, SemaGrow federates efficiently and effectively the access to the information that is stored in Cassandra (part of the original news text) and Strabon (RDF data).

5   Discussion and Conclusions

Table 2 outlines how the BDE Stack fulfills the initially derived requirements and indicates what measures support each requirement. In certain ways the BDE platform was inspired by the LOD2 Stack [1], which however used Debian packaging as deployment technology, since Docker was not yet available. Table 1 gives an overview of how the BDE Stack compares to major Big Data distributions Hortonworks, Cloudera, MapR and BigTop. The Plug-and-play components row describes how customized workflows can be realized. Extensibility means whether its possible to add custom components. The scaling refers to the requirement to pay additional licensing fees for cluster deployments. The comparison shows, that the BDE platform is complementary in many ways and especially with its extensibility, adaptability, consequent Docker-containerization and deployment as well

Table 2. BDE platform measures for addressing the requirements.

Requirement	BDE platform measure
R1 Simplifying use	M1.1 Integration of Web UIs for the overall platform and individual components
	M1.2 Visual Big Data workflow authoring and monitoring
	M1.3 Unified and integrated semantic representation of data
R2 Easing deployment	M2.1 Dockerization of components
	M2.2 Deployment orchestration using Docker Swarm
	M2.3 Support for various deployment schemes (individual machine, cloud, cluster)
R3 Managing heterogeneity	M3.1 Ingestion of heterogeneous data through RDF mapping and transformation
	M3.2 Direct operation of analytical algorithms for Big Data components on top of RDF data representations
	M3.3 Extensibility of the platform with custom, domain specific components
R4 Improving scalability	M4.1 Federated, parallelized query execution with Ontario and Semagrow
	M4.2 Pushing down of analytical queries on semantic representations to optimized Big Data analytics components with SANSA

its semantic layer goes far beyond the state-of-the-art. However, the interplay and integration with other Big Data platforms (especially Apache BigTop) is not only possible but increasingly advancing with related standardization efforts such as ODPi (https://odpi.org).

The advantages of semantics in distributed architectures appeal to both the Semantic Web community and to the Big Data processing community. The contribution the Semantic Web is SPARQL query processing that can scale over more voluminous query responses than what is currently possible. This makes Semantic Web approaches and representations a viable solution for new domains and applications, such as Earth observation (cf. Subsect. 4.2), where not only the underlying data but also the volume of the response is often orders of magnitude larger than the implicit limits of current SPARQL endpoints. For the Big Data processing community, the ability to join results from heterogeneous data stores allows integrating data and metadata. As an example, in the BigDataEurope SC5 climate modelling pilot numerical data in Hive and its provenance metadata in a triple store is joined in order to provide filters that refer not only to the data itself but also to provenance metadata regarding this data's origin.

References

1. Auer, S., Bryl, V., Tramp, S. (eds.): Linked Open Data – Creating Knowledge Out of Interlinked Data. LNCS, vol. 8661. Springer, Cham (2014)
2. Charalambidis, A., Troumpoukis, A., Konstantopoulos, S.: Semagrow: optimizing federated SPARQL queries. In: SEMANTiCS (2015)
3. Big Data Europe. WP2 deliverable: Report on interest groups workshops III (2016)
4. Big Data Europe. WP6 deliverable: Pilot evaluation and community specific assessment (2016)
5. Grady, R.B.: Practical Software Metrics for Project Management and Process Improvement. Prentice Hall, Upper Saddle River (1992)
6. Konstantopoulos, S., Charalambidis, A., Mouchakis, G., Troumpoukis, A., Jakobitch, J., Karkaletsis, V.: Semantic web technologies and big data infrastructures: SPARQL federated querying of heterogeneous big data stores. In: ISWC Demos and Posters Track (2016)
7. Kyzirakos, K., Karpathiotakis, M., Koubarakis, M.: Strabon: a semantic geospatial DBMS. In: Cudré-Mauroux, P., Heflin, J., Sirin, E., Tudorache, T., Euzenat, J., Hauswirth, M., Parreira, J.X., Hendler, J., Schreiber, G., Bernstein, A., Blomqvist, E. (eds.) ISWC 2012. LNCS, vol. 7649, pp. 295–311. Springer, Heidelberg (2012). doi:10.1007/978-3-642-35176-1_19
8. Kyzirakos, K., Vlachopoulos, I., Savva, D., Manegold, S., Koubarakis, M.: Geotriples: a tool for publishing geospatial data as RDF graphs using R2RML mappings. In: ISWC Posters & Demonstrations Track (2014)
9. Nikolaou, C., Dogani, K., Bereta, K., Garbis, G., Karpathiotakis, M., Kyzirakos, K., Koubarakis, M.: Sextant: visualizing time-evolving linked geospatial data. J. Web Sem. 35, 35–52 (2015)
10. Williams, A.J., Harland, L., Groth, P., Pettifer, S., Chichester, C., Willighagen, E.L., Evelo, C.T., Blomberg, N., Ecker, G., Goble, C., Mons, B.: Open PHACTS: semantic interoperability for drug discovery. Drug Discov. Today 17(21–22), 1188–1198 (2012)

Spatially Cohesive Service Discovery and Dynamic Service Handover for Distributed IoT Environments

Kyeong-Deok Baek and In-Young Ko[✉]

School of Computing, Korea Advanced Institute of Science and Technology,
Daejeon, Republic of Korea
{kyeongdeok.baek,iko}@kaist.ac.kr

Abstract. The proliferation of the Internet of Things (IoT) enables the provision of diverse services that utilize IoT resources distributed in ad-hoc network environments. This has resulted in a new challenge, the issue of how to efficiently and dynamically discover appropriate IoT services that are necessary to accomplish a user task in the vicinity of the user. In this paper, we propose a service discovery method that finds IoT services from a user's surrounding environment in a spatially cohesive manner so that the interactions among the services can be efficiently carried out, and the outcome of service coordination can be effectively delivered to the user. In addition, to ensure a certain Quality of user Experience (QoE) level for the user task, we develop a service handover approach that dynamically switches from one IoT resource to an alternative one to provide services in a stable manner when the degradation of the spatial cohesiveness of the services is monitored. The spatio-cohesive service discovery and dynamic service handover algorithms are evaluated by simulating a mobile ad-hoc network (MANET) based IoT environment. Then, various service discovery strategies are implemented on this simulation environment, and several options for the service discovery and handover algorithms are tested. The simulation results show that compared to various baseline approaches, the proposed approach results in a significant improvement in the spatial cohesiveness of the services discovered for user tasks. The results also show that the approach efficiently adapts to dynamically changing distributed IoT environments.

Keywords: Internet of things · Distributed service discovery · Service handover · Spatio-cohesive service coordination · Task-oriented computing

1 Introduction

Recent years have witnessed a wide spread proliferation of Internet of Things (IoT) and this has [1] enabled the provision and deployment of diverse services that utilize IoT resources in urban environments. For effective collection of data and management of IoT services, recent commercial IoT-based systems, such as Microsoft Azure IoT Hub[1],

---

[1] https://azure.microsoft.com/en-us/services/iot-hub/.

© Springer International Publishing AG 2017
J. Cabot et al. (Eds.): ICWE 2017, LNCS 10360, pp. 60–78, 2017.
DOI: 10.1007/978-3-319-60131-1_4

have adopted the cloud computing model. However, owing to a rapid increase in the number of IoT resources [2] in recent years, a centralized cloud computing model no longer remains scalable. Therefore, recent studies have suggested distributed computing models based on the Mobile Ad-hoc Network (MANET) [3] to address the scalability issue [4]. In this distributed model, as can be seen in Fig. 1, IoT resources are accessible directly via a MANET, which allows for IoT resources to interact with each other in order to provide services to users in a more efficient and flexible manner without the need to deploy infrastructure. In addition, composite services can be developed by integrating IoT and other types of services such as Web and cloud services together to accomplish user tasks.

For user-centric provision of IoT services, we developed a task-based service provision framework in our previous research [5]. In this framework, a *task* is a description of a user's demand, which entails the services required to provide necessary functionalities. A *service* defines a set of functionalities that are necessary to support an activity for a user task. A service is realized through a service instance that requires a set of *resources*, which are IoT devices that provide certain primitive functionalities. In this paper, we assume that IoT resources have enough computational power and networking capabilities to form a MANET without any infrastructure. However, even if an IoT resource is constrained by limited computation or networking capabilities, a *service gateway* may be used to enable the resource to join a MANET.

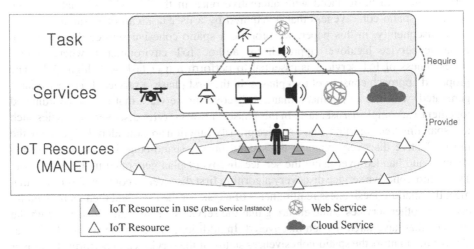

Fig. 1. Distributed IoT Environment Using MANET

Such a distributed IoT environment, however, raises a new challenge to efficiently and dynamically discovering appropriate IoT resources, which provides necessary IoT services to accomplish a user task, in the vicinity of the user. To address this issue, several studies have been undertaken in the service-oriented architecture and Web services domains [6]. In particular, mobility, location dependency, and composability of distributed services have been considered as important issues to address to enable effective service provision in practical IoT environments [7]. First, owing to mobility of users and IoT resources, the availability of IoT services frequently changes. Therefore, a discovery algorithm need to dynamically discover valid IoT services that are necessary to perform user tasks. Second, unlike traditional Web services, the spatial location of a user and IoT resources affects the effectiveness of delivering IoT services to users. This implies that, as many studies point out [6, 7], to deliver the outputs of IoT services to a user in an effective and efficient manner, corresponding IoT resources need to be located close to the user. Third, various services in an IoT environment need to coordinate with each other to perform complex user tasks, and therefore, cooperative IoT resources need to be located cohesively in a space where the user is located.

Even though there have been many studies on evaluating locational context of IoT services [6], none of them have focused on ensuring IoT resources that are involved with service coordination to be spatially cohesive to each other, to accomplish user tasks effectively. In this paper, we define *spatio-cohesive services* as services that utilize IoT resources that are located cohesively in a user's vicinity with the distance among the services short enough such that the outcome of the service coordination can be effectively delivered to the user. The spatio-cohesiveness of IoT services is a critical factor to maximize the Quality of user Experience (QoE) of a user task. Therefore, when degradation of spatio-cohesiveness of IoT services is detected, some of the services need to be replaced with alternative ones in the MANET so that a certain degree of spatio-cohesiveness that is required by a user task can be ensured.

Consequently, in this paper, we propose a spatio-cohesive service discovery and dynamic service handover approach for ad-hoc IoT environments, where spatio-cohesiveness of IoT services is essential to perform a user task with high QoE. The proposed approach consists of two phases. In the first phase, a service discovery plan is generated, which is a systematic plan for discovering services that can be coordinated together to accomplish a user task. In this phase, various service discovery strategies such as a spanning tree algorithm and a service priority function are considered to generate the most effective discovery plan. In the second phase, spatio-cohesive services are discovered and handed over using the service discovery plan generated in the first phase. The spatio-cohesive service discovery algorithm first discovers a core set of services that meet the spatio-cohesiveness requirements of a task for the user, and then incrementally discovers other services while ensuring that the services are spatially cohesive with the services that have already been discovered. In addition, the dynamic service handover algorithm monitors the spatio-cohesiveness status of the services and performs handover when some of the services are beyond the requirement boundary.

We evaluated the spatio-cohesive service discovery and handover algorithms by simulating a distributed IoT environment. Further, we implemented the service discovery strategies on the simulation environment and tested two options for allowing concurrent service discovery and discovering multiple candidate services with the service discovery plans. Because there is no research on spatio-cohesiveness among services, we implemented baseline as a basic algorithm that does not consider spatial distance among services. Consequently, we observed that the proposed algorithms show improvements in terms of spatio-cohesiveness of services discovered for user tasks.

The remainder of this paper is organized as follows. Section 2 discusses related work on service discovery in dynamic environments. Section 3 explains the spatio-cohesive service discovery and dynamic service handover method. Section 4 outlines the evaluation conducted of the proposed approach and analyzes the simulation results obtained. Section 5 introduces our testbed implementation and demo scenario. Finally, Sect. 6 concludes this paper.

2 Related Work

Service discovery in service computing environments can be classified into two types: proactive and reactive service discovery [8]. Reactive service discovery algorithms discover services according to a user's request, while proactive service discovery algorithms discover services before a user's request based on advertisement from IoT resources. In distributed IoT environments, it is necessary to discover services in a reactive manner according to the requirements of user tasks because of scalability. Furthermore, in IoT environments, a service discovery often requires the discovery of IoT resources that are necessary to provide the service [8]. Therefore, service discovery in this paper is closely associated with the discovery of the required IoT resources. In particular, the service discovery process in IoT environments includes the selection of the best IoT resources among candidate resources that can provide required services.

There have been various studies on scalable discovery of required services for performing user tasks in MANET environments [8, 9]. A MANET is an infrastructure-less environment, which can enable resources with high mobility to be connected with each other in a scalable manner. However, owing to the infrastructure-less characteristics of MANET, we cannot maintain a centralized service registry to discover services. Therefore, service discovery in a MANET needs to be performed in a decentralized manner [8]. Although this makes service discovery in a MANET more complex than a centralized environment, it enables heterogeneous and mobile IoT services to be discovered in a robust manner.

To improve the efficiency of service discovery in a MANET, some studies use a cross-layer method that merges service discovery into network capabilities [8, 10]. Their work focuses on optimizing network protocols for service discovery rather than improving QoE that users can observe.

One recent study points out that it is important to coordinate IoT services based on a goal-oriented framework to effectively deal with complex demands from users [4]. In particular, they state that it is essential to find multiple services that need to participate in service coordination to accomplish a user goal. However, in a dynamic MANET environment, where the availability of services changes often, it is challenging to discover multiple services within a user's vicinity. Furthermore, providing failure-resilient service discovery is also a challenging issue owing to the highly dynamic nature of MANET [4], and diverse contexts of services to consider [6, 11].

With regard to delivering services to users while ensuring a high QoE, context-aware service discovery has been considered as an important issue in service provision [6, 11]. In these studies, contextual information such as a user's preferences and location are utilized to increase the integrated quality of discovered services from users' perspectives and in terms of the efficiency of network architecture. In particular, the location of services is a critical context that affects QoE in performing user tasks [12]. For example, in a recent study, the goal was to optimize the integrated quality of discovered services by considering the relative locations of the services within a net-work [13]. In this research, they use a genetic algorithm to improve the quality of discovered services in an iterative manner. However, this approach requires frequent verification of the status of the network, and therefore it may incur considerable communication effort among services that are distributed over a dynamic network environment like MANET.

3   Spatially Cohesive Service Discovery and Dynamic Service Handover

Figure 2 shows the main phases and overall process of the spatio-cohesive service discovery and dynamic service handover algorithms. In the first phase, the spatio-cohesiveness requirements are extracted and a plan for the service discovery and handover is generated. A *spatio-cohesiveness requirement* is a locational dependency between two services or between a user and a service, which should be located cohesively to each other. After the extraction, the requirements are represented in a graph, and the service discovery plan is instantiated on the graph, using various dis-covery strategies that guides traversal on the graph. In the second phase, IoT services for a user task in an IoT environment are discovered in a spatially cohesive manner based on the service discovery plan generated. The dynamic service handover step is

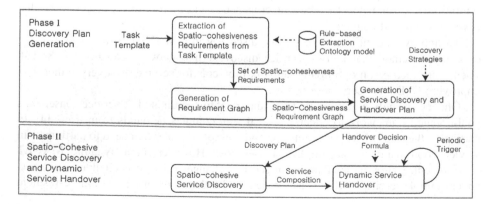

Fig. 2. The spatio-cohesive service discovery and handover process

periodically activated to monitor any changes in the spatio-cohesiveness requirements of the services, and to dynamically hand over some of the service instances that are not spatially cohesive to other services into alternative ones.

3.1 Problem Definition

The spatio-cohesive service discovery and dynamic service handover problem that this work tries to solve is defined formally as follows. For a given task, T, the problem is to find a proper set of IoT resources, which are necessary to provide a set of required services, S_T, of the task. Each service, s $\in S_T$, has its own service description, des_s, that contains the information about the required capabilities and constraints. A time series, TS, which is a sequence of time instances, τ_i, indicates a topology change of the MANET. Time instance τ_0 represents the time when the task starts its execution.

The MANET at time τ is represented as a graph, $N(\tau) = (D \cup \{u\}, E(\tau))$. The vertices of the graph include the set of IoT resources, D, which provide capabilities for services; and a user, u, that consumes the services. Each vertex v has function $v.\,\mathrm{coord}(\tau)$ that returns the physical coordinate of the node at time τ. A service that can be provided by utilizing an IoT resource, $d \in D$, is represented as $d_{service}$. An edge set of the network at time, τ, is defined as $E(\tau)$, where $E(\tau)$ is a set of direct links that forms MANET among IoT resources and user. The elements of $E(\tau)$ change as the connections among the IoT resources and user change because of the mobility.

In this work, we assume that there is only one user consuming the services in the environment. Because the focus of this work is on service discovery for user tasks in a spatially cohesive manner, the issues of simultaneous service discovery and the resource allocation problem with multiple users will be covered in future work.

A *spatio-cohesiveness requirement* is defined as (m, n, l), where m and n are a service or a user (i.e., m, n $\in S_T \cup \{u\}$), and they need to be located spatially close to each other within an upper limit, l. R_T is a set of spatio-cohesiveness requirements for a task, T, that can be automatically extracted from the service descriptions in the task template. In addition, spatio-cohesiveness requirements can be extracted from a user preference or can be manually provided by a user (Table 1).

Table 1. Summary of Formal Notations

Name	Notation
Service set for task T	S_T
Time instance	$\tau_i \in TS$
Network	$N(\tau) = (D \cup \{u\}, E(\tau))$
Vertex position	$v.\mathrm{coord}(\tau), v \in D \cup \{u\}$
Capability of IoT resource d	$d_{service}$
User	u
Connection	$(d_i, d_j)\ or\ (u, d_i) \in E(\tau)$
Spatio-cohesiveness requirement	$r_i = (m, n, l) \in R_T, for\, m, n \in S_T \cup \{u\}$
Candidate resource selection	$C(\tau) \subseteq 2^D$
Spatio-cohesive resource selection	$c_{SC}(\tau) \in C(\tau)$

The problem is to find the spatio-cohesive solution $c_{SC}(\tau) \in C(\tau)$, where $C(\tau) \subseteq 2^D$ is a set of candidate IoT resources that provide the services of the task at time τ. The spatial distance between the resources that provide the services in each service pair (services m and n), or between a resource that provides a service, m, and the user, n, of set c is defined as function $distance_c(m, n)$, and needs to be minimized for each spatio-cohesiveness requirement of the task. Moreover, achievement of the requirement is the ratio of distance and limit value, and will be 0 if distance exceed the limit. Thus, measurement of the achievement of a spatio-cohesiveness requirement, $r_i(c)$, with a candidate IoT resource set, c, can be measured as follows:

$$r_i(c) = \max\left(1 - distance_c(r_i.m.coord(\tau), r_i.n.coord(\tau))/_{r_i.l}, 0\right) \qquad (1)$$

In addition, according to the definition of the connectivity of a network [14], the spatio-cohesiveness of a candidate set of IoT resources can be measured as the average of the minimum distance of a service that causes failure to one of its spatio-cohesiveness requirements. Therefore, the spatio-cohesiveness objective function, $R_T(c)$, of a candidate resource set, c, can be defined as follows:

$$R_T(c) = \sum_{s \in S_{T,u}} \min_{r_i=(s,s_k,l) \in R_T} (r_i(c))/|S_T \cup \{u\}| \qquad (2)$$

The spatio-cohesiveness of a candidate IoT resource set is a value between zero and one. A low value indicates that there are many failures of meeting the requirements by IoT resources, while a high value indicates that most of the spatio-cohesiveness requirements are strongly achieved by the resources.

The spatio-cohesive service discovery problem is to find a set of resources that provide the spatio-cohesive services at the initial time, i.e., $c_{SC}(\tau_0)$, and the dynamic service handover problem is to find a new set of resources that provide spatio-cohesive services when there is a change in the IoT MANET, i.e., $c_{SC}(\tau_i)(i > 0)$. Furthermore, the dynamic service handover problem has an additional objective function, $handover(c(\tau_{i-1}), c(\tau_i))$, which is for identifying the number of handovers from $c(\tau_{i-1})$ to $c(\tau_i)$. The objective function for service handover can be defined as follows:

$$handover(c(\tau_{i-1}), c(\tau_i)) = |c(\tau_i) - c(\tau_{i-1})| \qquad (3)$$

By minimizing the number of IoT resources to be handed over, we can reduce the overhead of service handovers in accordance with the changes in the IoT MANET. Therefore, the following defines the problem of finding spatio-cohesive services for a user task, while minimizing service handover cost throughout the lifecycle of the task:

$$c_{SC}(\tau_i) = \left\{ \begin{array}{ll} \min_{c \in C(\tau_i)}(R_T(c)) & (i = 0) \\ \min_{c \in C(\tau_i)}(R_T(c), handover(c(\tau_i) - c_{SC}(\tau_{i-1}))) & (i > 0) \end{array} \right\} \qquad (4)$$

Because there are multiple objective functions for each spatio-cohesiveness requirement with some constraints, this problem is a constrained multi-objective optimization problem. To check if the objective functions included in $R_T(c)$ are conflicting with each other, let us consider an IoT resource, d_i, which is selected for the task. The discovery algorithm search for IoT resources that are necessary to provide the services that are spatially cohesive to the services of d_i. If we assume that all the resources and the user are distributed over the MANET randomly and uniformly, the resources are expected to be distributed in all directions from d_i. Therefore, choosing an alternative resource d_i' instead of d_i will make one of the spatio-cohesiveness requirement achievements decrease, while making another requirement achievement increase.

3.2 Discovery Plan Generation

Spatio-Cohesiveness Requirement Extraction. Studies on representing the functionality and the QoS of Web services semantically [12, 15] show that it is possible to infer the QoS requirements of a service in a semantic manner based on the semantic descriptions of the service. Similarly, the spatio-cohesiveness requirements of a user task can be extracted automatically from a task template. Therefore, as shown in Fig. 3, we define a brief ontology model to represent the spatio-cohesiveness requirements of a user task. The ontology model is developed using Web Ontology Language (OWL) and represented as a Unified Modeling Language (UML) class diagram. As shown in Fig. 3, a Task has its own *Task Template*, which is composed of multiple *Services*. A service can have some *Service Properties* representing the quality attributes and constraints of the service. In addition, a service can be classified into one or more *Service Category*. The services of a task and the *User* of the task can be combined to form a *Service Boundary*, by which a spatio-cohesiveness requirement can be specified for the task. Finally, a set of *Spatio-cohesiveness Requirements* are extracted from task template.

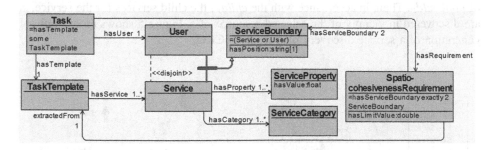

Fig. 3. Ontology model of spatio-cohesiveness requirements

Spatio-cohesiveness requirements that are represented based on this model can be extracted using a rule-based reasoning method. Figure 4 shows an example rule represented in Semantic Web Rule Language (SWRL). This rule states that if a task requires an audio service with volume property less than 30 dB, the distance between the user and the audio service should be less than 5 m:

Task(?t) ^hasTemplate(?t, ?tmpl)
^hasUser(?t, ?u) ^hasService(?tmpl, ?s)   hasRequirement(?t, ?req)
^hasCategory(?s, Audio)   ⟶   ^hasServiceBoundary(?req, ?u)
^hasProperty(?s, Volume)   ^hasServiceBoundary(?req, ?s)
^swrlb:lessThan(?v, 30)   ^hasLimitValue(?req, 5)

Fig. 4. SWRL rule for extracting spatio-cohesiveness requirements

Spatio-cohesiveness Requirement Graph Generation. In our approach, we represent the spatio-cohesiveness requirements as a graph called a *spatio-cohesiveness requirement graph (SCRG)*, in which the user and services are represented as vertices ($S_T \cup \{u\}$), and the spatio-cohesiveness requirements among them are represented as edges (R_T). We define the vertex that represents a user as the root of a SCRG. Figure 5 shows an example in which a SCRG is generated. The label on each edge represents the upper limit of spatio-cohesiveness requirement. The requirement graph is an abstraction of spatial positions and relationship of services. It does not represent the functional dependencies among services. In this paper, we assume that the user and all services in a SCRG are connected.

Discovery Plan Generation. On the basis of the SCRG generated in the previous stage, we now make a service discovery plan to discover services in a systematic manner. A service discovery plan for a task, T, is defined as follows:

$$P_T = (child_P, nonChild_P),$$

where $child_P$ is a subset of the spatio-cohesiveness requirements, and indicates the path in the requirement graph to discover a set of child services from its parent service. $nonChild_P$ is a complementary set of $child_P$. Therefore, the IoT resource that is selected for a service, s, is evaluated as to whether it meets the spatio-cohesiveness requirements of $nonChild_P$. Then, in accordance with the $child_P$, the child services for the service, s, are discovered in vicinity of the selected resource of s. Figure 5 shows an example of generation of a service discovery plan from a SCRG.

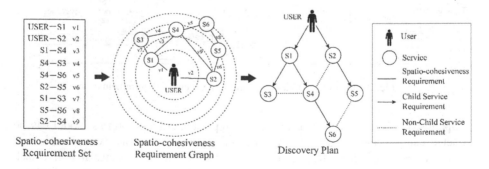

Fig. 5. Example of building a service discovery plan

Generation of the $child_P$ set is done by constructing a spanning tree of the SCRG. Therefore, our service discovery strategy is composed of a spanning tree algorithm and a service priority function. The spanning tree algorithm is for constructing the $child_P$, and the service priority function is for assigning a priority to each service based on their importance. This allows an algorithm to discover high-priority services effectively among child services of a service. Various spanning tree algorithms, such as Depth-First Search (DFS), Breadth-First Search (BFS), and Minimum Spanning Tree (MST) (e.g., the Prim's algorithm), can be used. For the service priority function, we can use an algorithm to calculate the centrality of a service in the requirement graph, which is a well-known indicator of a vertex's importance in the graph, or a requirement boundary from a service node to its parent service for strict prioritization.

3.3 Spatio-Cohesive Service Discovery and Handover

Spatio-Cohesive Service Discovery. Based on the service discovery plan built in the previous stage, the spatio-cohesive service discovery algorithm finds the IoT resources necessary to realize the services of a user task from the child services of the user node. Then, discovered services recursively finds the resources for the child services. For each resource, service discovery agent is deployed that allows discovery of child services in vicinity of the resource, as user requests. By the service discovery agent, user requests the resource of discovered service to discover its child services in vicinity of parent service. Therefore, service discovery request is propagated from parent service to child service, according to discovery plan.

During the service discovery process, for a service, there may be no available resource for its child service that meets the spatio-cohesiveness requirement. In this case, the algorithm re-discovers the resource of the parent service recursively until the service discovery of the parent service to its child service is successfully performed.

In this algorithm, there are two options that enable concurrent discovery of services, and the finding of multiple candidate resources for a service. Using these options, it is possible to make an efficient tradeoff between the failure rate and service discovery latency according to the conditions of the environment such as the dynamicity level, and the density of the deployed services.

Algorithm 1 shows the spatio-cohesive service discovery algorithm that is executed for each discovered service recursively. The service discovery is initially triggered by a user, and propagated to IoT resources. The inputs to the algorithm are a task, T, which includes own template, and a set of services, S_T, that need to be discovered for the task. A service discovery plan that is generated for the task is extracted from the task template, and the non-child services of the service to check the spatio-cohesiveness requirement are found based on the plan (Lines 2–3). For each non-child service, if it is already discovered and does not meet the requirement, the resource tries to discover the service of itself again from its parent (Lines 4–9). For simplicity, during the rediscovery of a service, current resource of the service is disabled, and an alternative resource is discovered (Lines 29–34). This makes the failed resource invisible in the further process.

If all the spatio-cohesiveness requirements of the non-child services are met, service discovery agent of the resource starts discovering child services of its associated service (Lines 10–16). Discovery of a child service is achieved by finding a set of candidate resources that can realize the child service in the vicinity of its parent service (Line 20). After a set of candidate resources is discovered, the resource selects the closest resource that can provide the child service, and starts discovery process from selected resource (Lines 23–26). If no available candidate resource is found for a child service, the resource tries to discover the parent service again (Lines 21–22). In addition, if the option is not allowed to discover the child services concurrently, it waits until the previous discovery session finishes, then proceeds to discovery of the next (Lines 13–15).

Algorithm 1. Spatio-cohesive Service Discovery Algorithm

```
 1: procedure SERVICEDISCOVERYTRIGGER(task T, service S)
 2:     nonChildSet = T.discoveryPlan.nonChildOf(S)
 3:     for each service nonChild in nonChildSet do
 4:         if isServiceDiscovered(nonChild) and not isRequirementMeet(S, nonChild) then
 5:             ReDiscovery(T, S)
 6:             Return
 7:         end if
 8:     end for
 9:     childList = T.discoveryPlan.childOf(S)
10:     for each service child in childList do
11:         ServiceDiscovery(T, S, child)
12:         if not allowConcurrency then
13:             waitUntilFinished(child)
14:         end if
15:     end for
16: end procedure
17:
18: procedure SERVICEDISCOVERY(task T, service parent, service child)
19:     candidateSet = findCandidateSet(T, isRequirementMeet, parent, child)
20:     if isEmpty(candidateSet) then
21:         ReDiscovery(T, parent)
22:     else
23:         child.setResource(candidateSet.extractMin())
24:         ServiceDiscoveryTrigger(T, child)
25:     end if
26: end procedure
27:
28: procedure REDISCOVERY(task T, service S)
29:     disableResource(S.getResource())
30:     T.resetDiscoveryState(S)
31:     parent = T.discoveryPlan.findParentOfService(S)
32:     ServiceDiscovery(T, parent, S)
33: end procedure
```

Algorithm 2 is for finding a set of candidate IoT resources by service discovery agent of the resource. The corresponding resource broadcasts the request of the child service (Line 3), then it finds candidates based on the replies from other resources (Lines 4–10). If a replied resource can be used to provide a required child service and meets the spatio-cohesiveness requirement, the service discovery agent adds the resource to the candidate set (Lines 4–7). If the option is not allowed to find multiple candidate resources, it returns the candidate resource that is found first (Lines 8–10).

Algorithm 2. Algorithm to Find Candidate IoT Resources for a Service

1: **procedure** FINDCANDIDATESET(task T, function condition, service $parent$, service $child$)
2:     $candidateSet = emptySet$
3:     $parent$.broadcastDiscoveryPacket($child$)
4:     **while** $time < delay$ and $candidate = $ getDiscoveryResponse() **do**
5:         **if** condition(T, $parent$, $candidate$) **then**
6:             $candidateSet$.insert($candidate$)
7:         **end if**
8:         **if** not $allowMultipleCandidates$ **then**
9:             break
10:        **end if**
11:    **end while**
12:    Return $candidateSet$
13: **end procedure**

Dynamic Service Handover. The major difference between the traditional handover and the service handover in an IoT environment is that the service handover requires consideration of a set of services to perform a user task, while the traditional handover considers only a single access point. Moreover, it is necessary to consider not only the case of handing over between a user's mobile device and a resource, but also the case of handing over between resources. In our approach, we utilize the service discovery plan generated for a user task to perform service handover in a dynamic manner.

Algorithm 3 shows the dynamic service handover algorithm that has a structure similar to that of the spatio-cohesive service discovery algorithm. The service handover is triggered by a user periodically to check whether the resources selected for a task meet the spatio-cohesiveness requirements, and propagated to services similar with the service discovery algorithm. When handover process is triggered at a service by its parent service, associated service discovery agent on the resource finds a set of candidate resources for its child services by using a handover decision rule with hysteresis margin and threshold (Lines 27–34). The handover decision rule calculates the spatio-cohesiveness between resource of the parent service and the candidate resource, and between resource of the parent service and the current resource, according to the spatio-cohesiveness requirement of the parent service and the child service. Unlike the service discovery algorithm, empty candidate set indicates that there is no better resource available for the child service in the vicinity of the parent service.

Algorithm 3. Dynamic Service Handover Algorithm

1: **procedure** SERVICEHANDOVERTRIGGER(task T, service S)
2:    $nonChildSet = T.discoveryPlan.$nonChildOf($S$)
3:    **for** each service $nonChild$ in $nonChildSet$ **do**
4:        **if** not isRequirementMeet(S, $nonChild$) **then**
5:            ServiceHandover(T, S)
6:        **end if**
7:    **end for**
8:    $childList = T.discoveryPlan.$childOf($S$)
9:    **for** each service $child$ in $childList$ **do**
10:       ServiceHandover(T, $child$)
11:       **if** not $allowConcurrency$ **then**
12:           waitUntilHandoverEnds($child$)
13:       **end if**
14:   **end for**
15: **end procedure**
16:
17: **procedure** SERVICEHANDOVER(task T, service S)
18:    $parent = $findParentOfService($S$)
19:    $candidateSet = $findCandidateSet($T$, handoverDecision, $parent$, S)
20:    **if** not isEmpty($candidateSet$) **then**
21:        S.setResource($candidateSet$.extractMin())
22:    **end if**
23:    ServiceHandoverTrigger(T, S)
24: **end procedure**
25:
26: **procedure** HANDOVERDECISION(task T, service S, resource $candidate$)
27:    $parent = T.discoveryPlan.$findParentOfService($S$)
28:    $req = T.$findRequirement($parent$, S)
29:    $currentResource = S.$getCurrentResource()
30:    $currentObj = $getDistance($parent$, $currentResource$)$/req.$getLimit()
31:    $candidateObj = $getDistance($parent$, $candidate$)$/req.$getLimit()
32:    Return $currentObj > threshold$ and $currentObj - margin > candidateObj$
33: **end procedure**

---

4   Evaluation

4.1   Evaluation Setting

We performed a number of simulations using the NS3 simulator[2] to evaluate the spatio-cohesive service discovery and dynamic handover algorithms. The NS3 simulator is one of the most well-known simulators for network environments. It provides various modules to simulate IoT environments that we exploited in this work. The simulations were conducted on a 64-bit Ubuntu 14.04 LTS operating system running on the virtual machine by Oracle VirtualBox 5.0.20 on Windows 10 Pro with an Intel i7-3770 processor. The size of the memory allocated for a virtual machine was 4 GB. In the simulation environment, 200 resource nodes were deployed in a 100 m × 100 m rectangular area in a uniform manner. Thus, each resource covered a 7 m × 7 m area,

---

[2] https://www.nsnam.org.

which is reasonable to deploying and managing resources in an IoT environment. Moreover, approximately half of the resources were generated as mobile nodes using the RandomDirection2d mobility model. We selected this model because it can reflect the behavior of mobile resources that move around in an IoT environment in an effective manner. Static resources were generated based on the ConstantPosition mobility model, which produces fixed position. The resource nodes communicated with each other through IEEE 802.11ac Wi-Fi channels with the VhtMcs0 physical mode, the most commonly used mode. In addition, for the Wi-Fi channels, we used the constant-speed propagation delay model, and the Cost-Hata propagation loss model, which are designed to simulate urban environments [16].

The number of services for a user task was set to 10 in order to simulate the situation of performing user task with medium complexity [4]. In addition, the velocity of the mobile resources was randomly set in the range 1 m/s to 10 m/s, which represents the walking and running speed of pedestrians in an IoT environment. For each service discovery strategy, we repeated the simulation 50 times, 20 s each. The simulation time set up for the experiments is much shorter than the time that usually takes to perform a user task in an IoT environment. However, in our simulation environment, 20 s were enough to test various situations that require dynamic service handovers. Before the simulation was initiated, the set of spatio-cohesiveness requirements, the SCRG, and service discovery plan of the simulation were generated based on a discovery strategy. At the beginning of each simulation, the spatio-cohesive service discovery is triggered by the user node. Once the service discovery process was completed, a series of service handover events were triggered one by one after a short delay.

4.2   Evaluation Results

Figure 6 shows the measurement of the spatio-cohesiveness metric over time. We tested the effectiveness of using the BFS and DFS strategies and compared their results against the case of discovering services without using the handover algorithm, and the case of not using a discovery plan. Without using a discovery plan, the service discovery process needed to find all of the services by the user node. Therefore, the spatio-cohesiveness requirements between a pair of services cannot be met, and this results in a poor spatio-cohesiveness condition. In the case of discovering services without using the handover algorithm, the result shows a continuous decrease in the spatio-cohesiveness of the services as time passes because of the mobility of resources. In contrast, the proposed BFS and DFS based approaches that use the service handover algorithm show high spatio-cohesiveness results under the condition of having mobile resources. As shown in Fig. 6, the BFS based approach and the DFS based approach do not have much difference.

Figures 7 and 8(a) compares the cases of using the options to allow concurrent service discovery and to discover multiple candidate services for the spatio-cohesive service discovery and dynamic service handover algorithms. Each box shown in the figure represents the simulation result generated by combining a service discovery strategy and an option. In the figure, the boxes show the number of service handovers

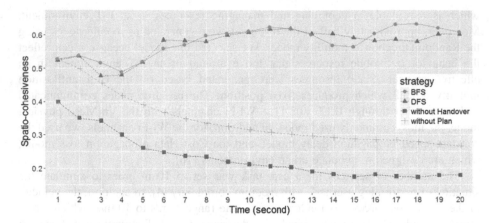

Fig. 6. Spatio-cohesiveness results compared to baseline approaches

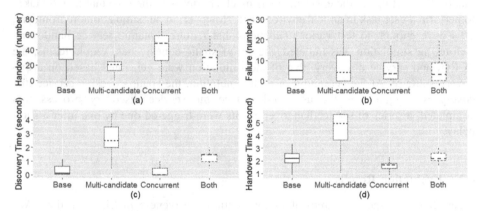

Fig. 7. Number of handovers (a), and number of failures (b), discovery time (c) and handover time (d) of using different combinations of service discovery strategies and options

made during the simulation (a), and the number of failures to meet spatio-cohesiveness requirements (b), and the time spent for the service discovery (c), and the time spent for the service handover (d), for each algorithm option.

The results show that choosing the option to allow multiple service discoveries effectively reduces the number of service handovers. These results are tightly related to the stability of the service discovery and handover algorithms. Considering multiple candidates during the process increases the stability of the algorithms (a less number of handovers means a more stable set of services), but makes them more time-consuming.

Choosing the option to allow concurrent service discovery resulted in the service discovery and handover time being less than that of the case in which this option was disabled. However, allowing concurrent discovery did not contribute significantly to improvement of the stability of the algorithms. Especially, allowing concurrent discovery effectively reduces the service discovery and handover time to find multiple candidate services when both options are used together.

Figure 8(a) compares the cases of using the options in terms of the spatio-cohesiveness. As shown in the figure, the case of enabling both options result in higher spatio-cohesiveness than other cases. Therefore, if the time constraint of service discovery and handover is not too hard, using both options of the algorithms can improve the stability and spatio-cohesiveness of the results.

Figure 8(b) compares the results of using the different service priority functions discussed in Sect. 3. For this experiment, service discovery plans were generated using BFS and three different service priority functions—random, degree centrality in a SCRG, and the limit value of the spatio-cohesiveness requirement of a service and its parent service. As can be seen in the figure, use of the service priority functions helps to improve spatio-cohesiveness of the services discovered especially during the service handover process (after the first time instance). However, the spatio-cohesiveness results of using different service priority functions are not significantly different. The case in which the limit value of the spatio-cohesiveness requirement is used results more steady spatio-cohesiveness of the services.

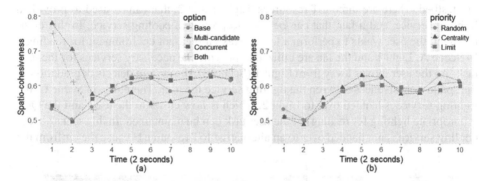

Fig. 8. Spatio-cohesiveness results of using different options (a), and different priority functions (b)

4.3 Threats to Validity

For the simulation, we did not consider the spatio-cohesiveness requirement extraction phase, and the simulation was conducted with randomly generated spatio-cohesiveness requirements. This was done because the spatio-cohesiveness requirement ontology model is not populated with practical task definitions yet. During the random generation process, we set the upper limit of requirements as 30 m to 45 m that covers about one-third of the entire environment area. Therefore, less than one-third of the IoT resources that are in the network are expected to spatially close within the upper limit. The lower bound of the limit seems high. However, because we consider mobile resources in dynamic IoT environments, such a setting might be appropriate.

5 Testbed Implementation

In this study, we implemented a testbed in which multiple IoT resources form a MANET. Because not many commercial IoT resources are programmable, we developed service gateways using Node-Red[3] on a Raspberry Pi 2 model B[4] that controls the power of IoT resources via Wemo[5] Insight model F7C029de. In addition, an Android application is developed to discover services by using the proposed algorithm as well as to perform user tasks on a mobile device (smartphone). In this testbed, once a user enters the MANET area, the user's mobile device automatically discovers the IoT resources that are necessary for performing the user task that the user selected from the available tasks in the testbed.

To measure the distance between resources, we use the Received Signal Strength Indicator (RSSI), which is commonly used for distance estimation. Therefore, while coordinating services for a task, if a user's mobile device detects degradation of the RSSI from an IoT resource that is used for providing a service, it discovers an alternative resource that meets the spatio-cohesiveness requirement of the task.

Figure 9 shows photos of the service-handover demonstration that we performed in our testbed. There are three IoT resources, Light 1 and 2, that can be used to provide a lighting service, and a fan, that can be used to provide a cooling service. In this environment, the user wants to perform a task, "Make a pleasant environment for reading." In Scene A, Light 1 and the fan are utilized to provide the necessary services for the user task. As the user moves away from Light 1 (Scene B), the system detects degradation in the spatio-cohesiveness between the lighting service the user. Therefore, in Scene C, the lighting service is handed over to Light 2, which is now closer to the user, and the QoE (appropriate lighting for reading) of the user task can be maintained. In this scenario, the cooling service is not handed over to an alternative IoT resource because the air from the

Fig. 9. Demonstration of a service-handover scenario in the testbed

---

[3] https://nodered.org/.

[4] https://www.raspberrypi.org/products/raspberry-pi-2-model-b/.

[5] http://www.wemo.com/.

fan can still reach the user in Scene C, and the QoE requirement (making the place cool) of continuously performing the user task can be met.

6 Conclusion

To perform a user task in an IoT environment, appropriate IoT services need to be discovered in a spatially cohesive manner in the vicinity of the user. Furthermore, to perform the user task in a stable manner under the highly dynamic MANET-based IoT environments, it is necessary to dynamically handover between services that utilize different IoT resources. In this paper, we proposed a service discovery method that finds IoT services from a user's surrounding environment in a spatially cohesive manner so that the interactions among the services can be effectively done, and the outcome of service coordination can be effectively delivered to the user. In addition, we developed a service handover approach to dynamically switching from an IoT resource to an alternative one to provide services in a stable manner when the degradation of the spatio-cohesiveness of the services is monitored.

The evaluation of the spatio-cohesive service discovery and dynamic service handover algorithms in an IoT environment was performed using the NS3 simulator. During the experiments, we observed that service discovery and handover using a discovery plan contributes to improving the spatio-cohesiveness level of the services discovered for user tasks. Furthermore, we found that enabling multiple candidate selections and concurrent service discovery are effective in improving the service discovery performance. In addition, the priority function used in building the discovery plans effectively increases the adaptability of the algorithms for dynamic IoT environments.

The main contribution of our work lies in its consideration of the spatial inter-relationship among services and user to discover effective services to improve user experience of performing a task, which is regarded as an important issue of mobile service discovery. Based on the spatial dependencies among the services, we organized the services to be discovered and handed over systematically, which improves efficiency and effectiveness of performing a user task. Moreover, our service handover algorithm reflects complex handover situations that are associated with the QoE of services.

In future work, we plan to extend our ontology model to represent diverse spatio-cohesiveness requirements, and QoE conditions. In addition, we will extend the service discovery and handover algorithms to deal with unconnected SCRG in order to support any type of service coordination for user tasks. Moreover, because the spatio-cohesiveness in service coordination should be considered as one of the critical factors of QoE, studies on the aspects of the spatio-cohesiveness requirements will be conducted.

Acknowledgment. This work was supported by the Basic Science Research Program through the National Research Foundation of Korea (NRF) funded by the Ministry of Science, ICT and Future Planning (2016R1A2B4007585).

References

1. Gubbi, J., et al.: Internet of things (IoT): a vision, architectural elements, and future directions. Future Gener. Comput. Syst. **29**(7), 1645–1660 (2013)
2. http://www.zdnet.com/article/internet-of-things-market-to-hit-7-1-trillion-by-2020-idc/
3. Corson, S., Macker, J.: Mobile ad hoc networking (MANET): routing protocol performance issues and evaluation considerations. No. RFC 2501 (1998)
4. Groba, C., Clarke, S.: Opportunistic service composition in dynamic ad hoc environments. IEEE Trans. Serv. Comput. **7**(4), 642–653 (2014)
5. Jimenez-Molina, A., Ko, I.-Y.: Spontaneous task composition in urban computing environments based on social, spatial, and temporal aspects. Eng. Appl. Artif. Intell. **24** (8), 1446–1460 (2011)
6. Truong, H.-L., Dustdar, S.: A survey on context-aware web service systems. Int. J. Web Inf. Syst. **5**(1), 5–31 (2009)
7. Elgazzar, K., Hassanein, H.S., Martin, P.: Daas: cloud-based mobile web service discovery. Pervasive Mob. Comput. **13**, 67–84 (2014)
8. Meshkova, E., et al.: A survey on resource discovery mechanisms, peer-to-peer and service discovery frameworks. Comput. Netw. **52**(11), 2097–2128 (2008)
9. Sailhan, F., Issarny, V.: Scalable service discovery for MANET. In: Third IEEE International Conference on Pervasive Computing and Communications. IEEE (2005)
10. Athanaileas, S.E., Ververidis, C.N., Polyzos, G.C.: Optimized service selection for manets using an aodv-based service discovery protocol. In: MEDHOCNET 2006, Corfu, Greece, June 2007
11. Surobhi, N.A., Jamalipour, A.: A context-aware M2 M-based middleware for service selection in mobile ad-hoc networks. IEEE Trans. Parallel Distrib. Syst. **25**(12), 3056–3065 (2014)
12. Chen, X., et al.: Web service recommendation via exploiting location and QoS information. IEEE Trans. Parallel Distrib. Syst. **25**(7), 1913–1924 (2014)
13. Klein, A., Ishikawa, F., Honiden, S.: Towards network-aware service composition in the cloud. Proceedings of the 21st International Conference on World Wide Web. ACM (2012)
14. Brandes, U., Erlebach, T.: Network Analysis: Methodological Foundations. LNCS, vol. 3418. Springer, Heidelberg (2005)
15. Ma, Q., et al.: A semantic QoS-aware discovery framework for web services. In: IEEE International Conference on Web Services, ICWS 2008 (2008)
16. Stoffers, M., Riley, G.: Comparing the NS-3 propagation models. In: 2012 IEEE 20th International Symposium on Modeling, Analysis and Simulation of Computer and Telecommunication Systems. IEEE (2012)

ALMOsT.js: An Agile Model to Model and Model to Text Transformation Framework

Carlo Bernaschina(⊠)

Politecnico di Milano, Piazza Leonardo da Vinci, 32, 20133 Milano, Italy
`carlo.bernaschina@polimi.it`

Abstract. Model Driven Development (MDD) requires model-to-model and/or model-to-text transformations to produce application code from high level descriptions. Creating such transformations is in itself a complex task, which requires mastering meta-modeling, ad hoc transformation languages, and custom development tools. This paper presents ALMOsT.js, an agile, in-browser framework for the rapid prototyping of MDD transformations, which lowers the technical skills required for Web and mobile developers to start be proficient with modeling and code generation. ALMOsT.js is shown at work in the creation of IFMLedit.org, a browser-based, online/offline environment for the MDD specification and rapid prototyping of web and mobile applications.

Keywords: Agile development · Model-driven development · Computer aided software engineering

1 Introduction

Model Driven Development (MDD) is the branch of software engineering that advocates the use of *models*, i.e., abstract representations of a system, and of *model transformations* as key ingredients of software development [21]. With MDD, developers use a general purpose (e.g. UML [4]) or domain specific (e.g., IFML [24]) modeling language to portray the essential aspects of a system, under one or more perspectives, and use (or build) suitable chains of *transformations* to progressively refine the models into executable code. General purpose languages, such as UML, aim at spanning the entire spectrum of software applications under all possible perspectives and thus are large and complex. It is hardly the case that developers can afford building their own MDD development environment, including such aspects as model editing, verification and code generation, because such a task could easily exceed the effort of the actual application to be delivered. Therefore, general purpose modeling languages must be supported by dedicated tools, which in turn present, despite the claims of interoperability, problems of so-called vendor lock-in. On the other hand, domain specific modeling languages [18] have been proposed to overcome the complexity and steep learning curve of general purpose approaches; their abstractions do not pretend to be universal, but embody domain knowledge that make the language speak in

© Springer International Publishing AG 2017
J. Cabot et al. (Eds.): ICWE 2017, LNCS 10360, pp. 79–97, 2017.
DOI: 10.1007/978-3-319-60131-1_5

terms closer to the intuition of the developer and of the application stakeholders. On the negative side, though, domain specific modeling languages have a more restricted user base, which prevents in many cases the development of industrial-strength supporting tools. For this class of languages, the need arises of creating and maintaining the tool chain necessary for bridging the gap between model specification and the real application deployment. Such a tool chain revolves around one or more domain specific *model transformations*. A such transformation is a procedure that takes in input a model, plus some additional domain or technical knowledge, and outputs a more refined model (yielding, at the end of the chain, the source code, which can be seen as the most concrete model). Creating model transformations is not itself an easy task. It is analogous to the construction of a compiler, which requires mastering very specific abilities, including meta-modeling (i.e., the specification of the grammar of the input language), definition of language semantics, and implementation of mapping rules. Presently, model transformation languages, such as ATL [17] and OMG QVT [3], and their support development environments, such as the ATL Integrated Environment (IDE) for Eclipse[1], are tools for the experts. They require specific skills regarding the language syntax and semantics and programming style. In many cases, organizations create and evolve a supporting MDD environment alongside their applications or product lines. Developers recognize recurring patterns of abstraction in their requirements and designs and tend to factor them out and promote them into abstractions of an emerging domain specific modeling language. Then, they try to boost reuse by generating code from the instantiation of their own abstractions. It is easy to see that this way of affording development requires the application of effective software engineering practices not only to the application, but also to its supporting MDD environment. Specifically, the well-known principles of agility should be seamlessly applied to both focuses of development, especially in the today's scenarios where scarcity of time and resources and high technological variability are the norm.

To address the agile development of MDD support tools, this paper presents **ALMOsT.js**, an agile, in-browser framework for the rapid prototyping of MDD transformations, which lowers the technical skills required for developers to start be proficient with modeling and code generation. The contributions of the paper can be summarized as follows:

- We describe ALMOsT.js, an MDD development framework that exploits well-know technical standards (JavaScript, Node.js and the JointJS GUI library) to help developers edit models in their formalism of choice and create transformations quickly and without complex meta-modeling steps.
- We discuss how ALMOsT.js can be used with an agile methodology, by rapidly prototyping, verifying, and evolving model editors and transformations.
- We illustrate how in ALMOsT.js developers specify transformations with a simple rule language that extends JavaScript naturally.
- We evaluate the effectiveness of ALMOsT.js in a case study in which the framework has been applied to a real world MDD scenario: the construction

---

[1] http://www.eclipse.org/atl/.

of an agile development environment for the OMG Interaction Flow Modeling Language (IFML) [24], called IFMLEdit.org[2]. ALMOsT.js has been used to create an IFML model editor, a model to model transformation mapping IFML to Place Chart Nets (PCNs) [20] for verifying the correctness of models via analysis and simulation, and two model to text transformations for automatically generating Web and cross-platform mobile code.

The paper is organized as follows: Sect. 2 surveys the work on agile MDD and transformation tools; Sect. 3 presents the architecture and components of the framework and reports on its application to the case study of IFML.org; finally, Sect. 4 draws the conclusions and gives an outlook on future work.

2 Related Work

Agile software development [2] is an incremental and iterative approach based on principles that aim at increasing productivity and adherence to requirements, while keeping the process as lightweight as possible. Agile Model Driven Development has been advocated as a promising approach [6], which has not yet fully expressed its potential [15,23]. Its idea is to organize the MDD process in ways that take advantage of the agile development principles:

1. Enabling an incremental and iterative development cycle by using toolchains able to test and validate even incomplete models [19].
2. Applying a Test-Driven Development, a distinctive feature of extreme programming [9], to MDD. [5,26].
3. Merging agile workflows such as SCRUM [25] with MDD in novel methodologies, to achieve system-level agile processes [29].

The above mentioned approaches mainly focus on enhancing the development cycle of the final product, assuming a predefined set of modeling languages and transformations. The rapid evolution of new technologies and the short time-to-market required today highlight the need to integrate modeling languages and tools in the loop [27], co-evolving iteratively the modeled final product and the MDD tools and languages. Accordingly, ALMOsT.js aims to foster agile processes in MDD not only when the models and transformations have already been chosen and are stable, but also when developers need a lightweight development environment that can be exploited to deliver minimum viable products rapidly, for the model concepts and transformations as well as for the target applications.

Model transformations are supported by many languages [13,22] and implemented by an even greater number of tools [12]. We limit the discussion to the works that are more relevant to our approach. Several works focus on domain specific languages and on different aspects of model transformation, as extensively discussed in [10,11,13]. The proposed transformation languages employ different styles:

---

[2] The environment is online and can be used at http://ifmledit.org/.

1. **Declarative:** transformations exploit mappings between elements of the input and of the output metamodel. An example is EMF Henshin [7], which focuses on model-to-model transformation using triple graph grammars.
2. **Imperative:** transformations are programmed as a sequence of operations. An example is Kermeta [14] an imperative programming language able to perform model transformations.
3. **Hybrid:** transformations mixing the declarative and imperative style. An example is ATLAS Transformation Language (ATL) [16] where declarative mappings can be extended by imperative constructs. The ATL transformation engine also automatically identifies the optimal rules execution order.

Some languages focus on particular features like Epsilon Transformation Language (ETL) [22], which is a model-to-model transformation language that can handle more than one source and target model at a time. Model-to-text transformation languages, such as Xpand, the imperative language part of OpenArchitectureWare, are generally template-based. Code is embedded inside plain text, making it easy to convert a source or configuration file into a template.

As illustrated in Sect. 3, with ALMOsT.js we aimed at quite a different objective, trading completeness for simplicity: reusing a popular, general purpose language, in which many developers are already proficient, to build a lightweight MDD environment whereby developers can design domain specific languages and the associated transformations rapidly and without resorting to external tools.

3 The ALMOsT.js Framework

In this section, we describe the architecture of ALMOsT.js (AgiLe MOdel Transformations), its rule language, and showcase its usage for customizing model editors and for building model to model and model to text transformations. For concreteness, the examples of data objects and rules are drawn from a simplified running example, and further expanded in a case study described in Sect. 3.6.

3.1 Requirements

Before introducing the architecture and use of ALMOsT.js, we pinpoint the requirements for its development that make the resulting environment amenable for use within an agile software development methodology. Here, we focus on the core characteristic of agile development methodologies, such as SCRUM [25], which commands developers to create minimum viable products rapidly and evolve them via frequent iterations (sprints). As a reference usage scenario for ALMOsT.js we imagine a software team, who has elicited the requirements for an application or product line and decides to exploit MDD in its development, by progressively factoring out domain abstractions from requirement and building a support environment for creating models, verifying them and generating code, in an iterative way. This scenario does not rule out the complementary case in which a tool company wants to implement rapidly and evolve over time a support environment for an existing standard or proprietary MDD language.

Under the above mentioned drivers, the requirements at the base of ALMOsT.js can be summarized as follows:

1. **No installation.** It must be possible for the team to use the framework instantly, with no installations.
2. **No new language.** It must be possible to start using the environment without learning languages that are not normally employed for application development.
3. **Fast start-up.** It must be possible to create a minimum viable model editor and model transformation in a very short time (e.g., less than one day).
4. **Parallel development.** It must be possible to work in team on different aspects of the same sprint, e.g., by adding a new concept to the model editor, and the corresponding generative rules to the model to model transformation and model to text transformation in parallel.
5. **Customized output.** There must be an easy and standard path to turn the (possibly prototypical) generated code into a complete version, by adding (manually) the missing aspects.
6. **Customized generation.** It must be possible to tailor the non-functional aspects of the generated code easily, e.g., by adding graphics and sample data collections for early validation of the product with stakeholders.

3.2   Architecture

In this section we show how we addressed requirements #1 (No installation) and #2 (No new language) in the design of the ALMOsT.js architecture. ALMOsT.js has the simple organization illustrated in Fig. 1. The implementation has been realized entirely in JavaScript, which is also the only language needed by developers to plug their domain models and transformations into ALMOsT.js, in observance of the above mentioned requirements.

At the core of the framework, the **Data Model** comprises the domain model objects, encoded according to the minimalist structure explained in Sect. 3.3. Objects of the data model are created and manipulated by the **Model Editor**, a GUI component, running in the browser, with the usual functions for creating and modifying projects consisting of MDD diagrams, and for storing, retrieving, manipulating and annotating models and their constituents. Alternative Model Editors can be plugged-in, as far as they are able to create objects of the Data Model, making ALMOsT.js independent of the model editing GUI (more details on the Model Editor are provided in Sect. 3.5).

The objects of the data model are also read and written by the model transformations, encoded as **Transformation Rules** in the **Model2Model** and **Model2Text** components. To avoid inventing yet another transformation language, transformation rules are coded as pure JavaScript functions, according to the prototypical signatures illustrated in Sect. 3.4. This choice also simplified the implementation of the **Rule Engine**, which employs the standard function call mechanism of JavaScript to orchestrate the execution of rules. The Rule Engine exploits the stateless and side effects free nature of model transformation and

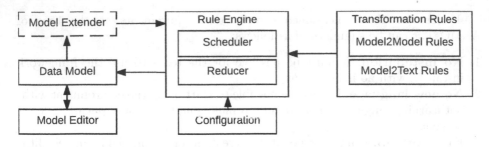

Fig. 1. Architecture of ALMOsT.js

employs a functional approach. Once configured, it acts as a main loop iterating one simple function, which takes as input a model object (described in Sect. 3.3) and returns as output the aggregated results of applying all the relevant rules to it. The Rule Engine function is internally organized in two sub-functions: the **Scheduler** is responsible of matching the input rules against the model objects (see Sect. 3.4), schedules rule execution, and collects their outputs. At each round of the main loop of the Rule Engine, the Scheduler manages the matching of the rules to the objects of the model and determines the rules to fire in the current round. The order of rule execution is pseudo-random: it depends on the order of definition of rules and on the order of creation of the model objects. All the output elements produced by the execution of the rules are grouped by a key (normally a property name of some model element) and passed to the Reducer for aggregation. The **Reducer** is the sub-function of the Rule Engine responsible of aggregating the results created by the execution of the Scheduler. If multiple rules generate outputs with the same key (i.e., multiple values for the same property of an object), then a *conflict* arises [28]. A typical case occurs when rules add multiple sub-elements or relationships to the same model element. The Reducer addresses the conflicts by merging output values referred to the same key into one resulting value. The merging policy is defined in the configuration of the Rule Engine; the available options are:

- **first:** (default) the first value, in terms of creation order, is used.
- **last:** the last value is used.
- **concat:** all the values are used and stored in an array.

The following fragment of the configuration of the Rule Engine specifies that conflicts related to the *name* property of an output object are treated with the `first` conflict resolution policy, whereas those related to the *children* property are treated with the `concat` policy.

```
{
    "name": "first",
    "children": "concat"
}
```

More details on how the Rule Engine can be configured to obtain different types of transformations are reported in Sect. 3.6.

3.3   Data Model

As shown in Fig. 1, the data model underpins all the components of ALMOsT.js. The artifacts of the data model can be regarded as a minimalist subset of the entities of the Essential Meta Object Facility (EMOF) OMG metamodel standard [1]. To cope with requirements #2 (No new language) and #3 (Fast start-up) the concepts of the data model are exposed as plain JSON objects, which facilitates creation, storage, manipulation, and migration between components. For example, developers can use model editors of their choice, provided that they can generate the JSON objects of the data model. Each MDD project must include a *model* object, which represents the entry point of the domain model; this is a root container object with two members:

1. **elements:** an array containing all the elements in the model.
2. **relations:** an array containing all the relations between elements in the model.

```
{
    "elements": [],
    "relations": []
}
```

Developers can start from this minimal assumption to fill a model object with their own model concepts (elements and relations). Even if ALMOsT.js does not require any particular structure for elements or relations, we propose a simple format, which facilitates element type identification, relations navigation, and separation of concerns between the data that are a proper part of the elements (attributes) and the descriptive information associated with them (metadata). According to this proposed format, each entry in the **elements** array must be an object (in JSON) endowed with the following properties:

1. **id:** a string that uniquely identifies the element in the model. It is used as a reference by relations.
2. **type:** a string that identifies the type of the element. No formal type system is assumed, type names are completely user-defined. Specifying types for model elements helps rule condition expressions to locate matching elements by simply checking their type name (see the examples of rule conditions in Sect. 3.4).
3. **attributes:** an array of objects, whose members are domain specific properties of the enclosing element.
4. **metadata:** an array of objects, whose members are descriptive properties of the enclosing element. Rules (see Sect. 3.4) should not take any decision based on metadata and should use them just as optional information that helps the output generation. Examples of metadata are graphical information (e.g., position and size of a model element in the graphical model editor) or debug information.

The following JSON object illustrates the representation of a model element: the *Mails* object of type *ifml. ViewContainer.*

```
{
    "id": "mails",
    "type": "ifml.ViewContainer",
    "attributes": {
        "name": "Mails",
        "landmark": true,
    },
    "metadata": {
        "graphics": {
            "position": { "x": 100, "y": 50},
            "size": { "width": 160, "height": 90}
        }
    }
}
```

Each entry in the **relations** array is a (relation) object, endowed with the mandatory **type** member, which describes the semantics of the relation. As for elements, no assumption is made on the available types of relations, which can be defined by the developer. The other members of a relation object must be references to elements in the **elements** array.

The following JSON object represents a hierarchy relationship between two model elements, *mails* and *mails-list*.

```
{
    "type": "hierarchy",
    "parent": "mails",
    "child": "mails-list"
}
```

3.4   Transformation Rules

As per requirement #2 (No new language), we did not create a custom rule language; the entire tool, including the rule language, is based on the JavaScript programming language. All transformations are expressed as Condition-Action (CA) rules [28], defined as pairs of plain JavaScript functions. Rule definitions are stored in JavaScript source format in the **Model to Model Rules** and **Model to Text Rules** components of Fig. 1. A CA rule simply consists of:

1. **Condition:** A function verifying if the current rule can be applied. It must return a Boolean value (or anything that can be interpreted as a Boolean value) informing the Rule Engine that the rule is activated and can fire.
2. **Action:** A function responsible of mapping the inputs to a series of outputs. The output of an action function must be a JSON object. Each member of these objects is treated separately and passed to the **Reducer** component for conflict resolution and aggregation (as explained in Sect. 3.2).

The condition part of a rule may match specific elements or relations in the model, or a whole model. Three type of rules can be identified.

1. **Model rules.** The input is the whole model. These rules can be used for setup purposes, e.g., for inserting fixed outputs into a model. The example below shows the declaration of a model rule.

```
createRule(
    // Condition function
    function (model) { return model.elements.length > 0; },
    // Action function
    function (model) {
        return {
            project: { type: "folder", name: "myProject" }
        };
    }
);
```

The (model to text) rule contains a condition function that receives in input a model object and tests if it contains elements. The action part simply returns a piece of the JSON configuration that specifies a fixed-name project folder.

2. **Element rules.** An element rule has a condition function that matches a specific model element and maps it (and possibly other related ones) into others elements and/or relations. Note that element rules can be used to identify patterns in the source model, by writing a condition part that traverses the model graph starting from the matched element. The following declaration illustrates the structure of an element rule.

```
createRule(
    function (element, model) {
        return element.type === "ifml.ViewContainer";
    },
    function (element, model) {
        return {
            elements: [
                { id: element.id, type: "pcn.PlaceChart",
                    attributes: {name: element.name} }
            ],
            relations: []
        };
    }
);
```

The (model to model) rule is activated by a match with an element of a given type (`ifml.ViewContainer`) and produces as output a model element of another type (`pcn.PlaceChart`), with the same Id and name of the input element.

Relation Rules. A relation rule has a condition that matches a specific relation and can be used to map relations into other relations and/or elements. Also relation rules can be used to identify patterns in the model, by writing conditions that traverse the model graph starting from the matched relation. An example of relation rule is as follows.

```
createRule(
    function (relation, model) {
        return relation.type === "hierarchy";
    },
    function (relation, model) {
        var id = relation.child + "-init";
        return {
            elements: [
                { id: id, type: "pcn.Transition", attributes: {} }
            ],
            relations: [
                { type: "source",
                  transition: id, source: relation.parent },
                { type: "target",
                  transition: id, target: relation.child },
            ]
        };
    }
);
```

The above rule matches an input relation object (of type `hierarchy`) and creates as outputs one element (of type `pcn.Transition`) and two relations associated with it, one of type `source`, associated with the parent of the matched relations, and one of type `target`, associated with the child of the matched relation[3].

---

[3] The rule transforms a hierarchical nesting relationship between IFML ViewContainers into an initialization PCN transition, which connects the place associated with the enclosing container to that associated to the enclosed one.

A Complete Example. To illustrate transformations in ALMOsT.js, we provide a small, yet realistic, example of model to model and model to text transformation: the well-known mapping of Entity-Relationship concepts to the relational model and to the SQL Data Definition Language [8]. Specifically, the two rules below respectively map an N:M relationship of the Entity-Relationship model into a bridge table of the relational model and into the SQL code for generating such table.

```
createRule(
    function (element, model) { // custom type checking function
        return model.isNMRelation(element);
    },
    function (relation, model) {
        var role1 = model.getRole1(relation), // first entity
            role2 = model.getRole2(relation), // second entity
            id = relation.id,
            // generate ids for relational elements
            id1 = id + '-ref-' + role1.id, // role1 column id
            id2 = id + '-ref-' + role2.id; // role2 column id

        return {
            elements: [
                // create bridge table
                { id: id, type: 'ER.Table',
                  attributes: {name: relation.attributes.name } },
                // create column referencing the 1st role table
                { id: id1, type: 'ER.Column',
                  attributes: {name: role1.attributes.name } },
                    // create  column referencing the 2nd role table
                { id: id2, type: 'ER.Column',
                  attributes: {name: role2.attributes.name } },
            ],
            relations: [
                // relate columns with table
                {type: 'ER.ColumnOfTable', table: id, column: id1 },
                {type: 'ER.ColumnOfTable', table: id, column: id2 },
            ]
        };
    }
)
```

```
createRule(
    function (element, model) {
        return model.isNMRelation(element);
    },
    function (relation, model) {
        var role1 = model.getRole1(relation), // first entity
            role2 = model.getRole2(relation), // second entity
            id = relation.id,
            name = relation.attributes.name, // name of the table
            results = {  // will contain the SQL source code
                project: { children: id }
            };
        results[id] = { type: 'file', name: name + '.sql',
        content:
            // CREATE TABLE statement composition
            'CREATE TABLE ' + name + ' ( ' +
            // add  column referencing the 1st table
            role1.attributes.name + ' int,' +
            // add  column referencing the 2nd table
            role1.attributes.name + ' int );'
        };
        return results;
    }
)
```

Note that the two rules can be added to the same project independently, so realizing requirement #4 (Parallel development). More generally, parallel work also on rules of the same model transformation is fostered by the blackboard structure of the Rule Engine: rules fired by the Scheduler produce independent outputs, which are merged by the Reducer based on simple configurable policies. This organization minimizes the coupling between rules and simplifies the control flow of the transformation. Finally, the design, implementation and debugging of model transformation rules does not require a visual model editor, because all inputs and outputs are (manually editable) JSON objects.

More examples of transformation rules are reported Sect. 3.6, where we describe how we validated ALMOsT.js in a real world project.

3.5   Model Editor

The starting point of the MDD practices is model editing, which can be useful also in absence of code generation, e.g., for documentation. We decided not to tie ALMOsT.js to a specific editor, but to make the architecture easily integrated with any editor of choice, provided that it respects the technical requirements of being executable in the browser and produce and consume JSON objects. In the case study of Sect. 3.6, the Model Editor is built on top of the JointJS diagram library. It manages models in the format compliant to the specifications illustrated in Sect. 3.3; inside the editor, model objects are enriched with meta-data in order to fit JointJS internal representation requirements. Developers can customize the appearance and behavior of both types of elements managed by the Model Editor, **entities** and **links**. The introduction of a new element in the editor requires different workflows for entities and links.

For entity-type elements, customization typically requires to:

1. Define the graphical representation of the element (in the popular SVG format).
2. Define the element type and register it in the list of available types.
3. (optional) Define the editable attributes.
4. (optional) Define its list of valid super- and sub-element types.
5. (optional) Define the default outgoing link type and constraints.

For link-type elements:

1. Define the graphical representation of the link.
2. Define and register the link type.
3. (optional) Define the editable attributes.
4. (optional) Define the valid source and target element types.

All the above customization aspects are specified declaratively, without programming. Following the philosophy of ALMOsT.js, developers are free to maintain a 1 to 1 relation between graphical elements in the editor and elements in data the model, or specify a custom mapping rule.

3.6   Case Study: IFMLEdit.org

ALMOsT.js has been applied to the construction of IFMLEdit.org [4], a browser-based tool that allows developers to model, analyze and rapidly prototype Web and mobile applications, following the MDD paradigm with the OMG Interaction Flow Modeling Language [24]. The purpose of the project is many-fold: (1) foster the adoption of MDD and IFML by offering an open source, free online tool whereby developers can quickly familiarize with the methodology and standard. (2) Implement the mapping from IFML to a formal language (namely Place Chart Nets [20]), to study in-depth the semantics of the language and thus support the development of other code generation tools and the portability of projects across tools; (3) offer to students of web engineering courses a simple environment for the hands-on experimentation with the main practices of MDD.

At the core of IFMLedit.org there are three different model transformations: a model to model transformation from IFML to PCN, for supporting semantic analysis, and two model to text transformations, respectively implementing code generation for the Web and for the Cordova cross-platform mobile environment.

The Data Model of IFMLEdit.org comprises five types of elements corresponding to the core concepts of IFML: *ViewContainer*, *ViewComponent*, *Event*, *NavigationFlow* and *DataFlow*, and two super-types: *ViewElement* which comprises *ViewContainers* and *ViewComponents* and *Flow* which comprises *NavigationFlows* and *DataFlows*.

---

[4] http://ifmledit.org.

The Data Model also comprises three types of relations: *hierarchy, source* and *target*, equipped with helper functions for easing model navigation in rules: *hierarchy* is navigable through *getParent* and *getChildren, source* is navigable through *getSource* and *getOutbounds* and *target* is navigable through *getTarget* and *getInbounds*. Other custom helper functions facilitate graph traversal by rules; examples are *getAncestors* and *getDescendants*, which support the traversal of trees of *hierarchy* relations.

Model to Model Transformation. IFMLEdit.org allows tool builders to analyze and simulate the semantics of their models, thanks to a set of mapping rules from IFML to PCN implemented with ALMOsT.js. Figure 2 shows the input IFML model (left) and the corresponding output PCN model (right), generated by the transformation rules. The PCN diagram expresses formally the dynamics of the application interface specified by the IFML diagram, using the classic concepts of Petri Nets, i.e., places, transitions, arcs and tokens[5].

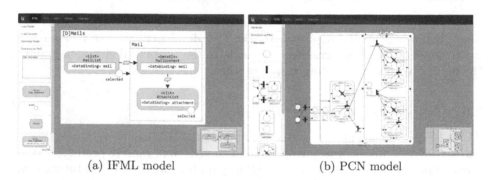

(a) IFML model   (b) PCN model

Fig. 2. IFML (left) to PCN (right) transformation

Each rule of the model to model transformation is responsible of generating a part of the PCN diagram and thus creates an output object with a single member called *model*, following the structure presented in Sect. 3.3 (i.e., possessing only the two *elements* and *relations* members). The Reducer is configured so to concatenate the list of elements and relations generated by the rules.

Overall, the transformation from IFML to PCN consists of 19 rules, which address all the aspects of the language.

Model to Text Transformations. To support requirement #5 (Customized Output), IFMLEdit.org allows the developer to simulate in the browser or download the modeled application. This function has required two model to text transformations for generating the Web and mobile code. For brevity, we illustrate only the Web version (the mobile transformation rules are quite similar).

---

[5] PCNs add to Petri Nets advanced modularization constructs, which make complex diagrams particularly readable.

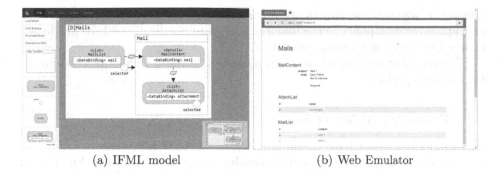

(a) IFML model   (b) Web Emulator

Fig. 3. IFML (left) to Web Code (right) transformation

The output of the model to text (for the web) transformation is a Node.js project following the MVC design pattern. The generated project can be run inside the browser[6], as shown in Fig. 3 or executed at the server side; the output artifacts are organized in a file system like structure and compressed after the execution of the transformation rules into a zip file, to facilitate download and installation.

Each model to text rule generates a subset of the folders and files that constitute the project. Model rules generate constant folders and files (e.g., system folders, configurations files, HTTP server initialization scripts, etc.) and textual files containing sample data collections for filling the interface with content (repositories). Each element rule generates the source code of a controller (JavaScript code) and of a view (Pug template). The following piece of JSON shows the output of a model to text rule that creates a package folder inside the project folder, and inserts in it the source code corresponding to some matched input element (for example, the source code of a Pug view, if the marched input element is an IFML ViewContainer of ViewComponent).

```
{
    "project": {
        "name": "project", "isFolder": true, "children": ["package"]
    },
    "package": {
        "name": "package.json", "content": "...Source File Content..."
    }
}
```

The Reducer is configured to concatenate the property values of each output object. In this way, for example, each model to text rule can simply generate code dealing with a specific aspect of the matched element (e.g., Java code or JSON configuration) as a value of the **children** property of the folder that must

---

[6] To support client-side execution of the web project, a web worker is instantiated when IFMLEdit.org is accessed, which emulates a Node.js environment and the HTTP protocol.

contain it, without worrying about other aspects treated by distinct rules. The concat aggregation policy of the Reducer takes care of collecting all the produced source files under the appropriate folder.

The model to text transformations consist of 22 rules, 12 for the web and 10 for the mobile platform.

3.7 Extending ALMOsT.js

To support requirement #6 (Custom generation), ALMOsT.js can be extended in a variety of ways. The **Model Extender** component, shown in Fig. 1, can be optionally used to enrich the Data Model, described in Sect. 3.3, with custom utility functions that facilitate rule implementation. Template engines like EJS[7] or Pug[8] (formerly known as Jade) can be easily incorporated in the action part of model to text rules. This facility supports: requirement #2 (No new language), because developers can reuse their favorite web platforms; requirement #4 (Parallel development) because graph walking algorithms are separated from actual code generation; and requirement #6 (Custom generation), because rules can use different output standards adopting the most appropriate frameworks, libraries, web languages or platforms.

3.8 Discussion and Limitations

The experience of implementing IFMLEdit.org demonstrated the suitability of ALMOsT.js in an agile MDD context. The case study addressed an already existing MDD language (IFML), but we deem that the lessons learned apply also to the case in which the development team designs its own domain specific abstractions and generation rules.

The development of the IFMLEdit.org project followed a SCRUM iterative cycle. The element of IFML were added one at a time to the online tool, developing in parallel the Model Editor customization, the model to model rules, and the model to text rules.

The first sprint featured a zero application consisting of just an empty ViewContainer and was realized in 24 h. Each subsequent sprint added further language elements, in the following order: multiple ViewContainers, NavigationFlows between ViewContainers, Landmark ViewContainers, Nested AND ViewContainers, ViewComponents, NavigationFlows between ViewComponents, DataFlows, XOR ViewContainers, arbitrarily nested ViewContainers and Actions. Each sprint took on average less than 2 days to complete. The most complex sprint (adding arbitrarily nested AND and XOR ViewContainers) took 5 days, mostly allocated to the model to model transformation rules, which were necessary to sort out the innumerable semantic issues that arise with such an advanced design patterns.

---

[7] http://www.embeddedjs.com/.
[8] http://www.pugjs.org.

However, the objective of simplicity, prominent for achieving agility of use, also implied trade-offs with respect to the expressive power and capabilities of the rule language and engine. In particular, the following limitations can be observed:

1. No advanced mechanisms are offered to exercise finer control over the execution of rules, such as execution priorities.
2. No higher order transformations are supported, because rules cannot match other rules but only model objects.
3. Rule modularization is not supported, because rules of the same transformation cannot be divided into packages, e.g., clustering rules according to different purposes. If a complex transformation has to be split into separated stages or focuses, it must be formulated as a set of sub-transformations.

4   Conclusions and Future Work

This paper presented ALMOsT.js, a framework for supporting MDD in agile methodologies. The design philosophy of the framework stemmed from 6 requirements, which could be summarized in one: keep it simple.

The usability of ALMOsT.js has been validated in the implementation of a complex MDD project, featuring one model to model transformation and two model to text transformations, applied to the full spectrum of features of a real-world MDA standard language (IFML).

The future work will focus on the experimentation and further assessment of ALMOsT.js in the industry, with two scenarios: companies that do not yet use MDD in their practices, to understand if the agile approach of ALMOsT.js lowers the reluctance of traditional developers towards modeling; companies already applying in-house domain specific models and code generation techniques, to understand the added value of ALMOsT.js lightweight, single-language approach and browser-based architecture. The second line of future work will address early-stage MDD education, where we will assess the impact of hands-on experimentation in introductory software engineering and web engineering courses at the Bachelor level.

References

1. Meta Object Facility Version 2.5.1. http://www.omg.org/spec/MOF/2.5.1/. Accessed 10 Jan 2017
2. Principles behind the Agile Manifesto. http://agilemanifesto.org/principles.html
3. Query View Transformation V 1.3. http://www.omg.org/spec/QVT/1.3/. Accessed 10 Jan 2017
4. UML unified modeling language. www.uml.org/. Accessed 10 Jan 2017
5. Alpaslan, G., Kalipsiz, O.: Model driven web application development with agile practices. CoRR abs/1610.03335 (2016). http://arxiv.org/abs/1610.03335
6. Ambler, S.W.: Agile model driven development is good enough. IEEE Softw. **20**(5), 71–73 (2003). http://dx.doi.org/10.1109/MS.2003.1231156

7. Arendt, T., Biermann, E., Jurack, S., Krause, C., Taentzer, G.: Henshin: advanced concepts and tools for in-place EMF model transformations. In: Petriu, D.C., Rouquette, N., Haugen, Ø. (eds.) MODELS 2010. LNCS, vol. 6394, pp. 121–135. Springer, Heidelberg (2010). doi:10.1007/978-3-642-16145-2_9

8. Atzeni, P., Ceri, S., Paraboschi, S., Torlone, R.: Database Systems - Concepts, Languages and Architectures. McGraw-Hill Book Company, New York (1999)

9. Beck, K.: Extreme Programming Explained: Embrace Change. Addison-Wesley Longman Publishing Co., Inc., Boston (2000)

10. Biehl, M.: Literature Study on Model Transformations. Technical report, ISRN/KTH/MMK/R-10/07-SE, Royal Institute of Technology, July 2010

11. Calegari, D., Szasz, N.: Verification of model transformations: a survey of the state-of-the-art. Electron. Notes Theoret. Comput. Sci. **292**, 5–25 (2013). http://www.sciencedirect.com/science/article/pii/S1571066113000042. Proceedings of the XXXVIII Latin American Conference in Informatics (CLEI)

12. Cetinkaya, D., Verbraeck, A.: Metamodeling and model transformations in modeling and simulation. In: Proceedings of the Winter Simulation Conference, WSC 2011, pp. 3048–3058. (2011). http://dl.acm.org/citation.cfm?id=2431518.2431880

13. Czarnecki, K., Helsen, S.: Feature-based survey of model transformation approaches. IBM Syst. J. **45**(3), 621–645 (2006). http://dx.doi.org/10.1147/sj.453.0621

14. Falleri, J.R., Huchard, M., Nebut, C.: Towards a traceability framework for model transformations in kermeta. In: Aagedal, J., Neple, T. (eds.) ECMDA-TW 2006: ECMDA Traceability Workshop, pp. 31–40. Sintef ICT, Norway, Bilbao (Spain), July 2006. https://hal-lirmm.ccsd.cnrs.fr/lirmm-00102855

15. Hebig, R., Bendraou, R.: On the need to study the impact of model driven engineering on software processes. In: Proceedings of the 2014 International Conference on Software and System Process, ICSSP 2014, NY, USA, pp. 164–168. (2014). http://doi.acm.org/10.1145/2600821.2600846

16. Jouault, F., Allilaire, F., Bézivin, J., Kurtev, I.: ATL: a model transformation tool. Sci. Comput. Program. **72**(1–2), 31–39 (2008). http://dx.doi.org/10.1016/j.scico.2007.08.002

17. Jouault, F., Kurtev, I.: Transforming models with ATL. In: Bruel, J.-M. (ed.) MODELS 2005. LNCS, vol. 3844, pp. 128–138. Springer, Heidelberg (2006). doi:10.1007/11663430_14

18. Kelly, S., Tolvanen, J.: Domain-Specific Modeling - Enabling Full Code Generation. Wiley (2008). http://eu.wiley.com/WileyCDA/WileyTitle/productCd-0470036664.html

19. Kirby Jr., J.: Model-driven agile development of reactive multi-agent systems. In: Proceedings of the 30th Annual International Computer Software and Applications Conference, COMPSAC 2006, vol. 02, pp. 297–302 (2006). http://dx.doi.org/10.1109/COMPSAC.2006.144

20. Kishinevsky, M., Cortadella, J., Kondratyev, A., Lavagno, L., Taubin, A., Yakovlev, A.: Coupling asynchrony and interrupts: place Chart Nets. In: Azéma, P., Balbo, G. (eds.) ICATPN 1997. LNCS, vol. 1248, pp. 328–347. Springer, Heidelberg (1997). doi:10.1007/3-540-63139-9_44

21. Kleppe, A., Warmer, J., Bast, W.: MDA explained - the Model Driven Architecture: practice and promise. Addison Wesley Object Technology Series. Addison-Wesley (2003). http://www.informit.com/store/mda-explained-the-model-driven-architecture-practice-9780321194428

22. Kolovos, D.S., Paige, R.F., Polack, F.A.C.: The Epsilon transformation language. In: Vallecillo, A., Gray, J., Pierantonio, A. (eds.) ICMT 2008. LNCS, vol. 5063, pp. 46–60. Springer, Heidelberg (2008). doi:10.1007/978-3-540-69927-9_4
23. Matinnejad, R.: Agile model driven development: an intelligent compromise. In: Proceedings of the 2011 Ninth International Conference on Software Engineering Research, Management and Applications, SERA 2011, pp. 197–202 (2011). http://dx.doi.org/10.1109/SERA.2011.17
24. OMG: Interaction flow modeling language (IFML), version 1.0. (2015). http://www.omg.org/spec/IFML/1.0/
25. Schwaber, K., Beedle, M.: Agile Software Development with Scrum. Prentice Hall, Upper Saddle River (2002)
26. Stahl, T., Voelter, M., Czarnecki, K.: Model-driven software development: technology, engineering, management (2006)
27. Vaupel, S., Strüber, D., Rieger, F., Taentzer, G.: Agile bottom-up development of domain-specific ides for model-driven development. In: Proceedings of the Workshop on Flexible Model Driven Engineering co-located with ACM/IEEE 18th International Conference on Model Driven Engineering Languages & Systems (MoDELS 2015), Ottawa, Canada, 29 September 2015, pp. 12–21 (2015). http://ceur-ws.org/Vol-1470/FlexMDE15_paper_4.pdf
28. Widom, J., Ceri, S. (eds.): Active Database Systems: Triggers and Rules For Advanced Database Processing. Morgan Kaufmann, Burlington (1996)
29. Zhang, Y., Patel, S.: Agile model-driven development in practice. IEEE Softw. 28(2), 84–91 (2011). http://dx.doi.org/10.1109/MS.2010.85

A Big Data Analysis Framework
for Model-Based Web User Behavior Analytics

Carlo Bernaschina, Marco Brambilla, Andrea Mauri[✉], and Eric Umuhoza

Politecnico di Milano, Piazza Leonardo da Vinci 32, 20133 Milan, Italy
{carlo.bernaschina,marco.brambilla,andrea.mauri,eric.umuhoza}@polimi.it

Abstract. While basic Web analytics tools are widespread and provide statistics about website navigation, no approaches exist for merging such statistics with information about the Web application structure, content and semantics. Current analytics tools only analyze the user interaction at page level in terms of page views, entry and landing page, page views per visit, and so on. We show the advantages of combining Web application models with runtime navigation logs, at the purpose of deepening the understanding of users behaviour. We propose a model-driven approach that combines user interaction modeling (based on the IFML standard), full code generation of the designed application, user tracking at runtime through logging of runtime component execution and user activities, integration with page content details, generation of integrated schema-less data streams, and application of large-scale analytics and visualization tools for big data, by applying both traditional data visualization techniques and direct representation of statistics on visual models of the Web application.

1 Introduction

In recent years, the software language engineering community has put more and more emphasis on the design and experimentation of languages that cover the requirement specification [5,6], design [30], and verification/validation [10,14–16, 19] of software artifacts. Some of these experiences have also spawned commercial products and thus have been applied in industrial settings, with excellent results.

On the other side, a completely different line of work is ongoing regarding the usage analysis of Web applications with the aim of extracting leads to optimizing the user experience. Indeed, with the increasing need to meet customer preferences and to understand customer behavior, Web analytics has become the tool of choice towards taking informed business and interaction design decisions. Several tools exist that support analysis of Web server logs and extract information on application usage (based on data mining, machine learning, and big data analysis; or simpler pragmatic approaches in the UX field, like A/B split testing). However, those tools are unaware of the design structure and the actual content managed by the application. Therefore, understanding the relation between the output of the quantitative log analysis, the structure of the application, and the displayed content is not an easy task, especially for large and complex systems.

© Springer International Publishing AG 2017
J. Cabot et al. (Eds.): ICWE 2017, LNCS 10360, pp. 98–114, 2017.
DOI: 10.1007/978-3-319-60131-1_6

Some analytics tools, like Google Analytics [22], can take into account the page content but this implies a high development overhead which yields to high risk of errors and maintainability.

In this paper we propose a model-driven engineering approach that combines user interaction models with user tracking information and details about the visualized content in the pages. Our claim is that integration of appropriately designed modeling languages for user interaction development and Web log analytics approaches has high potential of delivering valuable insights to designers and decision makers on the continuous improvement process of applications. We show our full development cycle from visual application modeling (through the OMG's standard Interaction Flow Modeling Language, IFML [8,24]), to full code generation, user interaction log collection and integration with content and semantics of visualized data, down to display of results both through traditional data visualization techniques and with direct representation of statistics on visual models of Web applications represented with the OMG's IFML standard. Our proposal can be applied to applications developed using any model-driven development approach and generating runtime usage logs containing references to the conceptual model elements.

The paper is organized as follows: Sect. 2 introduces some background on model-driven development of Web applications, the IFML language and the WebRatio toolsuite; Sect. 3 describes our method for merging logs, enriching them with information coming from models, and applying analysis over them. Section 4 presents a case study that demonstrate the approach at work; Sect. 5 reviews the related works; and Sect. 6 summarizes the results and concludes.

2   MDD of User Interaction with IFML and WebRatio

This work focuses on the general concept of model-driven integration of user interaction designs with user navigation logs. However, to make things working in practice, we exploit as background: a specific modeling language, namely IFML, and a specific toolsuite, called WebRatio.

2.1   The IFML Language

The Interaction Flow Modeling Language (IFML) [24] is an international standard adopted by the Object Management Group (OMG) that supports the platform independent description of graphical user interfaces for applications accessed or deployed on such systems as desktop computers, laptop computers, PDAs, mobile phones, and tablets. An IFML model supports the following design perspectives:

- The *view structure specification*, which consists of the definition of view containers, their nesting relationships, their visibility, and their reachability;
- The *view content specification*, which consists of the definition of view components, i.e., content and data entry elements contained within view containers;

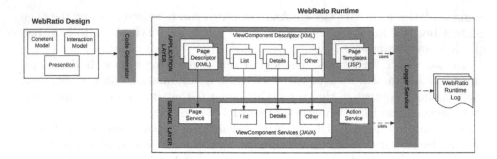

Fig. 1. WebRatio architecture: design time, generation time, and runtime environments.

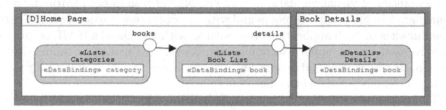

Fig. 2. Example of IFML diagram.

- The *events specification*, which consists of the definition of events that may affect the state of the user interface. Events can be produced by the user's interaction, by the application, or by an external system;
- The *event transition specification*, which consists of the definition of the effect of an event on the user interface;
- The *parameter binding specification*, which consists of the definition of the input-output dependencies between view components and between view components and actions; and
- The reference to *actions* triggered by the user's events.

IFML concepts can be stereotyped to describe more precise behaviours. For instance, one could define specific stereotypes for describing web pages (a specific kind of view container); forms, lists and details (specific kinds of view component); submission or selection events; and so on. By exploiting this extensibility feature, we defined custom extensions of IFML for covering Web [2] and mobile [3,9,28] applications.

Figure 2 shows an example of an IFML diagram: the *Home Page* ViewContainer contains a ViewComponent displaying a list of book categories. The user can click on the *books* event and thus selects a category in order to see the books belonging to it in the *Book List* ViewComponent. By clicking on a book (through *details* event) the user navigates to the *Book Details* page, where he can see the details of the selected book.

2.2 WebRatio Implementation of IFML

WebRatio Web Platform [1] is a model-driven development tool which implements the Web-extended version of IFML. The tool supports developers in the specification of the domain model and of the interaction flow model for web applications. The tool features model checking and full code generation that produces application code executable on top of any platform conforming to the J2EE specifications. Figure 1 represents an high level view of the WebRatio architecture, which supports developers ad design time, code generation time, and execution time.

Design time. WebRatio provides four main integrated modeling and development environments, supporting respectively: the modeling of IFML diagrams for the specification of the user interaction; the modeling of UML class diagrams (or ER diagrams) for the content design; the graphical layout template and style design environment; and the specification and development of the modeling level and execution level of custom IFML elements defined as extensions of IFML. The WebRatio models are saved as XML documents.

Code generation. Based on the input provided through the modeling and development environments, WebRatio provides code generators which transform the specifications of the application into concrete, executable implementations. The code generators are implemented using ANT, XSLT, and Groovy. The generated code consists of a Java EE code covering both front-end of back-end of web applications. The generated components include:

- The configuration file of the Controller which contains the navigation control flow logic;
- The action classes which are invoked by the Controller and, in turn, invoke the runtime services;
- The XML configuration files (called runtime descriptors) of the runtime services. In particular, every IFML ViewComponent in the the model produces one runtime descriptor configuring the behavior of a generic runtime service. For example, a specific List ViewComponent generates a runtime descriptor containing, among other things, the SQL code for extracting the content of the index from the data repository;
- The server-side templates for dynamically building the actual pages of the application.

Execution time. The WebRatio run-time framework consists of object oriented components and services for organizing the business tier, clustered in three main layers: the service layer, the application layer, and the logging layer.

The service layer is deployed once and for all for every application and does not actually need code generation, because it comprises general purpose services. Therefore, at runtime one single service class is deployed for each type of component, which is then instantiated with the smart service creation approach described next. The deployed services are:

- a Page Service, that is in charge of calculating the whole page structure for every page of the site;
- a set of ViewComponent Services (one for each component type), in charge of executing the logic of each view component;
- a set of Action Services (one for each action type), in charge of executing the business logic of each IFML action;

These services get configured based on the XML configuration files (descriptors) allocated at the application layer: the code generator produces one JSP template and XML descriptor for each Page in the IFML model; and one XML descriptor for each IFML Action and ViewComponent. The execution logic features a emphsmart service creation policy: services that implement ViewComponents or business actions are created upon request, cached, and reused across multiple requests, i.e., for evey designed IFML element, which gets configured through the XML descriptors. Access to the information stored in the XML ViewComponent descriptors is granted by standard parsing tools. This behaviour is tracked by the *Activity log*: a set of pre-built functions for logging each service is provided. The logging service can be extended with user-defined modules, so as to log the desired data.

In our approach, we are more concerned with the *logging service*. It represents a valuable asset, as it allows the logging of the execution of all application objects and shows the queries executed to fulfill the HTTP page requests. This information, along with the HTTP access log data available on the Web server, and information from conceptual models are the inputs of the proposed user interaction analysis framework.

3   Modeling, Integration, Analysis, and Visualization

In this section we describe the steps of our method, as summarized in Fig. 3.

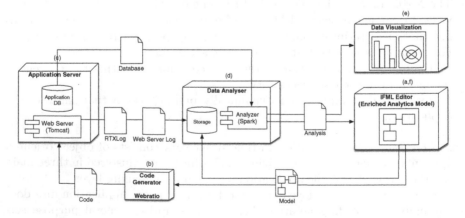

Fig. 3. Architectural overview of the approach and flow of artifacts.

3.1 Application Modeling and Generation

The first phase of our approach consists on modeling the Web application using a visual modeling editor (Fig. 3(a)) supporting the IFML standard presented in Sect. 2. A code generator (Fig. 3(b)) takes in input the IFML application models and deploys the web application on a web server (Fig. 3(c)), ready to be executed.

3.2 Analysis

The second phase (Fig. 3(d)), is concerned with two main aspects: (i) the integration of the website log, the database containing the information displayed in the site pages, and the Web application models specified in IFML; and (ii) the analysis of the resulting schema-less enriched stream. The outcome is a rich analytics able to provide deep insights regarding the behavior the user visiting the website. Since this phase is the core step of our approach, we provide full details about inputs, process and output.

Inputs. Four pieces of information are consumed in this phase:

1. *The Application Model*: The IFML model of the deployed web application, as designed in the first step of the process (see Sect. 2.1).
2. *The Application Server Log*: The common access log produced by the server where the website is deployed. The access log records all the requests processed by the application server in the standard Extended Common Log Format. Listing 1.1 shows an example of a Web server log line.

Listing 1.1. Example of Application Log line.

```
0:0:0:0:0:0:0:1 - - [22/Jun/2016:11:10:52 +0200] "GET /
    BookStore/page21.do?kcond4.att60=11&sp=page14&link=
    ln67&var6=true&var10=false&fllbck=.sv1&cbck=
    wrReq32387 HTTP/1.1" 200 21973
```

3. *The Runtime Components Log*: The runtime component log (RTX) that stores events and data produced and consumed by the application runtime for serving page requests. It traces the history of the ViewContainers, ViewComponents, and operations that are executed, along with the executed queries. Moreover, the RTX log contains the data about the requested pages and their contents. Furthermore, all logged information contains a reference to the relevant application model elements and to the consumed data in the database. Each log line has the following structure (exemplified in Listing 1.2):
   - *Timestamp*, the time when the logged instruction is written in the log;
   - *Log Level*, the level to which the log line belongs;
   - *Host*, the host sending the request, i.e., the IP address of the client computer used by user to navigate the application;
   - *Java* class, the Java class executed;
   - *Session ID*, the User session identifier;
   - *Model element*, the model element executed;
   - *Message*, the log message.

Listing 1.2 shows an example of RTX log line.

Listing 1.2. Example of WebRatio RTX log line.

```
22  Jun  2016  11:10:51,761  DEBUG  [http-bio-8080-exec-5]  (
    com.webratio.rtx.core.ServiceProvider:45)  -  [119354
    A67C7C0177D4A7F411E75BCDE7][page21][pwu6Block]
    Creating  service:  WEB-INF/descr/pwu6Block.descr
```

4. *Database*: Detailed reference to the database used to populate the requested pages.

While the application model is given at design time and we can assume it as persistent, the information contained in the logs and in the database changes as the users visit the website. The model can be changed in order to add or evolve features in the web application, but in that case the whole analysis process will be restarted.

Process. The analyzer (Fig. 3(d)) integrates the logs with the IFML model and database content. More precisely the following steps, summarized by Fig. 4, are performed:

- The logs, the models and are transformed in a tabular format in order to be easily queried and manipulated (Fig. 4(a)).
- The domain instances are imported from the database of the web application (Fig. 4(a)). In order to be domain independent and support any type of data, our process takes all the tables present in the database and create a unique table with the structure: `table_name,attribute_name,value`.
- The *Runtime Component Log* is joined with the *Application Server Log* using the session id and the timestamp (Fig. 4(b)). This *Enriched Log* shows, for each page requested by a user, the corresponding model element computed for serving the page content. Moreover it contains information regarding the link (if any) used for navigating the website. At this stage of the process we know all the elements involved in the users navigation, but only limited to their id. For instance, by imaging a possible user navigation in the model shown in Fig. 2, we may know that a user went to *page2* from *page1* through link *ln66* and he was shown elements with oid *3* of table *t1* by the component *d2*.
- The *EnrichedLog* is joined with the *Application Model* using the id of the model elements present in both inputs to obtain a *GlobalLog* (Fig. 4(c)). This allow to add to the *EnrichedLog* information regarding to the model elements involved in the user interaction. For instance now we may know that the user went to the *Book Details page* from the *Home Page* through link named *Details* and he was shown the elements with oid *3* of table *book* by the component *Details*.
- Finally the *GlobalLog* is joined wit the *DatabaseInstances* using the name of the tables and the oid of the instances (Fig. 4(d)). With this final step

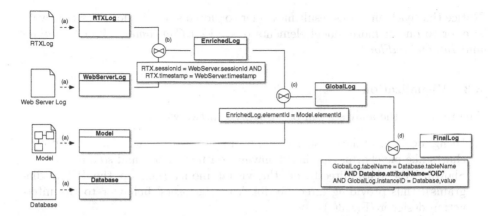

Fig. 4. Steps for the integration of the logs with the models and the database instances

we are able to assign an actual value to the oids present in the *GlobalLog*. To continue with the previous example, in addition with the information regarding the navigation, we know that the users looked the book *Lord of the Rings*.

From this global, denormalized view one can generate any desired behaviour analysis. In particular every statistics is computed with one or more SQL-like queries on the *FinalLog*. We describe in details how the analytics are computed in Sect. 4.

Output. With our approach we can compute three kinds of user behaviour analytics:

1. **Navigation-based analytics:** comprehends information regarding how the users navigate the website (e.g., average time spent on a page, number of requests for a page, links followed, and so on). These are the typical metrics computed also by other tools.
2. **Content-based analytics:** comprehends information regarding the domain entities involved in the user interaction, their types and their semantics. For example, in the case of an e-commerce website for selling books, these metrics may include the top-k visualized books, the top-k clicked authors and so on.
3. **Structure-based analytics:** comprehends information regarding the kind of widget, visualization, or even navigation pattern used in the user interaction model. For instance, analytics may include metrics like: top-k elements clicked by users when shown in a map throughout the site, or top-k elements clicked when shown in the first three positions of a list, or top-k elements clicked when an attribute of type image is shown in the page versus an attribute of type currency.

Notice that each analytics result may refer to global statistics for the whole website, or to one or more model elements (e.g., *ViewCompoment*, *ViewContainer* and *InteractionFlow*).

3.3 Visualization

The results of the analyses can be visualized in two ways:

- Using a canonical data visualization tool (Fig. 3(e)), which shows traditional charts like pie charts, bar charts, navigation flow charts and so on;
- Showing the data directly on the visual model (i.e., on the IFML diagrams), thus providing direct feedback of the users' behavior to the interaction designer (Fig. 3(f)).

Our main contribution on the visualization side focuses on the latter; in particular we devised three means of visualization:

- **Color:** the analytics is shown through the change of color of the corresponding model element (e.g., *ViewComponent* or *ViewContainer*) based on some quantitative analysis result associated to it. This is useful for statistics that refer to many model elements and provides a immediate bird's-eye view of the user behavior for the whole website: for instance, one may color the pages in the IFML diagram based on the number of page views achieved by each of them. For instance, Fig. 6(a) shows an IFML diagram with colored pages based on the number of visits per page (from green to red, where green means higher number of visits);
- **Label:** the analytics is shown with a label on the corresponding model element. For instance, one may display the number of clicks that a widget or a link has received; or the most clicked content instance(s) in a list, such as the top clicked book;
- **Properties:** the analytics is shown in a separate property panel, detailing the data about the corresponding model element. This is useful when many statistics or multi-valued results need to be displayed for the same element. For instance, Fig. 6(b) shows the list of top-10 visited elements in a list.

4   Case Study

In order to validate our approach we conducted a case study where we modeled mall e-commerce-like website selling books. The application (the IFML model is shown in Fig. 5) provides information on books such as their title, authors, price, category, description, reviews, comments and rating, and allows to navigate and purchase goods (the actual payment phase is not covered).

In our implementation, we rely on the code generators of WebRatio (www.webratio.com) which produces Java EE applications and deploys them on any servlet container. We simulated the behavior of users visiting the website and then we computed statistics related to their interaction. The entire process of

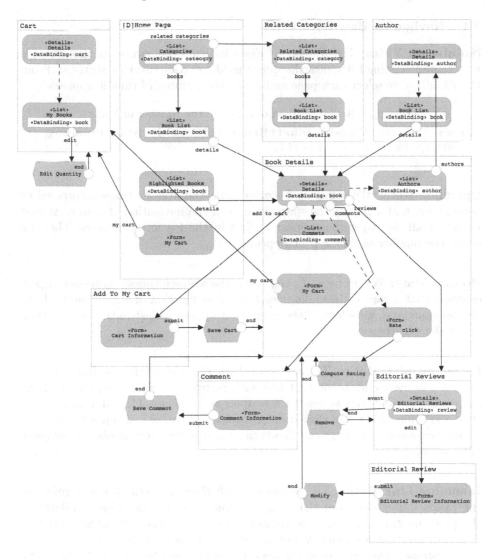

Fig. 5. IFML model of the application used in our case study

analysis is performed on the Microsoft Azure platform, using an Apache Spark Cluster as processing engine and technology for the processing and computation of the required analysis. Scala was used as a programming language to write the Spark application. We computed both navigation based and content based statistics. In the following we describe how the statistics are computed starting from the data obtained with the analysis step described in Sect. 3[1].

---

[1] The actual queries are omitted since they are quite long and do not fit in the paper.

4.1   Navigation-Based Statistics

Number of Visits. The *Number of Visits* indicate the number of times a page was visited. Listing 1.3 shows the query used for compute this statistic. From the *FinalLog* we select each page and count the number of time it appears.

Listing 1.3. Queries used to compute the Number of Visits

```
select  RequestedPageId ,count( RequestedPageId ) from  FinalLog
    group by  RequestedPageId ;
```

Entrance Rate: We define as *Entrance Rate* the average times, expressed as percentage, that a page is the starting point of the navigation of a user. At first we select all the pages that appear as first in the log for each users. Then we count the number of time a page appears.

Bounce Rate: We define as *Bounce Rate* the average times, expressed as percentage, that a page is the ending point of the navigation of a user. Similarly to the previous case, we select all the pages that appear as last in the log and then count the occurrences.

Outgoing Link Ratio. The *Outgoing Link Ratio* statistic of a link refers to the ratio of users clicking on that link with respect to the others present in the same page. As first step we select the couples `sourcePage,linkId`, that indicate that a specific *linkId* was clicked on a specific *sourcePage*. Then we count the occurrence grouped by page and divide the value by the total number of outgoing navigation.

Incoming Link Ratio. The *Incoming Link Ratio* statistic of a link refers to the ratio of user that uses that link to come in a specific page. Similarly to the previous statistic, at first we compute the couples `requestedPage,linkId`, that indicate that a specific *linkId* was clicked in order to arrive on a specific *requestedPage*. Then we count the occurrence grouped by page and divide the value by the total number of incoming navigation.

Average Staying Time: The *Average Staying Time* indicate the number of milliseconds that the users, in average, spent on a given page. The computation of this statistic is composed by four steps: (1) we select all the tuples in the form `time,sessionId,sourcePage,requestedPage` ordered by *time*. (2) Next, for each *sessionId* we identified the timestamp of the next request of a different page (e.g., *sourcePage!=requestedPage*). (3) Then the residence time is computed by making the difference of the two timestamps. (4) Finally the average is computed.

4.2 Content-Based Statistics

Top 10 Visualized Instances. For each *ViewComponent* we compute the top 10 visualized instances as follows: for each visualized instance (from the *Final-Log*) we assign value 1 as *occurrence* to the pair `ViewComponent,Instance`. Then, we compute the sum of occurrences of each pair. The results are grouped by *ViewComponent* and ordered in a descendant order on the number of occurrences.

Top 10 Clicked Instances. For each *ViewComponent* of type *List* we computed the top 10 clicked instances. This statistic was extracted as follows: for each clicked instance (from the *FinalLog*) we assign value 1 as *occurrence* of the instance. Then, we compute the sum of occurrences of each instance. The results are grouped by *ViewComponent* and ordered in a descendant order on the number of occurrences.

4.3 Visualization

As stated in Sect. 3 we focused on a novel type of visualization, that is showing the analytics information directly on the model. Figure 6 shows some examples: Fig. 6(a) show the Average Staying Time as background color of the pages (green means high staying time, red low), Fig. 6(b) shows the Top 5 displayed books on a *ViewCompent*, while Fig. 6(c) show the Outgoing Link Ratio of the HomePage link.

5 Related Works

At the best of our knowledge this work is the first one which attempts to combine the information coming from the application models with runtime Web logs of the application execution, at the purpose of deepening the understanding of how the users consume the provided content. The works we found closer to our approach are: (i) a *Web Quality Analyzer* proposed by Fraternali et al. [17], a framework which integrates the design-time conceptual schemas of the application and the usage data collected at runtime to analyze the quality of website. Their approach focuses on quality analysis while ours studies the user interactions; (ii) in [25], Salini et al. proposed an approach that exploits the data coming from web usage analytics of existing web applications to generate the mobile navigation model of their corresponding mobile applications; and (iii) a model-driven approach for tracking information concerning the user navigation over websites proposed by Marco Winckler and Florence Pontico [32].

Web Log Analysis. Web Analytics is the measurement, collection, analysis and reporting of Internet data for the purposes of understanding and optimizing Web usage [11]. The Web analytics process relies on defining goals, defining metrics, collecting data, analyzing data and implementing changes [31]. So far, several

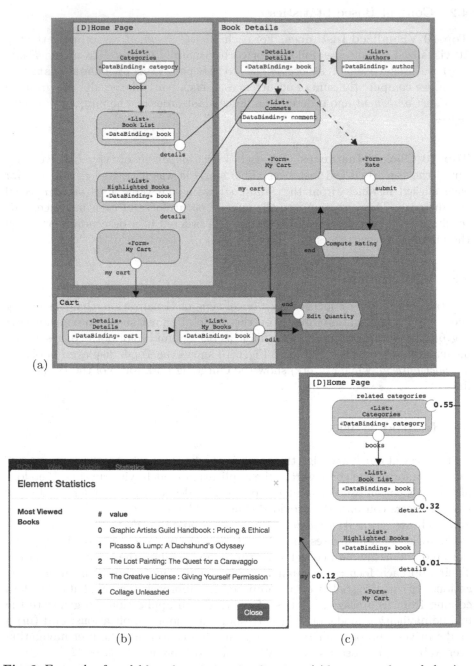

Fig. 6. Example of model-based statistics visualization: *(a)* heat-map of user behavior on the IFML diagram; *(b)* fine grained statistics related to a selected model element; *(c)* user navigational behavior on the IFML model as percentage of users clicking on each outgoing navigation flow. (Color figure online)

tools which support that process have been proposed. Among them we can cite Google Analytics, Yahoo Web Analytics, Compete, and ClickTale. All those tools provide the most important metrics: Unique Visitors, Visits/Sessions and Page Views [11].

However, those tools provide analysis mainly at web page level and they do not take into account the content of the visited pages. There exist some analytics tools, like Google Analytics [22], which provide a way to track the user interactions at content level but at a high cost of hard coding which yields to high risk of errors and maintainability of the web application. Furthermore, those event tracking mechanisms support a small subset of user interactions. Our approach combines the application models with runtime Web logs of the application execution, at the purpose of deepening the understanding of how the users consume the provided content and thus help analysts and decision makers to take more informed decisions on the website design and evolution.

The log analysis field in general, and Web log analysis in particular, has attracted the research community due to several advantages offered by the tracking the interactions of the users with an information access systems. Agosti et al. [4] have presented the current challenges and trends of this field.

Conceptual Modeling of WebApplications. This work is widely related to a large corpus of researches that address conceptual modeling of Web applications. Among them we can cite: (i) Araneus [23], a modeling proposal for Web applications that allows one to represent the hypertext organization, with nested relations and inter-page links; (ii) W2000, formerly HDM [7], which introduced a notion of model-based design, clearly separating the activities of authoring in-the-large (hypertext schema design) and authoring in-the-small (page and content production); (iii) OO-HDM [26], a UML-based approach for modeling and implementing Web application interfaces; (iv) OO-HMETHOD [20], based on UML interaction diagrams; (v) Hera [29], a model-driven design approach and specification framework focusing on the development of context-dependent or personalized Web information system; (vi) Web Application Extension for UML (WAE) [13], a UML extension for describing Web application interfaces and the client-server interactions; (vii) WebDSL [21], a domain-specific language consisting of a core language with constructs to define entities, pages and business logic, plus extensions.

Tools for Conceptual Modeling. Several design tools has been developed in the past. In the work [18] the authors developed Papyrus, a open-source graphical modeling tool for building UML models. This tool also allows the definition of DSLs through a very advanced implementation of UML profiles. In [12] the authors proposed *beContent*, a model-driven framework for developing and maintaining rich web applications. It allows to design the application, to deploy it on different platforms and to co-evolve it in order to adapt its components according to changes performed on the model. Syriani et al. [27] proposed a web framework that allows the definition, management and transformation of models. It runs entirely on the cloud and does not require any installation.

6   Conclusion

The main contribution of this work is to show the feasibility and advantages of a model-driven approach that blends design-time information and runtime execution data of websites for generating insightful analyses of user behavior. Thanks to the proposed information fusion, new kinds of analysis can be performed, which highlight the role and meaning of visualized data in the Web pages. New analytics results can be generated based on the displayed objects, their categorization and their properties. Results of the analysis can be immediately displayed directly on the visual models of the applications, thus making it immediate for a designer to spot possible problems or advantages of a given user interaction design choice. As a future work, we plan to: experiment scalability aspects on a cloud-based distributed solution; study additional Web-specific analyses (for instance, by combining the analysis with variant-based approaches such as A/B testing for checking how users react to different versions of the interface or of the rendered content); and explore a different tool to enhance the visualizations. As a limitation, this method cannot be applied to every existing web application, the web application must have been created using the WebRatio platform. A more general way to cover a wider range of web applications can be used through covering more modeling languages. Another point is that all the steps from analyzing to visualizing can be automatized to enable obtaining results easily, so that users who are not familiar with big data analysis technologies can obtain results easily.

References

1. WebRatio. http://www.webratio.com
2. Acerbis, R., Bongio, A., Brambilla, M., Butti, S.: Model-driven development based on omg's ifml with webratio web and mobile platform. In: Cimiano, P., Frasincar, F., Houben, G.-J., Schwabe, D. (eds.) ICWE 2015. LNCS, vol. 9114, pp. 605–608. Springer, Cham (2015). doi:10.1007/978-3-319-19890-3_39
3. Acerbis, R., Bongio, A., Brambilla, M., Butti, S.: Model-driven development of cross-platform mobile applications with WebRatio and IFML. In: 2nd ACM International Conference on Mobile Software Engineering and Systems, MOBILESoft 2015, Florence, Italy, 16–17 May 2015, pp. 170–171 (2015). http://dx.doi.org/10.1109/MobileSoft.2015.49
4. Agosti, M., Crivellari, F., Di Nunzio, G.M.: Web log analysis: a review of a decade of studies about information acquisition, inspection and interpretation of user interaction. Data Min. Knowl. Discov. 24(3), 663–696 (2012)
5. Ameller, D., Franch, X., Gómez, C., Araujo, J., Svensson, R.B., Biffl, S., Cabot, J., Cortellessa, V., Daneva, M., Fernández, D.M., et al.: Handling non-functional requirements in model-driven development: an ongoing industrial survey. In: 2015 IEEE 23rd International Requirements Engineering Conference (RE), pp. 208–213. IEEE (2015)
6. Amyot, D., Mussbacher, G.: User requirements notation: the first ten years, the next ten years. J. Softw. 6(5), 747–768 (2011)

7. Baresi, L., Garzotto, F., Paolini, P.: From web sites to web applications: new issues for conceptual modeling. In: Liddle, S.W., Mayr, H.C., Thalheim, B. (eds.) ER 2000. LNCS, vol. 1921, pp. 89–100. Springer, Heidelberg (2000). doi:10.1007/3-540-45394-6_9

8. Brambilla, M., Fraternali, P.: Interaction Flow Modeling Language Model-Driven UI Engineering of Web and Mobile Apps with IFML. The OMG Press, Morgan-Kaufmann, Burlington (2014)

9. Brambilla, M., Mauri, A., Umuhoza, E.: Extending the interaction flow modeling language (IFML) for model driven development of mobile applications front end. In: Awan, I., Younas, M., Franch, X., Quer, C. (eds.) MobiWIS 2014. LNCS, vol. 8640, pp. 176–191. Springer, Cham (2014). doi:10.1007/978-3-319-10359-4_15

10. Breu, R., Chimiak-Opoka, J.: Towards systematic model assessment. In: Akoka, J., Liddle, S.W., Song, I.-Y., Bertolotto, M., Comyn-Wattiau, I., Heuvel, W.-J., Kolp, M., Trujillo, J., Kop, C., Mayr, H.C. (eds.) ER 2005. LNCS, vol. 3770, pp. 398–409. Springer, Heidelberg (2005). doi:10.1007/11568346_43

11. Burby, J., Brown, A., et al.: Web analytics definitions (2007)

12. Cicchetti, A., Ruscio, D., Eramo, R., Maccarrone, F., Pierantonio, A.: beContent: a model-driven platform for designing and maintaining web applications. In: Gaedke, M., Grossniklaus, M., Díaz, O. (eds.) ICWE 2009. LNCS, vol. 5648, pp. 518–522. Springer, Heidelberg (2009). doi:10.1007/978-3-642-02818-2_52

13. Conallen, J.: Building Web Applications with UML. Addison Wesley, Boston (2002)

14. Cordeiro, L., Fischer, B.: Verifying multi-threaded software using SMT-based context-bounded model checking. In: Proceedings of the 33rd International Conference on Software Engineering, ICSE 2011, NY, USA, pp. 331–340 (2011). http://doi.acm.org/10.1145/1985793.1985839

15. Di Rocco, J., Di Ruscio, D., Iovino, L., Pierantonio, A.: Mining metrics for understanding metamodel characteristics. In: Proceedings of the 6th International Workshop on Modeling in Software Engineering, MiSE 2014, NY, USA, pp. 55–60 (2014). http://doi.acm.org/10.1145/2593770.2593774

16. Di Ruscio, D., Pelliccione, P.: A model-driven approach to detect faults in FOSS systems. J. Softw. Evol. Proc. 27(4), 294–318 (2015). http://dx.doi.org/10.1002/smr.1716

17. Fraternali, P., Lanzi, P.L., Matera, M., Maurino, A.: Model-driven web usage analysis for the evaluation of web application quality. J. Web Eng. 3(2), 124–152 (2004). http://www.rintonpress.com/xjwe3/jwe-3-2/124-152.pdf

18. Gérard, S., Dumoulin, C., Tessier, P., Selic, B.: 19 Papyrus: a UML2 tool for domain-specific language modeling. In: Giese, H., Karsai, G., Lee, E., Rumpe, B., Schätz, B. (eds.) MBEERTS 2007. LNCS, vol. 6100, pp. 361–368. Springer, Heidelberg (2010). doi:10.1007/978-3-642-16277-0_19

19. Giese, H., Tichy, M., Burmester, S., Schäfer, W., Flake, S.: Towards the compositional verification of real-time uml designs. SIGSOFT Softw. Eng. Notes 28(5), 38–47 (2003). http://doi.acm.org/10.1145/949952.940078

20. Gómez, J., Cachero, C., Pastor, O., Pastor, O.: Conceptual modeling of device-independent web applications, pp. 26–39 (2001)

21. Groenewegen, D.M., Hemel, Z., Kats, L.C.L., Visser, E.: Webdsl: a domain-specific language for dynamic web applications. In: OOPSLA Companion, pp. 779–780 (2008)

22. Ledford, J.L., Teixeira, J., Tyler, M.E.: Google Analytics. Wiley, New York (2011)

23. Mecca, G., Merialdo, P., Atzeni, P., Crescenzi, V., Crescenzi, V.: The (short) araneus guide to web-site development. In: WebDB (Informal Proceedings), pp. 13–18 (1999)

24. (OMG), O.M.G., Brambilla, M., Fraternali, P.: IFML: Interaction Flow Modeling Language. http://www.ifml.org
25. Salini, A., Malavolta, I., Rossi, F.: Leveraging web analytics for automatically generating mobile navigation models. In: 2016 IEEE International Conference on Mobile Services (MS), pp. 103–110. IEEE (2016)
26. Schwabe, D., Rossi, G., Rossi, G.: The object-oriented hypermedia design model, pp. 45–46 (1995)
27. Syriani, E., Vangheluwe, H., Mannadiar, R., Hansen, C., Van Mierlo, S., Ergin, H.: Atompm: a web-based modeling environment. In: Demos/Posters/StudentResearch@ MoDELS, pp. 21–25 (2013)
28. Umuhoza, E., Brambilla, M., Cabot, J., Bongio, A., et al.: Automatic code generation for cross-platform, multi-device mobile apps: some reflections from an industrial experience. In: Proceedings of the 3rd International Workshop on Mobile Development Lifecycle, pp. 37–44. ACM (2015)
29. Vdovják, R., Frăsincar, F., Houben, G.J., Barna, P.: Engineering semantic web information systems in Hera. J. Web Eng. 1(1–2), 3–26 (2003)
30. Völter, M., Benz, S., Dietrich, C., Engelmann, B., Helander, M., Kats, L.C.L., Visser, E., Wachsmuth, G.: DSL Engineering - Designing, Implementing and Using Domain-Specific Languages. dslbook.org (2013). http://www.dslbook.org
31. Waisberg, D., Kaushik, A.: Web analytics 2.0: empowering customer centricity. Original Search Engine Mark. J. 2(1), 5–11 (2009)
32. Winckler, M., Pontico, F.: A model-driven architecture for logging navigation. In: Workshop on Logging Traces of Web Activity: Workshop on the Mechanics of Data Collection, Co-located with 15th International World Wide Web Conference (WWW 2006), Edinburgh, Scotland (2006)

From Search Engines to Augmented Search Services: An End-User Development Approach

Gabriela Bosetti[1]([✉]) , Sergio Firmenich[1,2] ,
Alejandro Fernandez[1] , Marco Winckler[3] ,
and Gustavo Rossi[1,2]

[1] LIFIA, CIC, Facultad de Informática, Universidad Nacional de La Plata,
La Plata, Argentina
{gabriela.bosetti,sergio.firmenich,
alejandro.fernandez,gustavo}@lifia.info.unlp.edu.ar
[2] CONICET, La Plata, Argentina
[3] ICS-IRIT, Université Toulouse III, Toulouse, France
winckler@irit.fr

Abstract. The World Wide Web is a vast and continuously changing source of information where searching is a frequent, and sometimes critical, user task. Searching is not always the user's primary goal but an ancillary task that is performed to find complementary information allowing to complete another task. In this paper, we explore primary and/or ancillary search tasks and propose an approach for simplifying the user interaction during search tasks. Rather than focusing on dedicated search engines, our approach allows the user to abstract search engines already provided by Web applications into pervasive search services that will be available for performing searches from any other Web site. We also propose allowing users to manage the way in which the search results are presented and some possible interactions. In order to illustrate the feasibility of this approach, we have built a support tool based on a plug-in architecture that allows users to integrate new search services (created by themselves by means of visual tools) and execute them in the context of both kinds of searches. A case study illustrates the use of such tool. We also present the results of two evaluations that demonstrate the feasibility of the approach and the benefits in its use.

Keywords: Web search · Web augmentation · Client-side adaptation

1 Introduction

Searching is one of most frequent user task over the Web; popular rankings like Alexa's[1] can be used for supporting this claim. We can observe that most popular Web sites mainly support this task: Google, YouTube, Baidu, Yahoo, Wikipedia, Amazon are part of the leading group at the top 10 of the ranking, dismissing Facebook and QQ, and repetitions of Google under a different top-level domain.

---

[1] http://www.alexa.com/topsites/global; 0 accessed on January 17, 2017.

© Springer International Publishing AG 2017
J. Cabot et al. (Eds.): ICWE 2017, LNCS 10360, pp. 115–133, 2017.
DOI: 10.1007/978-3-319-60131-1_7

From the perspective of standard information retrieval [3, 14] a search is often treated as a task, driven by an information need that is formulated by a user as a query. The query is processed by a dedicated system able to select the documents that (according to certain rules) better match the user's query; a refinement process might be used to create new queries or to improve the obtained results. Nonetheless, users play a major role in the search process, therefore the understanding of user activities is essential to improve the tools supporting the search tasks [11, 14].

Web browsers and search engines are the primary tools people use to access the vast quantity of information available online [16]. A careful analysis of user activity over Web browsers shows that, quite often, users have to combine search engines of Web sites and browsing activities to find information that fulfils their needs [3]. It is also quite well known from empirical studies [1] that users maintain several tabs or windows open in the browser to track the context of multiple searches for information that interest them. It is interesting to notice that these activities might serve to accomplish two types of user goals [10, 20]: to start an investigation about a particular information (primary search task) and/or to find complementary information providing details about the information currently displayed on a Web site (secondary or ancillary search task). While opening new windows or tabs in the browser might help users to run multiple primary searches in parallel, the same actions create an articulatory distance between the main user's focus of interest [6], as illustrated in Fig. 1. The latter shows a task model matching two primary searches for finding the Web site of ICWE2016 and WWW2016, respectively, from the Google's search engine. Figure 2, in turn, shows the recursive pattern of the ancillary search tasks. There, a user visiting the ICWE2016 Web site performs an ancillary search on DBLP for getting a list of related articles for a concrete author, and navigates to Google to search for the online file. Both sites offered complimentary information that allows him to understand better the program of the ICWE2016 Web site and decide to attend or not.

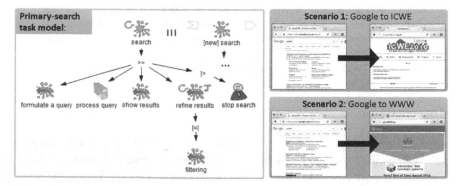

Fig. 1. Primary search task model and two possible scenarios

Today, users can type their search queries directly on the browser's interface components, without having to visit the Web site of a search engine. Safari lets users specify the search engine the user wants to use, like Google, Yahoo! or Bing. Mozilla Firefox also allows users to choose a particular search provider when the browser detects that the current Web page provides one. However, such feature works only for well-known Web sites, as IMDB. In all the cases, search results started from the browser's UI are still displayed in a new tab or window.

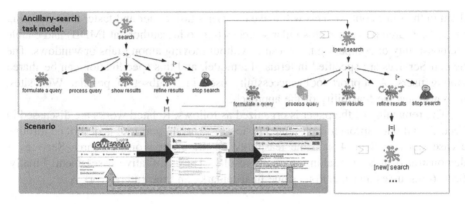

Fig. 2. Ancillary search task model and scenario

There are generic search engines based on Web content scrapers, as Google, and domain-specific ones, as IMDB. The first ones cover a broad scope and their goal is to provide entries to data sources/Web sites containing the information that users are looking for. In contrast, domain-specific search engines are focused in data with a narrow scope but more accurate if users want specialized details about objects in a concrete domain [15]. Indeed, most of ancillary search tasks are triggered because users need complementary information that is frequently domain-specific. A taxonomy of online page-views according to different categories is presented in [12]. The paper reports that users navigate among different kind of pages and use to revisit pages of the same kind and structure. This fact may also suggest that structured objects may take advantage of the already existing content and behavior on the Web, since they share a same domain, and that was the purpose of [8].

In this paper, we propose an innovative approach for dealing with ancillary search tasks while navigating the Web by means of Web Augmentation [17]. The approach is aimed at allowing users to: (i) create custom queries to generic and/or domain-specific search engines; (ii) trigger queries as ancillary search tasks for Web sites they are visiting; and (iii) display the results of the search in the context of the user task. For that, custom user searches are deployed as search services that can be accessed from the Web browser whenever the user is navigating the Web. An overview of this idea is shown in Fig. 3. In this case, we show how our approach would better support the scenario presented in Fig. 2. Instead of requiring the user to move from the current Web page he is visiting, we propose to obtain search results transparently and show

Fig. 3. A scenario for the search approach

them in the same context. This will make the user's task easier and faster than repeating the same search process with similar searches (e.g. other authors at IMDB) and enable the possibility of comparing their results without moving among tabs or windows. The Search Services are specified in terms of a model, and this specification can be shared among users. A support tool, successfully used in a subset of popular Web sites, demonstrates the feasibility of the approach.

The remaining of the paper is organized as follows. Related works are discussed in Sect. 2, and our approach is presented in Sect. 3. The support tool is described through a case study in Sect. 4, and the feasibility on the abstraction of Search Services is demonstrated in Sect. 5 along with the results of a quantitative assessment. Finally, Sect. 6 discusses and mentions some final remarks and the future work.

2 Related Works

Web search has been a target of research from the beginnings of the Web. Several works have contributed in various aspects, as scraping, semantic search or collaborative search. Although these research lines are very active, the design of user interfaces for searching was also studied. For instance, concerning information presentation, focus + context visualizations [4] proposed to maintain the object of primary interest presented in detail, meanwhile, other contextual information may be available. This is clearly related to the idea of ancillary searches; a deep analysis of this kind of search was recently presented in [20]. An inside-in approach for complementary information is presented in [18], conceiving its presentation in the light of the principle of "Overview first, zoom and filter, then details-on-demand". The paper also argues that information may require being presented in different ways, through visualizations that effectively cover different *task-domain information actions*. In the following sections, we propose an approach in which end-users are the ones managing how resulting information may be visualized. To achieve this, it is necessary to analyze search results.

Web search engines return a result page with a list of Web documents (URIs) matching the search criteria. Results are usually presented as a list of page titles and one or two sentences taken from the content. Recent advances in Web search interfaces provide, for a predefined set of element types, such as movies, recipes or addresses, rich snippets that help the user in recognizing relevant features of each result element (e.g., movies playing soon or reviews for a given movie). In some cases, these rich snippets include the data the user is looking for. It is possible because Web site creators include structured data in Web pages that computers can recognize and interpret, and that can be used to create applications. Viewing the Web as a repository of structured, interconnected data that can be queried is the ultimate goal of the Semantic Web [2]. However, end users do not always have the means to exploit or add information to the Semantic Web. The tools presented here let them extract and use structured data from any Web site, with no need of training on Web development.

There are also models describing how people deal with information and their informational needs [19] and, consequently, providing context to our approach. Eight characteristics (features) adequately describe the information behavior in various disciplines [7]. Such characteristics are: starting (begin the information seeking), chaining (following connections to related content), browsing (semi-structure browsing), differentiating (to filter information), monitoring (keeping up-to-date), extracting (selectively identifying relevant material in an information source), verifying (checking for accuracy), and ending (final searches and tidying up). The definition of Search Services allows the extraction of relevant features in Web content, putting the user in control of what are the relevant properties of information objects, to use them later for differentiation. Moreover, when users select which information repositories will be defined as search services, they conduct an initial quality filter (verification by provenance).

Being able to mouse-select elements in the primary window to launch in-situ searches is a means to support search start and information chaining activities. When the results are displayed in overlay mode, these connections become explicit and persistent, as pointed in [20]. It presents an approach for providing end users with the possibility of executing ancillary searches. However, while in [20] developers need to implement a *broker* at server-side, in charge of retrieving external and already existing services, we provide EUD support at client-side, with no need for server-side components nor textual programming skills for users. Additionally, we allow defining the semantic and structure of the results to be retrieved, without restraints on what kind of element to consider. This suggests a same structure of information for all end users using our tool, which can be later used for querying the common repository or using similar visualizations for the objects sharing a same class.

Finally, research on information seeking behavior has focused mostly on how individuals seek information. However, in many contexts, this process involves collaboration [9]. The Search Service definition tool we propose stores service definitions in a local storage. Users can export and import definitions, thus supporting simple sharing via e-mail and instant messaging. MacLean and colleagues found this simple form of collaborative tailoring via customization files and email sharing an effective mechanism to foster a culture of end-user tailoring [13].

3 End-User Driven Search Services

A vast number of searching scenarios may allow appreciating that the integration support offered by Web applications and Web browsers, in combination, may jeopardize the user experience, since the user may need to perform repeatedly extra operations to obtain the desired information (such as open a new tab, enter the URL of the target Web searcher, etc.). Furthermore, if the user is performing an ancillary search, he might refine the query or go back to the original information context to do it.

This paper presents an approach based on a flexible architecture that allows end users to customize the way they perform Web searches. It follows an inside-in approach, which enables users to perform ancillary searches without leaving their current context. While other searches are performed in relation to a specific goal in mind, the ancillary search is nested among other tasks and is aimed at providing complementary information to the user's current context. The solution we propose allows dealing with ancillary search tasks while navigating the Web by means of Web Augmentation [17], reducing the user's efforts and, therefore, the gulf of execution and the evaluation gulf, as explained in [20].

The searching process we target entails the following steps: (1) define a query; (2) select a search engine; (3) enter the query and trigger the search; and (4) inspect and interact with the results. Our goal is to improve the user experience with search tasks, particularly with the steps 2–4. For the first step, we comply with the query language imposed by the underlying search engine. To better support the aforementioned steps, our approach propose a tool allowing to:

- Trigger searches from the current Web page for reducing the interaction required to perform a search in any foreign search engine.
- Transform search results (DOM elements) into domain objects with specific semantic and structure.
- Integrate the resulting domain instances in the current Web site for further visualization and interaction.

In order to achieve these objectives, we propose:

- Allowing users to encapsulate existing Web applications' search engines into pervasive Search Services. Given that not all the Web applications supporting searches provide an API, we propose to reproduce automatically the UI interaction required to perform a search. It implies that users must select the UI search engine components to create Search Services.
- Integrating the new Search Services with the Web browser search mechanism for ancillary searches. Users should be able to use the created Search Services from any other existing Web site.
- Displaying results in the context of the current Web site, to enable different ways of visualizations supporting primary and ancillary searches. It is done by parsing the DOM, extracting the search results from it, and creating domain object instances.

A simple scenario is presented in Fig. 3, where a user is navigating the accepted articles on the ICWE2016's Web site. At some point, he requires seeing other publications of a particular author. Certainly, this secondary information requirement would be better satisfied by using domain specific Web applications (such as DBLP or Google Scholar) instead of a generic searcher (such as Google or Bing). In this setting and using our tool, the user would be able to trigger the search from the current Web site by highlighting the author's name and selecting the desired Search Service (e.g. DBLP). At the right of the same figure, results are listed in an overlay popup whose content follows a table structure. It is built automatically given the domain object specification that was made when the user created the Search Service. There is also a toolbar where options for filtering and ordering results are available (if these were defined for the

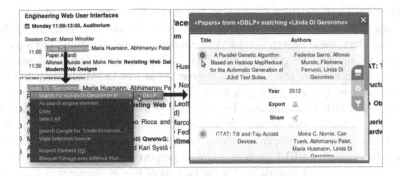

Fig. 4. Example of use of a Search Service

triggered Search Service). Furthermore, a third option allows the user to change the visualization, in case that a table structure does not fit the user's needs.

3.1 The Approach in a Nutshell

We have designed and developed an architecture for supporting the creation of Search Services and the integration tool needed for letting end-users to use these new services from any Web site they are visiting. The architecture has three layers: (i) end-user support tool, (ii) current search results, (iii) model layer, as it shown in Fig. 5.

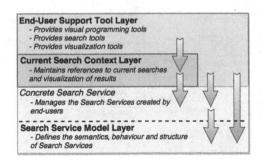

Fig. 5. Search service architecture overview

This diagram refines the relationships between layers' components:

1. *The Search Service Model.* This layer supports the creation of Search Services, which are based on the search engines that Web applications already provide. Note that this is focused mainly on the creation of Search Services for those Web applications that, despite providing a search engine, do not provide an API. In this sense, our approach enables the creation of a service (API alike) based on how users would use these search engines. A Web search engine's interface is usually composed of an input and submit button, filtering and ordering options, and some

mechanism for paginating the results. We propose to abstract all these UI components (DOM elements), wrapping them with objects that conform a Search Service. Then, these Search Services can emulate the user behavior and retrieve the corresponding results given a particular query specification. To provide an API-alike mechanism, results are not interpreted just as DOM elements but also as abstractions of the underlying domain objects. E.g. if the search engine being abstracted is DBLP, results may be wrapped by the *Paper* domain object, which could be populated with domain properties such as *title* or *authors*. Even more, this object may also have properties whose values are taken from another DOM (obtained from another URL), such as a *bibtext* property; we explain this in Sect. 3.3. All these concepts are materialized at the bottom of Fig. 6. As we will show later, it is convenient to provide a semantic layer on the top of the search results because it allows the creation of visualizers that go beyond presenting the raw results.

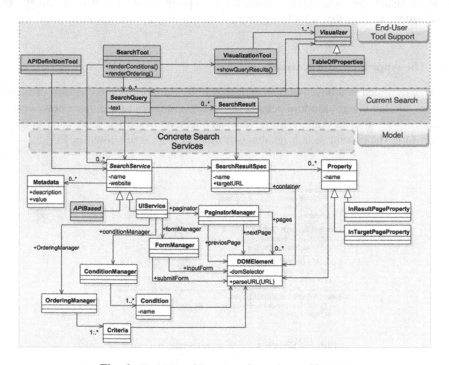

Fig. 6. End-User driven search service architecture

2. *End-User Tool Support.* Search Services can be incrementally specified by using our visual tool, selecting the key DOM elements related to a Web search engine. The tool is part of the End-User Support Tool layer, where other two components coexist. One is the *SearchTool*, which creates the corresponding menus for the existing *SearchService* objects and allow users to perform searches from the current Web site. The *VisualizationTool* takes the search results and allows users to interact with them in different ways by selecting a particular *Visualizer*.

3. *Current Search Context.* The middle layer represents the current *Query* and its results. When the user wants to perform a search task, a *SearchQuery* is created (there are basically two strategies for doing that: text selection and text input). When the search results are retrieved, these are materialized as instances of *SearchResult*; the Visualizer presents them considering the corresponding *SearchResultSpec;* this specification describes results by means of properties whose values are also obtained from the DOM where actually the results are, or from the DOM obtained from the *targetURL*, which is basically the real domain object's Web page.

3.2   Search Service Architecture: Flexibility, Compliance and Extensibility

In this subsection, we explain two important aspects of the architecture. Flexibility and compliance are related to how the Search Service may be mapped to different search engines provided by Web applications. Extensibility is related to how the whole toolkit may be extended with further components for searching and visualizing results.

Flexibility and Search Engine Coverage

There are some components presented in Fig. 6 that deserves a bit more explanation:

- *Properties*: these are defined for the *SearchResultSpec*. Consider a Search Service is being defined in the e-commerce domain, and properties such as *price* and *availability* could be defined. However, although there is some information obtainable from the DOM presenting the results, more information could be available in the site; as the *technical description* or *shipping costs*. Both kind of properties are extracted from different DOMs, and the behavior for retrieving both DOMs is different. This is the reason why we separate two kind of properties. On the one hand, *InResultPageProperty* allows users to define properties whose values will be extracted from the search result's page. In the other hand, *InTargetPageProperty* let users to extract values for further properties. Our toolkit knows how to reach the actual result's Web page because the attribute *targetURL* is defined, mandatorily, in *SearchResultSpec*. With this in mind, we can note that when using our approach for an ancillary search, as in Fig. 4, the listed properties could be obtained from different Web pages, but this will be transparent from the users' point of view.
- *Ordering and Filtering*: To better reproduce the power of original Web search engines, we have contemplated ordering and filtering features. Some Web sites, such as DBLP, let the users filter results according to some criteria, such as "Journal only", which is applied by clicking an anchor on the search page. Our model contemplates this kind of filters through the *ConditionManager* and *Condition* classes. Ordering and filtering functionality, with roots in the original Web application, will be available in the menu displayed at the right of Fig. 4. Most of the search engines we analyzed offer ordering by clicking anchors or buttons, and these options are available from the same main menu in our Visualizers.
- *Search Execution Strategies:* Search engines spread on the Web have different configurations concerning the involved UI components and the interaction design for

executing a search. Most of them have an input element to write the query but not all of them have a trigger (e.g. anchor, button) and that makes it necessary to have different strategies for carrying it out. We have implemented three alternatives, but they can be extended and integrated in the tool. We covered sites requiring to write a query into an input and handling the query execution by one of these alternatives: (1) clicking on a trigger to load the page with the results; (2) clicking a DOM element but loading the results through ajax-call; (3) listening to some input-related event (e.g. keypress, blur, change) and loading the results by ajax-call. Subclasses implementing this behavior are *WriteAndClickToReload*, *WriteAndClickForAjaxCall* and *WriteForAjaxCall* respectively. The strategy is automatically assigned when the user creates a search service, and each strategy can know if it is applicable based on the components the user has defined and the success on retrieving new results, so this is not a concern for the end user.

Extensibility

As shown in Sect. 5, the Search Service Model is compliant with most of the Web sites' search engines that we analyzed. However, beyond this model and the Search Service specification tool, our approach is extensible in two ways. On the one hand, the model is extensible by means of the creation of services based on the existence of application APIs. On the other hand, the end-user support tool is extensible by the creation of new Visualizers. We next explain both APIBased and Visualizer extension points.

- *APIBased Search Services*: Some Web applications offer APIs for retrieving information from their databases (Twitter, Facebook, etc.). In these cases, it is very common that APIs expressivity goes much further than what is possible to do with our UI-based Search Service approach. With this extension point, developers could create new Search Services (using application's API) to be integrated later by end-users.
- *Visualizers*: as shown in Fig. 4, search results become domain objects whose properties are listed in the default Visualizer (*TableOfProperties*), which create a table where columns represent each property and rows each object. However, further visualizers could be developed and integrated into our toolkit. E.g., a new Visualizer (*GroupByPropertyValue*) could allow users to select a property and group the already obtained results according to the value of this property. Consider the example of Fig. 4; it would be possible to group the author's articles within *journal* or *conference* boxes. Beyond this, other visualizers could be focused on calculating information and visualizing more processed information. For instance, the users could be interested in seeing quickly the evolution of the author production in a chart showing the amount of articles per venue, instead of just displaying each article.

It is important to note that these two extension points require advance JavaScript programming skills to be developed, but once created, they can be installed and used by end-users, who can configure the parameters of such new visualizers according to specific properties defined for the *SearchResultSpec* of a given *SearchService*.

4   Tool Support and Case Studies

In order to support our approach, we implemented our tool as a Firefox extension. It allows both the specification and execution of Search Services. The use of this tool for the first purpose is illustrated in Figs. 6 and 7, and for the second one in Fig. 8.

End users can define *UI-based Search Services* through our tool. A Search Service of such nature should be capable of automatically emulating a search that otherwise the user has to do manually (e.g. opening a new tab, navigating to the search engine of a Web site, typing a text, triggering a button, etc.). Once such Search Service is defined, the user can use it for performing ancillary searches by highlighting a text in any Web page he is navigating and choosing a service from which he wants to retrieve results. In this sense, the selected *UI-based Search Service* must know: in which *input control* the text should be entered, which *button* to trigger to perform the search, how to obtain more results, and how to interpret them. Filters and sorting mechanisms can also be defined, but they are not mandatory.

Consider Amaru, she is always surfing the Web and she uses to look for related books when she finds something (a topic, an author, etc.) of her interest. She is an active user of GoodReads and every time she finds some term of her interest, she copies it, opens a new tab in her browser, accesses GoodReads and performs a search with the copied text as keywords. In this setting, it would be very convenient for her to be able to carry out such searches from the same context in which she is reading the comments.

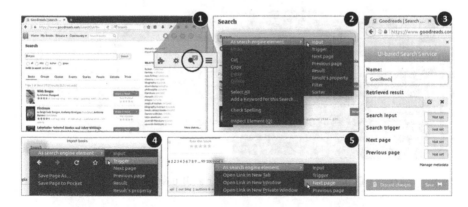

Fig. 7. Defining the input, trigger and pagination elements for a Search Service

Figure 7 shows how Amaru is starting to create a *UI-based Search Service* by selecting DOM elements from the Web site of *GoodReads*, concretely from its search engine[2]. To do so, she navigates to the Web page where the search engine is and enables the «Search Service definition mode» by clicking the highlighted button in step 1. In this concrete case, she should select, at least, the input (step 2) and trigger (step 4)

---

[2] https://www.goodreads.com/search?q=Borges.

controls, and also the one retrieving more results (step 5). The DOM elements defined as the UI-Search-Service controls are selected by right-clicking them.

As it is the first time she defines a Search Service and she has no other Services available in her personal account, the tool asks her to give it a name through a form opened in the sidebar (step 3). Otherwise, the tool should ask the user to select an existing UI-based Search Services for which it is starting to define the controls.

Then, she specifies the kind of results the Search Service will retrieve, as shown in Fig. 8. First, she selects an element in the DOM which represents the main container of the element he is expecting to have as a result. Such element is the highlighted one in step 6. When she chooses to define it as a "result" through the context menu options, a form is loaded in the sidebar (step 7), where she must complete some required data. In this case, she names this kind of results as *Book rating* and selects one of the available selectors generated in relation to the available XPaths for the selected DOM element. This will allow her to choose more than one instance of *Book rating* in the same context (as shown in steps 6 and 8, there are other instances in the background). In a similar way, she should define the properties of the results that are of her interest, so these will be displayed when an ancillary search is performed. For instance, in steps 8 and 9, Amaru is defining the *Title* property for a *Book rating* result of the *GoodReads* Search Service. She repeats the last two steps for defining also a *Rating* property.

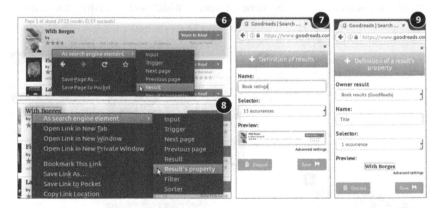

Fig. 8. Defining the expected results for a UI-based Search Service

After the mandatory elements of a *UI-based Search Service* have been defined (input, more-results trigger and the expected structure of the results), the Search Services becomes available in the browser's context menu whether the user has highlighted any text in any Web page. In step 1 of Fig. 9, Amaru has highlighted some text of a Wikipedia article and she is performing an ancillary search using it as keywords. At such point, she can also use other previously defined services, for instance, *Amazon* or *Google Books*. When she clicks one of the menu items, let us say *Amazon*, a draggable panel appears in the middle of the screen, presenting the results of the search for the highlighted keywords. Now, as shown in step 2, she can access related books to *Julio Cortázar* and see their *Title* and *Authors*. However, she can also access the

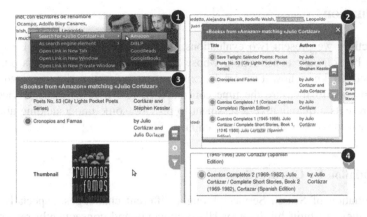

Fig. 9. Visualizing the results of ancillary searches

remaining properties (in this case, the *Thumbnail*) by clicking the «+» button at the left of each row, which will display a section with the data hidden due to the lack of space. There are three fixed buttons at the right of the panel that allow her to: (1) change the kind of visualization – she is using the default one –; (2) to configure some parameters of the selected visualization, as the order or the priority of the columns in the responsive layout; and (3) to apply filters if any was defined. At the bottom of the visualization, as shown in step 4, there are navigation buttons so she can get more results.

Note that the use of the Search Service is not exclusive for the Wikipedia Web site; it is always available in the context-menu of the browser, no matter the Web site the user is navigating. Multiple ancillary searches with the same or diverse Search Services can be performed in the same context, using different keywords, as shown in steps 5 and 6 of Fig. 10. This way, Amaru can search for a second time, by selecting *Rayuela*, one of the books of *Julio Cortázar* listed in the first ancillary search' results popup. This time she is using the Search Service she defined for *GoodReads*, and she is accessing information that was not present in the results of the *Amazon* Search API.

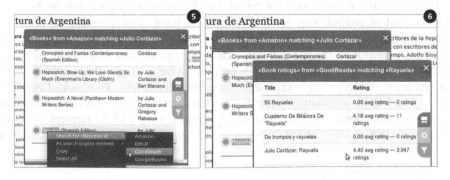

Fig. 10. Performing an ancillary search over the results of a previous one

5 Evaluation

In this section, we present two evaluations of our approach. Section 5.1 proposes a validation by software construction, whose main objectives are to stress our Search Service model to know how it covers the existing search functionality provided by Web applications and to measure the time consumed in real uses of these services. Section 5.2 presents a quantitative assessment to provide some understanding about how our approach influences the user interaction.

5.1 Validation by Construction

The instantiation of Search Services brought different challenges, especially considering that they are built by different UI components and consequently require a different kind of interactions for executing their associated behaviors. For instance, consider the 20 first sites in the top 500 by Alexa[3] meeting the following conditions: (1) the interface is not Chinese or Russian by default; (2) sites of with the same domain but different top-level domain are considered just once (e.g. Google.com and Google. co.in); (3) only consider one instance using the same search engine (e.g. Msn redirects to Bing); (4) do not consider the one with no search engine (t.co). The 20 sites are Google, Youtube, Facebook, Yahoo, Wikipedia, Amazon, Live, Vk, Twitter, Instagram, Reddit, Linkedin, Ebay, Bing, Tumblr, Netflix, Wordpress, Microsoft, Aliexpress and Blogspot. In this list, all sites have a search engine but they are executable by different means.

By analyzing the 20 sites, we can see that the involved UI controls differ in kind and quantity. There is no variation on the kind of DOM element used for entering the search's keywords, it is an input field, but it is different for the remaining controls involved in the search process. Moreover, there is a site hiding the input until the user clicks on a concrete element of the DOM (Live.com), and the search execution strategies (see Sect. 3.3.1) are not always the same: 11 sites use the *WriteAndClickToReload*; 5 the *WriteAndClickForAjaxCall*; 2 the *WriteForAjaxCall*. We successfully defined the service for those 18 of the list of 20 sites with such strategies; but the remaining 2 (Instagram and Live) required different ones that we are currently working on.

Back to the UI components, 17 of the full list of sites have a trigger element, but they differ in kind of control: they were 10 buttons, 6 inputs and 1 anchor. Concerning pagination, 14 sites have a control for retrieving the following elements, but just 12 of them have a control for the previous ones. This is due to the way they handle results in the presentation layer; Instagram and Wordpress have a single DOM element that attaches more results in the results area, expanding its height. Facebook, Live, Vk, Twitter and Tumblr automatically retrieve more results when the user scrolls down to the bottom of the page. The remaining site, Microsoft, have a DOM element with this purpose but clicking on it redirects to a specialized form for searching a concrete kind of

---

[3] http://www.alexa.com/topsites/global; 1.

results, where the results of the first page were included. However, this specialized searcher does not allow changing the search keywords; if you do that, you are redirected to the first results' page. Of the 14 sites with clickable paging elements, 8 of them cause the page to be reloaded while another 6 apply the changes via ajax-call. In this matter, sorting elements are present in 8 sites; 5 of them reload the document and 3 of them use ajax. Filter elements are defined for 16 sites, of which 11 reload and 5 use ajax.

The domain-specific abstraction of search results is another issue to face. Most sites in the list can retrieve more than one kind of result: news, images, videos, channels, playlists, people, pages, places, groups, applications, events, emails, etc. This is not just a problem for naming the kind of results a Search Service retrieves, but for choosing a selector (an XPath) to retrieve a concrete DOM element with a specific structure in the DOM (or all of them) whenever possible, since search engines often present different kinds of results with a different structure/style.

Regarding search performance, we logged the times for the 18 search engines in a 15-inch notebook, with a resolution of 1366×768 pixels. We cannot say the results are the same for other resolutions, since their UI elements may vary. The purpose of this analysis was to demonstrate that our tool allows the instantiation of the Search Services. We successfully defined a Search Service for each of the search engines of the aforementioned 20 sites, and we report below the times it took for executing an ancillary search. For each of them, we performed a search: (1) from exactly the same Web context[4], (2) searching for the same keywords (*Borges*) and (3) expecting to have results with just two textual properties: name and description. There were only two cases in which *Borges* did not produce any results, so the word was replaced by one of the same length: *Ubuntu*. For each test, we cleared the cached and offline Web content, the offline content and user data. We also reloaded the target Web page to augment, for making it sure that no class was already instantiated and giving the entire Search Services the same conditions before executing the search. The full search process took 7 s with a standard deviation of 5 s. The differences depended on the strategy and the response time of the search engines. E.g., there were 3 cases in which it took between 14 and 18 s, while in other 3 sites it took between 1 and 2 s. For a full list of logged times and a video, please visit the project's Web site[5].

This evaluation demonstrated that our approach is feasible, covering 18 of 20 analyzed sites by using our support tool and the 3 implemented strategies to date. The sites for the test were not conveniently chosen, but from a public Web traffic report. Other search engines using the same interaction strategies can be visually specified as Search Service by end users, and such strategies can be also extended by developers to cover more search engines. We also reported the average time in order to compare it against the estimation of times in the following subsection, regarding the time it takes a task to be performed manually or with the support of our tool.

---

[4] https://en.wikipedia.org/wiki/Argentine_literature.

[5] https://sites.google.com/site/webancillarysearches/testing-in-top-sites.

5.2   Quantitative Assessment Based on GOMS-KLM

We present in this section a quantitative assessment of search tasks based on the GOMS-Keystroke (KLM) model [5]. This formal model is used to evaluate the efficiency of interaction with a given software for a specific, very detailed, scenario. The resulting time is calculated by the summation of each user action, whose time are already known in this model. For example, the average time to perform the action *reach* with the mouse is 0.40 s or *click on a button* is 0.20 s. Thus, by providing a detailed scenario of user actions with the Web browser and Web applications, it is possible to use GOMS-KLM to predict performance.

We focus this quantitative assessment for the case of ancillary searches. By using GOMS-KLM, we have specified the traditional scenario for performing ancillary searches, shown in Fig. 2, in order to compare it against our approach, introduced in Fig. 3. We have split the whole scenario into six tasks: *(i)* Visit ICWE2016's Web site, *(ii)* select an author, *(iii)* search for that author in DBLP, *(iv)* select a resulting paper, *(v)* search for that paper in Google Scholar, *(vi)* point the mouse over the article's title. The results are shown in Table 1. Note that tasks *i, ii, iv* and *vi* are equivalents in both scenarios. However, the time required for tasks *iii* and *v*, the ones actually requiring to use the search engine provided by other Web sites, are quite faster using our approach. The whole scenario (involving two ancillary searches) takes 46.6 s without our approach. In contrast, by using Search Services, it takes just 18 s. It is interesting to note that performing a search using our approach and visualizing the results from any Web site would take only 1.5 s (this time would be always the same, regardless both the current Web site and the target Search Service).

Table 1.   GOMS-KLM results for both scenarios

Task	Time (without using SS)	Time (using SS)
1. Go to the ICWE web site	8.7 s	8.7 s
2. Select the target author	2.6 s	2.6 s
3. Perform the first ancillary Search (DBLP)	15.9 s	1.5 s
4. Select an article	2.6 s	2.6 s
5. Perform a second ancillary Search (Google Scholar)	15.7 s	1.5 s
6. Point the title of a target paper (Google Scholar)	1.1 s	1.1 s
Total	**46.6 s**	**18 s**
Define both Search Services	-	*39.2 s*

However, for using the Search Services of DBLP and Google Scholar, it is required that the user has previously defined them. With this in mind, we have defined also the GOMS-KLM scenarios related to the creation of both DBLP and Google Scholar Search Services. They were defined considering: *(i)* input search form, *(ii)* trigger search UI component, *(iii)* abstract the search result into the concept "*Paper*" with the property *title*, *(iv)* give a name to the Search Service and save it. All the tasks take 19.6 s.

The specification is the simpler, without filtering nor ordering options, since such features were not required for the contemplated scenarios. This means that the first time the user experiences the scenario in Fig. 3, he needs an extra time of almost 40 s. However, the definition is required just once and it can be avoided by installing an existing Search Service specification.

As a final comment, it is interesting to note that in the context of ancillary searches, the user would go back to the primary context, in this case, the ICWE's Web site. With our approach, it would not require further interaction because the user already is in ICWE's Web site; meanwhile in a traditional scenario it will require further interaction.

6 Conclusions and Future Works

In this paper, we presented an end-user driven approach for the customization of Web search tasks. The main objective is to get better support for ancillary searches, although the approach also reaches primary searches. Several contributions are made in this context. First, we propose an end-user support tool for the creation of Search Services based on the automatic execution of UI interaction required to perform searches in existing and third-party search engines. Second, we propose to transform search results into domain objects and take advantage of the semantic information as in [8]. Both together achieve a third contribution, which is the creation of new ways of interaction and visualization of results in-situ.

We fully supported our approach with already working tools, used in 18 existing Web applications as a way of validating our aims by software construction. Besides, we have shown that our approach is very convenient in terms of performance when users require complementary information for accomplishing their tasks.

Our approach still lacks a full end-user evaluation to measure the potential of adoption and the usefulness of in-situ visualizations of results. Another evaluation with end-users is necessary to demonstrate that the specification of Search Service is clearly doable without programming skills. However, beyond some usability issues, the abstraction of Search Engine components may not be a limitation because of the seamless observed in existing search engines and their common use nowadays.

Beyond further evaluations and the improvement of our tools (e.g. to support more strategies to cover more search engines), some other works are planned. First, although we foresee the usefulness of defining metadata for Search Services, we have not exploited it yet. For instance, this kind of information could be used for automatically perform searches in parallel given a particular context of information. Collaboration is another aspect to be addressed. So far, we allow users to share service specifications by sending their corresponding files, but we want to analyze how to reach this goal through collaborative techniques better. Finally, domain-specific visualizers could improve how end-users interact with the information obtained by our search services.

References

1. Aula, A., Jhaveri, N., Käki, M.: Information search and re-access strategies of experienced web users. In: WWW 2005 Proceedings of the 14th International Conference on World Wide Web, pp. 583–592 (2005)
2. Berners-Lee, T., Hendler, J., Lassila, O.: The semantic web. Sci. Am. **284**, 34–43 (2001). doi:10.1038/scientificamerican0501-34
3. Broder, A.: A taxonomy of web search. SIGIR Forum **36**, 3–10 (2002). doi:10.1145/792550. 792552
4. Card, S.K., Mackinlay, J.D., Shneiderman, B.: Focus + context. In: Readings in Information Visualization, pp. 306–309 (1999)
5. Card, S.K., Moran, T.P., Newell, A.: The Psychology of Human-Computer Interaction. L. Erlbaum Associates, Hillsdale (1983)
6. Cava, R., Freitas, C.M.D.S., Barboni, E., Palanque, P., Winckler, M.: Inside-in search: an alternative for performing ancillary search tasks on the web. In: 2014 9th Latin America Web Congress IEEE, pp. 91–99 (2014)
7. Ellis, D., Haugan, M.: Modelling the information seeking patterns of engineers and research scientists in an industrial environment. J. Doc. **53**, 384–403 (1997). doi:10.1108/ EUM0000000007204
8. Firmenich, S., Bosetti, G., Rossi, G., Winckler, M., Barbieri, T.: Abstracting and structuring web contents for supporting personal web experiences. In: Bozzon, A., Cudre-Maroux, P., Pautasso, C. (eds.) ICWE 2016. LNCS, vol. 9671, pp. 77–95. Springer, Cham (2016). doi:10.1007/978-3-319-38791-8_5
9. Golovchinsky, G., Qvarfordt, P., Pickens, J.: Collaborative information seeking. Computer (Long Beach Calif.) **42**, 47–51 (2009). doi:10.1109/MC.2009.73
10. Hearst, M.A.: User interfaces for search. In: Modern Information Retrieval the Concepts and Technology Behind Search Engines, pp. 21–56 (2010)
11. Hearst, M.A.: "Natural" search user interfaces. Commun. ACM **54**, 60–67 (2011). doi:10. 1145/2018396.2018414
12. Kumar, R., Tomkins, A.: A characterization of online search behavior. IEEE Data Eng. Bull. **32**, 1–9 (2009). doi:10.1145/1772690.1772748
13. MacLean, A., Carter, K., Lövstrand, L., Moran, T.: User-tailorable systems: pressing the issues with buttons. In: Proceedings of the SIGCHI Conference on Human Factors in Computing Systems Empower people - CHI 1990, pp. 175–182 (1990). doi:10.1145/97243. 97271
14. Marchionini, G.: Exploratory search: from finding to understanding. Commun. ACM **49**, 41 (2006). doi:10.1145/1121949.1121979
15. McCallum, A., Nigam, K., Rennie, J., Seymore, K.: A machine learning approach to building domain-specific search engines. In: Proceedings of the Sixth International Joint Conference on Artificial Intelligence, IJCAI 1999, pp. 662–667 (1999)
16. Morris, D., Morris, M.R., Venolia, G.: SearchBar: a search-centric web history for task resumption and information re-finding. In: Proceedings of the SIGCHI Conference on Human Factors in Computing Systems – CHI 2008, pp. 1207–1216 (2008)
17. Obal, C., Diaz, O.: The augmented web: rationales, opportunities and challenges on browser-side transcoding. ACM Trans. Web **9**, 1–30 (2015). doi:10.1145/2735633
18. Shneiderman, B.: The Eyes have it: a task by data type taxonomy for information visualizations. In: Proceedings of 1996 IEEE Symposium on Visual Language, pp. 336–343 (1996)

19. Wilson, T.D.: Models in information behaviour research. J. Doc. **55**, 249–270 (1999). doi:10.1108/EUM0000000007145
20. Winckler, M., Cava, R., Barboni, E., Palanque, P., Freitas, C.: Usability aspects of the inside-in approach for ancillary search tasks on the web. In: Abascal, J., Barbosa, S., Fetter, M., Gross, T., Palanque, P., Winckler, M. (eds.) INTERACT 2015. LNCS, vol. 9297, pp. 211–230. Springer, Cham (2015). doi:10.1007/978-3-319-22668-2_18

Temporal Analysis of Social Media Response to Live Events: The Milano Fashion Week

Marco Brambilla[✉], Stefano Ceri, Florian Daniel, and Gianmarco Donetti

Dipartimento di Elettronica, Informazione e Bioingegneria, Politecnico di Milano,
Piazza Leonardo da Vinci 32, 20133 Milan, Italy
{marco.brambilla,stefano.ceri,florian.daniel,
gianmarco.donetti}@polimi.it

Abstract. Social media response to catastrophic events, such as natural disasters or terrorist attacks, has received a lot of attention. However, social media are also extremely important in the context of planned events, such as fairs, exhibits, festivals, as they play an essential role in communicating them to fans, interest groups, and the general population. These kinds of events are geo-localized within a city or territory and are scheduled within a public calendar. We consider a specific scenario, the Milano Fashion Week (MFW), which is an important event in our city.

We focus our attention on the spreading of social content in *time*, measuring the delay of the event propagation. We build different clusters of stakeholders (fashion brands), we characterize several features of time propagation and we correlate it to the *popularity* of involved actors. We show that the clusters by time and popularity are loosely correlated, and therefore the time response cannot be easily inferred. This motivates the development of a predictor through supervised learning in order to anticipate the space cluster of a new brand.

1 Introduction

Thanks to the wide adoption of smartphones, which enable continuous sharing of information with our social network connections, the online response to popular real world events is becoming increasingly significant, not only in terms of volumes of contents shared in the social network itself, but also in terms of velocity in the spreading of the news about events with respect to the time and to the geographical dimension. It has been noted that social signals are at times faster than media news with highly impacting events, such as terrorist attacks or natural disasters.

This work deals more specifically with the problem of social media response to a scheduled and popular real world event, the Milano Fashion Week occurred from the 24th to the 29th February of 2016, analysing the behaviour of users who re-acted (or pro-acted) in relationship with each specific fashion show during the week.

MFW, established in 1958, is part of the global "Big Four fashion weeks", the others being located in Paris, London and New York [4]. The event is organized

© Springer International Publishing AG 2017
J. Cabot et al. (Eds.): ICWE 2017, LNCS 10360, pp. 134–150, 2017.
DOI: 10.1007/978-3-319-60131-1_8

by *Camera Nazionale della Moda Italiana*, who manages and fully co-ordinates about 170 shows, presentations and events, thus facilitating the work of show-rooms, buying-offices, press offices, and public relations firms. Camera Nazionale della Moda carries out essential functions like drawing up the calendar of the shows and presentations, managing the relations with the institutions, the press office and creation of special events. MFW represents the most important meeting between market operators in the fashion industry. Out of the 170 shows, we are interested only in the catwalk shows, which are the core of the fashion week. The whole set of catwalks includes a total of 73 brands; among them, 68 brands organize one single event, 4 brands organize 2 events, and 1 brand organizes 3 events.

We address three main research questions in the paper:

RQ1. Can we describe the temporal dimension of the dynamic reactions to live events related to a brand on social networks?
RQ2. Is the temporal dimensions a relevant and different aspect with respect to the popularity of a brand? Or the dynamics can be simply described once the level of popularity of the event is known?
RQ3. Given the features of the brand and events, can we predict what will be the reaction to the event on social networks?

We formulate our problem as the analysis of the response in time of scheduled, popular events on social media platforms. Our goal is to describe and characterize the time at which social media respond to the events which appear in the official calendar and are linked to specific brands. Informally, we observe either peaks of reactions which then quickly disappear, or instead slower reactions that however tend to remain observable for a longer time. Estimating the time latency of social responses to events is important for the brands, which could plan reinforcement actions more accurately, essentially by adding well-planned social actions so as to sustain their social presence over time.

Our approach is as follows. We start by establishing correspondences between social events and calendarized events by filtering them with the same brand identity. We then discuss the Granger causality between the two time series. We next normalize and discretize the social media response curves, and finally we build the clustering of the brands.

We then cluster brands by online popularity, and show that popularity does not precisely explain the difference in time reactions, although it is weakly associated with it. This motivates the development of a predictor of the time dispersion of the social signal of a brand which is based just on time-specific features; our predictor uses supervised learning in order to anticipate the cluster of a new, unclustered brand.

The paper is organized as follows: Sects. 2 present our approach to data collection and preparation; then Sect. 3 describes the data analysis leading to clustering of brands by time responses. Section 4 introduces brand popularity and shows the best matching between brands clustered by time response and popularity; then, Sect. 5 introduces the time class predictor for a new brand, Sect. 6 presents related work, and Sect. 7 concludes.

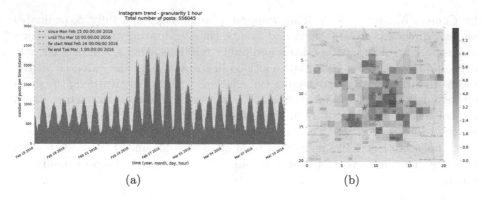

Fig. 1. Temporal overview for the three analyzed weeks of Instagram posts (a) and map representing the geographical distribution of MFW events (represented by red stars) and post density (b) - showing the spreading in the city of both the MFW events and the accumulated Instagram posts. (Color figure online)

2   Data Collection and Preparation

We initially extracted posts by invoking the social network APIs of Twitter and Instagram; for identifying the social reactions to MFW, we used a set of 21 hashtags and keywords provided by domain experts in the fashion sector, i.e., researchers of the *Fashion in Process* group (FIP) of Politecnico di Milano[1]. We focused on 3 weeks, before, during and after the event. In this way, we collected 106K tweets (out of which only 6.5% geolocated) and 556K Instagram posts (out of which 28% geolocated); eventually, we opted for considering only Instagram posts, as they represent a much richer source for the particular domain of Fashion with respect to Twitter. Figure 1 shows the temporal and geographical distribution of the posts.

We performed an initial analysis of the content, for associating each post with the corresponding event. In this specific scenario, the task was simple because each event was directly associated with a fashion brand, mentioned in the posts; the characterization of the brands was again provided by the FIP experts. For instance, for identifying the posts related to the Gucci catwalk, which was held on February 24th at 2:30 pm in Milano, Via Valtellina 17, we collected the posts containing the hashtags and keywords *#Gucci* and *Gucci*, filtering the posts through suitable regular expressions. This allowed us to collect 7718 Instagram posts related to that specific event.

3   Time Response Analysis

For studying the temporal dynamics of the social media reactions (RQ1), we consider two signals: the temporal series of number of posts related to a given

---

[1] http://www.fashioninprocess.com/.

brand B_i, and the one recording the presence of a live event for the brand B_i. The latter, which we name *calendar signal* for given brand B_i, is valid through the time intervals Δt of analysis and is defined as:

$$calendar\,Signal(B_i, \Delta t) = \sum_{e \in\, B_i} 1_e(\Delta t) \tag{1}$$

where $1_e(\Delta t)$ indicates the presence of the event e in the time window Δt. Intuitively, the signal has value 1 in the intervals when a live event of brand B_i is taking place, while it has value 0 when no events are taking place.

3.1 Granger Causality

We then focused on determining a causality relationship between the events and the follow-up posts. Specifically, we measured the Granger causality between two different time series, (a) specific events and (b) posts reacting to those events; this had to be done brand by brand, by suitable selections both from the calendar and from the dataset of social posts.

The Granger Causality test applies to two different time series; a time series is said to Granger-cause another time series if it can be shown, usually through a series of t-tests and F-tests, that the values of the first series provide statistically significant information about future values of the second series. In particular, we want to reject, with statistical significance, the null hypothesis of the test that the calendar signal does not Granger-cause the social media signal. We focus on the F-test, which evaluates the ratio of two scaled sums of squares reflecting different sources of variability, constructed so that the statistic tends to be greater when the null hypothesis is not true.

As an example, Fig. 2 shows the social response to the event of *Versace* on 26th February at 8 pm. Note a strong reaction in the social media relatively close to the scheduled events: indeed, we have a peak of about 180 Instagram posts in the time window starting when the event is just completed, and then the number of posts per time window decreases rapidly. Figure 3 shows the graph of the Granger causality for Versace. The graph shows that the maximum causality is achieved with a lag of 15 min from the event, and then decreases. We performed tests like these for all the brands that have one or more events in the Milano Fashion Week 2016 calendar, confirming Granger causality for all the brands. Figure 4 shows the normalized Granger Causality results for all the 68 analyzed brands.

3.2 Clustering

We then looked for similarity of the time response curves for posts. We normalized each peak at 1, since we were no longer interested in the statistical relevance, but just in the shape of the curves. We then considered a period of 5 h and applied k-means clustering in L-dimensional space, where $L = 20$ is the number of points we have collected for each curve (therefore, we discretized the curve using 15 min intervals).

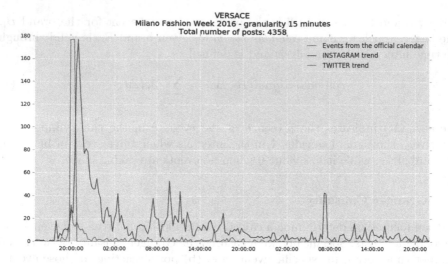

Fig. 2. Social media response to Versace's event of 26th February at 20:00. The granularity is 15 min. The red line represents the *calendarSignal* function and the other lines report the number of posts in the specific time window. The blue line is for Instagram, the green one is for Twitter. (Color figure online)

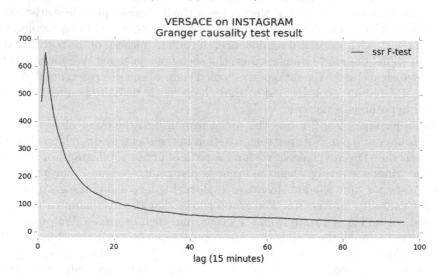

Fig. 3. Granger Causality test results for Versace between the *calendarSignal* and the number of Instagram posts in time.

For deciding the ideal number k of clusters which better describes the scenario, we computed clustering for $k \leq 15$, and then we computed inertia for each such choice. Inertia (or within-cluster sum-of-squares) is a measure of internal coherence; the inertia curve is monotonically decreasing, with the maximum

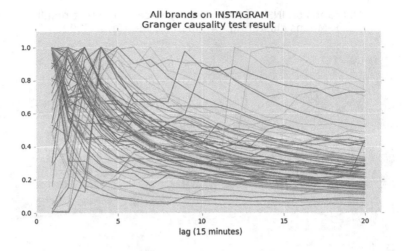

Fig. 4. Normalized Granger Causality test results for all the 65 analyzed brands.

Table 1. Information on the intra-cluster inertia and inertia gain when adding one additional cluster, after clustering the results with different values of k, for $3 \leq k \leq 5$.

k	Inertia	Inertia gain
3	23.82	12.39
4	17.92	5.90
5	14.87	3.05

value corresponding to just one global cluster and the minimum value equal to 0 when the number of clusters coincide with the number of elements.

Based on inertia values (partially reported in Table 1), we picked k equal to 4. The values of inertia justify our choice, as: $k = 3$ corresponds to 23.82, $k = 4$ to 17.92, $k = 5$ to 14.87, and the inertia is decreasing very slowly for $k \geq 4$. The resulting clusters are associated to a color code:

- *Yellow*, with high immediate response;
- *Red*, with lagged response peak at 15 min;
- *Green*, with lagged response peak at 45–60 min;
- *Blue*, with an initial significant response but with another lagged response peak after 3 h.

Figure 5 shows the clusters produced. We then selected the most representative elements for each cluster as the medoid, i.e., the element in the cluster that is the closest to the centroid of that cluster. The medoids shown in Fig. 6 are:

- *Costume National*, from the yellow group;
- *Trussardi*, from the red group;
- *Alberta Ferretti*, from the green group;
- *Emporio Armani*, from the blue group.

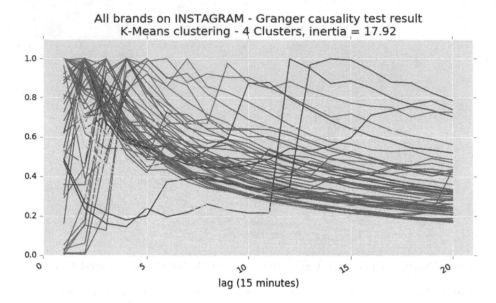

Fig. 5. Clusters produced by k-means clustering in 20-dimensional space with k = 4. (Color figure online)

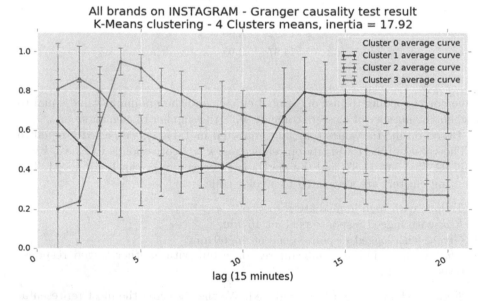

Fig. 6. Most representative elements of each cluster. (Color figure online)

4   Comparison with Brand Popularity

4.1   Popularity Analysis

We next turned to a simpler observation, the *brand popularity*, in order to evaluate the popularity of a brand and see if it relates to the above clusters; we focused

on 65 brands which were hosting fashion shows during MFW. In our analysis,
We extracted from our Twitter and Instagram datasets a classic set of popularity
features, related to each brand: the number of posts on Instagram, number of
likes collected on Instagram, number of comments collected on Instagram, number of posts on Twitter, number of likes collected on Twitter, number of retweets
collected on Twitter. We then performed a principal component analysis (PCA)
in order to find the best features, and not surprisingly we noticed that likes on
Instagram essentially dominate, as the 2 principal components are:

- Number of likes on Instagram (99.9% of total variance)
- Number of comments on Instagram (0.0025% of total variance)

In the end, we run k-means over these attributes, asking again for 4 different
clusters, in order to better compare our final results. Figure 7 shows the outcome
of this clustering. The groups could be described as following: from the red cluster
to the blue cluster we are going from the most unpopular brands to the most
popular ones, in the two social media of Twitter and Instagram. These results
were confirmed by our experts in fashion design.

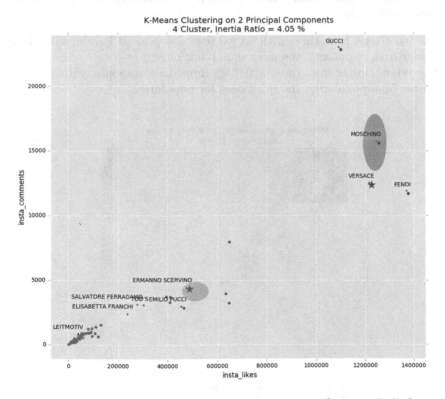

Fig. 7. Representation of popularity clusters with respect to the two principal components (comments and likes on Instagram, respectively). (Color figure online)

4.2   Time – Popularity Correlation

We then studied the correlation among the two clusterings, namely temporal behaviour and popularity (RQ2). We label one clustering result arbitrarily and then rename the other clustering labels with all the possible permutations in the set of adopted labels. For each re-labelling, we compute a measure of correlation between the two clustering results, assuming a correspondence between same-name-labels, and we take the best renaming permutation in terms of the specific measure adopted. We pick as statistic measure of validation the accuracy in juxtaposing one cluster to another one. We recall that we are in a multiclass case, then the accuracy will be the percent of true-matching between the two clustering results.

We visualize the best matching by using a confusion matrix, a table that visualizes the correlations of the two different results clustering. One clustering is taken on the rows, the other one on the columns, total correlation requires all diagonal elements to be one and all other elements to be zero; with partial correlation we measure the accuracy of the matrix as the sum of all correlation values appearing along the diagonal. The best accuracy we obtain is 41.54%, that produces the confusion matrix in Fig. 8. In a few words, the best correlation is obtained juxtaposing:

- The red cluster from time, with lagged response peak at 15 min, with the red cluster from popularity, the most unpopular ones;
- The yellow cluster from time, with high immediate response, with the yellow cluster from popularity, the third ones for popularity;

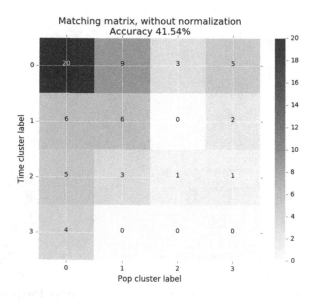

Fig. 8. Matching matrix comparing time response versus popularity response.

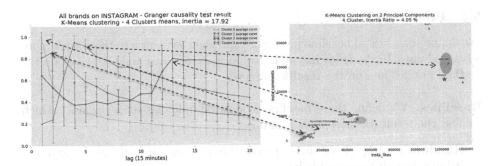

Fig. 9. Visual mapping between the temporal clusters and the popularity-based ones. (Color figure online)

- The green cluster from time, with lagged response peak at 45–60 min, with the blue cluster from popularity, the most popular ones;
- The blue cluster from time, with an initial significant response but with the lagged response peak at 3 h and 15 min, with the green cluster from popularity, the ones just below the most popular.

Figure 9 shows a gross-granularity association between temporal clusters and the popularity-based ones, to give an intuition on the best possible mapping between the two aspects. However, notice that in summary, time-based and popularity-based clustering results show low correlation. This highlights the fact that popularity alone is not a sufficient analysis dimension for guessing/understanding the social response in time of given brands.

5   Prediction

Once we realized that simple popularity metrics cannot be good proxies for describing the temporal dynamics of a brand, we focused on determining the best prediction model for estimating the membership of a brand in one of the four clusters determined before (RQ3). In practice, we consider as labels the set of 4 clusters and we aim at predicting the correct label for a given brand B.

5.1   Pre-processing Phase

For building the prediction, we consider only the 68 brands featuring one event during the MFW, to avoid aliasing due to the overlapping impact of multiple events. Subsequently, in the test phase we check the results of the prediction on all the 73 brands.

Some pre-processing of the data was performed in order to convert categorical variables into numerical ones. This avoids problems with the models that requires non-categorical features only. Other transformations have been applied to: (1) normalize continuous variables to a common range; and (2) convert variables with types allowing for many representations (such as date and time) to a unique, common format.

5.2 Prediction Models

We define some trivial baseline predictors and then we compare them with more advanced ones. In the following we introduce each of them and we report on the method applied and on the results obtained.

Baselines. We define three baseline models (simple classifiers) against which we test the results of more advanced ones:

1. **Random** strategy: it predicts the membership with a uniform random distribution to all the possible labels;
2. **Most frequent** strategy: it always predicts the membership to the most frequent label in the training set;
3. **Stratified** strategy: it generates label predictions according to the probability distribution of the labels in the training set.

Naive Bayes Classifier for Multivariate Bernoulli Models. This Naive Bayes classifier applies to data that is distributed according to multivariate Bernoulli distributions; i.e., we assume each feature to be a binary-valued variable. Therefore, samples are represented as binary-valued feature vectors. Variables can be naively binary or transformed into binary.

Logistic Regression. In regularized logistic regression, the probabilities describing the possible outcomes of a single trial are modeled using a logistic function. We implement a version that can fit a multiclass (with one-vs-rest paradigm) norm-2 penalized logistic regression with optional regularization C. This can be represented as an optimization problem that minimizes the following cost function:

$$\min_{w,C} \frac{1}{2} w^T w + C \sum_{i=1}^{m} \log(\exp(-y_i(\mathbf{X}_i^T w + c)) + 1). \tag{2}$$

Cross-Validated Logistic Regression. We also improve the basic Logistic Regression model performance through cross-validation. The function is given 10 values in log-scale between 0.0001 and 10000 for the parameter C, the best hyperparameter is selected by the Stratified K-fold cross-validator. Being a multiclass problem, we compute the hyperparameters for each class using the best scores got by doing a one-vs-rest in parallel across all folds and classes.

Support Vector Machine. Support vector machines are a set of supervised learning methods used for classification, regression and outlier detection.

We adopt a One-versus-rest strategy, thus the problem can be formulated at each iteration as a 2-class problem. Given training vectors $x_i \in R^p$, with $i = 1, ..., n$, in two classes, and a vector $y \in \{1, -1\}^n$, the Support Vector Classifier solves the following dual minimization problem:

$$\min_{a} \frac{1}{2} a^T \mathbf{Q} a - e^T a \tag{3}$$

subject to:

$$y^T a = 0, 0 \leq a_i \leq C, \ i = 1, ..., n. \tag{4}$$

where e is the vector of all ones, $C > 0$ is the upper bound, $\mathbf{Q}$ is an $n \times n$ positive semidefinite matrix, $\mathbf{Q}_{ij} = y_i y_j K \langle x_i, x_j \rangle$, where $K \langle x_i, x_j \rangle = \phi(x_i)^T \phi(x_i)$ is the kernel function. Training vectors are implicitly mapped into a higher dimensional space by the function ϕ. The decision function applied by SVM is therefore:

$$sign(\sum_{i=1}^{n} y_i a_i K \langle x_i, x \rangle + b) \tag{5}$$

Decision Tree. Decision Trees are a non-parametric supervised learning method used for classification and regression. The goal is to create a model that predicts the value of a target variable by learning simple decision rules inferred from the data features.

Random Forest. This is a perturb-and-combine technique specifically designed for trees. This means a diverse set of classifiers is created by introducing randomness in the classifier construction. The prediction of the ensemble is given as the averaged prediction of the individual classifiers. With random forest, we build each tree in the ensemble starting from a sample, drawn with replacement (i.e., a bootstrap sample) from the training set. In addition, when splitting a node during the construction of the tree, the split that is chosen is no longer the best split among all features. Instead, the split that is picked is the best split among a random subset of the features, hence yielding an overall better model.

5.3 Fitting the Models

We fit the models described in the previous section, in order to compare the results and pick the best one in the final classification step. The different models were fit with different features in order to find the best combination of input variables for the final classification and probability estimation. The final selected features are: $[x^{start\_minute}, x^{start\_hour}, x^{start\_day}, x^{end\_minute}, x^{end\_hour}, x^{end\_day}, x^{class}, x^{live}, x^{type}, x^{invitation}, x^{open}]$.

Table 2 reports the main performance coefficients of the models, covering: time performance indicators (time needed for performing the fitting and the prediction), weak indicators, and cross-validation indicators. Indeed, we applied Leave-One-Out Cross Validation (LOO-CV) for all the adopted models, in order to obtain better performance for the different models. This has been possible thanks to the relatively small size of the data-set.

More precisely, the columns in Table 2 describe:

1. the classifier adopted for the problem;
2. the time spent during the fitting phase, when training the model;
3. the time spent for the final prediction, for classifying the training samples;
4. the number of misclassified elements in the training (error);

Table 2. Performance of all the techniques adopted in the classification problem. From left to right: the classifier adopted, the time spent during the fitting phase, the time spent for the final prediction, the number of training errors, the log loss on the training, the number of errors by adopting leave-one-out cross validation, the average log loss on the test (one row at a time) with the standard error adopting leave-one-out cross validation, the average log loss on the training with standard error adopting leave-one-out cross validation.

Classifier	Time indicators (s.)		Weak indicators		Leave-One-Out Cross-Validation indicators				
	Fitting	Prediction	Errors	Log loss	Errors	Log loss test		Log loss training	
						Mean	Std	Mean	Std
Bernoulli NaiveBayes	0,0035	0,0015	22	1,39806	36	2,76323	4,17595	1,3901	0,0349
Logistic Regression	0,0225	0,0015	17	0,61651	33	0,76421	0,72045	0,6129	0,0123
Cross-Validated Log. Regr.	2,7521	0,0015	21	0,86406	32	0,56453	0,36066	0,8162	0,0666
Support Vector Machine	0,0190	0,0060	15	1,03000	38	0,49924	0,24064	0,9835	0,0444
Decision Tree	0,0035	0,0020	2	0,03798	37	17,03322	17,26817	0,0375	0,0044
Random Forest	1,1363	0,2067	2	0,25385	28	0,83452	0,79575	0,2507	0,0052
Dummy MostFrequent	0,0020	0,0020	36	17,03282	36	17,50636	17,26817	17,0328	0,2398
Dummy Stratified	0,0020	0,0015	48	20,81789	48	13,24806	16,79498	21,6853	1,8348
Uniform Distribution		0,0005		1,38629					
Random Distribution		0,0020		1,75860	60,47000			1,7586	0,0980

5. the log loss on the training;
6. the number of misclassified elements in the test adopting LOO-CV (error);
7. the average log loss on the test adopting LOO-CV;
8. the standard deviation of the log loss on the test adopting LOO-CV;
9. the average log loss on the training adopting LOO-CV;
10. the standard deviation of the log loss on the trainimg LOO-CV.

The number of errors for the random strategy is an average of the number of errors above 100 runs, in order to keep this measure more confident.

5.4   Results and Discussion

Execution Time. In terms of execution time, all these models are fast both in the training phase and in the prediction phase. This is also related to the fact that the dataset is relatively small. The only two techniques that are slower than the others are Cross-Validated Logistic Regression (due to the cross-validation phase that has to choose the best value for the parameter C) and Random Forest (because of the diverse set of classifiers created by introducing randomness in their construction, and because of the high number of different estimators which we set to 1000).

Evaluation of Training. Regarding the number of errors obtained when using as test set the training self itself, it is well known that this type of predictions leads to some significant overfitting; however, with small data-sets like ours, also this measure can be considered. For this performance indicator we have a predominance of the tree-based algorithms, i.e., Decision Trees and Random Forest. However, the respective performance accuracy is a clear example of over-fitting: indeed, decision-tree learners can often create over-complex trees that do not generalise the data well. In any case, each method is outperforming the baselines Dummy Most Frequent and Dummy Stratified.

From the *log loss test* columns, where again test and train set are perfectly overlapping, we can notice the overfitting of the tree-based learners, together with nice performance in probability estimation from the other models, too.

Evaluation of LOO-CV. Regarding the *Leave One Out Cross-Validation* indicators, as we said we have applied prediction one sample at a time, using all the remaining ones as training set. This avoids the overfitting problem and thus are more accurate and truthful performance indicators. As one can expect, the number of errors are increasing. For this indicator, the best model is the Random Forest, that seems to beat the overfitting problem of the decision trees, upon which it is based. over 73 (for every sample, it has to choose randomly among 4 possible target values), and Every model we adopted is outperforming the baseline methods.

Considering the probability estimation indicators, i.e., log loss, we immediately notice the overfitting of the Decision Tree, as expected from the previous analysis. Indeed, the average log loss on the test for this type of estimator is really close to the baselines strategies Most Frequent and Stratified. Together with the test log loss standard deviation, these measures are proving the high variance of the over-complex model built by the tree. The best model considering the log loss on the test samples (both in terms of mean and standard deviation) is therefore the Support Vector Machine, with the lowest mean of the log loss and also the lowest standard deviation. The remaining indicators of the log loss on the training sets underline once again the overfitting issue of the Decision Tree and fairly good performance of the other models.

Summary. Concluding, if we give more importance to the Leave-One-Out Cross-Validation indicators, there are two models that can be preferred to the others: the first one is the Random Forest, which has a really low number of errors (only 2 over 68) when considering all the data-set as training and test set, and maintains the lowest number of errors (28 over 68) among the other techniques in the classification phase of the Cross-Validation. Figure 10 reports the clustering of the Granger causality curves as obtained by the Random Forest predictor. The second is the Support Vector Machine, which has the best performance in probability estimation, as we can see from the log loss LOO-CV indicators. However, this technique reports a significant number of errors (38 errors) in the classification phase of the Cross-Validation.

6   Related Work

Various studies analyzed the impact of events on social media. We consider both works that deal with response to events on the social media in general, and works that focus specifically on the role of social media in the fashion domain.

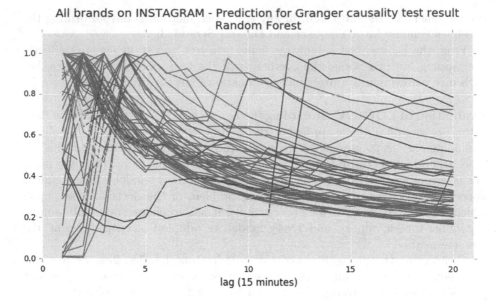

Fig. 10. Clusters produced by the Random Forest predictor in terms of Granger causality curves.

Social Media Event Response. The work [10] is concerned with 21 hot events which were widely discussed on Sina Weibo; it empirically analyzes their posting and reposting characteristics. In the work [2], by automatically identifying events and their associated user-contributed social media documents, the authors show how they can enable event browsing and search in a search engine. The work [3] underlines how user-contributed messages on social media sites such as Twitter have emerged as powerful, real-time means of information sharing on the Web. The authors explore approaches for analyzing the stream of Twitter messages to distinguish between messages about real-world events and non-event messages.

The focus of [7] is to detect events and related photos from Flickr, by exploiting the tags supplied by users. This task is hard because (1) Flickr data is noisy; (2) capturing the content of photos is not easy. They distinguish between tags of aperiodic events and those of periodic events; for both classes, event-related tags are clustered such that each cluster consists of tags with similar temporal and locational distribution patterns as well as with similar associated photos. Finally, for each tag cluster, photos corresponding to the represented event are extracted.

The problem of event summarization using tweets is well faced by [6], where the authors argue that for some highly structured and recurring events, such as sports, it is better to use sophisticated techniques to summarize the relevant tweets via Hidden Markov Models.

The paper [5], adding the information given by cell-phone traces, deals with the analysis of crowd mobility during special events. They analyze nearly 1 million cell-phone traces and associate their destinations with social events. They show that the origins of people attending an event are strongly correlated to the type of event.

Finally, [1] proposes a procedure consisting of a first collection phase of social network messages, a subsequent user query selection, and finally a clustering phase, for performing a geographic and temporal exploration of a collection of items, in order to reveal and map their latent spatio-temporal structure. Specifically, both several geo-temporal distance measures and a density-based geo-temporal clustering algorithm are proposed. The paper aims at discovering the spatio-temporal periodic and non-periodic characteristics of events occurring in specific geographic areas.

Social Media Analysis for Fashion. The work [12] presents a qualitative analysis on the influence of social media platforms on different behaviors of fashion brand marketing. They analyze their styles and strategies of advertisement. The authors employ both linguistic and computer vision techniques. The study [11] set out to identify attributes of social media marketing (SMM) activities and examine the relationships among those perceived activities, value equity, relationship equity, brand equity, customer equity, and purchase intention through a structural equation model.

The findings of [14] show that different drivers influence the number of likes and the number of comments to fashion posts. Namely, vivid and interactive brand post characteristics enhance the number of likes. An analysis in [8] was conducted during a 2011 Victoria's Secret Fashion Show that reference the show. Although the majority were idiosyncratic remarks, many tweets contain evidence of social status comparisons to the fashion models.

The article [9], based on two studies of the fashion industry, examines one of its key institutions, London Fashion Week (LFW). The authors argue that this event is a materialization of the field of fashion. They examine how LFW renders visible the boundaries, relational positions, capital and habitus at play in the field, reproducing critical divisions within it.

Finally, [13] develops a motion capture system using two cameras that is capable of estimating a constrained set of human postures in real time. They first obtain a 3D shape model of a person to be tracked and create a posture dictionary consisting of many posture examples.

7   Conclusions

We discussed how social content spreads in time as a response to live events, focusing on the Milano Fashion Week. We demonstrated that brands can be clustered into 4 classes of increasingly prolonged responses, from ones showing a peak which is close in time to the event, to ones which do not show sharp peaks but rather a smoothed behavior; we also showed that brand popularity alone is not sufficient for explaining such a difference. Estimating the time latency of social responses to events is important for the brands, which could plan reinforcement actions more accurately, essentially by adding few but well-planned social actions so as to sustain their social presence over time. Our future work is to attempt to correlate spreading in time with other features beyond brand popularity, e.g. by studying the profiles of each brand's social networks and specifically of Instagram.

Acknowledgments. This work is partly funded by the research project FaST – Fashion Sensing Technology (ID 187010) funded by Regione Lombardia Smart Fashion and Design call. We thank the FashionInProcess group (http://www.fashioninprocess.com) of Politecnico di Milano, and especially Paola Bertola, Chiara Colombi and Federica Vacca, who supported us in the definition of the domain-specific knowledge related to the event.

References

1. Arcaini, P., Bordogna, G., Ienco, D., Sterlacchini, S.: User-driven geo-temporal density-based exploration of periodic and not periodic events reported in social networks. Inf. Sci. **340**, 122–143 (2016)
2. Becker, H., Naaman, M., Gravano, L.: Learning Similarity Metrics for Event Identification in Social Media (2010)
3. Becker, H., Naaman, M., Gravano, L.: Beyond Trending Topics: Real-World Event Identication on Twitter (2011)
4. Bradford, J.: Fashion Journalism (2014)
5. Calabrese, F., Pereira, F.C., Lorenzo, G., Liu, L., Ratti, C.: The geography of taste: analyzing cell-phone mobility and social events. In: Floréen, P., Krüger, A., Spasojevic, M. (eds.) Pervasive 2010. LNCS, vol. 6030, pp. 22–37. Springer, Heidelberg (2010). doi:10.1007/978-3-642-12654-3_2
6. Chakrabarti, D., Punera, K.: Event Summarization using Tweets. Yahoo! Research (2011)
7. Chen, L., Roy, A.: Event Detection from Flickr Data through Wavelet-based Spatial Analysis (2009)
8. Chrisler, J., Fung, K., Lopez, A., Gorman, J.: Suffering by comparison: Twitter users' reactions to the Victoria's Secret Fashion Show. Body Image **10**, 648–652 (2013)
9. Entwistle, J., Rocamora, A.: The field of fashion materialized: a study of London fashion week. J. Sociol. **40**, 735–751 (2006)
10. Guan, W., Gao, H., Yang, M., Li, Y., Ma, H., Qian, W., Cao, Z., Yang, X.: Analyzing user behavior of the micro-blogging website Sina Weibo during hot social events. Physica A **395**, 340–351 (2014)
11. Kim, A., Ko, E.: Do social media marketing activities enhance customer equity? An empirical study of luxury fashion brand. J. Bus. Res. **65**, 1480–1486 (2011)
12. Manikonda, L., Venkatesan, R., Kambhampati, S., Li, B.: Trending Chic: Analyzing the Influence of Social Media on Fashion Brands. Department of Computer Science, Arizona State University (2016)
13. Okada, R., Stenger, B., Ike, T., Kondo, N.: Virtual Fashion Show Using Real-Time Markerless Motion Capture. Corporate Research & Development Center, Toshiba Corporation (2006)
14. de Vries, L., Gensler, S., Leeflang, P.: Popularity of brand posts on brand fan pages: an investigation of the effects of social media marketing. J. Interact. Mark. **26**, 83–91 (2012)

Trading Off Popularity for Diversity in the Results Sets of Keyword Queries on Linked Data

Ananya Dass[(✉)] and Dimitri Theodoratos[(✉)]

New Jersey Institute of Technology, Newark, USA
{ad292,dth}@njit.edu

Abstract. Keyword search is the most popular technique for querying the ever growing repositories of RDF graph data on the Web. However, keyword queries are ambiguous. As a consequence, they typically produce on linked data a huge number of candidate results corresponding to a plethora of alternative query interpretations. Current approaches ignore the diversity of the result interpretations and might fail to satisfy the users who are looking for less popular results. In this paper, we propose a novel approach for keyword search result diversification on RDF graphs. Our approach instead of diversifying the query results per se, diversifies the interpretations of the query (i.e., pattern graphs). We model the problem as an optimization problem aiming at selecting k pattern graphs which maximize an objective function balancing relevance and diversity. We devise metrics to assess the relevance and diversity of a set of pattern graphs, and we design a greedy heuristic algorithm to generate a relevant and diverse list of k pattern graphs for a given keyword query. The experimental results show the effectiveness of our approach and proposed metrics and also the efficiency of our algorithm.

1 Introduction

Keyword search is the most popular technique for querying RDF data on the Web because it frees the user from knowing a complex structured query language (e.g., SPARQL) and allows querying the data without having full or partial knowledge of its structure/schema. The convenience and flexibility of keyword search comes with a cost. Keyword queries are ambiguous. As a consequence, there is usually a huge number of candidate results of which very few are relevant to the user intent. Several approaches try to exploit structural or semantic characteristics of the data and/or query results in order to filter out irrelevant results. A better technique ranks the results in descending order of their estimated relevance [7,13]. The relevance is usually estimated based on scoring functions which employ statistics-based IR-style metrics for flat documents (e.g., tf*idf or PageRank) adapted to the structural characteristics of the data. Ranking can be complemented with top-k processing, wherein gains can be achieved in the processing time by avoiding the computation of results which are not expected to be in the top-k positions [20].

© Springer International Publishing AG 2017
J. Cabot et al. (Eds.): ICWE 2017, LNCS 10360, pp. 151–170, 2017.
DOI: 10.1007/978-3-319-60131-1_9

Even though, the statistics-based metrics can be effective in returning the most popular results, they fail to capture the diversity of the result set and may dissatisfy the users who look for less popular results [11,12]. For instance, a user issuing the query "python" could be interested in searching about the snake "python", the "Python" programming language, or the Monty "Python" comedy group. If results are returned to the user based on the most plausible interpretation of the query (in this case "Python" programming language) then there is an inherent risk of leaving the user who is interested in "python" snake or in Monty "Python" comedy group. This problem is known as the over-specialization problem [18]. Diversifying the results retrieved for a keyword query could be a meaningful solution to this problem. By introducing diversity in the result set, the search mechanism can maximize the user's chance of finding at least one of the retrieved results relevant to her intent [6]. Additionally, even if a keyword query has a single, clearly defined interpretation, it can still be under-specified to some extent. For example, a user searching for "apple electronics" may be interested in laptops, desktops, or the best selling apple electronics, or sale on apple electronics, or service centers for apple electronics. Therefore, another motive for diversifying search results is to cover different aspects of the entire result space and enable the user to explore and find desired results [12].

Search result diversification is a well-studied problem in Information Retrieval and Recommendation Systems [12,14]. However, the problem of diversifying the results of keyword search over RDF graph data has remained under-addressed. In recent years, there is a proliferation of RDF repositories on the Web, and keyword search is commonly used for retrieving data from these repositories. While ranking ensures that the most popular results of a given keyword query are ranked on top, it is often the case that the top results tend to be homogeneous, making it difficult for users interested in less popular aspects to find results relevant to their intent. Thus, result diversity can play a big role in ensuring that the users get a broad view of the different aspects of the results and in satisfying a maximum number of users who are interested in different interpretations of the query.

Our Approach. In this paper, we propose a novel technique for diversifying keyword search results on RDF graph data. We formulate the diversification problem as an optimization problem over pattern graphs. Pattern graphs are structured queries which cluster together results with the same structural and semantic characteristics and represent alternate interpretations of a keyword query. By diversifying pattern graphs instead of query results, we address the data scalability problem of diversification since pattern graphs can be computed efficiently by exploiting a structural summary of the RDF data without exhaustively computing the query results. Further, by diversifying pattern graphs we diversify the alternative interpretations of the query. Given a positive integer k, our diversification approach aims at selecting a k-size set of pattern graphs which maximizes the number of pattern graphs which are relevant to at least one user intent. In order to do so, our approach trades off popularity for diversity.

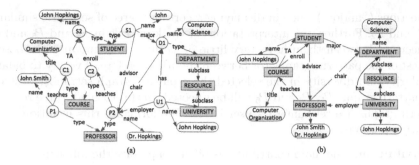

Fig. 1. (a) An RDF graph D, (b) The Structural Summary S of D.

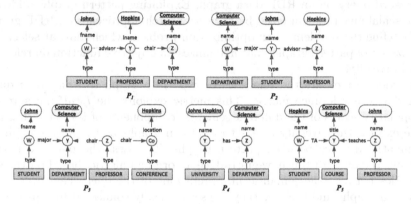

Fig. 2. Patterns graphs of the query $Q = \{$Johns, Hopkins, Computer, Science$\}$.

As an example, consider the RDF data graph D of Fig. 1(a), and its structural summary S in Fig. 1(b). Consider also the keyword query $Q = \{$Johns, Hopkins, Computer, Science$\}$ on D. Figure 2 shows five pattern graphs of Q computed over S. These pattern graphs are alternative interpretations of Q. For instance, the pattern graph P_1 interprets "Johns" as a student advised by Professor "Hopkins" who is the chair of the "Computer Science" department. Among those pattern graphs P_1, P_2, P_4 and P_5 provide meaningful interpretations while, pattern graph P_3 does not seem to be relevant to a user intent. Let us assume that the pattern graphs are ranked on popularity in descending order as follows: P_1, P_2, P_3, P_4, P_5. Let us also assume that we are required to return only three pattern graphs. If we select the relevant pattern graphs based on popularity, we are going to return the list (P_1, P_2, P_3). Therefore, we are returning only two pattern graphs which are relevant to a user intent. We can now try to also diversify the pattern graphs. We can do so by taking into account semantic dissimilarities between them. For instance, P_1 and P_2 interpret the keywords in the same way: "Johns" is a student, "Hopkins" is a professor and "Computer Science" is a department and they link these interpretations in a quite similar way. Pattern graph P_3 shares three out of four keyword interpretations with P_1 and P_2 (for the keywords "Johns", "Computer", and "Science") and some connections ("major"), therein displaying a certain degree of semantic similarity to P_1 and P_2. Further, the interpretations of the keywords in P_4 and P_5 and the connections between them have very little or nothing in common with those of the rest of the patterns graphs (or between them). Therefore, if we try to balance popularity and diversity in the selected three pattern graph set, most probably the pattern graphs P_1, P_4 and P_5 will be selected. In this case, all three returned pattern graphs are relevant to a user intent. We formalize this intuition in the following sections.

Contribution. The main contributions of the paper are the following:

- We formalize the problem of diversifying the pattern graphs returned by a keyword query on an RDF data graph. Exploiting pattern graphs addresses the scalability problem faced by keyword search approaches on RDF graphs. We define the problem as an optimization problem which aims at selecting a k-size set of pattern graphs that maximizes an objective function on relevance and diversity.
- In order to measure the relevance of a pattern graph to a keyword query, we devise a relevance metric. This metric exploits the *tf*idf* measure and popularity scores of the different semantic components of the pattern graph.
- We express the diversity of a set of pattern graphs as the average pairwise semantic distance between pattern graphs. To assess the pairwise pattern graphs semantic distance, we introduce an original metric. based on the similarities between the semantic interpretations of the query keywords in the pattern graphs and the way they are semantically connected but also on the dissimilarity between the concepts involved in the pattern graphs.
- To cope with a high complexity of the diversification problem, we design a greedy heuristic algorithm for computing a list of top-k pattern graphs trading off popularity for diversity.

We ran extensive experiments to evaluate and fine-tune the effectiveness of the approach and the proposed metrics and the efficiency of the algorithm.

2 Data Model and Pattern Graph Computation

Data Model. The Resource Description Framework (RDF) provides a framework for representing information about Web resources in a graph form. The RDF vocabulary includes elements that can be broadly classified into Classes, Properties, Entities and Relationships. All the elements are resources. Similarly to [7,8], our data model is an RDF graph defined as follows:

Definition 1 (RDF Graph). An *RDF graph* is a quadruple $G = (V, E, L, l)$:

V is a finite set of vertices, which is the union of three disjoint sets: V_E (representing entities), V_C (representing classes) and V_V (representing values).

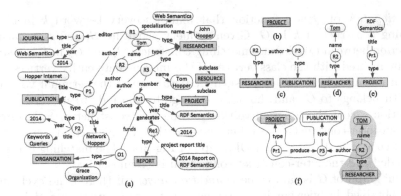

Fig. 3. (a) An RDF graph, (b), (c), (d) and (e) class, relationship, value and property matching constructs, respectively, (f) inter-construct connection and result graph.

E is a finite set of directed edges, which is the union of four disjoint sets: E_R (inter-entity edges called *Relationship* edges which represent entity relationships), E_P (entity to value edges called *Property* edges which represent property assignments), E_T (entity to class edges called *type* edges which represent entity to class membership) and E_S (class to class edges called *subclass* edges which represent class-subclass relationship).

L is a finite set of labels that includes the labels "type" and "subclass".

l is a function from $V_C \cup V_V \cup E_R \cup E_P$ to L. That is, l assigns labels to class and value vertices and to relationship and property edges.

Entity and class vertex and edge labels are Universal Resource Identifiers (URIs). Vertices are identified by IDs which in the case of entities and classes are URIs. Every entity belongs to a class. Figure 3(a) shows an example RDF graph. For simplicity, vertex and edge identifiers are not shown in this example.

Query Language Semantics. A *query* Q on an RDF graph G is a set of keywords. A *keyword instance* of a keyword k in Q is a vertex or edge label in G containing k. The *answer* of Q on G is a set of result graphs of Q on G. Each result graph is a minimal subgraph of G involving at least one instance of every keyword in Q and is formally defined below. In order to facilitate the interpretation of the semantics of the keyword instances, every instance of a keyword in Q is matched against a small subgraph of G which involves this keyword instance and the corresponding class vertices. This subgraph is called *matching construct*. Figures 3(b), (c), (d) and (e) show a class, relationship, value and property matching construct, respectively, for different keyword instances in the RDF graph of Fig. 3(a). Underlined labels in a matching construct denote the keyword instances. Each matching construct provides information about the semantic context of the keyword instance under consideration. For instance, the matching construct of Fig. 3(d) shows that Tom is the name of an entity $R2$ of type Researcher.

A *signature* of Q is a function that matches every keyword k in Q to a matching construct of k in G. Given a query signature S, an *inter-construct connection* between two distinct matching constructs C_1 and C_2 in S is a simple path augmented with the class vertices of the intermediate entity vertices in the path (if not already in the path) such that: (a) one of the terminal vertices in the path belongs to C_1 and the other belongs to C_2, and (b) no vertex in the connection except the terminal vertices belong to a construct in S. Figure 3(f) shows an inter-construct connection between the matching constructs for keywords Project and Tom in the RDF graph of Fig. 3(a). The matching constructs are shaded and the inter-construct connection is circumscribed.

A subgraph of G is said to be *connection acyclic* if there is no cycle in the graph obtained by viewing its matching constructs as vertices and its inter-construct connections between them as edges. Given a signature S for Q on G, a *result graph* of S on G is a connected, connection acyclic subgraph of G which contains only the matching constructs in S and possibly inter-construct connections between them. A *result graph* for Q on G is a result graph for a signature of Q on G. Figure 3(f) shows a result graph for the query {Project, Tom} on the RDF graph of Fig. 3(a).

The Structural Summary and Pattern Graphs. In order to construct pattern graphs we use the structural summary of the RDF graph as in [10,20]. Intuitively, the structural summary is a graph that summarizes the RDF graph.

Definition 2 (Structural Summary). The *structural summary* of an RDF graph G is a vertex and edge labeled graph constructed from G as follows:

1. Merge every class vertex and its entity vertices into one vertex labeled by the class vertex label and remove all the type edges from G.
2. Merge all the value vertices which are connected with a property edge labeled by the same label to the same class vertex into one vertex labeled by the union of the labels of these value vertices. Merge also the corresponding edges into one edge labeled by their label.

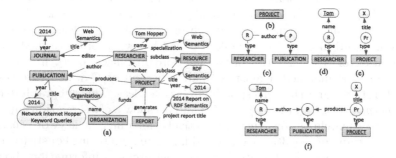

Fig. 4. (a) Structural Summary G', (b), (c), (d) and (e) are matching constructs for keywords in the keyword query Q1={Tom, author, Project, title} on G' (f) Pattern Graph of Q on G'.

3. Merge all the relationship edges between the same class vertices which are labeled by the same label into one edge with that label.

Figure 4(a) shows the structural summary for the RDF graph G of Fig. 3(a). Similarly to matching constructs on the data graph we define matching constructs on the structural summary. Since the structural summary does not have entity vertices, a matching construct on a structural summary possess one distinct entity variable vertex labeled by a distinct variable for every class vertex and a distinct value variable for every value vertex label which does not contain a keyword instance. Figure 4(b), (c), (d), and (e) show the class, relationship, value and property matching constructs for the keywords "Project", "author", "Tom", and "title", respectively, on the structural summary of Fig. 4(a).

Pattern graphs are the subgraphs of the structural summary, strictly consisting of one matching construct for every keyword in the query Q and the connections between them without these connections forming a cycle.

Definition 3 (Pattern Graph). A *(result) pattern graph* for a keyword query Q is a graph similar to a result graph for Q, with the following two exceptions:

(a) The labels of the entity vertices in the result graph, if any, are replaced by distinct variables in the pattern graph. These variables are called *entity variables* and they range over entity labels.

(b) The labels of the value vertices are replaced by distinct variables whenever these labels are not the keyword instances in the result graph. These variables are called *value variables* and they range over value labels in the RDF graph.

Figure 4(f) shows an example of a pattern graph, for the keyword query $Q = \{$Tom, author, project, title$\}$ on the RDF graph of Fig. 3(a). This pattern graph is computed over the structural summary of Fig. 4(a) combing the matching constructs of Fig. 4(b), (c), (d) and (e). Labels R, P, and Pr are entity variables and X is a value variable.

Given a keyword query Q over an RDF data graph G, we first find all the matching constructs for all the keywords in Q on the structural summary G' and then generate all the pattern graphs on G' for all possible signatures of Q.

3   Balancing Relevance and Diversity

We provide in this section a formal definition of the problem we address and then elaborate on its components: how to assess the relevance and the diversity of sets of pattern graphs.

3.1   Problem Statement

Our goal is to provide the user with a set of pattern graphs which is relevant and diverse. To this end, we define the problem as an optimization problem. Let

G denote an RDF data graph, Q be a keyword query on G, $\mathcal{P}$ be the set of pattern graphs of Q on G and k be a positive integer. Given a subset $\mathcal{S}$ of $\mathcal{P}$, let $relevance(\mathcal{S}, Q)$ denote the relevance of S with respect to Q, and $diversity(\mathcal{S})$ denote the diversity of set $\mathcal{S}$. We aim at selecting a subset $\mathcal{S}$ of $\mathcal{P}$ which maximizes the objective function $\alpha * relevance(\mathcal{S}, Q) + (1 - \alpha) * diversity(\mathcal{S})$, where $\alpha \in [0, 1]$, is a parameter which tunes the importance of relevance and diversity. In other words,

$$\mathcal{S} \in \arg \max\nolimits_{\mathcal{S}' \subseteq \mathcal{S},\, |\mathcal{S}'| = k}(\alpha * relevance(\mathcal{S}', Q) + (1 - \alpha) * diversity(\mathcal{S}'))$$

The tuning parameter α allows to give more importance to the relevance or diversity of the pattern graph set to be selected. If $\alpha = 1$, the selected pattern graph set will have the most relevant pattern graphs without considering diversity. If $\alpha = 0$, the pattern graph set will be selected solely based on its diversity.

We assume that the relevance of one pattern graph is independent of the relevance of another pattern graph to Q. The relevance of a set of pattern graphs $\mathcal{S}'$ of size k to a keyword query Q is the average relevance of its pattern graphs:

$$relevance(\mathcal{S}', Q) = 1/k * \left(\sum\nolimits_{P \in \mathcal{S}'} relevance(P, Q)\right)$$

where $relevance(P, Q) \in [0, 1]$ and denotes the relevance of pattern graph P to Q. The diversity of $\mathcal{S}'$ is defined as:

$$diversity(\mathcal{S}') = \sum\nolimits_{P_i, P_j \in \mathcal{S}', P_i \neq P_j} dist(P_i, P_j)/k(k-1)$$

where $dist(P_i, P_j)$ denotes the semantic distance between pattern graphs P_i and P_j; $dist(P_i, P_j) \in [0, 1]$. Dividing the sum by the total number of pattern graph pairs, normalizes $diversity(\mathcal{S}')$ in the $[0, 1]$ range. We define in the next sections metrics for assessing $relevance(P, Q)$ and $dist(P_i, P_j)$.

3.2   Assessing the Relevance of a Pattern Graph

Our approach exploits statistical information for the popularity (frequency) of the class and value vertices and the property and relationship edges of the pattern graphs in the RDF graph. In doing so, it also takes into account structural and semantic information of the pattern graphs. In this sense, two edges with the same label are different if they involve entity variables of different types. For assessing the popularity of value vertex vertices with keyword instances in the pattern graph, we employ the well known $tf*idf$ metric of Information Retrieval (IR) adapted to the syntactic and semantic features of the RDF data.

Consider a pattern graph P over an RDF data graph G. Let $C_1, \ldots, C_n$ be the class vertex labels in P. Let also $|V_{C_i}|$ denote the number of entities of type C_i in the RDF graph G, and $|V_E|$ denote the total number of entities in G. The popularity of the class vertices of P is given by the formula:

$$pop_c(P) = 1/n * \left(\sum\nolimits_{C_i \in \{C_1, \ldots, C_n\}} |V_{C_i}|/|V_E|\right)$$

Let $P_1, \ldots, P_m$ denote the distinct (owner class vertex, property edge label) pairs in P. Let also $|E_{P_i}|$ denote the number of property edges complying with P_i in the RDF graph G, and $|E_P|$ denote the total number of property edges in G. The popularity of the property edges of P is defined as:

$$pop_p(P) = 1/m * \left(\sum\nolimits_{P_i \in \{P_1, \ldots, P_m\}} |E_{P_i}|/|E_P|\right)$$

Let $R_1, \ldots, R_u$ denote the distinct (domain class vertex, relationship edge label, range class vertex) triples in P. Let also $|E_{R_i}|$ denote the number of relationship edges complying with R_i in the RDF graph G, and $|E_R|$ denote the total number of relationship edges in G. The popularity of the relationship edges of P is given by the formula:

$$pop_r(P) = 1/u * \left(\sum\nolimits_{R_i \in \{R_1, \ldots, R_u\}} |E_{R_i}|/|E_R|\right)$$

For defining the popularity of value vertices with keyword instances in a pattern graph, we modify the *tf*idf* metric so that it applies to RDF graphs. The metric *tf*idf* (term frequency, inverse document frequency) used in IR reflects how important a term is to a document in a corpus of documents. *tf(t, d)* denotes the frequency of a term t in a document d while *idf(t)* is the logarithmically scaled inverse fraction of the documents that contain the term. In the context of an RDF graph G, the set of property edges in G which have the same label L and are incident to entity vertices of a type C correspond to a document. This set of property edges is denoted by $E(C, L)$. Given a keyword k_i and a set of property edges $E(C, L)$, let $E(k_i, C, L)$ be the subset of $E(C, L)$ which contains only those property edges whose value comprises k_i. Then:

$$tf(k_i, E(C, L)) = |E(k_i, C, L)|/|E(C, L)|$$

Let W denote the set of all property edge sets $E(C, L)$ in G. For a given keyword k_i, let W_i be the subset of W consisting of those property edge sets $E(C, L)$ such that $tf(k_i, E(C, L)) > 0$ (that is, property edge sets where k_i occurs in the value of at least one of their property edges). Then:

$$idf(k_i) = log(|W|/|W_i|)$$

Let $k_1, \ldots, k_j$ denote the keywords which appear in the labels of value vertices in a pattern graph P. Note that, multiple keywords can appear in the label of a value vertex in P. Let v_i denote the value vertex whose label contains the keyword k_i and L_i is the label of the property edge connecting v_i to an entity variable vertex of type C in P. This means that k_i also appears in the values for the set of property edges for the (C_i, L_i) pair. Then, the popularity of value vertices containing keywords in P is given by the formula:

$$pop_v(P) = 1/j * \left(\sum\nolimits_{k_i \in \{k_1, \ldots, k_j\}} tf(k_i, (C_i, L_i)) * idf(k_i)\right)$$

We define the relevance of pattern graph P to keyword query Q as the sum of the popularity of the components of P as follows:

$$relevance(P,Q) = 1/4 * \left(\sum_{i \in \{c,p,r,v\}} pop_i(P)\right)$$

Clearly, the values of $pop_i(P)$ are in the range $[0,1]$. By dividing the sum by 4 we guarantee that $relevance(P,Q)$ also ranges between 0 and 1.

3.3 Assessing the Semantic Distance Between Two Pattern Graphs

In order to measure the semantic distance of two pattern graphs, we consider both structural and semantic features of the pattern graphs.

The first factor we consider in assessing the distance of two pattern graphs is the similarity of their matching constructs. Remember that the matching constructs are small graphs that involve only a single keyword instance and provide a context for interpreting the keywords. Given a pattern graph P for a keyword query $Q = \{k_1, \ldots k_n\}$, let $mc(P)$ denote the set of matching constructs of Q—one for every keyword in Q. The larger the number of keywords which are interpreted in the same way in the two pattern graphs, the more similar the pattern graphs are. The similarity of the matching constructs in the two pattern graphs is given by the formula

$$mc\_sim(P_1, P_2) = (|mc(P_1) \cap mc(P_2)|)/n$$

where n is the number of keywords in Q. Note that $n = |mc(P_1)| = |mc(P_2)|$. Clearly, $mc\_sim(P_1, P_2) = 1$ if P_1 and P_2 share the same matching constructs, and $mc\_sim(P_1, P_2) = 0$ if they have no common matching constructs.

For instance, Fig. 5 shows 5 pattern graphs of a query with 5 keywords. Intuitively, P_2 and P_3 are more similar to P_1 than P_4 and P_5 because P_4 and P_5 interpret the keyword semantics differently. Metric mc_sim catches this intuition since $mc\_sim(P_1, P_2) = mc\_sim(P_1, P_3) = 5/5$ while $mc\_sim(P_1, P_4) = mc\_sim(P_1, P_5) = 4/5$.

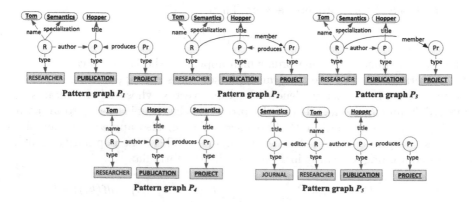

Fig. 5. Pattern graphs of Q={Tom, semantics, publication, hopper, project}.

Although P_2 and P_3 have the same common matching constructs with P_1, P_2 looks more similar to P_1 than P_3 does. Therefore, the second factor we consider is to what extent matching constructs for the same keywords are connected in the same way in the two pattern graphs. The higher the number of pairs of keywords in P_1 and P_2 whose matching constructs are connected in the same way in the two pattern graphs, the more similar P_1 and P_2 are. Of course, if the matching constructs of two keywords are not the same in P_1 and P_2, their connections cannot be the same in the two pattern graphs and this pair of keywords does not contribute to the similarity of P_1 and P_2. We define a connection between two keywords k_i and k_j of Q in a pattern graph P of Q as a graph consisting of the matching constructs of k_i and k_j, respectively, and a simple path between these matching constructs in P augmented with type edges and class vertices for every entity variable vertex in the path. Let z be the number of unordered pairs of query keywords which have the same connection in the two pattern graphs. The similarity of the keyword pair connections in P_1 and P_2 is given by:

$$conn\_sim(P_1, P_2) = \frac{z}{(n(n-1))/2}$$

where n is the number of keywords in Q. The denominator reflects the number of unordered keyword pairs for the keywords in Q. Similarly to $mc\_sim(P_1, P_2)$, $conn\_sim(P_1, P_2)$ ranges between $[0,1]$, with 1 indicating that the matching constructs for all the keywords and all the connections between them are the same.

In the example of Fig. 5 both pattern graphs P_2 and P_3 have five common matching constructs with P_1. However, $conn\_sim(P_1, P_2) = 6/10$ and $conn\_sim(P_1, P_3) = 4/10$. Intuitively, P_2 looks more similar to P_1 than P_3 to P_1.

Measuring the similarity of two pattern graphs P_1 and P_2 based solely on the similarity of matching constructs and matching construct connections, $mc\_sim(P_1, P_2)$ and $conn\_sim(P_1, P_2)$, cannot entirely capture their semantic closeness. Compare, for instance, the pattern graphs P_4 and P_5 with the pattern graph P_1 in Fig. 3. Both P_4 and P_5 have 4 keyword matching constructs and 6 pairs of matching construct connections in common with P_1. However, our intuition suggests that P_5 is less similar (more dissimilar) to P_1 than P_4 as it has the class vertex (concept) "Journal" which does not appear in P_1. In contrast, P_1 and P_4 have the same class vertices. Therefore, we introduce the metric of concept dissimilarity to capture the dissimilarity of two pattern graphs. Let $c(P)$ denote the set of class vertices in a pattern graph P. Given two pattern graphs P_1 and P_2 of a keyword query,

$$concept\_dsim(P_1, P_2) = \frac{|(c(P_1) \cup c(P_2)) - (c(P_1) \cap c(P_2))|}{|c(P_1) \cup c(P_2)|}$$

$concept\_dsim(P_1, P_2)$ ranges between 0 (when P_1 and P_2 have all their class vertices in common) and 1 (when P_1 and P_2 do not have common class vertices). The higher the value of $concept\_dsim(P_1, P_2)$, the more distant the pattern graphs P_1 and P_2 are.

Taking into account all the factors, we define the distance $dist(P_1, P_2)$ of two pattern graphs P_1 and P_2 as shown next

$$dist(P_1, P_2) = \frac{1 - [(mc\_sim(P_1, P_2) + conn\_sim(P_1, P_2))/2 - concept\_dsim(P_1, P_2)]}{2}$$

Note that $concept\_dsim(P_1, P_2)$ is considered with a negative sign since it expresses dissimilarity.

$dist(P_1, P_2) = 0$ when the two pattern graphs are the same and $dist(P_1, P_2) = 1$ when $concept\_dsim(P_1, P_2) = 1$.

4   Algorithm

In this section, we present an algorithm for the problem of balancing the relevance and diversity of sets of pattern graphs stated in Sect. 3.1. Exhaustively generating all size-k subsets of a set of pattern graphs and computing their relevance and diversity in order to find an optimal one has exponential complexity in the number of the pattern graphs. In fact, different versions of the diversification

---

Algorithm 1. PGDiversification (Pattern Graph Diversification)

Input: $Q = \{k_1, \ldots, k_n\}$: a keyword query with n keywords,
        S: structural summary of the data graph,
        α: tuning factor, k: size of the output list.
Output: $\mathcal{P}_{div}$: set of diversified pattern graphs of size k.
1: **for all** $k_i \in Q$ **do**
2:     $L_i \leftarrow \{$set of all matching constructs of k_i on $S\}$;
3: $\mathcal{P} \leftarrow ComputePatternGraphs(\{\times_{i=1}^{n} L_i\}, S)$;
4: $\mathcal{P} \leftarrow SortByRelevance(\mathcal{P})$;
5: $\mathcal{P}_{div} \leftarrow \mathcal{P}[0]$;
6: $i = 1$;
7: **while** $i < k$ **do**
8:     $j = i$;
9:     $NextIndex = -1$;
10:    $NextScore = 0$;
11:    **while** $j \leqslant |\mathcal{P}|$ **do**
12:       $distance = 0$;
13:       **for all** $p_i \in \mathcal{P}_{div}$ **do**
14:         $distance = distance + dist(p_i, \mathcal{P}[j])$
15:       $CurrentScore = \alpha * relevance(\mathcal{P}[j], Q) + (1 - \alpha) * distance/|\mathcal{P}_{div}|$;
16:       **if** $CurrentScore > NextScore$ **then**
17:         $NextScore = CurrentScore$;
18:         $NextIndex = j$;
19:       $j = j + 1$;
20:    $\mathcal{P}_{div}.add(\mathcal{P}[NextIndex])$;
21:    $swapPGs(\mathcal{P}[i], \mathcal{P}[NextIndex])$;
22:    $i = i + 1$;

---

problem have previously been shown to be NP-hard [1,5,12,14]. Therefore, we design a heuristic algorithm, called *PGDiversification*, which greedily selects a new pattern graphs at every iteration and incrementally computes the relevance and diversity of pattern graph sets. Algorithm *PGDiversification* takes as input a keyword query Q, the structural summary S of an RDF graph, the tuning parameter α, and a positive integer k. The output is a subset of the set of pattern graphs of Q on S of size k.

The algorithm starts by finding all the matching constructs of the keywords in query Q on S (lines 1–2) and then generates the set $\mathcal{P}$ of pattern graphs for all possible signatures of Q (line 3). The pattern graphs are generated as r-radius Steiner graphs [16]. Different algorithms can be used for generating the pattern graphs (for instance, in [8,20]). The way the pattern graphs are computed is orthogonal to our approach for pattern graph diversification. The pattern graphs of $\mathcal{P}$ are ranked in descending order of their relevance (line 4). The variable $\mathcal{P}_{div}$ represents the output set of size k which is a subset of the set of pattern graphs $\mathcal{P}$. Initially, the set $\mathcal{P}_{div}$ contains a pattern graph with the highest relevance (line 5). Subsequently, at every iteration, a pattern graph is chosen for inclusion in $\mathcal{P}_{div}$ so that the new $\mathcal{P}_{div}$ set maximizes the objective function (line 8–22). The process terminates when $|\mathcal{P}_{div}| = k$.

5   Experimental Results

We implemented our approach and ran experiments to examine: (a) the effectiveness of our distance metric in assessing the semantic distance of pattern graphs, and the quality of our approach in retrieving a maximum number of relevant pattern graphs, (b) the quality of the approximation of the greedy heuristic algorithm, and (c) the efficiency of the *PGDiversificiation* algorithm in computing the set of pattern graphs that trades off relevance and diversity.

5.1   Datasets and Queries

We used the DBLP[1] and Jamendo[2] real datasets for our experiments. DBLP is a bibliography database of 600 MB of size, containing 8.5 M triples. Jamendo is a repository of Creative Commons licensed music of 85 MB of size, containing 1.1 M triples. The experiments were conducted on a standalone machine with an Intel i7-5600U@2.60GHz processor and 8GB memory. We experimented with a large number of queries and Table 1 reports on 10 of them for each dataset in the interest of space.

5.2   Effectiveness Results

Effectiveness of the Distance Metric. We first want to examine the quality of our distance metric. To this end, for each of the queries in Table 1, we select five

---

[1] https://datahub.io/dataset/l3s-dblp.
[2] http://dbtune.org/jamendo/.

Table 1. Keyword Queries on the Jamendo and DBLP Datasets

Keyword queries on Jamendo		Keyword queries on DBLP	
Q ID	Keywords	Q ID	Keywords
J1	document, teenage, fantasie	D1	concatenable, aspectisation, oliver
J2	nuts, spy4, lemonade	D2	dataflow, quantization
J3	divergence, obsession, lyrics	D3	donatella, intermittent, congestion
J4	reflection, record	D4	balvinder, coscheduling, article
J5	document, cool, divergence	D5	springer, inproceedings
J6	cicada, performance	D6	skogstad, tensorial, morphology
J7	extraordinary, blissful, madness	D7	hierarchical, hybridization
J8	awesome, passion, spy4	D8	person, tolga, coscheduling
J9	guitarist, lemonade	D9	charles, peephole, inproceedings
J10	disgusting, revenge, fantasie	D10	tolga, forward, normalizability

of their pattern graphs. We ask three expert users to score the semantic similarity of each one of these five pattern graphs with another pattern graph (the pattern graph with the highest relevance). The scores are integers in the range $[0, 3]$. A score of 0 denotes that the two pattern graphs are totally dissimilar. The ground truth is determined by majority vote. We also use our distance metric $dist(P_1, P_2)$ to rank the five pattern graphs in descending order of their distance from the most relevant pattern graph. We assess the quality of the ranking based on $dist(P_1, P_2)$ using the *normalized Discounted Cumulative gain* (nDCG) metric which is defined as follows. The *discounted cumulative gain* (DCG) for position n in a ranked list is given by the following formula:

$$DCG_n = \sum_{i=1}^{n} \frac{2^{rel_i} - 1}{log_2(i + 1)}$$

In order to take into account equivalent classes of pattern graphs in the ranked lists (that is, pattern graphs which have the same rank), we have extended nDCG by introducing minimum, maximum and average values for it. The nDCG_{max} value of a ranked list RL_e with equivalence classes corresponds to the nDCG value of a strictly ranked (that is, without equivalence classes) list obtained from RL_e by ranking the pattern graphs in the equivalence classes correctly (that is, in compliance with the scores given by the expert users). The nDCG_{min} value of RL_e is defined analogously. The nDCG_{avg} value of RL_e is the average nDCG value over all strictly ranked lists obtained from RL_e by ranking the pattern graphs in the equivalence classes in all possible ways. The nDCG values range between 0 and 1. Figure 6(a) and (b) show the nDCG_{min}, nDCG_{max} and nDCG_{avg} values for the queries on the two datasets. As one can see, all the values are very close to 1. This implies that our distance metric successfully assesses the semantic similarity of two pattern graphs.

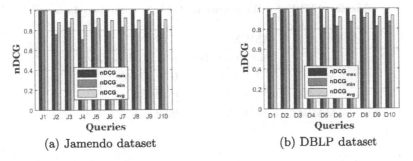

(a) Jamendo dataset

(b) DBLP dataset

Fig. 6. nDCG$_{max}$, nDCG$_{min}$ and nDCG$_{avg}$ for the queries on the two datasets

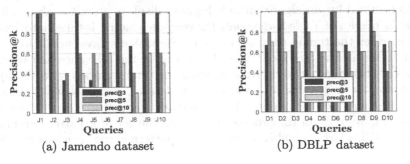

(a) Jamendo dataset

(b) DBLP dataset

Fig. 7. Prec@k for $k = 3$, 5 and 10, for the queries of Table 1 on the two datasets based solely on the relevance metric ($\alpha = 1$).

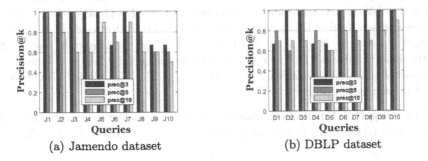

(a) Jamendo dataset

(b) DBLP dataset

Fig. 8. Prec@k for $k = 3$, 5 and 10, for the queries of Table 1 on the two datasets based on the relevance and diversity metrics ($\alpha = 0.5$).

Effectiveness of the Approach. In order to evaluate the quality of the approach in retrieving relevant results, we measure the relevant results retrieved by Algorithm *PGDiversification* for different queries when only our relevance metric, and when our metric which balances relevance and diversity is taken into account. Three expert users characterize the retrieved pattern graphs as relevant or not to the query based on whether the pattern graphs express meaningful interpretations of the query and ground truth is determined by majority vote.

Table 2. MAP@k for the queries of Table 1 on the two datasets

	JAMENDO		DBLP	
	Rel.	Rel. & Div.	Rel.	Rel. & Div.
MAP@10	0.53	0.62	0.51	0.66
MAP@5	0.72	0.77	0.72	0.84
MAP@3	0.86	0.86	0.82	0.91

The quality of our approach on a query is expressed by *precision@k* (*prec@k*), which is the ratio of the number of relevant pattern graphs in the set of k pattern graphs returned by our algorithm to k. Figure 7 displays *prec@k* for $k = 3$, 5 and 10 for the queries of Table 1 when only the relevance metric is taken into account (that is, when the tuning parameter $\alpha = 1$). Figure 8 displays *prec@k* for $k = 3$, 5 and 10 for all the queries of Table 1 when both the relevance and diversity metrics are taken into account (that is, when the tuning parameter $\alpha = 0.5$). As we can see, for each value of k and for each dataset, *precision@k* is the same or better in most of the queries. This observation demonstrates the benefit of introducing diversity in the process of selecting the set of k pattern graphs as, when $\alpha = 0.5$, a larger number of the selected pattern graphs can satisfy at least one user.

We also measure the mean average precision at k (MAP@k) for two different values of the tuning parameter: $\alpha = 1$ and $\alpha = 0.5$. MAP@k is the mean of the average precision at k for the queries of Table 1. The values of MAP@k are shown in Table 2 and summarize the measurements displayed in Figs. 7 and 8. As one can see, taking into account diversity ($\alpha = 0.5$) improves MAP@k in all cases. For instance, it improves MAP@10 by 17% on the Jamendo and by 29% on the DBLP dataset. The increase is more pronounced when k is larger. This is expected since a larger k offers more chances to the algorithm to diversify the selected pattern graph set.

Calibrating the Tuning Parameter. In order to select a good value for the tuning parameter α we ran experiments to measure $MAP@k$ for $k = 3$, 5, and

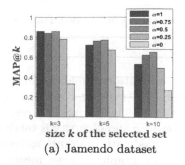

(a) Jamendo dataset

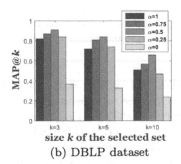

(b) DBLP dataset

Fig. 9. MAP@k for $k = 3$, 5 and 10, for the queries of Table 1 with different α values.

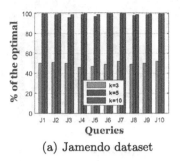
(a) Jamendo dataset

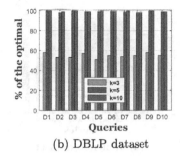
(b) DBLP dataset

Fig. 10. Closeness (%) of the objective function values computed by algorithm *PGDiversification* to the optimal ones, for $k = 3, 5, 10$, for the queries on the two datasets.

10, for all the queries of Table 1, varying α from 0 to 1 in increments of 0.25. The results are depicted in Fig. 9. As one can see, the highest values for $MAP@k$ are displayed for $\alpha = 0.5$ for all values of k. Therefore, α has been set to 0.5 in all other experiments. Reasonably, the lowest values for $MAP@k$ are displayed for $\alpha = 0$, since all the pattern graphs (with the exception of the initial most popular pattern graph) are selected based on the semantic distance metric and the popularity metric, which is a good criterion for relevance, did not contribute to the selection of relevant pattern graphs.

5.3 Quality of the Approximation by the HeuRistic Algorithm

We also wanted to evaluate the quality of the approximation of the optimal solution by the heuristic greedy algorithm. We computed the optimal solution and measured the values of the objective function for $k = 3, 5$, and 10, by running an algorithm which exhaustively enumerates all combinations of k pattern graphs from the generated set of candidate pattern graphs. This was possible since the queries were selected so that they do not generate on the structural summary a huge number of candidate pattern graphs. Figure 10 displays the closeness of the objective function values produced by our algorithm as a percentage to the optimal ones for $k = 3, 5$, and 10, for all the queries of Table 1. Not surprisingly, the closeness is around 50% for $k = 3$ since the selected pattern graph set is always initialized with the most popular pattern graph which might not be part of the optimal solution and the next pattern graph selections are based on their distance from previously selected ones. However, it moves very close to 100% already for $k = 5$ for all queries on both datasets, and it is perfect or almost perfect for $k = 10$. These results suggest that the greedy heuristic produces a good approximation of the optimal solution.

5.4 Efficiency Results

We ran Algorithm *PGDiversification* for all the queries of Table 1. The execution time is reported in Fig. 11 for pattern graphs sets of size $k = 5$ and tuning

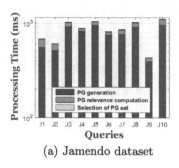

(a) Jamendo dataset

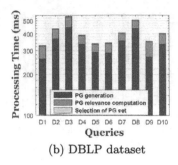

(b) DBLP dataset

Fig. 11. Processing time of *PGDiversification* for the queries on the two datasets.

parameter $\alpha = 0.5$. The execution time for each query consists of three components: (a) the generation of the pattern graphs using the structural summary, (b) the computation of the relevance of the pattern graphs and the selection of one with the highest relevance, and (c) the application of the greedy heuristic for generating the list of k pattern graphs which is both relevant and diversified. One can see that the execution time is dominated by the pattern graph generation process. This is expected since computing the pattern graphs requires accessing the database for finding all the keyword matching constructs.

Overall, our approach displays interactive execution times ($\leqslant 1$ s) on both datasets. This performance demonstrates that our approach succeeds in improving the number of relevant pattern graphs selected without suffering from the scalability issue of keyword search approaches.

6 Related Work

The goal of the diversification problem is to solve the over-specialization problem where a highly homogenous set of results is returned to the user due to relevance-based ranking and/or personalization. Result diversification is a way to minimize user dissatisfaction by providing a diverse set of results. In general, the diversification problem is defined as selecting a subset of the retrieved result set with k results such that the diversity among these k results is maximized. In [4], the concept of maximal marginal relevance (MMR) is used to tradeoff between relevance and novelty. In [14] an axiomatic approach for result diversification is adopted. Reference [12] is a review of different definitions for diversity, and of algorithms and evaluation metrics for diversification. It categorizes diversity definitions as content-based [22], novelty-based [21] and coverage-based [1]. Most of these approaches perform diversification as a post-processing or re-ranking step of candidate result retrieval which can incur a huge computation cost since the number of candidate results can be extremely large for a keyword query. In contrast, our diversification process is part of a query disambiguation phase which takes place before extracting any search results. We compute pattern graphs corresponding to alternate interpretations of a given keyword query since they offer clear semantics and quality information for diversification. This way, we also avoid the computation overhead of computing all relevant results.

A number of papers study search result diversification on structured and semi-structured databases [2,11,15,17]. However, there are only few contributions on diversifying keyword search results on RDF data. Previous techniques cannot be directly applied in this context as the semantic information in RDF data graphs requires different criteria and methods. [19] addresses diversification of entity search results using a categorization technique to cluster similar entities. [3,9] addresses diversification issues of keyword query results on RDF data, but do not consider relevance aspects. In this paper, we exploit structural and semantic characteristics of pattern graphs for capturing their relevance to a query, and also the similarity and dissimilarity between pattern graphs.

7 Conclusion

We presented a novel technique which exploits pattern graphs of a keyword query instead of the query results for trading off popularity for diversity in the result sets of keyword queries on RDF data graph. Diversification of pattern graphs addresses the scalability problem of keyword search approaches on large data graphs and also allows diversifying the interpretations of the keyword query. We introduced metrics to assess the relevance of a pattern graph and the semantic distance between patterns graphs. We also designed a greedy heuristic algorithm which maximizes an objective function on pattern graph sets balancing popularity and diversity. Our extensive experimental results show the feasibility of our approach in terms of the efficiency and the effectiveness of the algorithm.

References

1. Agrawal, R., Gollapudi, S., Halverson, A., Ieong, S.: Diversifying search results. In: WSDM, pp. 5–14. ACM (2009)
2. Aksoy, C., Dass, A., Theodoratos, D., Wu, X.: Diversification of keyword query result patterns. In: Cui, B., Zhang, N., Xu, J., Lian, X., Liu, D. (eds.) WAIM 2016. LNCS, vol. 9659, pp. 171–183. Springer, Cham (2016). doi:10.1007/978-3-319-39958-4_14
3. Bikakis, N., Giannopoulos, G., Liagouris, J., Skoutas, D., Dalamagas, T., Sellis, T.: RDivF: diversifying keyword search on RDF graphs. In: Aalberg, T., Papatheodorou, C., Dobreva, M., Tsakonas, G., Farrugia, C.J. (eds.) TPDL 2013. LNCS, vol. 8092, pp. 413–416. Springer, Heidelberg (2013). doi:10.1007/978-3-642-40501-3_49
4. Carbonell, J., Goldstein, J.: The use of MMR, diversity-based reranking for reordering documents and producing summaries. In: SIGIR, pp. 335–336 (1998)
5. Carterette, B.: An analysis of NP-completeness in novelty and diversity ranking. Inf. Retrieval 14(1), 89–106 (2011)
6. Chen, H., Karger, D.R.: Less is more: probabilistic models for retrieving fewer relevant documents. In: SIGIR, pp. 429–436. ACM (2006)

7. Dass, A., Aksoy, C., Dimitriou, A., Theodoratos, D.: Exploiting semantic result clustering to support keyword search on linked data. In: Benatallah, B., Bestavros, A., Manolopoulos, Y., Vakali, A., Zhang, Y. (eds.) WISE 2014. LNCS, vol. 8786, pp. 448–463. Springer, Cham (2014). doi:10.1007/978-3-319-11749-2_34

8. Dass, A., Aksoy, C., Dimitriou, A., Theodoratos, D.: Keyword pattern graph relaxation for selective result space expansion on linked data. In: Cimiano, P., Frasincar, F., Houben, G.-J., Schwabe, D. (eds.) ICWE 2015. LNCS, vol. 9114, pp. 287–306. Springer, Cham (2015). doi:10.1007/978-3-319-19890-3_19

9. Dass, A., Aksoy, C., Dimitriou, A., Theodoratos, D., Wu, X.: Diversifying the results of keyword queries on linked data. In: Cellary, W., Mokbel, M.F., Wang, J., Wang, H., Zhou, R., Zhang, Y. (eds.) WISE 2016. LNCS, vol. 10041, pp. 199–207. Springer, Cham (2016). doi:10.1007/978-3-319-48740-3_14

10. Dass, A., Dimitriou, A., Aksoy, C., Theodoratos, D.: Incorporating Cohesiveness into keyword search on linked data. In: Wang, J., Cellary, W., Wang, D., Wang, H., Chen, S.-C., Li, T., Zhang, Y. (eds.) WISE 2015. LNCS, vol. 9419, pp. 47–62. Springer, Cham (2015). doi:10.1007/978-3-319-26187-4_4

11. Demidova, E., Fankhauser, P., Zhou, X., Nejdl, W.: DivQ: diversification for keyword search over structured databases. In: SIGIR, pp. 331–338. ACM (2010)

12. Drosou, M., Pitoura, E.: Search result diversification. ACM SIGMOD Rec. 39(1), 41–47 (2010)

13. Elbassuoni, S., Ramanath, M., Schenkel, R., Weikum, G.: Searching RDF graphs with SPARQL and keywords. IEEE Data Eng. Bull. 33(1), 16–24 (2010)

14. Gollapudi, S., Sharma, A.: An axiomatic approach for result diversification. In: WWW, pp. 381–390. ACM (2009)

15. Hasan, M., Mueen, A., Tsotras, V., Keogh, E.: Diversifying query results on semi-structured data. In: CIKM, pp. 2099–2103. ACM (2012)

16. Li, G., et al.: Ease: an effective 3-in-1 keyword search method for unstructured, semi-structured and structured data. In: SIGMOD, pp. 903–914 (2008)

17. Li, J., Liu, C., Yu, J.X.: Context-based diversification for keyword queries over XML data. Proc. KDE 27(3), 660–672 (2015)

18. Radlinski, F., Dumais, S.: Improving personalized web search using result diversification. In: SIGIR, pp. 691–692. ACM (2006)

19. Ruotsalo, T., Frosterus, M.: Semantic entity search diversification. In: ICSC, pp. 32–39 (2013)

20. Tran, T., Wang, H., Rudolph, S., Cimiano, P.: Top-k exploration of query candidates for efficient keyword search on graph-shaped (RDF) data. In: ICDE (2009)

21. Zhang, M., Hurley, N.: Avoiding monotony: improving the diversity of recommendation lists. In Recommender Systems, pp. 123–130 (2008)

22. Ziegler, C.-N., McNee, S.M., Konstan, J.A., Lausen, G.: Improving recommendation lists through topic diversification. In: WWW, pp. 22–32. ACM (2005)

The Qanary Ecosystem: Getting New Insights by Composing Question Answering Pipelines

Dennis Diefenbach[1(✉)], Kuldeep Singh[2,3], Andreas Both[4], Didier Cherix[5], Christoph Lange[2,3], and Sören Auer[2,3]

[1] Laboratoire Hubert Curien, Saint Etienne, France
dennis.diefenbach@univ-st-etienne.fr
[2] Fraunhofer IAIS, Sankt Augustin, Germany
kuldeep.singh@iais.fraunhofer.de, {langec,auer}@cs.uni-bonn.de
[3] University of Bonn, Bonn, Germany
[4] DATEV eG, Nuremberg, Germany
contact@andreasboth.de
[5] FLAVIA IT-Management GmbH, Kassel, Germany
didier.cherix@gmail.com

Abstract. The field of Question Answering (QA) is very multi-disciplinary as it requires expertise from a large number of areas such as natural language processing (NLP), artificial intelligence, machine learning, information retrieval, speech recognition and semantic technologies. In the past years a large number of QA systems were proposed using approaches from different fields and focusing on particular tasks in the QA process. Unfortunately, most of these systems cannot be easily reused, extended, and results cannot be easily reproduced since the systems are mostly implemented in a monolithic fashion, lack standardized interfaces and are often not open source or available as Web services. To address these issues we developed the knowledge-based Qanary methodology for choreographing QA pipelines distributed over the Web. Qanary employs the qa vocabulary as an exchange format for typical QA components. As a result, QA systems can be built using the Qanary methodology in a simpler, more flexible and standardized way while becoming knowledge-driven instead of being process-oriented. This paper presents the components and services that are integrated using the qa vocabulary and the Qanary methodology within the Qanary ecosystem. Moreover, we show how the Qanary ecosystem can be used to analyse QA processes to detect weaknesses and research gaps. We illustrate this by focusing on the Entity Linking (EL) task w.r.t. textual natural language input, which is a fundamental step in most QA processes. Additionally, we contribute the first EL benchmark for QA, as open source. Our main goal is to show how the research community can use Qanary to gain new insights into QA processes.

Keywords: Semantic web · Software reusability · Question answering · Service composition · Semantic search · Ontologies · Annotation model

© Springer International Publishing AG 2017
J. Cabot et al. (Eds.): ICWE 2017, LNCS 10360, pp. 171–189, 2017.
DOI: 10.1007/978-3-319-60131-1_10

1 Introduction

The amount of data, information, and knowledge available on the Web and within enterprise environments is increasing constantly. Especially in enterprise environments a strong trend to better connected data can be observed, leading to interlinked and accessible data unlocking the company's information for intense data analytics and information retrieval. Novel interfaces are required for enabling users to retrieve information in such scenarios and interact with it. Natural language interfaces are being considered to bridge the gap between large amounts of (semi-structured) data and users' needs. Recent industrial applications show the capabilities and advantages of natural language interfaces in the field of Question Answering (QA). These include *Apple Siri*[1], *Microsoft Cortana*[2], and "Ok Google"[3]. However, these proprietary platforms do not facilitate experimentation with cutting-edge research approaches, they offer only limited interfaces for integrating third-party components and they are generally not open, reusable and extensible by developers and the research community.

Several QA systems have been developed recently in the research community, for example, [5,8,11,17]. These systems perform well in specific domains, but their reusability for further research is limited because of their focus on specific technologies, applications or datasets. As a result, creating new QA systems is currently still cumbersome and inefficient. Particularly, the research community is not empowered to focus on improving particular components of the QA process, as developing a new QA system and integrating a component is extremely resource-consuming. Some first steps for developing flexible, modular QA systems have started to address this challenge, e.g., [9,13]. However, these approaches lack several key properties required for constructing QA systems in a community effort as they are, for example, bound to a particular technology environment and have rather static interfaces, which do not support the evolution of the inter-component data exchange models. For this reason we presented the qa vocabulary [18] as a flexible and extensible data model for QA systems. Based on the vocabulary, we developed the Qanary [3] methodology for integrating components into QA systems; it is independent from programming languages, agnostic to domains and datasets, as well as enabled for components on any granularity level within the QA process.

This work presents the Qanary *ecosystem*: the components and services currently implemented around the qa vocabulary by using the Qanary methodology. We present a general workflow that can be used to construct and particularly analyze as well as optimize future QA systems in a community effort using the Qanary ecosystem. It can be broken down into two phases: (1) the identification and integration of existing state-of-the-art approaches to solve a particular task in the QA pipeline, and (2) the derivation of benchmarks for sub-tasks of a QA process from well-known QA benchmarks such as the Question Answering over

---

[1] http://www.apple.com/ios/siri/.

[2] http://windows.microsoft.com/en-us/windows-10/getstarted-what-is-cortana.

[3] https://support.google.com/websearch/answer/2940021?hl=en.

Linked Data (QALD) challenge[4]. Hence, the described approach is dedicated to support the engineering process to build components for a QA system and the system by itself, by using the knowledge-driven approach for flexible component integration and quality evaluations. In this paper, we show this workflow applied to the task of EL, which is key in the QA process. Therefore, we consider components dedicated to the tasks of named entity identification/recognition (NER) and named entity disambiguation (NED), which are integrated into the Qanary ecosystem. The included components are the NER tool of *DBpedia Spotlight* [14], the *Stanford NER tool* [10] and the *Federated knOwledge eXtraction Framework* (FOX) [20] as well as the NED components *Agnostic Disambiguation of Named Entities Using Linked Open Data* (AGDISTIS) [22] and the named entity disambiguator of DBpedia Spotlight. In addition two combined approaches for NER and NED are also provided as components: *IBM Alchemy*[5] and *Lucene Linker* – a component that we implemented following the idea of the QA system *SINA* [17]. Moreover, we devised a benchmark for entity linking (EL) based on the well-known Question Answering over Linked Data (QALD) challenge. Our contribution here has three aspects. First, we provide researchers with a tool for comparing NED and NER w.r.t. QA, thus enabling them to compare their components with the state-of-the-art just by implementing a Qanary wrapper around their novel functionality. Second, we provide the results of comparing existing tools, i.e., an expressive benchmark for the quality of entity linking components w.r.t. natural language questions, thus enabling the QA community to gain new insights into QA processes. Third, we compute a list of questions that are completely annotated w.r.t. the entity linking process. Hence, researchers investigating a processing step of a QA system that comes after entity linking can reuse these annotations to create an environment for conveniently testing and continuously improving their components.

As a result, the QA community is empowered to easily reuse entity linking functionality for QA systems (or for the development of other tools depending on named entities) and reuse a profound benchmark for QA systems both for the evaluation of new entity linking components and as input for components active in the subsequent processing steps of a QA system (e.g., relation detection or query computation). However, the entity linking functionality and experiments presented in this paper are just a proof that Qanary's knowledge-driven and component-driven approach as well as the previously described general workflow provides key advantages particularly in contrast to existing systems and other benchmarks.

The next section describes related work. Section 3 gives an overview of our recent work which laid the groundwork for the Qanary ecosystem. Section 4 gives an overview of the components and services that are available in the Qanary ecosystem. Section 5 describes how the Qanary ecosystem can be used to gain new insights into QA processes w.r.t. the EL task. Section 6 concludes and points to future research areas.

---

[4] http://greententacle.techfak.uni-bielefeld.de/cunger/qald.
[5] http://www.alchemyapi.com/.

2   Related Work

In the context of QA, a large number of systems and frameworks have been developed in the last years. This can be observed for example from the number of QA systems (> 20 in the last 5 years) that were evaluated against the QALD benchmark. Many QA systems use similar techniques. For example, there are services for named entity identification (NER) and disambiguation (NED) such as DBpedia Spotlight [14] and Stanford NER [10], which are reused across several QA systems. These reasons led to the idea of developing component-based frameworks that make parts of QA systems reusable. We are aware of three frameworks that attempt to provide a reusable architecture for QA systems. The first is QALL-ME [9] which provides a reusable architecture skeleton for building multilingual QA systems. The second is openQA [13], which provides a mechanism to combine different QA systems and evaluate their performance using the QALD-3 benchmark. The third is the Open KnowledgeBase and Question-Answering (OKBQA) challenge[6]. It is a community effort to develop a QA system that defines rigid JSON interfaces between the components. Differently from these works we do not propose a rigid skeleton for the QA pipeline, instead we allow multiple levels of granularity and enable the community to develop new types of pipelines.

Recognizing named entities in a text and linking them to a knowledge base is an **essential task** in QA. DBpedia Spotlight [14], Stanford NER [10], FOX [20], and Alchemy API are a few of the tools dedicated to such tasks. Furthermore, tools such as DBpedia Spotlight, AGDISTIS [22], Alchemy API etc. not only identify information units in text queries but also point every named entity to a knowledge resource for disambiguation.

We are not aware of any work that has tried to compare in a systematic way existing approaches that tackle sub-processes of QA pipelines, for example EL. Atdag and Labatut [1] compare a few NER tools applied to bibliographic text, whereas researchers in [16] present NERD, a framework for evaluating NER tools in the context of Web data where a wrapper of NER/NED services was implemented but the independent registration of new services is not possible. Platforms such as GERBIL [23] and GERBIL for QA[7] offer benchmarks for EL tools and full QA systems and they generate persistent URIs for experiment results. This enables third-party evaluations and citable URIs. Their main goal is not to gain new insights into the underlying processes but only to generate one final metric that is publishable. For example, they do not generate a summary indicating in which cases the corresponding tool succeeded or failed.

In contrast, the Qanary reference implementation is a full-featured framework for component-oriented QA process creation, which is *additionally* enabled to support benchmarking of the included distributed components. We give a comparison of the described tools in Table 1.

---

[6] http://www.okbqa.org/.
[7] https://github.com/TortugaAttack/gerbil-qa.

Table 1. Overview of tools related to benchmarks in the field of question answering in comparison to the benchmark functionality of Qanary.

Property vs. Tool	NERD	Gerbil	Gerbil-qa	Qanary
Support for analyzing NER/NED task	■	■	□	■
Support for analyzing QA process quality	□	□	☑	■
Third party evaluations	□	■	■	■
Fine-grained information	■	□	□	■
Traceability of intermediate results	☑	□	□	■
Computation of citable URIs	□	■	■	□

3 The qa Vocabulary and the Qanary Methodology

To advance the QA process, researchers are combining different technologies to optimize their approaches. However, reusability and extensibility of QA components and systems remains a major hurdle. There are many components and services, which are provided as standalone implementations but can be useful in QA processes (e.g., the previously mentioned DBpedia Spotlight, AGDISTIS etc.), but there has so far not been a methodology to integrate them within QA pipelines. Instead substantial programming efforts had to be invested as each component provides its own API or integration method.

To address this challenge, and to promote reusability and extensibility of QA components, we introduced the qa vocabulary [18]. This vocabulary can represent information that is generated during the execution of a QA pipeline when processing a question given as speech or text input. Consider, for example, the question "When was Barack Obama born?". Typical information generated by components of a QA pipeline are the positions of named entities (NE) (such as "Barack Obama"), the ontological relations used to express the relational phrase in the question (that "born" refers to dbo:birthPlace[8]), the expected answer type (here: a date), the generated SPARQL query, the language of the question and possible ontologies that can be used to answer it.

The rationale of qa is that all these pieces of information can be expressed as annotations to the question. Hence, these exposed pieces of information can be provided as an (RDF) knowledge base containing the full descriptive knowledge about the currently given question.

qa is built on top of the *Web Annotation Data Model* (WADM)[9], a vocabulary to express annotations. The basic constructs of the WADM are annotations with at least a target indicating what is described and a body indicating the description.

---

[8] PREFIX dbo: <http://dbpedia.org/ontology/>.
[9] W3C Candidate Recommendation 2016-09-06, http://www.w3.org/TR/annotation -model.

```
PREFIX oa:  <http://www.w3.org/ns/oa#>
<anno>    a                oa:Annotation ;
          oa:hasTarget    <target>       ;
          oa:hasBody      <body>        .
```

In qa, a question is assumed to be exposed at some URI (e.g. URIQuestion that can be internal and does not need to be public) and is of type qa:Question. Similarly other QA concepts (qa:Answer, qa:Dataset, etc.) are defined in the vocabulary; please see [18] for further details. As a result, when using qa, the knowledge of the QA system is now representable independently from a particular programming language or implementation paradigm because everything is represented as direct or indirect annotations of a resource of type qa:Question. The qa vocabulary is published at the persistent URI https://w3id.org/wdaqua/qanary# under the CC0 1.0 license[10].

The qa vocabulary led to the Qanary methodology [3] for implementing processes operating on top of the knowledge about the question currently processed within a QA system, leading to the possibility of easy-to-reuse QA components. All the knowledge related to questions, answers and intermediate results is stored in a central Knowledge Base (KB). The knowledge is represented in terms of the qa vocabulary in the form of annotations of the relevant parts of the question.

Within Qanary the components all implement the same service interface. Therefore, all components can be integrated into a QA system without manual engineering effort. Via it's service interface a component receives information about the KB (i.e., the endpoint) storing the knowledge about the currently processed question of the user. Hence, the common process within all components is organized as follows:

1. A component fetches the required knowledge via (SPARQL) queries from the KB. In this way, it gains access to all the data required for its particular process.
2. The custom component process is started, computing new insights of the user's question.
3. Finally, the component pushes the results back to the KB (using SPARQL).

Therefore, after each process step (i.e., component interaction), the KB should be enriched with new knowledge (i.e., new annotations of the currently processed user's question). This way the KB keeps track of all the information generated in the QA process even if the QA process is not predefined or not even known. A typical QA pipeline consists of several steps such as NER, NED, relation identification, semantic analysis, query computation and result ranking. Most recently we provided a reference implementation of Qanary [19]. We call this implementation *message-driven*; it follows the architectural pattern that we have previously described for search engines in [4]. The processing steps might be implemented in different components with dedicated technology provided by distinguished research groups. The message-driven implementation of Qanary [4]

---

[10] https://creativecommons.org/publicdomain/zero/1.0/.

laid foundations for the QA ecosystem. The advantage of such an ecosystem is that it combines different approaches, functionality, and advances in the QA community under a single umbrella.

4 The Qanary Ecosystem

The Qanary ecosystem consists of a variety of components and services that can be used during a QA process. We describe in the following what components and services are available.

The Qanary ecosystem includes various components covering a broad field tasks within QA systems. This includes different components performing NER like FOX [20] and Stanford NER [10] and components computing NED such as DBpedia Spotlight and AGDISTIS [22]. Also industrial services such as the Alchemy API are part of the ecosystem.

Furthermore, Qanary includes a language detection module [15] to identify the language of a textual question. A baseline automatic speech recognition component is also included in the reference implementation. It translates audio input into natural language texts and is based on Kaldi[11]. Additionally it should be noted that a monolithic QA system component [7] was developed in the course of the WDAqua project[12] and is integrated in Qanary. Additional external QA components are included in the ecosystem. In particular, Qanary includes two components from the OKBQA challenge[13] namely the template generation and disambiguation component. All components are implemented following the REST principles. Hence, these tools/approaches become easy to reuse and can now be invoked via transparent interfaces. To make it easy to integrate a new component we have created a Maven archetype that generates a template for a new Qanary component[14]. The main services are encapsulated in the *Qanary Pipeline*. It provides, for example, a service registry. After being started, each component registers itself to the *Qanary Pipeline* central component following the local configuration[15] of the component. Moreover, the *Qanary Pipeline* provides several web interfaces for machine and also human interaction (e.g., for assigning a URI to a textual question, retrieving information about a previous QA process, etc.). Particularly, as each component automatically registers itself to the *Qanary Pipeline*, a new QA system can be created and executed just by on-demand configuration (a concrete one is shown in Fig. 1). Hence, the reference implementation already provides the features required for QA systems using components distributed over the Web.

---

[11] http://kaldi-asr.org.
[12] http://www.wdaqua.eu.
[13] http://www.okbqa.org/.
[14] github.com/WDAqua/Qanary/wiki/How-do-I-integrate-a-new-component-in-Qanary%3F.
[15] The configuration property `spring.boot.admin.url` defines the endpoint of the central component (and can be injected dynamically).

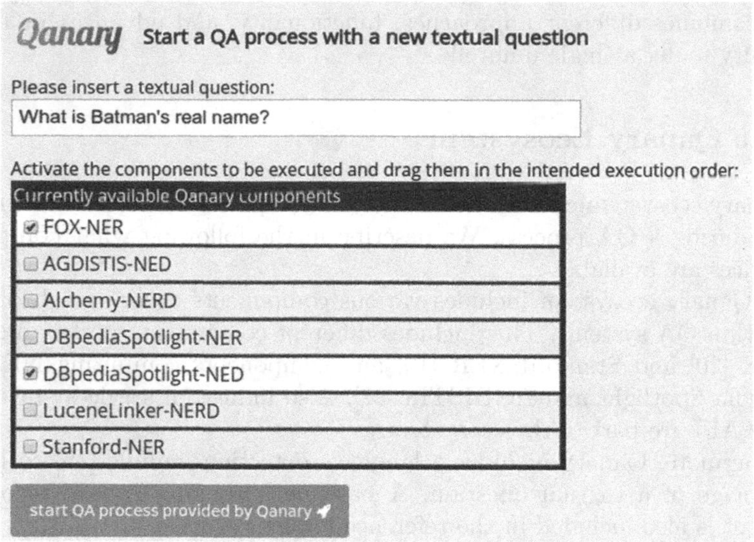

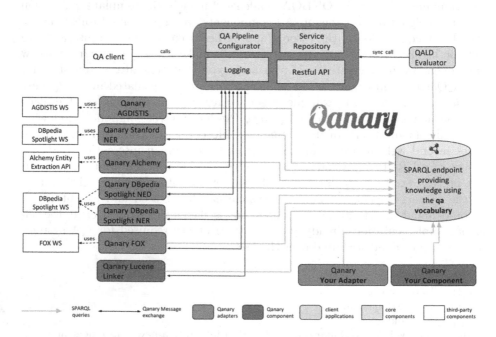

Qanary Start a QA process with a new textual question

Please insert a textual question:

What is Batman's real name?

Activate the components to be executed and drag them in the intended execution order:

Currently available Qanary components

☑ FOX-NER

☐ AGDISTIS-NED

☐ Alchemy-NERD

☐ DBpediaSpotlight-NER

☑ DBpediaSpotlight-NED

☐ LuceneLinker-NERD

☐ Stanford-NER

start QA process provided by Qanary ✈

Fig. 1. Snapshot of the Web interface for defining a textual question and a sequence of components to process it (here only NED/NER components where registered).

Fig. 2. The Qanary reference architecture implementation highlighting the NER/NED components.

An additional interface allows for benchmarking a QA system created on demand using Gerbil for QA[16], thus allowing third-party evaluation and citeable URIs. Figure 2 illustrates the complete reference architecture of Qanary and a few of its components. Additional services include a user interface called Trill [6] for a fully working QA system. A running demo can be found at http://www.wdaqua.eu/qa. The code is maintained in the repository at https://github.com/WDAqua/Qanary under the MIT License[17].

5   Gaining New Insights into the QA Process: The EL Task

To show how Qanary can be used to gain new insights into QA processes we focus here on the EL task. We present the qa vocabulary used to represent the information produced by NER and NED tools. Moreover, we describe the components of the Qanary ecosystem that are integrated using the Qanary methodology and that can be used for the EL task. We describe how we constructed a benchmark for EL out of QALD. The analysis of the benchmark will show: what are the best tools to tackle QALD, where are current research gaps, and for which questions do single tools fail and why.

Finally, we present a new dataset that can be used as a gold standard for a sub-task of the QA process.

The following workflow is not restricted to the EL task but can be applied to any other sub-task of the QA process to gain new insights into QA processes.

5.1   The Qanary Vocabulary for the EL Task

The qa vocabulary is designed to be extensible so as not to constrain the creativity of the QA community developers. All information that can possibly be generated and that might need to be shared across QA components can be expressed using new annotations. This principle follows the understanding that standards that allow communication between QA components must be defined by the community. Taking into consideration the state-of-the-art (e.g., [2, 10, 22]), the qa vocabulary was extended with standard concepts for NER and NED representations. This in particular uniforms the representation of the input and output of every integrated component, making it easy to compare and analyze the integrated tools. Note that this does not only hold for tools that can be used for EL but for every tool integrated into the Qanary ecosystem.

To describe an entity spotted within a question we introduced a dedicated annotation:

```
qa:AnnotationOfSpotInstance a owl:Class;
    rdfs:subClassOf qa:AnnotationOfQuestion.
```

---

[16] http://gerbil-qa.aksw.org.
[17] https://opensource.org/licenses/MIT.

If in the question "When was Barack Obama born?" a spotter detects "Barack Obama" as an NE, this fact can be expressed by the following annotation, where oa:SpecificResource and oa:hasSelector are concepts of the WADM to select a part of a text.

```
<anno1>   a   qa:AnnotationOfSpotInstance .
<anno1>   oa:hasTarget [
   a oa:SpecificResource ;
      oa:hasSource      <URIQuestion>;
      oa:hasSelector    [
        a oa:TextPositionSelector;
        oa:start"9"^^xsd:nonNegativeInteger;
        oa:end   "21"^^xsd:nonNegativeInteger
        ]
] .
```

For named entities, we define the new concept qa:NamedEntity and a corresponding annotation subclass (i.e., annotations of questions whose body is an instance of qa:NamedEntity):

```
qa:NamedEntity            a owl:Class ;
qa:AnnotationOfInstance a owl:Class ;
   owl:equivalentClass [
      a                      owl:Restriction ;
      owl:onProperty         oa:hasBody ;
      owl:someValuesFrom qa:NamedEntity
      ] ;
   rdfs:subClassOf   qa:AnnotationOfQuestion .
```

If an NED tool detects in the question "When was Barack Obama born?" that the text "Barack Obama" refers to dbr:Barack_Obama[18], then this can be expressed (using oa:hasBody) as:

```
<anno1>   a   qa:AnnotationOfInstance ;
  oa:hasTarget [
   a                   oa:SpecificResource ;
   oa:hasSource      <URIQuestion> ;
   oa:hasSelector  [
      a               oa:TextPositionSelector ;
      oa:start"9"^^xsd:nonNegativeInteger;
      oa:end   "21"^^xsd:nonNegativeInteger
   ]
   ] ;
  oa:hasBody   dbr:Barack_Obama .
```

Note that using annotations provides many benefits thanks to the inclusion of additional metadata such as the creator of the information, the time and a trust score. However, this information is omitted here for readability.

5.2 Reusable NER and NED Components

The following components integrated into the Qanary ecosystem solve the task of NER, NED or the combined task of EL.

---

[18] PREFIX dbr: <http://dbpedia.org/resource/>.

- **Stanford NER (NER)** is a standard NLP tool that can be used to spot entities for any ontology, but only for languages where a model is available (currently English, German, Spanish and Chinese) [10].
- **FOX (NER)** integrates four different NER tools (including the Stanford NER tool) using ensemble learning [20].
- **DBpedia Spotlight spotter (NER)** uses lexicalizations, i.e., ways to express named entities, that are available directly in DBpedia [14].
- **DBpedia Spotlight disambiguator (NED)**, the NED part of DBpedia Spotlight, disambiguates entities by using statistics extracted from Wikipedia texts [14].
- **AGDISTIS (NED)** is a NED tool that uses the graph structure of an ontology to disambiguate entities [22].
- **ALCHEMY (NER + NED)**: Alchemy API[19] is a commercial service (owned by IBM) exposing several text analysis tools as web services.
- **Lucene Linker (NER + NED)** is a component that we implemented following the idea of the SINA QA system [17], which employs information retrieval methods.

Note that thanks to the integration as Qanary components all these tools can be interwoven, i.e., each NER tool can be combined with each NED tool just by configuration.

5.3 A QALD-based Benchmark for EL in QA

To compare the different entity linking approaches, we created a benchmark based on the QALD (Question Answering over Linked Data) benchmark used for evaluating complete QA systems. The QALD-6 training set[20], which is the recent successor of QALD-5 [21], contains 350 questions, including the questions from previous QALD challenges. For each question, it contains a SPARQL query that retrieves the corresponding answers. For example, the following SPARQL query corresponds to the question "Which soccer players were born on Malta?".

```
PREFIX dbr: <http://dbpedia.org/resource/>
PREFIX dbo: <http://dbpedia.org/ontology/>
SELECT DISTINCT ?uri WHERE {
    ?uri    a               dbo:SoccerPlayer.
    ?uri    dbo:birthPlace  dbr:Malta.
}
```

EL tools should provide functionality to interlink the named entities present in the question with DBpedia (or other data), i.e., they should be able to identify "Malta" and link to it the resource http://dbpedia.org/resource/Malta. Our benchmark compares the URIs generated by an EL tool with the resource URIs in the SPARQL query (i.e., those in the http://dbpedia.org/resource/

---

[19] http://alchemyapi.com.

[20] Training Questions of Task 1: http://qald.sebastianwalter.org/index.php?x=challenge&q=6.

$$\text{Precision}(q) = \frac{\text{\# correct URIs retrieved by the EL configuration for } q}{\text{\# URIs retrieved by the EL configuration identifying NE in } q}$$

$$\text{Recall}(q) = \frac{\text{\# correct URIs retrieved by the EL configuration for } q}{\text{\# gold standard answers for } q}$$

$$F_1\text{-measure}(q) = 2 \times \frac{\text{Precision}(q) \times \text{Recall}(q)}{\text{Precision}(q) + \text{Recall}(q)}$$

Fig. 3. Metrics used in the EL benchmark

namespace)[21], which are obviously required for answering the question. Hence the gold standard for each question is given by all resource URIs in the SPARQL query.[22]

The metrics for a question q are calculated as defined in the QALD benchmark and are reported in Fig. 3[23]. The metrics over all questions are defined as the average of the metrics over the single questions. The corresponding benchmark component is available at https://github.com/WDAqua/Qanary.

Note that this procedure can be generalized and applied to many subprocesses of a QA pipeline. For example, one might establish a benchmark to recognize relations or classes, a benchmark to identify the type of the SPARQL query required to implement a question (i.e., a *SELECT* or an *ASK* query), a benchmark for identifying the answer type (i.e., list, single resource, ...) and so on.

We used our benchmarking resource described above to evaluate EL tools. We have identified different strategies to annotate entities in questions. These include using the spotters Stanford NER, FOX, DBpedia Spotlight Spotter, the NED tools AGDISTIS, and the DBpedia Spotlight disambiguator, as well as the (monolithic w.r.t. NER and NED) EL tools Alchemy and Lucene Linker. Each of them is implemented as an independent Qanary component, as presented in Sect. 5.2. According to the Qanary methodology the computed knowledge about a given question is represented in terms of the qa vocabulary and can be interpreted by the benchmark component.

For the benchmark all three NER components are combined with each of the two NED components. All questions of QALD-6 are processed by each of the six

---

[21] Note that resources of the type http://dbpedia.org/ontology/* would match a DBpedia property or class and therefore are not considered here.

[22] This definition of the gold standard ignores the *order* of the URIs. In practice this definition rarely causes problems, but in theory one could construct counter-examples that pinpoint the limitations. Imagine the question "What German actors were not born in Germany?", and imagine that the word "German" got linked to the entity dbr:Germany and "Germany" to dbr:Germans - clearly a wrong linking, but "correct" w.r.t. our gold standard definitions. However, in QALD (release 6) there are no questions in which such a mix-up is likely to happen.

[23] In the corner cases where the number of system answers or the number of gold standard answers is zero we follow the same rules that are used in the QALD evaluation; see https://github.com/ag-sc/QALD/blob/master/6/scripts/evaluation.rb.

resulting configurations, and by the two monolithic tools. The benchmark was executed exclusively using the service interface of the *Qanary Pipeline*.

Table 2 shows the benchmark results. The "Fully detected" column indicates the number of questions q where some resources were expected and the EL configuration achieved $\text{Recall}(q) = 1$. The column "Correctly Annotated" indicates for how many questions we obtained $\text{Precision}(q) = \text{Recall}(q) = 1$. Finally, the table shows for each configuration the precision and recall metrics over all questions.

Table 2. Benchmark of the QALD-6 data using the Qanary reference implementation.

Pipeline configuration	Fully detected	Correctly annotated	Precision	Recall	F$_1$-measure
StanfordNER + AGDISTIS	200	195	0.76	0.59	0.59
StanfordNER + Spotlight disamb	209	189	0.77	0.62	0.61
FOX + AGDISTIS	189	186	0.83	0.56	0.56
FOX + Spotlight disambiguator	199	192	0.86	0.59	0.58
Spotlight Spotter + AGDISTIS	209	204	0.75	0.62	0.62
Spotlight spotter + disambiguator	242	**213**	0.76	0.71	**0.68**
Lucene linker	**272**	0	0.01	**0.78**	0.03
Alchemy	143	139	**0.91**	0.42	0.42

5.4 Discussion

We presented the Qanary methodology, which allows to interweave the analysed tools. Thanks to the qa vocabulary we can collect (from the SPARQL endpoint) the results produced by every configuration. A detailed overview showing for each question if the pipeline configurations lead to a recall resp. F-measure of 1 can be found at:

- https://raw.githubusercontent.com/WDAqua/Qanary/master/ICWE-results/
  - Recall_1.csv and
  - F-measure_1.csv.

We analysed both this data and the results presented in Table 2 to draw some conclusions on the performance of the used tools with respect to QALD.

For some QALD-6 questions none of the pipeline configurations is able to find the required resources, for example:

– *Q1:* "Give me all cosmonauts." with the following resources requested in the SPARQL query: dbr:Russia, dbr:Soviet_Union. For this question one should be able to understand that cosmonauts are astronauts born either in Russia or in the Soviet Union. Detecting such resources would require a deep understanding of the question. Q201 is similar: "Give me all taikonauts.".

– *Q13:* "Are tree frogs a type of amphibian?"; requested resources: dbr:Hylidae, dbr:Amphibian.
  The problem here is that the scientific name of "tree frogs" is Hylidae and there is no such information in the ontology except in the free text of the Wikipedia abstract.

– *Q311:* "Who killed John Lennon?"; requested resource: dbr:Death_of_John_Lennon.
  The problem is that one would probably assume that the information is encoded in the ontology as a triple like "John Lennon", "killed by", "Mark David Chapman" but this is not the case. Even if in the question the actual NE is "John Lennon", DBpedia happens to encode the requested information in the resource "Death of John Lennon". A similar case is Q316 ("Which types of grapes grow in Oregon?"), where the resource dbr:Oregon_wine is searched.

Spotter comparison. An unexpected result is that FOX as a spotter has a lower recall than the Stanford NER tool, even though FOX also includes the results of Stanford NER. This can be seen from comparing the recall of the configurations that combine these tools with AGDISTIS or the DBpedia Spotlight disambiguator. The reason is that, for example, in Q101 ("Which German cities have more than 250,000 inhabitants?") the word "German" is tagged by the Stanford NER tool as "MISC" (miscellaneous). However, FOX only supports the tags "PERS" (person), "ORG" (organisation), and "LOC" (location). This explains why FOX has a lower recall but a higher precision than Stanford NER.

The spotters based on NLP (e.g., Stanford NER and FOX) perform worse than the DBpedia Spotlight Spotter, which is mainly based on vocabulary matching. Syntactic features do not suffice to identify "Prodigy" in Q114 ("Give me all members of Prodigy.") or "proinsulin" in Q12 ("Is proinsulin a protein?"). Moreover, there are cases like Q109 ("Give me a list of all bandleaders that play trumpet."), where bandleaders is not an NE in the NLP sense but is modeled as a resource in the DBpedia *ontology*. Similarly, in Q133 ("Give me all Australian nonprofit organizations."), the resource dbr:Australia is expected for the adjective "Australian".

NED comparison. The results show that the DBpedia Spotlight disambiguator performs better than AGDISTIS w.r.t. QA. AGDISTIS works on co-occurrences of NE. These occur often in longer texts but are rare in questions. If only one NE is spotted, AGDISTIS can only decide based on the popularity of the resources but not on the context as DBpedia Spotlight does.

EL comparison. The best spotter, the DBpedia Spotlight spotter, and the best NED, the DBpedia Spotlight disambiguator, also perform best in the EL task.

Only the Lucene Linker has a higher recall but must be followed by a disambiguation tool in the next step to increase precision. The Alchemy API shows the lowest recall.

Our evaluation does not permit the conclusion that the combination of DBpedia Spotlight spotter and disambiguator should be recommended in general. The best choice may depend on the questions and on the particular form of dataset. The DBpedia Spotlight disambiguator, for example, is tightly connected to Wikipedia; even its algorithm cannot be ported to other ontologies. Alchemy, despite showing a very low F_1-score and recall, could be a useful resource for QA over other datasets, such as Freebase or Yago. This is one of the many reasons that makes Qanary in general a valuable resource. For a new QA scenario, Qanary empowers developers to quickly combine existing tools and more easily determine the best configuration. Moreover, a detailed analysis of the configurations can help to detect the main problems of the different strategies to further improve the complete QA process or just individual components. Hence, using Qanary provides insights on the quality of each component w.r.t. the current use case, leading to an optimization of the system based on the given data.

A combination of tools solving the same task in an ensemble learning approach is now possible and is recommended as the benchmark results already indicate. Note that such an analysis is not possible using existing benchmarking tools such as Gerbil or Gerbil for QA since they only provide a final overall score. On the other hand such an analysis is needed to detect existing research gaps and push the advancement of QA further. Hence, following the Qanary methodology the research community is enabled to develop new QA processes and components in a joint engineering effort and to validate the given quality metrics within the specific QA scenario. This again proves the potential impact of Qanary within the engineering process of QA systems.

```
PREFIX qa:   <http://www.wdaqua.eu/qa#>
PREFIX oa:   <http://www.w3.org/ns/openannotation/core/>
PREFIX xsd:  <http://www.w3.org/2001/XMLSchema#>
PREFIX dbr:  <http://dbpedia.org/resource/>
PREFIX git:  <https://github.com/dbpedia-spotlight/>
<anno1> a                qa:AnnotationOfInstance ;
        oa:annotatedAt  "2016-04-30T15:00:43.687+02:00"^^xsd:dateTime ;
        oa:hasTarget [
                a               oa:SpecificResource ;
                oa:Selector  [
                        a            oa:TextPositionSelector ;
                        oa:end   4 ;
                        oa:start 21
                     ] ;
                oa:hasSource   <URIQuestion>
                ] .
        oa:hasBody      dbr:Margaret_Thatcher ;
        oa:annotatedBy git:dbpedia-spotlight .
```

Fig. 4. Example data of Q178: "Was Margaret Thatcher a chemist?"

5.5   Dataset of Annotated Questions for Processing in QA systems

We provide a new dataset with questions of the QALD-6 benchmark, which are completely annotated with disambiguated named entities (DBpedia resource URIs) computed by applying our benchmarking to the EL configurations described in Sect. 5.3. This dataset contains 267 questions (out of 350 questions in the QALD-6 training set) because the components could not annotate the rest. A Turtle file, representing the results in terms of the qa vocabulary, is published at https://github.com/WDAqua/Qanary under the CC0 license. A typical fragment of the provided data is provided in Fig. 4.

We imagine the following usage scenarios for this dataset. It can be used as input for steps in a QA process *following* the EL task that require annotated named entities, such as relation detection or query computation. Consequently, QA components that depend on the output of the EL task can now be evaluated without depending on concrete EL components (and without the results being influenced by possible flaws). Hence, in conjunction with the SPARQL queries already defined in QALD-6, we established a new gold standard for evaluating parts of a QA process. We also provide the component for computing this dataset (cf., Sect. 5.3); it can be extended if improved EL configurations are available or when a new version of the QALD benchmark is released.

6   Conclusion and Future Work

We have presented the status of the Qanary ecosystem, which includes a variety of components and services that can be used by the research community. These include typical components for sub-tasks of a QA pipeline as well as a number of related services.

Since all messages exchanged between components are expressed using the qa vocabulary, all information generated during a QA process can be easily kept. Thanks to this uniform message format it is now possible to easily compare existing tools and integrate new ones. Moreover, the Qanary methodology allows to integrate independent, distributed components, implemented in different programming languages in a loosely-coupled fashion. This allows the creation of comprehensive QA systems in a community effort.

Driven by the demand for better QA technology, we propose a general workflow to develop future QA systems. It mainly breaks down into two parts: (1) the identification and integration of existing state-of-the-art approaches to solve a particular sub-task in the QA pipeline, and (2) the derivation of a benchmark from benchmarks for QA such as QALD. Additionally a new gold standard for the sub-task can be provided. In contrast to other approaches the qa vocabulary allows to analyse a QA process. Hence, full traceability of the information used in the QA process is ensured, enabling, for example, the optimization of the assigned components. Additionally, the Qanary methodology allows to create such processes in a flexible way. This allows researchers to focus on particular tasks taking advantage of the results of the research community and contributing to it directly in a reusable way.

We have demonstrated this workflow in the case of EL. This way we realized a set of reusable components as well as the first benchmark for EL in the context of QA. All together we have shown how Qanary can be used to gain deep insights in QA processes. While having such insights the engineering process can be steered efficiently towards the improvement of the QA components. Hence, the presented engineering approach is particularly well suited for experimental and innovation-driven approaches (e.g., used by research communities).

The Qanary ecosystem is maintained and used within the WDAqua ITN project[24] (2015–2018 [12]), where Qanary is the reference architecture for new components. All artifacts are published under permissive open licenses: MIT for the software, CC0 for the datasets and the vocabulary.

One of our future task is to populate the Qanary ecosystem with any component significant to the QA community. According to the literature, the tasks of relation and answer type detection are of particular relevance, but not yet sufficiently covered by existing components. Additionally, as Qanary provides easy access to different implementations having the same purpose, ensemble learning components for all steps within a QA process are becoming possible and will increase the flexibility as well as boost overall QA quality. Hence, our overall goal is to provide a fully-featured ecosystem for creating QA components and concurrently supporting the measuring and improvement of particular QA systems w.r.t. the considered use cases. This aim provides several research challenges, e.g., the (semi-)automatic creation of self-optimizing QA systems.

Acknowledgments. Parts of this work received funding from the European Union's Horizon 2020 research and innovation programme under the Marie Skłodowska-Curie grant agreement No. 642795, project: Answering Questions using Web Data (WDAqua). We would like to thank Elena Demidova for proof-reading.

References

1. Atdag, S., Labatut, V.: A comparison of named entity recognition tools applied to biographical texts. In: 2nd International Conference on Systems and Computer Science (ICSCS) (2013)
2. Auer, S., Bizer, C., Kobilarov, G., Lehmann, J., Cyganiak, R., Ives, Z.: DBpedia: a nucleus for a web of open data. In: Aberer, K., Choi, K.-S., Noy, N., Allemang, D., Lee, K.-I., Nixon, L., Golbeck, J., Mika, P., Maynard, D., Mizoguchi, R., Schreiber, G., Cudré-Mauroux, P. (eds.) ASWC/ISWC -2007. LNCS, vol. 4825, pp. 722–735. Springer, Heidelberg (2007). doi:10.1007/978-3-540-76298-0_52
3. Both, A., Diefenbach, D., Singh, K., Shekarpour, S., Cherix, D., Lange, C.: Qanary – a methodology for vocabulary-driven open question answering systems. In: The Semantic Web. Latest Advances and New Domains: 13th International Conference, ESWC 2016, Heraklion, Crete, Greece, 29 May–2 June 2016, Proceedings (2016)
4. Both, A., Ngonga Ngomo, A.-C., Usbeck, R., Lukovnikov, D., Lemke, C., Speicher, M.: A service-oriented search framework for full text, geospatial and semantic search. In: Proceedings of the 10th International Conference on Semantic Systems, SEM 2014, pp. 65–72. ACM (2014)

---

[24] http://wdaqua.eu.

5. Cabrio, E., Cojan, J., Aprosio, A.P., Magnini, B., Lavelli, A., Gandon, F.: QAKiS: an open domain QA system based on relational patterns. In: Glimm, B., Huynh, D. (eds.) Proceedings of the ISWC 2012 Posters & Demonstrations Track, vol. 914, CEUR Workshop Proceedings (2012). CEUR-WS.org

6. Diefenbach, D., Amjad, S., Both, A., Singh, K., Maret, P.: Trill: a reusable front-end for QA systems. In: ESWC P&D (2017)

7. Diefenbach, D., Singh, K., Maret, P.: Wdaqua-core0: a question answering component for the research community. In: ESWC, 7th Open Challenge on Question Answering over Linked Data (QALD-7) (2017)

8. Dima, C.: Answering natural language questions with intui3. In: CLEF (Working Notes) (2014)

9. Ferrández, Ó., Spurk, C., Kouylekov, M., Dornescu, I., Ferrández, S., Negri, M., Izquierdo, R., Tomás, D., Orasan, C., Neumann, G., Magnini, B., González, J.L.V.: The QALL-ME framework: a specifiable-domain multilingual question answering architecture. Web Semant. Sci. Serv. Agents World Wide Web 9(2), 137–145 (2011)

10. Finkel, J.R., Grenager, T., Manning, C.: Incorporating non-local information into information extraction systems by Gibbs sampling. In: Proceedings of the 43rd Annual Meeting on Association for Computational Linguistics, ACL 2005 (2005)

11. Freitas, A., Oliveira, J., Curry, E., O'Riain, S., da Silva, J.: Treo: combining entity-search, spreading activation and semantic relatedness for querying linked data. In: 1st Workshop on Question Answering over Linked Data (QALD-2011) (2011)

12. Lytra, I., Vidal, M.-E., Lange, C., Auer, S., Demidova, E.: WDAqua - answering questions using web data. In: Mannens, E., Dragoni, M., Nixon, L., Corcho, O. (eds.) EU Project Networking (2016)

13. Marx, E., Usbeck, R., Ngonga Ngomo, A., Höffner, K., Lehmann, J., Auer, S.: Towards an open question answering architecture. In: 10th International Conference on Semantic Systems (2014)

14. Mendes, P.N., Jakob, M., García-Silva, A., Bizer, C.: DBpedia spotlight: shedding light on the web of documents. In: Proceedings of the 7th International Conference on Semantic Systems, I-Semantics 2011 (2011)

15. Nakatani, S.: Language detection library for Java (2010). https://github.com/shuyo/language-detection

16. Rizzo, G., Troncy, R.: NERD: a framework for unifying named entity recognition and disambiguation extraction tools. In: 13th Conference of the European Chapter of the Association for Computational Linguistics (2012)

17. Shekarpour, S., Marx, E., Ngonga Ngomo, A.-C., Auer, S.: SINA: semantic interpretation of user queries for question answering on interlinked data. Web Semant. Sci. Serv. Agents WWW 30, 39–51 (2015)

18. Singh, K., Both, A., Diefenbach, D., Shekarpour, S.: Towards a message-driven vocabulary for promoting the interoperability of question answering systems. In: 2016 IEEE Tenth International Conference on Semantic Computing (ICSC) (2016)

19. Singh, K., Both, A., Diefenbach, D., Shekarpour, S., Cherix, D., Lange, C.: Qanary-the fast track to create a question answering system with linked data technology. In: The Semantic Web: ESWC 2016 Satellite Events, Heraklion, Crete, Greece, 29 May–2 June 2016, Revised Selected Papers (2016)

20. Speck, R., Ngonga Ngomo, A.-C.: Ensemble learning for named entity recognition. In: Mika, P., Tudorache, T., Bernstein, A., Welty, C., Knoblock, C., Vrandečić, D., Groth, P., Noy, N., Janowicz, K., Goble, C. (eds.) ISWC 2014. LNCS, vol. 8796, pp. 519–534. Springer, Cham (2014). doi:10.1007/978-3-319-11964-9_33

21. Unger, C., Forascu, C., Lopez, V., Ngonga Ngomo, A., Cabrio, E., Cimiano, P., Walter, S.: Question answering over linked data (QALD-5). In: CLEF (Working Notes) (2015)

22. Usbeck, R., Ngonga Ngomo, A.-C., Röder, M., Gerber, D., Coelho, S.A., Auer, S., Both, A.: AGDISTIS - graph-based disambiguation of named entities using linked data. In: Mika, P., Tudorache, T., Bernstein, A., Welty, C., Knoblock, C., Vrandečić, D., Groth, P., Noy, N., Janowicz, K., Goble, C. (eds.) ISWC 2014. LNCS, vol. 8796, pp. 457–471. Springer, Cham (2014). doi:10.1007/978-3-319-11964-9_29

23. Usbeck, R., Röder, M., Ngonga Ngomo, A., Baron, C., Both, A., Brümmer, M., Ceccarelli, D., Cornolti, M., Cherix, D., Eickmann, B., Ferragina, P., Lemke, C., Moro, A., Navigli, R., Piccinno, F., Rizzo, G., Sack, H., Speck, R., Troncy, R., Waitelonis, J., Wesemann, L.: GERBIL: general entity annotator benchmarking framework. In: 24th International Conference on World Wide Web (2015)

Improving Reliability of Crowdsourced Results by Detecting Crowd Workers with Multiple Identities

Ujwal Gadiraju[1(✉)] and Ricardo Kawase[2]

[1] L3S Research Center, Leibniz Universität Hannover,
Appelstr. 9a, Hannover, Germany
gadiraju@L3S.de
[2] mobile.de GmbH/eBay Inc., Berlin, Germany
rkawase@team.mobile.de

Abstract. Quality control in crowdsourcing marketplaces plays a vital role in ensuring useful outcomes. In this paper, we focus on tackling the issue of crowd workers participating in tasks multiple times using different `worker-ids` to maximize their earnings. Workers attempting to complete the same task repeatedly may not be harmful in cases where the aim of a requester is to gather data or annotations, wherein more contributions from a single worker are fruitful. However, in several cases where the outcomes are subjective, requesters prefer the participation of distinct crowd workers. We show that traditional means to identify unique crowd workers such as `worker-ids` and `ip-addresses` are not sufficient. To overcome this problem, we propose the use of *browser fingerprinting* in order to ascertain the unique identities of crowd workers in paid crowdsourcing microtasks. By using browser fingerprinting across 8 different crowdsourced tasks with varying task difficulty, we found that 6.18% of crowd workers participate in the same task more than once, using different `worker-ids` to avoid detection. Moreover, nearly 95% of such workers in our experiments pass gold-standard questions and are deemed to be *trustworthy*, significantly biasing the results thus produced.

Keywords: Crowdsourcing · Microtasks · Multiple identities · Quality control · Reliability

1 Introduction

With the ubiquity of the Internet these days, and the existing need for human intelligence, *crowdsourcing* has empowered millions of people around the globe by providing crowd workers an alternative source to earn their livelihood. A considerable number of real-world applications have showcased the value of this paradigm, ranging from mapping satellite imagery[1] to disaster relief and management

---

[1] http://www.digitalglobeblog.com/2014/03/10/missingmalayairjet/.

© Springer International Publishing AG 2017
J. Cabot et al. (Eds.): ICWE 2017, LNCS 10360, pp. 190–205, 2017.
DOI: 10.1007/978-3-319-60131-1_11

initiatives[2]. While inumerable examples of profitable crowdsourcing initiatives exist at present, ensuring high quality of results and inhibiting malicious activity are pivotal challenges.

In this work, we aim to tackle a specific kind of potentially malicious activity in paid crowdsourcing marketplaces. Several crowdsourced tasks often require participation from unique crowd workers. This is clearly apparent in surveys and other tasks that require subjective judgments from individuals. For instance, a requester[3] would not want multiple judgments from the same crowd worker in a task that gathers an opinion census of a newly launched product. However, through our experiments presented in this paper, we note that a significant number of workers tend to complete the same task multiple times (by using distinct worker-ids) in order to maximize their monetary gains. We define these workers as 'repeaters'. We reason that crowd workers who exhibit such behavior are primarily driven by monetary incentives. Recent work has shown that over the last 3 years surveys are one of the most prominent types of crowdsourced tasks, gaining wide popularity on Amazon's Mechanical Turk[4] (AMT) [1]. Hence, this is an important and timely problem to tackle. By posing as different workers (due to different worker-ids), the same individual can complete a given task any number of times within the task constraints. Workers benefit in the following two ways by doing so, with varying implications.

- *Completion Time.* Familiarity with the task due to repeated participation can result in workers requring much lesser time to complete a given task.
- *Monetary Rewards.* Workers consequently multiply the rewards attained on task completion.

Requesters on the other hand suffer from such repeated participation of workers in a given task in the following ways.

- In the best case of repeated participation, workers can complete tasks in a quick manner resulting in a reduced overall task completion time. If the repeated participation by workers is only motivated by an objective to maximize rewards by completing tasks with genuine effort, this can improve the results [5]. However in the contrasting case, workers with alternative intentions can sabotage a task repeatedly.
- Repeated participation by crowd workers in tasks where distinct workers are expected, implies that requesters bear costs without receiving qualitative returns on their investment. On detection of such activity in a post-processing manner, requesters may need to incur additional cost overheads to gather new judgments.

Existing methods on crowdsourcing platforms rely on user ip-addresses to prevent workers from participating in tasks multiple times if so specified by a

---

[2] http://www.mission4636.org/.
[3] A *requester* is one who deploys a task on a crowdsourcing platform in order to gather responses from the crowd.
[4] http://www.mturk.com/mturk/.

requester. However, crowd workers can change their IP addresses at will, thereby limiting the effectiveness of such methods. We present a novel method for quality assurance in paid microtask crowdsourcing. We propose the adoption of *browser fingerprinting* in order to identify crowd workers that participate in microtasks multiple times with distinct `worker-ids`. The concept of browser fingerprinting has evolved from device fingerprinting over the last decade, emulating the forensic essence of human fingerprints; the ability to uniquely identify different individuals.

The main contributions of this work are two-fold. First, by using browser fingerprints we expose the existence of the crowd workers who repeatedly complete a given task. We show that a substantial share of microtask participants are repeaters. Secondly, we show how current quality control mechanisms that rely on `worker-ids` and `ip-addresses` to restrict repeated participation are insufficient to determine the unique identiy of crowd workers.

2 Related Literature

We discuss related works in two different realms; (i) prior work related to browser fingerprinting and identifying multiple online identities, and (ii) those relevant to ensuring quality in crowdsourcing tasks.

2.1 Browser Fingerprinting and Identifying Multiple Online Identities

Over the last decade there have been advances in the reliable detection of unqiue web browsers. Eckersley investigated the version and configuration information that web browsers transmit upon request, in order to study the extent to which browsers are subject to device fingerprinting [2]. The variables considered in the hashing of browser fingerprints as prescribed by Eckersley included the following: the user agent string transmitted by HTTP, the HTTP accept headers, whether or not cookies are enabled, screen resolution, timezone, browser plugins, plugin versions and MIME types, system fonts and a partial supercookie test. The author showed that the distribution of browser fingerprints of users in their collection contained at least 18.1 bits of entropy, effectively meaning that in their experimental collection, only one in 286,777 other browsers shared its fingerprint.

Mowery and Shacham proposed the rendering of text and WebGL scenes to a `<canvas>` element, and thereby examining the pixels produced in order to tie a browser more closely to a user's operating system and hardware [14]. Mulazzani et al. proposed an efficient method to identify browsers by JavaScript engine fingerprinting [15]. In this paper, our proposal to use browser fingerprinting to detect the undesirable repeated participation of workers in crowdsourcing tasks, is inspired by these prior works. The contribution of our work in this context is the evidence we provide through rigorous experimentation, indicating the effectiveness of browser fingerprinting in improving the reliability of crowdsourcing results.

Prior works have also addressed the problem of identifying multiple identities in various online contexts. Gani et al. proposed a framework to detect multiple identities in social networks based on machine learning models and interaction between users [8]. Kafai et al. showed how online gamers use multiple accounts and identities in order to make more money or maximize rewards [10]. More recently, Yamak et al. proposed supervised machine learning algorithms to detect multiple identities of users in collaborative projects online [21]. In contrast to such previous works, in this paper we address the novel context of identifying *repeaters* in crowdsourcing microtasks.

2.2 Quality Assurance in Crowdsourcing

Several prior works have focused on methods from varying perspectives to improve the quality of crowdsourced work. Kittur et al. reflected on the measures required to ensure reliability in crowdsourced user measurements from the task design point of view [12]. Authors have also studied the effect of task pricing on the quality of results produced. Faradani et al. [4] focused on the duality between task completion time and pricing, and model quality as a tradeoff between these aspects. Wang et al. proposed a method to measure worker quality, based on which they determined the fair payment level for a worker [20]. Mason et al. showed that increasing monetary incentives of crowdsourced tasks attracts more workers but does not improve the quality of the results produced [13]. Oleson et al. proposed the usage of *gold-standard* questions to ensure reliability of responses and improve the quality of crowd work [16]. Eickhoff et al. proposed guidelines to inhibit spammers in crowdsourced tasks [3]. However, as shown by Gadiraju et al., the use of gold-standards alone are insufficient to curtail malicious activity in the crowd. The crowd consists of a significant number of *smart deceivers*, who take special precautions to avoid detection [7]. The authors studied the implications of task design as well as worker behavior on the quality of crowdsourced results. Other works have also investigated the motivation behind participation in crowdsourcing microtasks and the impact of motivation on performance of workers [11,18]. In contrast to these prior works, we study the problem of workers repeatedly participating in tasks using distinct worker-ids to maximize their earning and avoid the scrutiny of quality control mechanisms. This problem has not been explicitly addressed and studied in prior works. We propose to detect such crowd workers who repeatedly participate in tasks (thereby called *repeaters*), by using browser fingerprinting of workers.

Rzeszotarski and Kittur proposed *task fingerprinting*; collecting user activity logs through mousetracking in crowdsourcing tasks, in order to infer worker cognition and effectiveness [19]. While task fingerprints are extremely useful to model crowd workers and understand their cognitive processes, they cannot reliably be used to identify unique workers, since multiple workers can depict identical behavior in a given task. In a closely related application of browser fingerprinting, Rainer and Timmerer used browser fingerprinting in order to ensure unique participation in their experiments regarding QoE in multimedia streaming over HTTP [17]. However, the authors employ browser fingerprinting without measuring the actual effectiveness of the method. In this paper, we investigate the applicability of browser fingerprinting as a quality control mechanism in crowdsourcing microtasks.

3   Preliminary Validation Study – Multiple Accounts Usage by Crowd Workers

To first determine the legitimacy of multiple accounts usage by crowd workers on crowdsourcing platforms, we surveyed workers on CrowdFlower[5], a premier crowdsourcing platform.

3.1   Survey Design

CrowdFlower allows task requesters to restrict the participation of workers based on their reputation (in terms of *levels*, where `level-3` workers have the best reputation, followed by `level-2` and `level-1`). Thus, we considered the three different levels of worker participation by deploying identical surveys corresponding to each restriction. We collected responses from 100 crowd workers in each case, and rewarded them with 5 USD cents for responding to 5 questions in the survey. The workers were urged to respond honestly in the instructions, and the objective of the survey (i.e., to understand the usage of multiple accounts by workers) was conveyed accurately. The workers were first asked whether they used multiple accounts with different `worker-ids` to access more work. Then the workers were asked to select the types of tasks in which they typically used multiple accounts (if at all) among the following; *content access, content creation, information finding, interpretation and analysis, surveys, verification and validation* [6]. Workers were also asked the frequency with which they used multiple accounts with different `worker-ids` on a 5 point Likert-scale ranging from *1: Never* to *5: Always*. Finally, we asked workers how many active multiple accounts they used among the following options; $(1, 2, 3, 4, 5, > 5)$.

To verify the reliability of responses from the workers, we embedded a test question within the 5 questions in the survey. Here the workers were asked explicitly to enter the word 'COOL' in a corresponding text box. Those workers who failed to do so were deemed to be unreliable and we do not consider their responses in our analysis. We found 2 workers in each of `level-1`, `level-2`, and 1 worker in `level-3` who failed the test question.

3.2   Results

We found that 12.25% of crowd workers in `level-1`, 7.14% of the workers in `level-2` and 11.11% of workers in `level-3` claimed to use multiple accounts with different `worker-ids` to access more work.

---

[5] http://www.crowdflower.com/.

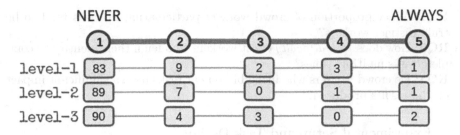

Fig. 1. Frequency (on a Likert-scale) with which workers corresponding to each of the 3 levels on CrowdFlower use different `worker-ids` to participate in tasks.

Figure 1 presents the frequency with which workers in each level use multiple accounts with different `worker-ids`. We note that although the vast majority of workers in each level claim to never use multiple `worker-ids`, few workers do indicate a moderate to high usage of such multiple accounts for participation.

We found that some workers in each of the three levels actively used more than one `worker-id` to participate in tasks, as shown in Table 1a. Those workers who use multiple accounts did not depict a signficant affinity towards a particular type of task for such repeated participation, as shown in Table 1b. This suggests that task type is not necessarily an important feature that facilitates or drives the repeated participation of workers.

Table 1. (a) Number of `worker-ids` actively used by workers, and (b) frequency of the usage of multiple `worker-ids` across different types of tasks corresponding to each of three levels on CrowdFlower.

(a)

# worker-ids	level-1	level-2	level-3
1	76	80	82
2	4	6	6
3	6	5	4
4	5	2	4
5	3	2	0
6 or more	4	3	3

(b)

Task Type	level-1	level-2	level-3
Content Access	4	4	1
Content Creation	3	3	3
Information Finding	8	7	3
Interpretation & Analysis	2	2	3
Surveys	9	8	5
Verification & Validation	5	7	6

Through surveying 300 crowd workers on CrowdFlower, we found evidence of the usage of multiple `worker-ids` by workers in order to access and complete more work, thereby maximizing their monetary rewards. As motivated earlier, this may however be an undesirable aspect depending on the task at hand.

4 Objectives and Methodology

By addressing the following research questions in this work, we propose the application of browser fingerprinting for improving quality assurance in crowdsourcing practice.

- **RQ1:** What proportion of crowd workers participating in tasks tend to be truly distinct workers?
- **RQ2:** How does *task difficulty* effect workers who feign their identity to complete tasks multiple times?
- **RQ3:** Do crowd workers who ineligibly repeat tasks have a significant impact on the results produced?

4.1 Experimental Setup and Task Design

Based on the responses of workers in the preliminary study, we did not find a correlation between the task type and the tendency of workers to use multiple `worker-ids`. With an aim to investigate the research questions stated earlier, we consider the task of logical reasoning. We first gather basic background information from the crowd workers through demographic questions. These are followed by 15 questions in the domain of *logical reasoning*. We used logical reasoning questions from $A + Click^6$. The logical reasoning questions were based on the Common Core Standards[7], which is a set of academic standards in mathematics and English. These learning objectives indicate what a student should know and be able to do at the end of each grade. We chose this setup since the progressing grade-level is a clear indicator of increasing difficulty in the logical reasoning questions. Such a setup would enable us to explore the impact of *task difficulty* on the repeated participation of workers.

In order to assess the impact of *task difficulty* among crowd workers, we deployed 8 microtasks that are designed similarly except for the difficulty level of the logical reasoning questions. Herein, we used graded questions from $A+Click$ to procure logical reasoning questions from the level of Grade 5 to Grade 12. An example question is presented in Fig. 2. We did not consider lower grades than the 5th, since initial experiments revealed that workers tend to perform with a 100% accuracy in those grades. In order to separate *trustworthy* workers (TW)[8] from *untrustworthy* workers (UW)[9], we intersperse attention-check questions (example shown in Fig. 3) as recommended by Gadiraju et al. [7].

Prior research has shown that having verifiable questions such as tags is a recommended way to design tasks and assess crowdsourced results [12]. The last two questions in the task asked crowd workers to provide as many tags as possible for two different pictures. Note that the order in which different questions were asked did not have an impact on the results reported in our work. Thus, we do not focus on this further. We paid the crowd workers according to a fixed hourly wage of 7.5 USD, for completing the tasks successfully. Corresponding to each of the 8 graded tasks that we deployed on CrowdFlower, we gathered 250 responses from independent crowd workers, resulting in a total of 2000 workers overall. We did not restrict the participation of workers based on the CrowdFlower *levels*.

---

[6] http://www.aplusclick.com/.

[7] http://www.corestandards.org/.

[8] Workers who correctly answer all 3 attention check questions embedded in the task.

[9] Workers who incorrectly answer at least 1 of the 3 attention check questions embedded in the task.

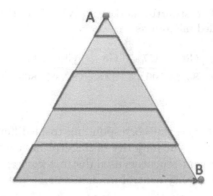

A bug goes from point A to point B along the indicated path. How many times does the insect turn left?

○ 9

○ 4

○ 10

○ 5

Fig. 2. An example logical reasoning question from $A + Click$ that was administered to crowd workers in the task corresponding to Grade 5.

This is an attention check question. Please select the third option.

○ Apple ○ Ball ○ Cat ○ Dog

Fig. 3. Attention-check questions to identify *untrustworthy* workers.

Finally, we extracted the browser fingerprints of crowd workers through a Javascript implementation[10]. As shown by Peter Eckersley [2], browser fingerprinting can anonymously identify a web browser with an accuracy of over 95%.

5   Results and Discussion

5.1   Can We Trust the Trustworthy Workers?

Table 2 presents the number of trustworthy workers (TW) determined by using the attention check questions across the different grades. We can clearly see that the percentage of TW is quite high. However, as we have motivated earlier in this paper, a repeater represents a breach in trust. If a particular task is designed to collect a limited number of responses from an individual worker, then each worker is eligible and expected to provide only those limited number of responses. Not respecting such clearly prescribed limits would amount to a 'violation of trust'.

Figure 4 presents the distribution between the number of distinct `worker-ids` corresponding to the distinct fingerprints in all the tasks. We observe a power-law

---

[10] http://valve.github.io/fingerprintjs/.

Table 2. Percentage of trustworthy workers (TW) out of 250 participating workers, across the different graded microtasks.

Grade	G5	G6	G7	G8	G9	G10	G11	G12
#TW	91.2%	86.4%	90.4%	82.8%	82.8%	86%	85.6%	87.6%

distribution with one fingerprint corresponding to 11 different `worker-ids`, and this gradually decreases to the majority of fingerprints corresponding to distinct `worker-ids`. It is clear that repeaters used distinct `worker-ids` in order to avoid detection by potential quality control mechanisms.

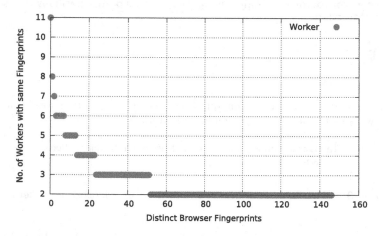

Fig. 4. Distribution of number of distinct `worker-ids` corresponding to each fingerprint across all grades (G5 through G12).

Having said that, Table 3a depicts the percentage of repeaters in each grade that are trustworthy (TW). We note that in each grade there are several repeaters, i.e., workers having different `worker-ids` with the same browser fingerprints. At first, this behavior might seem irrelevant; the average number of repeaters across all tasks is 6.18%. However, in some cases, these repeaters provide a substantial number of contributions (as also observed in Fig. 4). For example, nearly 18% of the total responses (contributions) collected in G6 correspond to answers from repeaters. We also observe that the fraction of repeaters in each grade are similar to our findings in the preliminary study.

Previous work by Gadiraju et al. proposed a classification of the most common type of untrustworthy workers [7]. The authors did not study *repeaters* as a particular case of untrustworthy workers. Under their proposed classification, repeaters would belong to the category of *Ineligible Workers* - due to violating the pre-requisite that a worker is not eligible to perform a task more than a certain number of times. However, repeaters circumvent their ineligibility by using

Table 3. (a) Percentage of trustworthy (TW) repeaters, and share of their contributions in each task, and (b) performance of all workers, repeaters alone and non-repeaters alone in each task.

(a)

Task	% TW Repeaters	% Contributions
G5	4.13	6.82
G6	8.78	17.48
G7	7.38	10.91
G8	5.35	13.43
G9	6.72	13.64
G10	5.86	10.29
G11	5.46	12.87
G12	3.71	7.58
All	6.18	11.63

(b)

Task	All Workers	Repeaters	Non-Repeaters
G5	77.61	75.71	77.88
G6	60.45	57.40	61.68
G7	57.36	60.30	56.61
G8	40.00	43.17	39.16
G9	39.93	35.60	41.28
G10	45.59	46.83	45.28
G11	32.97	33.18	32.91
G12	28.72	24.00	29.50

multiple `worker-ids` and do not demonstrate further untrustworthy characteristics. In fact, in terms of performance, we see little (non-significant) variations (see Table 3b). These results show that, despite the fact that *repeaters* are ineligible workers, they perform tasks with the diligence of an average trustworthy worker. This makes repeaters undetectable unless techniques such as fingerprints are employed.

Although no significant differences were found in terms of performance of the workers, the impact of repeaters becomes clear when one considers the demographics questions. Our demographics questions which included multiple choices for the age group (5 options), education (9 options), ethnicity (7 options) and gender (2 options), depict a significant change in all cases in the presence and absence of repeaters (with $p < 0.05$) in the distributions of at least one of the options provided. Based on these results, and considering that surveys are one of the most common types of crowdsourced tasks [1], we reflect on the susceptibility of surveys to the participation of repeaters, resulting in the generation of skewed and biased outcomes that can go unnoticed.

Finally, we found a moderately strong negative correlation between the difficulty-level of a task (an inherent function of progressive grades from G5 through G12) and the number of trustworthy `worker-ids` corresponding to repeaters (Pearson's $r = -0.3$). This suggests that the more difficult that a task is, the less often trustworthy workers tend to repeat it using different `worker-ids`. Thus, easier tasks that require less effort, or those which provide a better cost-benefit ratio, are more prone to attract repeaters.

5.2   The Case of Account Sharing Among Crowd Workers

Across the different tasks we found 21 cases where multiple (2 or more) browser fingerprints were associated with the same `worker-id` within the same task. Out of these 21 anomalies, in 7 cases the different fingerprints were associated with the same `ip-address`, suggesting that the workers switched or altered some browser configuration. In the other 14 cases, the different `ip-addresses` and the corresponding different fingerprints suggest that multiple workers have access to the same user account, and thereby correspond to the same worker-id. This can be attributed to scenarios where the users have different sessions with the same login through virtual machines (which is less likely), or it is a shared account where different persons work together using different devices. Although this is a breach in the quality control mechanisms for crowdsourced tasks, in this paper we focus on the more frequent case of workers participating repeatedly by using different worker-ids rather than multiple workers using the same worker-id.

5.3   Pruning Workers Using IP Addresses

We investigate the number of *repeaters* that can be detected relying solely on the worker ip-address. We detect one repeater in the task corresponding to G7 and another in G10. Thus, we note that using a worker's ip-address alone as a means to identify unique crowd workers is not sufficient.

5.4   The Privacy Perspective

Although the experiments in this work have been carried out after establishing user-consent, covertly tracking users as a means of their browser fingerprints can be considered to be an unsolicited intrusion of their privacy.

Having said that, we argue in favor of using *browser fingerprinting* to detect *repeaters* who attempt to maximize their monetary benefits by completing tasks mutliple times. Through our 8 crowdsourced tasks involving 2000 workers, we observed a significant participation of repeaters (6.18%). These repeaters account for over 13% of the total contributions by virtue of their repeated participation. Repeaters skew the purpose of requesters, especially in subjective types of tasks such as surveys. Due to the fact that these fingerprints are merely required to uniquely identify workers within a task, the user data used to generate browser fingerprints can be consequently discarded on task completion. Moreover, by using hashing functions to generate browser fingerprints, one does not need to store the underlying data representing browser characteristics such as agent strings, headers, plugin details, system fonts, cookie settings, and so forth. Due to these reasons, *browser fingerprinting* is a viable and effective method to prevent workers from violating task requirements in crowdsourced microtasks, thereby improving the quality of the results produced. Moreover, by relying solely on the hashed fingerprint, we can alleviate privacy concerns.

5.5   Caveats and Limitations

We acknowledge that the existing browser fingerprinting techniques are around 95% accurate. This means that there is room for a small percentage of errors. However, the elaborate hashing of various attributes that are considered for browser fingerprinting means that it is highly unlikely that two fingerprints will accidentally collide to be identical. Yet, a conservative approach can be the use of browser fingerprinting as a means to flag crowd workers for further scrutiny, rather than blocking potential *repeaters* immediately. Such an approach would also resonate with prior work that has called for less-aggresive means of dealing with sub-optimal or potentially malicious work.

Another limitation of this work stems from our inability to account for genuine explanations of repeated participation, as detected using browser fingerprinting. For example, a false positive could result from crowd workers working in Internet cafes [9], or family members sharing a computer.

6   Repeaters in Real-World Crowdsourcing Microtasks

We conducted an additional study to evaluate the occurrence of repeaters in real-world microtasks. We manually created a batch of 120 microtasks comprising of an equal distribution of all the different types. Table 4 presents some example tasks corresponding to each task type that were deployed. The different types were prescribed by a taxonomy proposed in previous work [6]. We deployed these tasks on CrowdFlower and collected 100 responses from distinct workers for each task, resulting in 12,000 human intelligence tasks (HITs). Once again we did not restrict participation of workers based on CrowdFlower *levels*. Table 4 also presents sample tasks that we created corresponding to each type; these tasks are noticeably designed to reflect real-world microtasks that have previously been deployed on crowdsourcing platforms such as Amazon's Mechanical Turk.

Table 4. Examples of different real-world microtasks that were deployed.

Task type	Sample tasks deployed
Content Access	Watch the following video
Content Creation	Transcribe the audio excerpt presented above
Information Finding	Find the middle-names of the following famous persons by searching on the Web
Survey	What is your age?
Interpretation & Analysis	Which of the following tweets has a neutral sentiment? Check all that apply
Verification & Validation	Choose the words which are synonyms of 'HAPPY' in the following list

6.1   Results – Distribution of Unique Fingerprints and worker-ids

As in case of the previous study, we used a JavaScript implementation to generate browser fingerprints corresponding to each worker participating in the tasks. We analyzed the browser fingerprints and found a power law distribution between the number of distinct worker-ids corresponding to each fingerprint across all 120 tasks, as shown in Fig. 5.

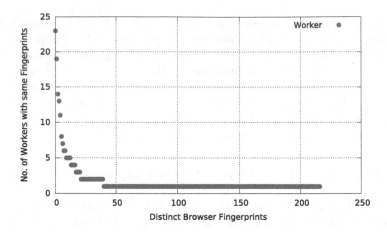

Fig. 5. Distribution of number of distinct `worker-ids` corresponding to each fingerprint across all 120 real-world microtasks of different types.

Once again we found that a significant portion of distinct browser fingerprints corresponded to more than one `worker-id` associated to the participating workers; nearly 18.5% of fingerprints corresponded to 2 or more `worker-ids`. This shows that the usage of multiple accounts with different `worker-ids` can be observed in real-world microtasks.

6.2   Evaluation of Repeaters

Due to the lack of a given groundtruth with respect to whether or not the multiple `worker-ids` associated with a unique fingerprint are a result of workers using multiple accounts, we interviewed a random selection of such workers for the purpose of evaluating the accuracy in identification of repeaters. We randomly selected 10 workers from the pool of 193 workers who corresponded to sharing the same fingerprint with at least one more worker in the pool. We contacted these workers via e-mail and recruited them to a follow-up 15 min Skype interview in return for 3 USD each. We promised to maintain the anonymity of workers and clarified the purpose of the interview beforehand. We carried out these interviews over 2 weeks following the completion of the tasks.

Workers were first asked about whether or not they participated in the tasks that they completed within this study. All 10 workers confirmed that they completed those tasks successfully. We then asked workers regarding the usage of multiple accounts to complete more tasks. 9/10 workers admitted to using multiple accounts to complete more work, and maximize their monetary rewards. Nearly all workers defended their actions since they claimed to have completed the tasks diligently each time they repeated it using a different *worker-id*. Through the interviews, it was apparent that workers were not aware of the unintentional consequences in skewing the reliability of results through their repeated participation. However, our findings suggest the high reliability of using browser fingerprinting to identify repeaters.

7   Conclusions and Future Work

In this paper we have showed that there are a significant number of repeaters that participate in crowdsourced tasks using distinct `worker-ids`. In the light of repeaters in crowdsourced microtasks, we present the following contributions and draw conclusions.

- Across 8 crowdsourced logical reasoning tasks with varying task difficulty and spanning 2000 workers, we have observed that over 13% of the workers are not distinct, but are a result of repeated participation from 6.18% of workers using different `worker-ids` (**RQ#1**). We found consistent results in further experiments using real-world microtasks.
- We found that there is a moderately high negative correlation between the task difficulty and the number of trustworthy (TW) repeaters. This means that with an increasing task difficulty the number of TW repeaters decreases (**RQ#2**). Thus, task requesters should be more prudent while deploying tasks that are relatively easy to complete; simple tasks have a greater propensity for repeated participation.
- Existing quality control mechanisms that rely on worker `ip-addresses` or `worker-ids` fail to detect repeaters. We have shown that browser fingerprinting can be used in order to identify repeaters in crowdsourced microtasks. Through our experimental tasks, we have found that repeaters significantly skew the demographic attributes within a given task, and thereby adversely affect the reliability of the results produced (**RQ#3**).

Our findings have important implications in crowdsourced tasks, especially when the tasks are subjective. It is vital to detect and prevent repeated participation of workers in a task in order to ensure reliable and unbiased results in crowdsourced microtasks. In the imminent future, we plan to investigate the usage of browser fingerprinting in tandem with other quality control mechanisms.

Acknowledgments. This research has been supported in part by the European Commission within the H2020-ICT-2015 Programme (AFEL project, Grant Agreement No. 687916).

References

1. Difallah, D.E., Catasta, M., Demartini, G., Ipeirotis, P.G., Cudré-Mauroux, P.: The dynamics of micro-task crowdsourcing: the case of amazon mturk. In: Proceedings of the 24th International Conference on World Wide Web, pp. 238–247. International World Wide Web Conferences Steering Committee (2015)
2. Eckersley, P.: How unique is your web browser? In: Atallah, M.J., Hopper, N.J. (eds.) PETS 2010. LNCS, vol. 6205, pp. 1–18. Springer, Heidelberg (2010). doi:10. 1007/978-3-642-14527-8_1

3. Eickhoff, C., de Vries, A.P.: Increasing cheat robustness of crowdsourcing tasks. Inf. Retrieval **16**(2), 121–137 (2013)
4. Faradani, S., Hartmann, B., Ipeirotis, P.G.: What's the right price? pricing tasks for finishing on time. In: Human Computation, Papers from the 2011 AAAI Workshop, San Francisco, California, USA, 8 August 2011
5. Gadiraju, U., Dietze, S.: Improving learning through achievement priming in crowdsourced information finding microtasks. In: Proceedings of the Seventh International Learning Analytics & Knowledge Conference, Vancouver, BC, Canada, pp. 105–114, 13–17 March 2017
6. Gadiraju, U., Kawase, R., Dietze, S.: A taxonomy of microtasks on the web. In: 25th ACM Conference on Hypertext and Social Media, HT 2014, Santiago, Chile, pp. 218–223, 1–4 September 2014
7. Gadiraju, U., Kawase, R., Dietze, S., Demartini, G.: Understanding malicious behavior in crowdsourcing platforms: the case of online surveys. In: Proceedings of the 33rd Annual ACM Conference on Human Factors in Computing Systems, CHI 2015, Seoul, Republic of Korea, pp. 1631–1640, 18–23 April 2015
8. Gani, K., Hacid, H., Skraba, R.: Towards multiple identity detection in social networks. In: Proceedings of the 21st International Conference on World Wide Web, pp. 503–504. ACM (2012)
9. Gawade, M., Vaish, R., Waihumbu, M.N., Davis, J.: Exploring employment opportunities through microtasks via cybercafes. In: 2012 IEEE Global Humanitarian Technology Conference, GHTC 2012, Seattle, WA, USA, pp. 77–82, 21–24 October 2012
10. Kafai, Y.B., Fields, D.A., Cook, M.: Your second selves: avatar designs and identity play in a teen virtual world. In: Proceedings of DIGRA, vol. 2007 (2007)
11. Kaufmann, N., Schulze, T., Veit, D.: More than fun and money, worker motivation in crowdsourcing-a study on mechanical turk. In: AMCIS, vol. 11, pp. 1–11 (2011)
12. Kittur, A., Chi, E.H., Suh, B.: Crowdsourcing user studies with mechanical turk. In: Proceedings of the SIGCHI Conference on Human Factors in Computing Systems, pp. 453–456. ACM (2008)
13. Mason, W.A., Watts, D.J.: Financial incentives and the "performance of crowds". In: Proceedings of the ACM SIGKDD Workshop on Human Computation, Paris, France, pp. 77–85, 28 June 2009
14. Mowery, K., Shacham, H.: Pixel perfect: fingerprinting canvas in html5. In: Proceedings of W2SP (2012)
15. Mulazzani, M., Reschl, P., Huber, M., Leithner, M., Schrittwieser, S., Weippl, E., Wien, F.: Fast and reliable browser identification with javascript engine fingerprinting. In: Web 2.0 Workshop on Security and Privacy (W2SP), vol. 5 (2013)
16. Oleson, D., Sorokin, A., Laughlin, G.P., Hester, V., Le, J., Biewald, L.: Programmatic gold: targeted and scalable quality assurance in crowdsourcing. In: Human Computation, Papers from the 2011 AAAI Workshop, San Francisco, California, USA, 8 August 2011
17. Rainer, B., Timmerer, C.: Quality of experience of web-based adaptive http streaming clients in real-world environments using crowdsourcing. In: Proceedings of the 2014 Workshop on Design, Quality and Deployment of Adaptive Video Streaming, pp. 19–24. ACM (2014)
18. Rogstadius, J., Kostakos, V., Kittur, A., Smus, B., Laredo, J., Vukovic, M.: An assessment of intrinsic and extrinsic motivation on task performance in crowdsourcing markets. In: Proceedings of the Fifth International Conference on Weblogs and Social Media, Barcelona, Catalonia, Spain, 17–21 July 2011

19. Rzeszotarski, J.M., Kittur, A.: Instrumenting the crowd: using implicit behavioral measures to predict task performance. In: Proceedings of the 24th Annual ACM Symposium on User Interface Software and Technology, Santa Barbara, CA, USA, pp. 13–22, 16–19 October 2011
20. Wang, J., Ipeirotis, P.G., Provost, F.: Quality-based pricing for crowdsourced workers (2013)
21. Yamak, Z., Saunier, J., Vercouter, L.: Detection of multiple identity manipulation in collaborative projects. In: Proceedings of the 25th International Conference Companion on World Wide Web, pp. 955–960. International World Wide Web Conferences Steering Committee (2016)

Maturity Model for Liquid Web Architectures

Andrea Gallidabino[✉] and Cesare Pautasso[✉]

Faculty of Informatics, University of Lugano (USI), Lugano, Switzerland
{andrea.gallidabino,cesare.pautasso}@usi.ch

Abstract. Liquid Web applications adapt to the set of connected devices and flow seamlessly between them following the user attention. As opposed to traditional centralised architectures, in which data and logic of the application resides entirely on a Web server, Liquid software needs decentralised or distributed architectures in order to achieve seamless application mobility between clients. By decomposing Web application architectures into layers, following the Model View Controller design pattern, we define a maturity model for Web application architectures evolving from classical *solid* applications deployed on single devices, to fully *liquid* applications deployed across multiple Web-enabled devices. The maturity model defines different levels based on where the application layers are deployed and how they migrate or synchronize their state across multiple devices. The goal of the maturity model described in this paper is to understand, control and describe how Web applications following the liquid user experience paradigm are designed and also provide Web developers with a gradual adoption path to evolve existing Web applications.

Keywords: Maturity model · Liquid web architecture · Decentralised web · Liquid software

1 Introduction

The metaphor of liquid software [17,34] illustrates the user experience when interacting with software deployed across multiple devices. Liquid software can (1) adapt the user interface to the *set of* devices being concurrently used to run the application; (2) seamlessly migrate a running application across devices and (3) synchronize the state of the application distributed across two or more devices, effectively breaking down the continuity boundaries that exist between devices in proximity both in physical space as well as in cyberspace.

Web applications were traditionally developed following a thin client architecture whereby most of the logic and the entire persistent state of the application would be executed and stored on a central Web server. They would offer only partial support for the liquid user experience in terms of the ability of migrating the application by simply sharing the URL pointing to the current state of the application and adapting the user interface by employing responsive Web design techniques [24].

© Springer International Publishing AG 2017
J. Cabot et al. (Eds.): ICWE 2017, LNCS 10360, pp. 206–224, 2017.
DOI: 10.1007/978-3-319-60131-1_12

As the Web technology platform has evolved with enhanced support for rich and thick client architectures and for protocols beyond HTTP to enable real-time push notifications and peer to peer (browser to browser) connections, it is time to revisit the architectural design space of contemporary Web applications to systematically study new deployment configurations and how these impact the liquid user experience.

In this paper we present a maturity model for liquid Web applications [25], based on multiple facets that determine the degree of liquidity of a Web application both in terms of which liquid user experience primitives are enabled as well as how these can be implemented with different performance and privacy guarantees. Each architectural configuration presents unique challenges and opportunities to deliver a liquid behavior under different constraints. For example, while it is relatively easy to synchronize the state of the application relying on a highly available, centralised master copy deployed in the Cloud, some privacy and latency issues may warrant considering more decentralised or distributed approaches to data management.

The maturity model is based on three orthogonal facets, each having three levels: (a) logic deployment (ultra-thin, thin, thick); (b) state storage (centralised, decentralised and distributed); (c) communication channel (HTTP, WebSockets and WebRTC). This paper provides a systematic discussion on the implications of the most significant architectural configurations on whether and how liquid Web applications can be built under the corresponding architectural constraints. Additionally, for each level, we survey existing Web development frameworks within the corresponding Web application architectures. As we are going to show, migration of Web applications can be achieved with all configurations, while cloning requires support for real-time synchronization that is only present in higher maturity levels.

2   Motivation

The relationship between users and their computing devices has evolved from multiple users sharing one expensive and large computer to the opposite, where multiple, cheap, mobile and Web-connected devices are owned by a single user. Web browsers are nowadays ubiquitous as they run on desktop computers, laptops, tablets, smart phones, digital cameras, smart televisions, cars, and – with some limitations – even kitchen refrigerators, watches and glasses. While responsive Web applications are designed to adapt to the different screen sizes and input/output capabilities of different devices, it is still challenging to develop rich Web applications which can seamlessly migrate across different user devices. For example, planning a trip on a large display and following the directions for the trip guided by the phone GPS; typing a short email by tapping on the phone screen but then as the email text grows longer deciding to continue the work on a computer with a real keyboard.

As users begin to use multiple devices concurrently – for example: to watch television while looking up information on their tablet, to play games across multiple telephones, to share pictures taken with personal devices and view them on a public display, to remotely control presentation slides from a watch, or to confirm a credit card transaction entered on a desktop computer using the fingerprint reader of the phone – only few Web applications fully take advantage of all available devices and distribute their user interface accordingly or allow users to re-arrange different user interface components at will.

Web developers and designers have successfully addressed the scalability challenge to serve Web applications to millions of users [1] and to personalize such applications to each user profile [6] (e.g., language, age, geographical location, regulatory constraints, etc.) and adapt it to the capabilities of their Web browsing device [24]. However, this has been under the assumption that users connect to use the Web application using one device at a time. Stateful Web applications which use cookies to establish a session with a particular Web browser make it difficult for users to switch devices in the middle of a browsing session [15]. Additionally, they may break when opening multiple tabs to run them and sometime assume that users logging in from different devices at the same time may indicate a security issue.

In this paper we provide a maturity model to assess how different Web application architectures can provide support for the liquid user experience for both sequential and parallel screening scenarios: – **Sequential screening**: users own more than one device, at any time they may decide to continue their work using another device. The application and the associated state seamlessly flow from one device to another; – **Parallel screening**: users own multiple devices and deploy the software on all of them. Users may decide to change the number of devices running the liquid application as well as to move components of the application from one device to another while keeping the state up to date.

Web developers can follow the maturity model to redesign, refactor and transform their applications to provide enhanced level of support for liquid behavior defining the following liquid UX primitives: – **Forward**: the ability of an application of redirecting the input/output of one device to another; – **Migrate**: the ability of moving a running application to another device; – **Fork**: the ability of creating a perfect copy of an existing application on another device. – **Clone**: the ability of creating a perfect copy of an existing application on another device, and keep the state of the original and the copy synchronised thereafter. Sequential screening can be achieved if an application defines either a migrate or fork primitive, while parallel screening can be supported either with clone or forwarding primitives.

3   Web Architecture Facets

Web applications comply with the client-server architectural style, in which persistent resources or services are provided by one server to multiple clients. Without loss of generality, we further describe Web application architectures using the Model-View-Controller (MVC) pattern, one of the most used design patterns in Web applications development [20]. In the MVC pattern Web applications are logically decomposed to manage separate concerns: data modeling and persistent storage, data processing and business logic, and data input/output and user interaction.

– **Model Layer** manages the persistent data of an application. The model of a Web application also includes any of its assets such as Web Pages, images, and scripts that need to be transferred to the clients. This layer requires some kind of data storage able to represent, organize and collect information: • in the server-side of an application it usually takes the form of a database such as relational databases like *Oracle* and *MySQL*, document oriented databases like *MongoDB*, or *CouchDB*, or other schemaless databases like *Redis* [8]; • in the client-side usually the file system of the device is used as storage, but due to the possibility of having clients running on heterogeneous devices and implemented using different programming languages, data storages can highly differ from client to client even in the same application. *WebSQL*, *IndexedDB*, *LocalStorage*, and *Cookies* are standard implementations of data storage APIs available in HTML5-compliant Web browsers.

– The **Controller Layer** consists of the logic of an application. The controller layer is a bridge between the model and view layers, it manipulates data and executes tasks received from either layer and forwards the results to the appropriate one. Depending on where the controller layer is deployed it can be implemented using different programming languages. In the server-side *PHP*, *ASP.NET*, and *Javascript* (using *Node.js*) are the most used programming languages, while in the client-side *Javascript* is the main option.

– The **View Layer** is the graphical user interface of an application, consisting of the visual representation of the data and information retrieved from the model layer and rendered into an interactive visualization.

Combining the client/server execution environment and the three MVC layers, we identify different deployment combinations. While the View Layer is constrained to run on the client, both Model and Controller Layer can be deployed on either side (or partitioned to run on both client and server). Additionally, we distinguish three alternative communication channels and protocols (HTTP, WebSockets and WebRTC) used to interconnect the layers of Web applications running on different devices.

In the following sections we discuss more in detail each facet which will be combined into the liquid Web application maturity model in Sect. 4.

3.1   Model Layer Deployment

Model layer deployment describes where the persistent state of the Web application is stored. We identify three levels based on whether data is logically centralised on the server or distributed towards the clients (Fig. 1):

Level 1 - Centralised - The model is stored using any data management solution that is solely deployed in the server-side. For scalability and availability purposes, the actual storage can be implemented using multiple virtual servers running in a Cloud data center. Conceptually, this is still a centralised solution

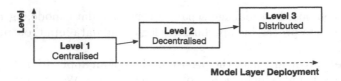

Fig. 1. Model layer deployment levels

as data is never managed by the client. The advantage is that no matter what client device is used to access it, the data will be readily available [32]. Users thus trade off the convenience of accessing "their" data anywhere with the loss of control over the actual location where it is stored and who else can access it. As clients always need to remotely request data from the server, there are also latency and availability implications to be considered. When multiple clients perform transactions to update shared resources, having a single master copy on the server helps to ensure consistency.

Some real world examples of centralised model layer deployments use databases created with *MySQL*, *MySQL Cluster*, or *Cassandra* [23].

Level 2 - Decentralised - The model layer is deployed both in the server and client-side of the Web application. Information stored in the server database is replicated or cached by the clients. Conversely, users may prefer to save the primary copy of their data in their own clients and use the server as a secondary backup.

Cookies are the simplest example of decentralised persistent storage on the Web. Web application using any technology mentioned in level 1 (e.g. *MySQL*) in combination with any HTML5 storage API (e.g. *localstorage, WebSQL*) falls in this category. *Apache CoudchDB* or *PouchDB* are databases that feature client-side caching with automatic synchronisation allowing offline availability of the retrieved data.

Decentralised approaches enhance: • data privacy, even though data must still be transmitted to the servers if there is not a direct communication channel between clients; • availability during offline operation, assuming the data has been prefetched by the client the Web application may still work while being disconnected from the server; • enhanced perceived performance when hitting data cached on the client.

Level 3 - Distributed - The model layer is distributed exclusively on the client-side of the Web application. There is no need to use the server to retrieve or store data, because clients completely own the state of the Web application. This positively impacts data privacy because the information of the users always remains on their devices and is never stored in a Web server outside of their control (e.g., in the Cloud).

Distributed model layer deployment can be achieved in a modern Web browser by using any combination of the storage APIs provided by the HTML5 standard, namely the *WebSQL, IndexedDB,* and *LocalStorage* APIs. On top of these technologies, or even using the File system of the devices running the client of the Web application it is possible to build distributed model layers able to automatically synchronise between clients (e.g. [36]).

3.2 Controller Layer Deployment

The Controller layer deployment determines where the Web application executes tasks and whether it can offload its workload. We define three levels with respect to the client thickness: (Fig. 2):

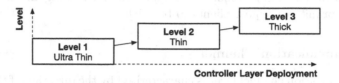

Fig. 2. Controller layer deployment levels

Level 1 - Ultra-Thin Client - In this level the controller layer of an application is deployed only on the server-side of the application. The only logic present on the client is the logic needed to retrieve content from the server and to display views when they are received from the server.

Primitive Web browsers that did not allow running scripts, such as *JavaScript* or *Java Applets* can be seen as Ultra-Thin clients. Ultra thin clients always display the view layer statically and cannot adapt it to the client's device. *Curling* pages on a Terminal is also an example of Ultra-Thin Client, in which the forwarded raw data is displayed. Web applications that do not require scripts to run in the client fall in this category as well.

Level 2 - Thin Client - The logic of an application is deployed on both server and client-side of the Web application. The server can offload part of the computations to the connected clients. The most offloaded task in Web applications is the creation of the views which is entrusted directly to the clients needing it, however in thin clients any simple task can be offloaded to the clients.

Whenever the client is thin, it is possible to make views responsive to the client's device. This allows the same applications to use a different look and feel in devices with different hardware specifications.

AngularJS, React, or *EmberJS* are some frameworks for isomorphic Web applications written in Javascript that require thin clients.

Level 3 - Thick Client - The logic of an application is entirely deployed on the client-side of the Web application. A big portion of the application computations are offloaded to the clients. As in level 2 clients compute the views they display. Additionally they execute computationally-heavy application-specific tasks that were not previously included in thin clients. The HTML5 WebWorker specification allows Web browser to run scripts in background, making it possible to develop complex client-side applications [7].

Thick clients can be aware of other connected clients, making it possible to adapt the view layer of the application on a set of devices instead of making it responsive to a single one. Complementary adaptive views can also automatically evolve in real-time if the application is able to propagate to all devices the knowledge of connections and disconnections of other clients.

Web 2.0 single-page applications generally require thick clients, any client described in level 2 can become a thick client if the entire controller logic layer is deployed in the client-side. Liquid Web applications featuring all the liquid user experience primitives require clients to be thick.

3.3   Communication Channel

The communication channel facet is characterised by the direction of the communication between the client and the server and whether clients can communicate directly. The levels shown in Fig. 3 are inclusive, whereby a higher level also includes all the features provided by the lower levels:

Fig. 3. Communication channels

Level 1 - Client-Server Pull - Clients are always the origin of all request-response interactions with the server. Clients request resources addressed by URIs and the server responds with the corresponding representations if they exist. On the Web, this kind of communication is implemented with the HTTP protocol.

Applications relying solely on the HTTP protocol cannot propagate state changes or events occurring on the server back to the clients in real-time. They can only simulate a quasi-real-time environment (with continuous polling). While the liquid migrate primitive can be implemented with HTTP only, cloning or forwarding cannot be implemented in level 1, because data synchronisation in liquid applications requires real-time notifications.

Level 2 - Client-Server Push - Similarly to the client-server pull level, clients are still the origin of the interaction with the server. However in level 2, clients open a two-way communication channel. In this level the server is allowed to propagate data and events to the connected clients immediately, meaning that it is possible to efficiently create real-time Web applications. The standard Web protocol used for implementing client server push is WebSocket. With Web-Socket it is possible to implement the liquid clone primitive since data synchronisation can happen in real-time. Liquid Web applications whose goal is to implement all possible liquid UX primitives need to consider at least a level 2 communication channel in the design of their architectures.

Level 3 - Peer-to-Peer - With the advent of the WebRTC protocol it is now possible to have peer-to-peer (P2P) communication among Web browsers. Architectures implementing level 3 communication channels still rely on the HTTP and WebSocket protocols for peer discovery purposes. Level 3 communication channels allow to lower the latency between clients, by potentially decreasing the number of hops in the communication, instead of propagating data relying on the server (client $\rightarrow$ server $\rightarrow$ client), in the best case it is possible to communicate directly between clients (client $\rightarrow$ client).

4   Maturity Model

Figure 4 shows the maturity model of liquid Web applications determined by combining the deployment configuration of their MVC layers across the server-side and client-side with the choice of the communication channels established between them. We identify five levels: 1. Web 1.0 Applications 2. Rich Web Applications 3. Real-time Web Applications 4. Hybrid Web Applications 5. Peer-to-peer Web Applications

The diagram also shows in which level is possible to use the *migrate/fork* and *clone/forward* liquid user experience primitives. Migrate and fork are possible in all levels, while clone and forward can only be achieved starting from level three.

Table 1 summarizes the different configurations per level as a combination of the facets explained in Sect. 3. The table also describes how the configurations affect the following quality attributes:

Latency or the proximity between two clients on the network: • *2 hops* means that whenever clients communicate to perform a liquid user experience primitive, the communication relies on an intermediary Web server (client $\rightarrow$ server $\rightarrow$ client); • *1 hop* means that clients – in the best case – can communicate directly with each other.

Liquid UX (primitive) Migrate/Fork can occur asynchronously (A) or synchronously (S). **Asynchronous** migration and fork of an application happens between two clients that cannot directly push the migrated state and logic to the target client, but they have to be stored in a central storage first; **Synchronous** migration and fork of an application can be implemented in systems in which clients can push migrated or forked state and logic to other clients without the need to store it in a central storage. **Clone/Forward** indicates in which configurations the liquid clone and forward operations are possible ✓ or not possible ✗.

View Adaptation describes which level of the view layer adaptability is possible to achieve in all configurations: *Static* means that the view does not

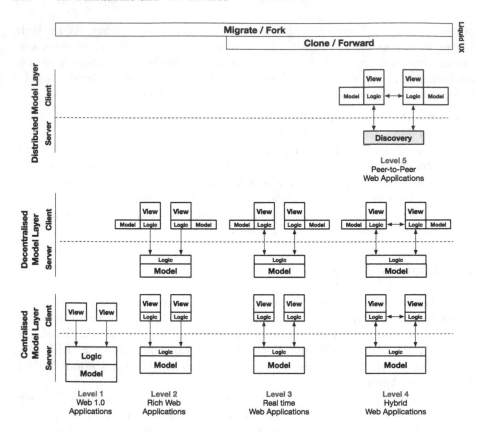

Fig. 4. Maturity model for Web architectures for centralised, decentralised and distributed model layer deployments. The controller layer is labeled as *Logic*.

Table 1. Maturity model: architectural configurations and quality attributes.

	Configuration			Quality attributes				
	Deployment		Channel	Latency	Liquid UX		View	Privacy
Level	Logic	Model		Hops	Migrate fork	Clone forward	Adaptation	Model
1	Centralized	Ultra-Thin	Pull	2	A	✗	Static	✗
2c	Centralized	Ultra-Thin	Pull	2	A	✗	Responsive	✗
2d	Decentralized	Thin	Pull	2	A	✗	Responsive	✗
3c	Centralized	Ultra-Thin	Push	2	A, S	✓	Complementary	✗
3d	Decentralized	Thin	Push	2	A, S	✓	Complementary	✓*
4c	Centralized	Ultra-Thin	P2P	1 or 2	A, S	✓	Complementary	✗
4d	Decentralized	Thin	P2P	1 or 2	A, S	✓	Complementary	✓*
5	Distributed	Thick	P2P	1	S	✓	Complementary	✓

dynamically adapt to the client hardware device capabilities, but it is displayed exactly as it was determined by the server; *Responsive* means that the view locally adapts to the client. *Complementary* means that it is possible to manually or automatically adapt the view to *set of* heterogeneous devices connected to the Web application.

Privacy describes if the users have control on their data by ensuring that it is exclusively stored in devices they own or trust: ✓ means that users are in control of their own data, ✗ means that the data is stored in untrusted devices, ✓* means that the data is stored in trusted devices, but is exchanged with or relayed across untrusted devices (e.g., a Web server running in the Cloud).

4.1   Level 0 - Solid Applications

All layers of a solid (or monolitic) application are deployed on the same machine, typically a standalone personal computing device, or a server to which multiple dumb terminal devices are attached. This architecture configuration predates the Web as intra-layer communication does not go through any Web protocol, but only happens locally within the same host using, e.g., local procedure calls or shared memory buffers.

Liquid migration can be achieved by the mean of input/output virtualisation of the clients, e.g. multiple users can access the operating system installed in the server by different screens. This architecture allows users to save their data on the server and access it from any screen. From the user perspective, switching terminal device amounts to successfully migrating their work from one screen to another. The *virtualised* client is therefore ultra-thin, where the view layer running on the server forwards the user interface input/output events and commands to the terminals connected to it.

The concept of the first Sun Ray [28] designed in 1997 can be considered as an example level 0, solid application architecture. The concept is designed to take advantage of stateless network computers whereby users authenticating with smart cards were instantaneously taken to their virtual desktops and could access their applications and data centrally managed on the server from anywhere.

4.2   Level 1 - Web 1.0 Applications

Level 1 Applications can be seen as the first generation of Web applications [4] built using the HTTP protocol. The logic and model layer are deployed on the server-side, while the view layers run client-side on Web browsers. In this level the content provided by the Web servers is static and cannot be changed by the clients. Web browsers retrieve content (e.g. Web pages written in HTML) by sending HTTP requests to Web servers addressed by URIs. Browsers display resources as they were sent from the server. The view layer is completely static, since there is no definition of a technology able to adapt the retrieved resources to different client rendering capabilities. At that time CSS media queries did not yet exist while the Extensible Stylesheet Language (XSL) did not provide any markup to adapt the content of a Web page to the device displaying it.

Level 1 supports asynchronous liquid migration by uploading resources to the server and then using their Uniform Resource Locator (URL) to retrieve the resource from another device. In this level cloning a liquid Web application is challenging to achieve, because data synchronisation does not happen in real-time, as clients can only resort to continuous HTTP polling. Likewise, migration in level 1 does not happen in real-time and requires the exchange or agreement on the URL addressing any resource that needs to be migrated between clients, which do not need to be available and connected at the same time.

This is the most basic architecture for implementing liquid applications that only need the liquid migration primitive, it does not provide any kind of view adaptation and cannot ensure data privacy (unless the Web server is owned and operated by the same organization owning the client devices). Synchronization between multiple clients can be achieved only by manually refreshing the Web page after sharing URLs via out-of-band channels, which does not fit with the real-time expectations of the liquid user experience.

4.3   Level 2 - Rich Web Applications

In Level 2 we consider rich Web applications [7] in which the controller layer is deployed both in the server and client-side. Level 2 architectures are the first ones able to have responsive views because the portion of the controller on the client-side can compute different views and do so based on the underlying hardware capabilities. Liquid migration is possible, but, like in level 1, shared URLs are needed to address and retrieve the resources representing the state to be shared among clients, since there is not a direct communication channel between clients. More in detail, after discovering the identifier of the resource being migrated, the Web application on the browser is manually refreshed to ensure the consistency of the displayed information with the model of the Web application. Again, cloning is hindered by the lack of real-time communication between clients attempting to immediately synchronize. The distance between two clients is always equal to two hops.

Depending on how the model layer is deployed on the clients we distinguish two different level 2 configurations: Level 2c - **centralised** - the model is deployed only in the server-side; Level 2d - **decentralised** - part of the model is stored in the client-side, in traditional Web applications it takes the form of *Cookies* or cached data, in modern rich Web application it takes various forms (e.g., local storage, service workers, WebSQL databases) as described in Sect. 3. Users of a liquid rich Web application can store their confidential data in their devices, nevertheless during a migration of the liquid Web application its state has to be transferred via the server, which may not be always owned or trusted by the user of the Web application.

The liquid user experience in this level is similar to the one in level 1, with the addition of support for a responsive view and the option of storing parts of the model locally on the client.

Example Level 2 Frameworks. CrowdAdapt [26] is a centralised level 2 framework for creating responsive Web pages. Web pages created with CrowdAdapt allow users to change the layout of the Web page as they desire and thereafter migrate their creations on other devices. In CrowdAdapt the controller layer provides the editing functionality and the automatic detection of the hardware specification of the device running the client. The users are able to choose between the layouts created by all the users of the Web application that better fit their needs.

PageTailor [5] is a decentralised level 2 framework with concepts similar to CrowdAdapt. Users can change the layout of Web pages using PageTailor and then reuse these layouts on subsequent visits. In this case layouts are not shared between multiple users as in CrowdAdapt.

4.4   Level 3 - Real-Time Web Applications

The deployment of the view and controller layers are the same as in level 2, however level 3 applications have access to client-server push communication channels. This make it possible for Web applications to synchronise data among clients and notify connections of new clients and detect disconnections of old devices. The liquid clone operation is implementable in level 3 because data can be synchronised in real-time between simultaneously connected devices. The awareness of the connected clients to the Web application allows to distribute the view layer among them. The complementary view implementable in level 3 Web applications increases the quality of the liquid user experience. Liquid migration, liquid cloning and complementary view control can happen at different granularity levels: **application level** - the Web application is monolithic and all devices receive all the assets and model of the whole application. Upon migration or cloning the new clients have a perfect copy of the whole application whose state is kept synchronised between them. Complementary view adaptation in this case can be implemented through *Web clipping* by concealing part of a view on all but one device. **component level** - in component-based Web applications clients receive only portions of the whole application. Liquid migration and cloning can be done at component level, thus moving and keeping synchronized only part of the application. In this granularity level complementary view development does not need *clipping*, because clients move or receive only the portions of the application they need and do not have to locally hide the components displayed on other devices.

The **decentralised** configuration of level 3 allows partial privacy on the data created by the users, as it can be stored only on trusted devices. During liquid migration or cloning on simultaneously connected devices there is no need to store any information on the server. However data sent between clients is still relayed through the WebSocket channels on the server, meaning that such data must be encrypted in order to ensure privacy. In the case of liquid migration on devices which are not simultaneously connected to the Web application, the entire model has to pass through the Web server regardless.

Example Level 3 Frameworks. Smart Composition [22] is a centralised level 3 framework that allow the creation of component-based (called widgets) multi-screen Web applications. By using a central *cross-device communication service* the infrastructure created by SmartComposition is able to compose distributed view layers among devices and keep the various components building the application synchronised.

Panelrama [38] is a centralised level 3 framework used to create distributed user interfaces using the concept of *panels*, Javascript objects defining pieces of user interface and logic. Panelrama provides an API to migrate and clone panels between devices and automatically create the complementary distribution of the view layer among the connected devices.

DireWolf [21] is a decentralised level 3 framework used to create multi-device mashups Web applications. Clients are aware of the connected devices in the application and can migrate widget-like components to any target device. Dire-Wolf offers the possibility to manage the device ownership, device information and specification, the widget state, and the application state of the whole application through its clients.

Liquid.js for DOM [35] is a decentralised level 3 framework based on *React.js* for component-based Web applications. By synchronising virtual DOMs between devices it is able to migrate and clone logic and model layers among the connected clients. It does not offer automatic cross-device complementary views. There also exists a level 4 Hybrid version of Liquid.js for DOM offering peer-to-peer data synchronisation between clients.

Bellucci *et al.* [3], Frosini *et al.* [11], and Raposo *et al.* [29] propose similar frameworks in which is possible to distribute and synchronise the view layer of the application on all connected devices.

4.5   Level 4 - Hybrid Web Application

Level 4 augments level 3 with the ability for clients to communicate directly with each other through P2P channels. The logic layer deployed in the client-side can send messages to other clients either directly with a single *hop* or through the server with two *hops*. Connected clients can send any kind of data between each other, including the entire assets of the application. Similarly to level 3, it is possible to have both asynchronous and synchronous migration and cloning operations among connected clients. Through the peer-to-peer channels decentralised hybrid Web applications can send confidential data directly among trusted clients without relaying any message through the server, ensuring privacy if confidential data does not need to be stored on the server.

Example Level 4 Frameworks. XD-MVC [19] is a decentralised level 4 framework for creating cross-device interfaces and automatic complementary adaptation views applications. XD-MVC implements migration at the application level and takes advantage of clipping off parts of the view layer in order to simulate migration between devices. Views can be annotated with rules about how they are expected to adapt to the set of connected devices. Given these rules the view is able to dynamically and automatically adapt to set of heterogeneous devices when a new client connects or disconnects.

PolyChrome [2] is a centralised level 4 framework for creating co-browsing applications with collaborative views spanning on multiple screens deployed on multiple devices. PolyChrome complementary view adaptation supports *stitching, replication, nesting,* and *overloading* layouts. Data synchronisation happens both through peer-to-peer and WebSocket channels. The framework creates components out of a legacy applications in order to be able to make a view span on multiple devices.

4.6 Level 5 - Peer-to-peer Web Applications

Peer-to-peer [31] Web architectures are at the highest level of the maturity model, the only one providing all quality attributes expected from liquid Web applications. Peer-to-peer Web applications allows connected clients to communicate with each other directly. When peers are linked with a fully-connected mesh, this amounts to the best case scenario with a latency of 1 hop. Indeed other topologies are possible, like rings, in which N connected clients are up to $N/2$ hops away, or stars, in which the hops number vary between 1 and 2, depending on which peers are communicating.

Level 5 applications allow synchronous migration, since connected clients can push the model and logic through the full-duplex peer-to-peer channel created with WebRTC at any time. Since there is no longer a central server available at all times asynchronous migration is not possible. Instead clients need to be online simultaneously in order to proceed with the migration. Since clients sense and propagate their availability across the peer to peer network, it is possible to have complementary view adaptation.

Data privacy is ensured in peer-to-peer Web applications because users are in full control of all devices storing and processing their data. Data is never stored in any server or Cloud storage platform. Also, data migrated or synchronised with another device is never sent through a Web server.

Level 5 Web applications allow strong mobility with direct model and logic transfer and synchronisation between peers, however this requires to a suitable discovery method. WebRTC, for example, uses a signaling server to initiate and establish the connection between clients. Clients first connect to the signaling server and then receive information on how they can join the rest of the peers. Once the topology is created and peers are connected, they are free to communicate among themselves and the signaling server is no longer involved.

Example Level 5 Frameworks. Liquid.js for Polymer [12] is a level 5 peer-to-peer framework which allows the creation of distributed component-based Web application built on top of the Polymer framework. Users instantiate any component provided by the Web application on their devices and share them directly with other users. If a peer does not own the assets of the component being sent to it, the peer will also receive the model of the component that is going to be migrated. Liquid.js allows to define strategies for creating different peer topologies. Currently Liquid.js for Polymer does not support automatic complementary view adaptation. However developers can build their own layout adaptations by using the API provided by the framework.

5   Related Work

Liquid software is named after a metaphor: as liquids adapt to the shape of their containers, *Liquid software* seamlessly adapts to the set of devices it is allowed to run on. In the Liquid Software Manifesto [34] Taivalsaari *et al.* make the case for liquid software, envisioning that the Web (thanks to its ubiquitous support) provides the most suitable platform for ensuring liquid applications can flow across heterogeneous devices.

The design space for liquid software is analysed in [13,14] where we defined twelve dimensions for designing liquid software and position existing technologies developed within the last two decades within the design space. In this paper we focus on the deployment configuration of Web applications designed according to the Model-View-Controller pattern and discuss more in detail how the configuration impacts the liquid Web application quality attributes.

Other fields close to liquid software concept are also working towards the definition and evaluation of technologies and architectures of frameworks which could be used to create applications able to seamlessly flow between multiple devices. Esenther [10] builds in 2002 a framework for real-time collaborative co-browser applications. Opera Unite (now discontinued) [27] was a Web browser extension allowing users to host social Web applications (e.g., photo sharing, social wall) on their Web browser. The goal of these efforts was to enable safe social networking whereby personal data would be exchanged directly between trusted devices.

Cross-device interfaces [18] study how to design collaborative environments spanning across multiple Web-connected devices. In the distributed user interfaces field, Santosa *et al.* [30] make a field study on the impact in the real world of the use of technologies enabling cross-device interactions. Given the responses of experts in the field, they collect and compare nine existing cloud-based data management software enabling cross-device collaboration between users.

The survey proposed by Elmqvist [9] discusses the state of the art of distributed user interfaces in the human computer interaction research area. Elmqvist summarizes how to achieve migration of the user interface and redirection of I/O of a device. The concept of forwarding used in this paper is similar to the concept of redirection used in the survey.

6   Conclusions

In this paper we described the maturity model of liquid Web architectures and provide examples of emerging frameworks and technologies across all five levels.

Since the term *liquid software* was introduced 21 years ago [17], we recognise that there has been an evolution in how software architectures are designed to bring seamless application mobility across multiple devices.

In the context of Web applications, the maturity model presented in this paper includes: 1. Web 1.0 applications; 2. Rich Web applications; 3. Real-time Web applications; 4. Hybrid Web applications; 5. Peer-to-peer Web applications. Each level's architectural configuration impacts the possible liquid user experience primitives. Most of the existing liquid Web application development frameworks [2,3,5,11,12,19,21,22,26,28,29,35,38] we surveyed are categorised by a level 3 (real-time) architecture, however we emphasize that higher level architectures are possible and they should be considered to deliver all the quality attributes that one would expect from a liquid Web application, in particular data privacy (with decentralized configurations) and a reduced latency between devices that do not need to communicate with or through a remote Web server all the time. We acknowledge that real-world liquid applications with multiple client implementations may span multiple levels in the maturity model. For the sake of simplicity and clarity we described the main five levels in the maturity model instead of all possible combinations thereof.

The choice of level and configuration should be implemented or which framework to use are important architectural decisions. Upgrading an architecture from a lower level to a higher one, or downgrading to lower levels fundamentally impact the design of the Web application and are likely to result in significant development costs. Still, over the history of the Web, application architectures have been gradually and steadily shifting towards the higher levels of the maturity model presented in this paper.

7   Future Work

As the number of devices connected to the Web and the average number of devices owned by one user increases [16], more frameworks will appear positioned across all levels of the maturity model targeting the creation of liquid Web application. An evaluation of the presented and future frameworks in terms of performance, scalability, and usability would allow developers to assess which framework is more suitable for executing liquid primitives in the sequential and parallel scenarios.

HTML5 standards are quickly evolving every year and new specification drafts are already defining new technologies that may be used to further extend and improve the liquid user experience provided by liquid applications. While we described five levels in our maturity model, we do not exclude that in the future higher levels will appear. For example, emerging technologies like Web Bluetooth (currently not yet a W3C standard) [37] aims to bring Bluetooth support in Web browsers which may be used to define a new maturity level in which there is no longer a need for a central server in order to perform client discovery and device pairing.

We based our description on the MVC design pattern adopted by traditional Web applications, however in the future it may become necessary to revisit the fundamental architectural abstraction and design principles of Web applications and study their interplay with a programmable world [33] of billions of heterogeneous interconnected devices.

Acknowledgements. This work is partially supported by the SNF with the "Fundamentals of Parallel Programming for PaaS Clouds" project (Nr. 153560).

References

1. Abbott, M.L., Fisher, M.T.: The Art of Scalability: Scalable Web Architecture, Processes, and Organizations for the Modern Enterprise. Pearson Education (2009)
2. Badam, S.K., Elmqvist, N.: Polychrome: a cross-device framework for collaborative web visualization. In: Proceedings of the Ninth ACM International Conference on Interactive Tabletops and Surfaces, pp. 109–118. ACM (2014)
3. Bellucci, F., Ghiani, G., Paternò, F., Santoro, C.: Engineering Javascript state persistence of web applications migrating across multiple devices. In: Proceedings of the 3rd ACM SIGCHI Symposium on Engineering Interactive Computing Systems, pp. 105–110. ACM (2011)
4. Berners-Lee, T., Fischetti, M., Foreword By-Dertouzos, M.L.: Weaving the Web: the Original Design and Ultimate Destiny of the World Wide Web by its Inventor. HarperInformation (2000)
5. Bila, N., Ronda, T., Mohomed, I., Truong, K.N., de Lara, E.: Pagetailor: reusable end-user customization for the mobile web. In: Proceedings of the 5th International Conference on Mobile Systems, Applications and Services, pp. 16–29. ACM (2007)
6. Brusilovsky, P., Maybury, M.T.: From adaptive hypermedia to the adaptive web. Commun. ACM **45**(5), 30–35 (2002)
7. Casteleyn, S., Garrig'os, I., Maz'on, J.N.: Ten years of rich internet applications: a systematic mapping study, and beyond. ACM Transactions on the Web (TWEB) **8**(3), 18 (2014)
8. DB-Engines: DB-Engines ranking (2017). http://db-engines.com/en/ranking
9. Elmqvist, N.: Distributed user interfaces: state of the art. In: Distributed User Interfaces, pp. 1–12. Springer, London (2011)
10. Esenther, A.W.: Instant co-browsing: lightweight real-time collaborative web browsing. In: Proceedings of WWW (2002)
11. Frosini, L., Manca, M., Paternò, F.: A framework for the development of distributed interactive applications. In: Proceedings of the 5th ACM SIGCHI Symposium on Engineering Interactive Computing Systems, pp. 249–254. ACM (2013)
12. Gallidabino, A., Pautasso, C.: Deploying stateful web components on multiple devices with liquid.js for Polymer. In: Proceedings of CBSE, pp. 85–90. IEEE (2016)
13. Gallidabino, A., Pautasso, C., Ilvonen, V., Mikkonen, T., Systä, K., Voutilainen, J.P., Taivalsaari, A.: On the architecture of liquid software: technology alternatives and design space. In: Proceedings of WICSA, pp. 122–127. IEEE (2016)
14. Gallidabino, A., Pautasso, C., Ilvonen, V., Mikkonen, T., Systä, K., Voutilainen, J.P., Taivalsaari, A.: Architecting liquid software. J. Web Eng. (2017)
15. Ghiani, G., Paternò, F., Santoro, C.: On-demand cross-device interface components migration. In: Proceedings of the 12th International Conference on Human Computer Interaction with Mobile Devices and Services, pp. 299–308. ACM (2010)

16. Google: The connected consumer (2015), http://www.google.com.sg/publicdata/explore?ds=dg8d1eetcqsb1_
17. Hartman, J., Manber, U., Peterson, L., Proebsting, T.: Liquid software: a new paradigm for networked systems. Technical report 96–11. University of Arizona (1996)
18. Husmann, M., Marcacci Rossi, N., Norrie, M.C.: Usage analysis of cross-device web applications. In: Proceedings of the 5th ACM International Symposium on Pervasive Displays, pp. 212–219. ACM (2016)
19. Husmann, M., Norrie, M.C.: XD-MVC: Support for cross-device development. In: 1st International Workshop on Interacting with Multi-Device Ecologies in the Wild (Cross-Surface 2015), ETH Zürich, Switzerland, Zürich (2015)
20. Jazayeri, M.: Some trends in web application development. In: Future of Software Engineering, FOSE 2007, pp. 199–213. IEEE (2007)
21. Kovachev, D., Renzel, D., Nicolaescu, P., Klamma, R.: DireWolf - distributing and migrating user interfaces for widget-based web applications. In: Daniel, F., Dolog, P., Li, Q. (eds.) ICWE 2013. LNCS, vol. 7977, pp. 99–113. Springer, Heidelberg (2013). doi:10.1007/978-3-642-39200-9_10
22. Krug, M., Wiedemann, F., Gaedke, M.: SmartComposition: a component-based approach for creating multi-screen mashups. In: Casteleyn, S., Rossi, G., Winckler, M. (eds.) ICWE 2014. LNCS, vol. 8541, pp. 236–253. Springer, Cham (2014). doi:10.1007/978-3-319-08245-5_14
23. Lakshman, A., Malik, P.: Cassandra: a decentralized structured storage system. ACM SIGOPS Operating Syst. Rev. 44(2), 35–40 (2010)
24. Marcotte, E.: Responsive Web Design. Editions Eyrolles (2011)
25. Mikkonen, T., Systä, K., Pautasso, C.: Towards liquid web applications. In: Cimiano, P., Frasincar, F., Houben, G.-J., Schwabe, D. (eds.) ICWE 2015. LNCS, vol. 9114, pp. 134–143. Springer, Cham (2015). doi:10.1007/978-3-319-19890-3_10
26. Nebeling, M., Speicher, M., Norrie, M.C.: CrowdAdapt: enabling crowdsourced web page adaptation for individual viewing conditions and preferences. In: Proceedings of the 5th ACM SIGCHI Symposium on Engineering Interactive Computing Systems, pp. 23–32. ACM (2013)
27. Opera: Opera Unite reinvents the Web (2009), http://www.operasoftware.com/press/releases/general/opera-unite-reinvents-the-web
28. Oracle: Sun Ray products (2016), http://www.oracle.com/technetwork/server-storage/sunrayproducts/overview/index.html
29. Raposo, M., Delgado, J.: Empowering the web user with a browserver. In: Quintela Varajão, J.E., Cruz-Cunha, M.M., Putnik, G.D., Trigo, A. (eds.) CENTERIS 2010. CCIS, vol. 109, pp. 71–80. Springer, Heidelberg (2010). doi:10.1007/978-3-642-16402-6_8
30. Santosa, S., Wigdor, D.: A field study of multi-device workflows in distributed workspaces. In: Proceedings of the 2013 ACM International Joint Conference on Pervasive and Ubiquitous Computing, pp. 63–72. ACM (2013)
31. Schollmeier, R.: A definition of peer-to-peer networking for the classification of peer-to-peer architectures and applications. In: Proceedings of the First International Conference on Peer-to-Peer Computing, pp. 101–102 (2001)
32. Sivasubramanian, S., Pierre, G., Van Steen, M., Alonso, G.: Analysis of caching and replication strategies for web applications. IEEE Internet Comput. 11(1), 60–66 (2007)
33. Taivalsaari, A., Mikkonen, T.: A roadmap to the programmable world: software challenges in the IoT era. IEEE Softw. 34(1), 72–80 (2017)

34. Taivalsaari, A., Mikkonen, T., Systa, K.: Liquid software manifesto: the era of multiple device ownership and its implications for software architecture. In: 2014 IEEE 38th Annual Computer Software and Applications Conference (COMPSAC), pp. 338–343. IEEE (2014)
35. Voutilainen, J.-P., Mikkonen, T., Systä, K.: Synchronizing application state using virtual DOM trees. In: Casteleyn, S., Dolog, P., Pautasso, C. (eds.) ICWE 2016. LNCS, vol. 9881, pp. 142–154. Springer, Cham (2016). doi:10.1007/978-3-319-46963-8_12
36. Wallis, M., Henskens, F., Hannaford, M.: A distributed content storage model for web applications. In: Proceedings of 2010 Second International Conference on Evolving Internet (INTERNET), pp. 98–106. IEEE (2010)
37. Web Bluetooth Community Group: Web bluetooth (2017), https://webbluetoothcg.github.io/web-bluetooth/
38. Yang, J., Wigdor, D.: Panelrama: enabling easy specification of cross-device web applications. In: Proceedings of the 32nd Annual ACM Conference on Human Factors in Computing Systems, pp. 2783–2792. ACM (2014)

Twisting Web Pages for Saving Energy

Eda Köksal[1], Yeliz Yeşilada[1(✉)], and Simon Harper[2]

[1] Middle East Technical University Northern Cyprus Campus,
Güzelyurt, Mersin 10, Turkey
edakoksal89@gmail.com, yyeliz@metu.edu.tr
[2] Schoool of Computer Science, University of Manchester, Manchester, UK
simon.harper@manchester.ac.uk

Abstract. Battery capacity (energy density) is increasing at around 3% per year. However, the increasing requirements of the mobile platform is placing higher demands on this capacity. In this case, there are three options: decrease our expectations of the mobile platform, increase the capacity and therefore size and weight of our batteries, or create energy saving solutions to extend battery-life with minimal effect on platform performance. Here we present a system called Twes+ which is inline with the last option and aims to transcode web pages for increasing battery-life when surfing-the-web without changing the look and feel. Our evaluation results show that there is a statistically significant energy saving when using our Twes+ transcoder. Our *redirect* service brings a 4.6% cumulative processor energy reduction, while *image* transcoding service, brings a 7% cumulative processor energy reduction. These savings equate to between a 40 to 60 min saving depending on the mobile device.

Keywords: Transcoding · Energy · Green computing · Mobile web

1 Open Data

Twes+[1] is open sourced at https://github.com/EdaKoksal/Twest.git. This repository includes materials used, raw measurements and statistical test results.

2 Introduction

Mobile device ownership is increasing worldwide especially in developing regions. For example, in 2014, the number of mobile Internet users exceeded the number of desktop users and in 2015, the number of connected devices per person rose to 3.42. 2020s prediction shows that this ratio will be doubled [34]. Although the popularity of mobile devices increases, these devices still have constraints such as the limited bandwidth of connection, device memory, processing power, battery size, etc. [16,28,39]. Further, battery capacity (energy density) is only increasing at around 3% per year [34]. On the other hand, web technologies evolve quickly

---

[1] Twes+ is pronounced as Twist

© Springer International Publishing AG 2017
J. Cabot et al. (Eds.): ICWE 2017, LNCS 10360, pp. 225–245, 2017.
DOI: 10.1007/978-3-319-60131-1_13

and these evolutions normally require increased energy capacity. Indeed, the total size and the total number of requests required between client and server are steadily increasing. For instance, total transfer size has increased from 831 KB to 2135 KB between 2012–2015 [3]. These changes and also the limitations of mobile devices can make their everyday usage inconvenient. Mobile devices connect to the web usually through a cellular connection or a Wi-Fi. These connections consume a serious amount of energy. For example, the battery life of an iPhone 7 is up to 240 h on standby, 12 h on the cellular connection and 14 h on Wi-Fi connection [46]. As a result of the constraints and the changes in the web, accessing the web from a mobile device can quickly drain the battery [19].

To address these limitations, this paper presents a system called Twes+ (*T*ranscoding *W*eb pages for *E*nergy *S*aving) that is proposed to save energy on the client-side. Twes+ aims to systematically transcode the source code of web pages on the proxy so that, during browsing, less energy is consumed on the client. Transcoding here means the process of transforming and adapting the source code of web pages into alternative forms to improve the user experience (UX) [4]. While Twes+ transcodes web pages, it takes three principles into account: the look&feel of the web page should not be changed, the changes should not depend on server-side, and the changes should not add extra energy consumption or settings to the client-side. When a web page is visually designed, it includes implicit information that guides the user around the web page [42]. Therefore, we want to preserve this implicit information. Even though some web pages have specific mobile versions, these typically include the subset of what is available on the desktop, so there is information loss. This can be addressed with ResponsiveWeb Design (RWD) but in practice not many sites are "fully responsive" and they can also remove functionality required on mobile devices [20]. Therefore, with this principle, we do not want to reduce the amount of information, and we do not want to change the look&feel. Regarding the second principle, even though there exist guidelines to develop web pages with energy awareness, very few pages take them into account [16,28]. Furthermore, there are many non-technical/programmer web developers, so expecting them to address these guidelines might not be realistic. By keeping Twes+ independent from the web server, the proposed adaptations can be applied to any web page systematically. That is why Twes+ is proposed as a proxy-based system. Regarding the third principle, the focus is to improve the energy consumption on the client, so the developed system should not add extra energy consumption to the client and also the user should not be asked to make configurations, etc. as users tend not to use such advance settings [27]. Therefore, Twes+ is implemented as a proxy which does not modify the look&feel of the page, not require any extra workload from the server side, not require the user to make any configurations and not consume extra energy on the client.

In the literature, many transcoding techniques are proposed (Sect. 3). Some of these aims to improve the UX for mobile or disabled web users, and because of the principles introduced above they are not appropriate for our purpose. Techniques have also been introduced to save energy such as compression servers, and are proven to save energy on the client-side [5,13,16,38,44]. However, there are still techniques that can improve energy but have not been investigated which include reducing the number of redirects and consolidating images. These are already proposed as guidelines but their effect on energy saving have not been demonstrated [12,16,17,28,30,44]. Therefore, Twes+ encodes two transcoding techniques: redirect transcoding and image transcoding service based on Internet Content Adaptation Protocol (ICAP) (Sect. 4). Therefore, the overall **contribution** of Twes+ is scientifically showing that these transcoding techniques can save energy as suggested by the unproven guidelines.

In order to assess the impact of Twes+, we conducted an experiment which investigates the following research questions: 1. "Does Twes+ save energy by applying redirect transcoding to reduce the number of redirects?" 2. "Does Twes+ save energy by reducing the number of requests/responses between client&server to retrieve images?". For redirect service, a controlled web page is used, and three different redirection cases are created. Case 1 is designed without any redirection. When the web page is requested, the content can be accessed from the original URL. Case 2 is designed with five redirections, as more than five redirections indicate an infinite redirection loop [18]. The web page content starts loading after the first five redirection requests/responses. Case 3 is designed with infinite redirection where the web page content cannot be accessed.

For image service, we used Alexa top 100 pages and assess the energy consumption with and without Twes+ (Sect. 5). For comparison, we measured the number of requests/responses and the cumulative energy of the processor. The experiment illustrates that Twes+ saves energy on the client side by reducing the number of redirect round trips and the number of image requests/responses without modifying the look&feel of the web pages. Twes+ was able to eliminate the infinite temporary redirect round trips and with Twes+ 4.6% reduction in cumulative processor energy is observed. With image transcoding service, on average 7% reduction in cumulative processor energy is observed by reducing the number of request/response round trips between the client and the server. It is also observed that this percentage is more than 10% with pages that include many images (Sect. 6). Based on these results, we also look at battery life improvements with Twes+ and our estimations show that Twes+ can provide significant battery life improvements, especially when the page has a lot of images and the number of redirection is above the given limit in the HTTP specification (Sect. 7). For example, in average Twes+ provides 7.6% power reduction and if we consider specific pages, for example ebay, we can say that Twes+ provides 40.27 min battery improvements when the cumulative processor energy is equal to the entire system energy consumption and 24.99 min improvement when the cumulative processor energy is equal to the 70% of the system energy consumption (assuming that 30% is used by other processes, and only ebay is processed). Although there are some limitations in the current version of Twes+ (Sect. 7), the overall results show that Twes+ saves energy by systematically transcoding the source code of web pages (Sect. 8).

3   Related Work

Web pages can be accessed in different ways on mobile devices including native applications, mobile web applications, widgets and browsers. Among these, browsers are cross-platform, and they can be used to access any web page. Most of the mobile browsers provide ways to control energy saving on mobile devices. For example, they can be used to block accessing the flash content, compress data, and transcode the visual look&feel of web pages. For instance, Google Chrome browser has data saver [13] on mobile device browsers for compression, Apple Safari has power saver on desktop browsers of laptops for blocking the flash content and Opera Mini browser has data saving mode [32] on mobile device browsers for transcoding the web pages to reduce the content, etc. These are simple but yet very useful techniques for saving energy, however more complex techniques can be explored that can potentially save even more energy.

When we consider the web architecture, we can see that energy consumption can be reduced in three ways: on the hardware side, network interaction and the software/application side. On the hardware side, there are studies that consider the ways of using renewable energy sources in the mobile devices themselves to improve the battery life. For instance, methods are introduced to integrate thermoelectric generator to the Central Processing Unit (CPU) heat pipe to use the waste heat of the processor and integrating photovoltaic unit to the devices to generate power from environmental illumination [2]. Intel Corporation developed a mobile processor that uses Core-Multi-Processor micro-architecture to achieve high performance with low power consumption [21]. Even though these are very crucial for saving energy on mobile devices, software and network can also be as important as hardware for saving energy.

There are also studies to improve energy consumption on the network side. For instance, the energy consumption of a customer is examined, and the results show that 0.83Wh/day for a terminal is consumed whereas 120Wh/day for the mobile network is consumed [15]. There are also studies to reduce the energy consumption of localization services such as a system called Senseless. Instead of only using the Global Positioning System (GPS), it combines accelerometer, GPS and 802.11 access point in location-aware services. Their early results show that their design can extend the battery life of mobile devices from nine hours to 22 h [6]. Another proposed optimisation for localisation is changing the host-centric network paradigms. The current host-centric approach is based on placing all intelligence at the host. By adopting an information-centric approach or by adapting a multi-access mobile networking, energy saving can be achieved. Briefly, the information-centric approach means the network connects the information consumers with information producers and distributors instead of nodes. The multi-access mobile networking approach means it uses surrounding networks together [34].

Compared to network and hardware based studies, there are limited studies on the software side. There are some studies on the protocol side. Google, for instance, has developed a protocol called SPeeDY Protocol (SPDY) which is proposed as a replacement of Hypertext Transfer Protocol (HTTP). It allows multiple requests in a Transmission Control Protocol (TCP) session, so the number of TCP sessions setup can decrease. In addition to this, the server and the client can push data to each other without a setup request. The main purpose of SPDY protocol is to reduce the latency [11]. In addition to studies in the protocol, there are also specific studies on mobile operating systems. For instance, for Android OS, one study shows that better compilers for Android can save energy [33]. There is also a study for energy saving on the mobile applications. This study gives guidelines to develop applications that considers the limitations of mobile applications by studying some micro-benchmarks. By writing the same micro-benchmark or common primitives or briefly code refactoring, the battery life improvement can be achieved [47]. In addition to these studies, the energy consumption by a browser to display components of a web page is also measured. For example, to download and display Apple page, the required energy is approximately 46 J–12 J of this 46 J is to download and render CSS files. To decrease the energy consumption, non-used functions are removed in the CSS file and five Joules have been saved [31]. Even though these are important studies to show how the energy is consumed by the browser, the results of such studies can be only be used to develop guidelines for web developers.

Besides these studies on the software side, some transcoding techniques have also been introduced in the literature either as a browser plugin, a client-side application or as a proxy based applications. Table 1 shows these techniques and our examination of them according to the principles discussed in the Introduction section: do they modify the look&feel of the page?, do they do transcoding on the server, client or proxy side? are there any studies to demonstrate that they save energy in the client?

There are some techniques that focus on adapting the content of web pages to address the screen size limitation and therefore they modify the look&feel. For example, some techniques provide a summary of the page [1,4,26,39], however even though this could be good for energy reduction, it also significantly reduces the content. Similarly, some techniques are introduced to simplify the page that can save energy [4,7,43]. Furthermore, there are also techniques that can potentially save energy, for example, techniques that rearrange the page content such that only the important parts are displayed to save energy [4,35,36,39], techniques that automatically add alternative texts to images that can replace images for saving energy [4,22,40], and color scheme changes so that the client spends less energy for rendering [4,22,40].

When we look at the location of the transcoding implementations exist in the literature, we see that there are some techniques already implemented as a proxy such as data compression and concatenation of external files. These studies focus on compressing data so to reduce the amount of data transferred [5,8,13,16,23] and they focus on concatenating external files such as stylesheets and scripts to reduce the amount of HTTP requests [16,38,44]. These studies demonstrate to improve energy consumption on mobile devices. There are also some techniques implemented on the proxy that affect the battery life negatively. For example [29] proposes image inlining but this requires both encoding on the server side and

Table 1. Transcoding methods [§P/wSN: People with Special Needs, S: Server-Side, C: Client-Side, P: Proxy-Side, *: Reverse-Proxy].

Technique	Reference	Location			Modify the look & Feel	Adaptation for P/wSN §	Screen size adaptation	Battery life improvement
		S	P	C				
Text magnification	[4]	✓	χ	✓	✓	✓	✓	?
Color scheme changes	[4,22,40]	χ	✓	✓	✓	✓	χ	?
Alternative text insertion	[4,22,40]	χ	✓	✓	✓	✓	✓	✓
Page rearrangement	[4,35,36,39]	✓	✓	χ	✓	✓	✓	?
Simplification	[4,7,43]	✓	✓	χ	✓	χ	✓	✓
Summarization	[1,4,26,39]	χ	✓	✓	✓	χ	✓	✓
Image consolidation	[12,16,17,24]	✓	✓*	χ	χ	χ	χ	?
Responsive image	[12,16,20,28,30]	✓	χ	✓	χ	χ	✓	?
Image inlining	[28,29]	✓	χ	χ	χ	χ	χ	χ
Data compression	[5,8,13,16,23]	✓	✓	✓	χ	χ	χ	✓
Concatenation of external files	[16,38,44]	✓	✓	χ	χ	χ	χ	✓
Reducing the number of redirects	[37,41,44]	✓	✓*	χ	χ	χ	χ	?
Expiration header	[16]	✓	✓	✓	χ	χ	χ	✓

decoding of these images on the client side. This was demonstrated to be six times slower than external linking on mobile devices.

Finally, when we look at Table 1, we see that there are still three techniques that can potentially improve the battery life and do not modify the look&feel of the page but they have not been technically investigated. These are image consolidation, responsive image and reducing the number of redirects. They are proposed as guidelines for developers to make their pages faster, but they have not been implemented as transcoding techniques. Although in Twes+, these promising techniques are implemented on a proxy, this paper only covers the image consolidation and the number of redirects reduction.

4   Twes+

Fig. 1 shows the software architecture of Twes+. Squid Caching Proxy is used as a proxy server (A in Fig. 1) which is a popular open source proxy server to cache the frequently accessed web pages. It can reduce the response time and the payload. Moreover, Squid Caching Proxy supports different content adaptation mechanisms. The ICAP mechanism is adopted because it can modify both the

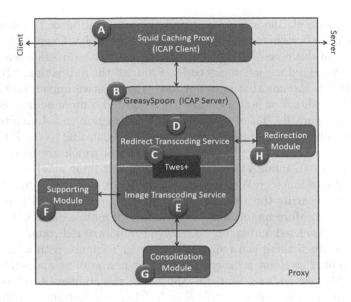

Fig. 1. Software architecture of Twes+.

request and response content. Moreover, it is not dependent on a single proxy server project or vendor [14].

The ICAP requires an HTTP proxy as a client (A) and an ICAP-server (B) to reside the adaptation algorithm [14]. The relation between the ICAP-server and the ICAP-client is many-to-many. The content of the web page comes from the ICAP-client to the ICAP-server for adaptation. In the Squid Caching Proxy-ICAP section [2], some ICAP-servers are suggested such as C-ICAP, ICAP-Server, POESIA, Traffic Spicer and GreasySpoon [14]. GreasySpoon (B) is adopted for Twes+ as it is an extensible open-source project [25]. Twes+ (C) resides on the GreasySpoon ICAP-server.

Twes+ mainly transcodes the response of a web page. When a web page is requested by the client and when the web server replies with a response, the data passes through the Squid Caching Proxy. Squid Caching Proxy forwards the response to the ICAP-server. Twes+ (C) resides on ICAP-Server (B) and contains two different adaptation services: (1) redirect transcoding service (D) and (2) image transcoding service (E). After the adaptation, the adapted response goes back from ICAP-Server to ICAP-Client and then from ICAP-Client to the client.

4.1   Redirect Transcoding Service

The goal of this service is to handle the temporary redirects. The content of an HTTP message is the status line, header and body. The status line shows

---

[2] http://wiki.squid-cache.org/Features/ICAP.

the HTTP protocol, the status code and the explanation of the status code. The status code is a three-digit integer and the most significant digit gives the classification of this code [18]. 3xx status codes are the redirection status codes. There are several redirection status codes. From all the redirection, "Temporary Redirect 307" is the one that does not have any positive impact on the client-side [18]. This redirection indicates that the requested content address is changed to another temporarily. There is also another redirection called infinite redirect loop and usually occurs because of a missing file. According to the HTTP specification, the client side should detect the redirects which are more than five as this indicates an infinite loop [18]. Although this limit is adjustable on the browsers, infinite loops create significant amount of network traffic. Moreover, each redirection carries the overhead of a request/response but without the content of the page. More importantly, the battery of mobile devices drains while the device is connected and trying to resolve temporary redirects.

Normally, when there is a redirect, the server response with a new address for the content. The client browser extracts this new address and sends another request to this new address. Statistics show that the number of web pages that includes redirection increased from 61% to 78% between December 2010 and January 2016 [3]. With Twes+, our goal is to eliminate these temporary and infinite redirects to save energy. When Twes+ redirect transcoding service receives the response, it checks the status code. If the status code indicates that there is a redirect, it starts the adaptation. The redirect transcoding service extracts the new address from the location field of the header and forwards this address to the redirection module (H in Fig. 1). There can be three cases: 1. there is no redirection; 2. the number of redirection is less than the limit which is recommended as five; 3. the number of redirection is more than the limit.

For case 1, redirect transcoding service forwards the responses without modification. For case 2, redirection module returns the new address to redirect transcoding service. The content of the web page is requested from this new address and forwarded to the client. For case 3, redirection module stops tracking after the fifth redirection. It returns negative to the redirect transcoding service. Redirect transcoding service updates the header and the body of the response as a warning message. This response is forwarded to the client. The results of the redirect transcoding module indicates that there is a reduction in the number of redirect round trips.

4.2 Image Transcoding Service

The image content of web pages are external resources except when they are included as data URI. External resources require request and response roundtrips to be added to a web page. The overall goal of image transcoding service is to combine images in order to reduce the number of requests and responses required to add images to web pages. In order to achieve this, this service includes a number of steps. The first request of the client passes through the proxy and reaches to the server. The server forwards the content and the HTML source is selected by Twes+ image transcoding service (E in Fig. 1). If the HTML source code of the page does not include an image tag, this file is transferred to the client without any modification. On the other hand, if there is an image tag, the transcoding process starts. On image transcoding service, the image tags are tracked to derive the essential data of the image. Image tags require minimum two attributes which are the location <src> and an alternate text <alt>. Although there are other attributes, the dimension attributes are controlled by Twes+. In addition to the individual image tags, there are arranging structures like frame tags that can include image tags inside them. These tags are processed as well. To sum up, Twes+ extracts image tags and retrieve the source, the width and the height. The default value of the width and the height are set zero in case they are not defined. After extracting these three attributes, they are forwarded to the supporting module (F in Fig. 1) of Twes+. This module downloads the image from the web server by using the given source attribute. The dimensions of the downloaded image are found out by ImageMagick[3]. The supporting module compares the required and the original dimensions. This two values might be equal or it might be the case that the second one is not specified. Therefore, we made an assumption that both the required and the original dimensions are the same. There is also a possibility that the required dimensions are defined but they are not the same as the original image. It means that the browser of the client needs to resize the image. However, this resizing can consume the battery of the client. Therefore, the resizing process is also done by the supporting module. When the required and the original dimensions are not equal, the supporting module applies the resizing to the original image. This way, the server-side does not change the original image while the client-side does not apply resizing.

In the current version of Twes+, the following image formats are supported PNG, JPEG/JPG, and GIF [3]. Since the goal is not to modify the look&feel of the web pages, the format of the images in the original web page is preserved. To be able to consolidate images, the name of the images on the web page are altered. The new name carries the name of the web page, the format of the image and a number based on the order of this format appears in the HTML. There are some possibilities that the supporting module returns negative. These cases are when the downloaded file is not an image or it is an animated GIF image. In order, not to have problems with the image formats that are not supported, only supported formats are considered in our module. Animated GIF images are not modified because they are the combination of several images and the dimensions do not have to be fixed. Responsive image technique is also applied for mobile devices by the supporting module.

After all the image tags are examined, the consolidation module (G in Fig. 1) is called. Consolidating all the images into one image is not a practical idea. If a web page contains too many images, this makes the concatenated image size too large and sending a large image is not safe in case of any network failure. In Twes+ implementation, every ten images of the same format are concatenated. The limit is ten because WildCards are used during the consolidation process and WildCards works on base ten. According to these concerns, the consolidation

---

[3] http://www.imagemagick.org.

(a) The First Image

(b) The Second Im-(c) The Third Image
age

Fig. 2. Images before concatenation

Fig. 3. Concatenated version of all images in Fig. 2 [Width:679 pixels Height:371 pixels]

module combines all the downloaded images. According to the order and the format of the image, the consolidation process is performed (see Fig. 2 for individual images and Fig. 3 for consolidated image from xkcd-Squirrelphone[4]). The consolidation module generates a decoded name with a keyword.

Images are concatenated vertically. The CSS is used to split images to access individual images. The cumulative height is used to determine the specific position of the images. Image tags are replaced with the new ones. The source is defined as a spacer.gif to hold a place in the layout for the image. The CSS styles with the location of the source are injected into the HTML file so an extra request/response is prevented. The modified HTML file is delivered to the client. The client-side requests consolidated image from the proxy. The *url_rewrite_program* feature of Squid Caching Proxy is used. When the requests are passing through the proxy, the keyword is tracked. If a request comes from the client that includes the keyword, the response is searched inside the proxy. Otherwise, the request is processed as it is.

5   Evaluation

In order to experimentally evaluate whether or not Twes+ can save energy, we conducted a study. This study aims to investigate the following two research questions:

1. *Redirect Transcoding Service:* The goal is to validate whether or not Twes+ redirection transcoding service can reduce the number of redirects and Twes+ can save energy. Therefore, the research question is *"Does Twes+ save energy by applying redirect transcoding to reduce the number of redirects?"*.
2. *Image Transcoding Service:* In order to validate whether or not Twes+ can save energy by reducing the number of image requests/responses with image consolidation technique. Thus, the research question is *"Does Twes+ save energy by reducing the number of requests/responses between client and server to retrieve images?"*.

---

[4] http://xkcd.com/1578/.

To address these questions, we followed different evaluation strategies for each of them. Regarding the first question, we mainly used a controlled web page and introduced different cases for measurement. Regarding the second question, we used real web pages and we did measurements with original and modified pages by Twes+.

5.1   Materials

For the redirect transcoding service experiments, we used the xkcd-Squirrelphone[5] web page as a controlled medium for measurement. In fact, we could use any web page for measurement as the content here is not relevant but since this was used in the literature [9] for other purposes, we decided to reuse this web page. We controlled the number of redirects by mainly creating different versions of this page by injecting controlled number of redirects. We mainly created three cases:

- Case 1: The control web page can be reached without any redirect.
- Case 2: After five redirects, the recommended action to the client-side is to take an action as infinite redirection. According to this suggestion, by using same web page and five redirects this case is designed [18]. When the client requests the web page, after five redirects web page content can be accessible.
- Case 3: With the same web page, over the server infinite redirection loop is designed. In this case, the client does not receive the content of the web page. After the fifth redirection, Twes+ changes the content to a warning message about the infinite redirect to the client.

Regarding the image transcoding service, we used web pages selected from the top 100 list of Alexa. We had to download these pages and serve them locally in order to do measurements. Some web pages we could not use because of the following reasons. Some sites have duplicates in different countries and more than one of those can be included in the top 100 list, therefore we eliminated replicated ones and kept only the first version in the list. For example, Google has duplicates for different countries. We could not do measurements for sites that require logins as we deleted the caching of browsers so we also eliminated those from the list downloaded. This was also a concern from the security perspective. Finally, when we downloaded web pages locally, we checked to see if their look&feel is modified, if it did, we also eliminated those. As a final check, we also wanted to make sure that we test pages that can be modified by Twes+ based on the supported image formats so if the page did not include the image formats supported by Twes+, we also eliminated those. At the end, we did the measurements for 16 web pages which are listed in Table 3.

5.2   Equipment

In order to perform the measurements, we need to control the overall architecture. Therefore, we used three identical computers as a client, a proxy and a

---

[5] http://xkcd.com/1578/

server (see Fig. 1). Their processor is Intel Core i7, model: 4770. The base and the turbo frequencies are 3.4 GHz and 3.9 GHz, respectively. Installed memory is 16 GB. The brightness of the screen is 50% and unified over all the measurements. The switch between these three desktops, Cisco Catalyst 2960-X Series is used and its speed is 80 Gbps (Gigabits per second).

5.3   Tools

The evaluation of Twes+ is done by measuring some key metrics with and without Twes+. These metrics are the number of requests/responses and the cumulative processor energy. Measurements are performed with Chrome Devtool and Intel Power Gadget. Devtool is used to measure the number of requests/responses. There are other browser base tools but the network throttling feature of Devtool is the main reason we preferred it over others. The network system is connected via wire. To be able to simulate regular 3G network, Devtool network throttling is used. The 3G network is preferred because it is more common for mobile devices and also it has a wider coverage. Moreover, it is able to give the number of requests, the load duration of the web page and the network timeline of the elements during the page load [10].

Intel Power Gadget is used to measure the energy consumed by the processor. It estimates power data by reading the data of registers and energy counters of the processor. It supports Intel Core processors from 2nd Generation up to 6th Generation Intel Core processors. It is able to measure the current package power consumption and the limit of the package power. In addition to power consumption, it gives the frequency of the package and the base frequency of installed processor.

5.4   Methodology

To be able to know the reduction in the number of redirects and in the number of image requests/responses, the overall number of requests/responses is measured. To observe the impact of Twes+, the energy consumed by the processor is measured with and without Twes+. During the measurements, the cache and the history of all the components (client/server/proxy) are cleared. Entire system works locally to prevent the influence of the Internet traffic. The system is tracked by using Intel Power Gadget and when there are no changes, the measurements are started and the measurements are repeated minimum of three times to ensure that there is no external effecting factors. While measuring the number of requests/responses, on the client only Chrome Devtool is left active. While measuring the power and the energy, on the client only Intel Power Gadget and the browser are left active. Although there is no limitation for the duration of the test for the number of requests/responses measurements, the duration of the power and the energy measurements is set as 30 seconds so the impact of starting and stopping the tools to measure is eliminated.

6   Results

We present our results based on the research questions introduced above.

"Does Twes+ save energy by applying redirect transcoding to reduce the number of redirects?"

For each of the three cases, the number of requests/responses, the load duration and the cumulative processor energy are measured with and without Twes+. These measurements are done when the web page is requested by the client. Table 2 shows the overall results. The required number of requests/responses for the control web page is seven. For case 1, the number of requests and responses are the same because there is no redirect. For case 2, the content is loaded after five redirects. Without Twes+, the number of requests/responses includes these redirects. With Twes+ it remains same as case 1. For case 3, the content cannot be reached because of the infinite number of redirects. The browser is prevented from interfering the redirects so the redirects never stop without Twes+. On the other hand, with Twes+ redirects are stopped. The client received a warning message from Twes+. By reducing the number of requests/responses, a reduction in the cumulative processor energy is observed for three cases. For case 1, this was dropped from 63.55 to 61.13 mWh, which means 3.8% decreased was observed on the original case. In case 2, this was dropped from 80.39 to 61.7, therefore 23.3% decrease was observed on the original case. In case 3, this was dropped from 76.02 to 72.5 which means 4.6% decrease observed (see Table 2). Finally, regarding the load duration, in case 1, we did not observe any difference (2.06 vs. 2.04) and in case 3, we could not measure for without Twes+ case and with Twes+, the loading finished in 0.576 s. In case 2, we observed a small increase in the load duration because of the process over Twes+. This process is detecting redirects, extracting the new location, tracking the remaining redirects and retrieving the content from the web server.

"Does Twes+ save energy by reducing the number of requests/responses between client and server?"

For each web page used in the experiment, the number of image requests/responses, the load duration and the cumulative processor energy are

Table 2. Total number of requests/responses and cumulative processor energy [w/:with, w/o:without]

	Number of requests/ responses		Load duration (s)		Cumulative processor energy (mWh)	
	w/o Twes+	w/Twes+	w/o Twes+	w/Twes+	w/o Twes+	w/Twes+
Case 1	7	7	2.06	2.04	63.55	61.13
Case 2	12	7	1.73	2.08	80.39	61.70
Case 3	69	2	*	0.576	76.02	72.50

measured with and without Twes+. The mean, median and the standard deviation of these results are also calculated. Table 3 shows that with Twes+, although Twes+ increases the page load duration, the number of total image requests/responses and the cumulative processor energy are lower than without Twes+. To determine whether or not the difference between with and without Twes+ is significant, the Paired-Dependent T-Test or its non-parametric alternative Wilcoxon Signed Rank Test is applied since same web pages are measured with and without Twes+ and the design is repeated measures. The normality of the difference between the results with and without Twes+ are checked by using ShapiroWilk (SW) test to decide either Paired-Dependent T-Test or its non-parametric alternative Wilcoxon Signed Rank Test is needed to be applied.

Table 3. Number of image requests/responses, the load duration(finish) and the cumulative processor energy [w/:with, w/o:without, SD: Standard Deviation]

Rank	Name	Number of image requests/responses		Load duration(s)		Cumulative processor energy (mWh)	
		w/o Twes+	w/Twes+	w/o Twes+	w/Twes+	w/o Twes+	w/Twes+
17	Ebay.com	165	28	25.57	30.15	76.18	66.75
28	Microsoft.com	30	5	17.92	20.6	93.72	89.02
29	AliExpress.com	192	31	84	96	76.23	77.03
33	Ask.com	6	5	3.4	3.68	67.22	63.62
43	Imgur.com	70	13	9.19	10.27	78.23	75.58
47	Imdb.com	61	30	10.46	12.49	73.22	67.58
50	Fc2.com	11	10	4.08	8.5	79.33	74.24
55	Stackoverflow.com	9	8	4.08	4.23	78.04	75.94
61	Odnoklassniki	56	18	21.43	23.4	77.34	75.37
72	Booking.com	37	26	7.52	8.36	73.74	63.96
75	NicoVideo.jp	43	16	7.43	21.61	80.65	72.74
76	Flipkart.com	149	29	21.48	28.95	83.38	72.18
85	BBC.co.uk	59	14	17.03	16.43	77.14	66.03
95	DailyMotion.com	45	9	10.93	12.29	75.43	67.32
96	Wikia.com	33	14	28.72	33.25	66.35	65.63
97	ChinaDaily.com.cn	108	87	36.39	29.65	89.29	80.80
	Mean	**67.13**	**21.44**	**19.35**	**22.49**	**77.84**	**72.11**
	Median	**50.50**	**15.00**	**13.98**	**18.52**	**77.24**	**72.46**
	SD	**56.90**	**19.69**	**19.78**	**21.83**	**6.92**	**6.92**

The result of the statistical tests presented in Table 4 shows that there is a significant difference between with and without Twes+ for the number of image requests/responses, the load duration and the cumulative energy. Interpretation of these results is that although there is a minor increase in the load duration of the web pages, there is a statistically significant reduction in the number of image requests/responses and this leads to energy saving with Twes+. To find the magnitude of effect, r value is calculated for the number of image requests/responses

Table 4. Statistical test differences with and without Twes+

	Number of image requests/responses	Load duration (s)	Cumulative processor energy (mWh)
Test	Wilcoxon	T-Test	T-Test
N	16	16	16
df		15	15
p	<.0005	.023	<.0005
t or z	-3.5198	2.532	-6.019
Eta Square or r	.6221	.3	0.71

and the results indicate a large effect size. Eta Squared statistics applied for the load duration and the cumulative processor energy and the results indicate respectively moderate and large effect size.

7 Discussion

Twes+ is a novel system that combines two different transcoding techniques to improve the energy consumption on mobile devices while browsing the web. These techniques are previously recommended as guidelines for developers but they are not implemented as Twes+ does. To our knowledge Twes+ is the only system that implements these techniques on the proxy. The experiments presented here also demonstrate that Twes+ saves energy, in particular it reduces the number of requests/responses between client and server and consequently it saves processor energy. Therefore, the overall contribution of Twes+ is to scientifically show that these suggested techniques as guidelines do save energy. Here we first discuss the results based on the two services included in the Twes+, and then we discuss the impact of energy saving on the battery life.

Twes+ redirect transcoding service is experimentally evaluated with three cases. The results show that the cumulative processor energy is almost the same with and without Twes+ while there is no redirect. On the other hand, Twes+ can eliminate the finite and infinite redirects of the web page and it leads to energy saving on the client-side. This reduction is because of Twes+ is removing the unnecessary requests/responses. Every redirect carries the overhead of the request and response without the content of the web page. Even though the browsers have a limit to prevent the infinite redirects, the round trip still occurs on the client-side. With Twes+, all the temporary finite or infinite redirects are removed.

To estimate the impact of this energy saving on the battery life of mobile devices, we made the following assumptions about the battery of the client device. The battery specification of a laptop (TOSHIBA Satellite P70) has the capacity of 4400 mAh and the voltage is 10.8 V. We estimated the battery life improvement for two cases: the cumulative processor energy is equal to the entire system energy consumption and we denote this as 100% and 70% of the entire system (in this case, we assume that 30% is used by other processes). Table 5 shows the battery life improvement estimations for the three cases of this transcoding service. The battery life of mobile devices varies from the tasks done over the device. With Twes+, this duration is extended for browsing by eliminating the redirects. If we assume that the cumulative processor energy is equal to the entire system energy, in case 2 the battery of the mobile device can stay 90.5 min longer. In real system, there are other components that consume energy such as screen, ethernet, etc. Therefore, if we assume that the processor energy is the 70% of the entire system energy consumption, in case 2 the battery life of the mobile device can stay 48.6 min longer.

Table 5. Battery life analysis for redirect transcoding service results

	Power reduction (%)	Battery life extension (100%) (min)	Battery life extension (70%) (min)
Case 1	2.83	10.9	7.4
Case 2	23.23	90.5	48.6
Case 3	4.33	14.3	9.6

We used a different strategy to evaluate Twes+ image transcoding service. We can say that we tested it in the wild. We picked actual web pages from Alexa 100. Instead we could also have a controlled page and do the measurements based on that, however we wanted to make sure we demonstrate the impact of Twes+ on real web pages. For some web pages as can be seen in Table 3, the reduction in the cumulative processor energy is low. This is mainly because the number of images is low. For example, StackOverflow page includes nine images requests/responses and with Twes+ this number reduces to eight. The number of transcoded images in these web pages is low because some of the images are inline images or background images or requested by the CSS files or the format of the image is not supported. However, in most pages, we observe 50% reduction in the number of requests/responses and also reduction in the cumulative processor energy. For example, the number of image requests/responses for Flipkart reduced from 149 to 29 images and the cumulative processor energy reduced from 83.38 mWh to 72.18 mWh (i.e. 13.4%).

To see the impact of this energy saving on the battery life of mobile devices, we followed the same assumption that the client works on battery with the specification given above. With the same ratios between the entire system energy and the processor energy consumption, Twes+ is able to extend the battery life (see Table 6). We see 7.6% power reduction and the mean value of the battery life extension is 26.04 min in case the cumulative processor energy is equal to the 100% of the entire system energy and 16.20 min in case the cumulative processor energy is equal to the 70%of the entire system energy. For example, although the number of images are low in StackOverflow page, with Twes+ there is 4.7% power reduction and 14.90 min (the cumulative processor energy consumption is equal to the entire system) or 9.94 min (the cumulative processor energy consumption is equal to 70% of the entire system) extension can be achieved. For Flipkart page, the battery life extension is 53.23 min for 100% case and 31.42 min for 70% case, and observed power reduction is 15.7%.

Table 6. Battery life analysis for image transcoding service results

		Power reduction (%)	Battery life extension (100%) (min)	Battery life extension (70%) (min)
17	Ebay	11.4	40.27	24.99
28	Microsoft	7.1	19.49	12.67
29	AliExpress	0.9	2.79	1.93
33	Ask	4.9	18.01	11.99
43	Imgur	0.9	2.66	1.85
47	Imdb	7.4	26.11	16.92
50	Fc2	5.0	15.98	10.62
55	Stackoverflow	4.7	14.90	9.94
61	Odnoklassniki	3.5	11.16	7.54
72	Booking	17.2	65.97	38.24
75	NicoVideo	9.2	30.08	19.13
76	Flipkart	15.7	53.23	31.42
85	BBC	14.8	53.88	32.13
95	DailyMotion	9.8	34.37	21.70
96	Wikia	1.5	5.54	3.82
97	ChinaDaily	7.5	22.21	14.39
	Mean	**7.6**	**26.04**	**16.20**
	Median	**7.3**	**20.85**	**13.53**
	SD	**5.14**	**19.10**	**11.04**

Besides these estimations, we also looked at the impact of image transcoding service of Twes+ on some popular mobile devices. The effect of 7% reduction is investigated for Samsung Galaxy S4-5-6-7 and iPhone 5-6s by using their battery specifications. On average, Internet usage time is estimated to be extended by 40.6 min(3G) and 48.3 min(Wi-Fi), and specifically for Samsung Galaxy S4 (33.6 - 3G, 42 - Wifi), S5 (46.2 - 3G, 50.4 - Wifi), S6 (42 - 3G, 50.4 - Wifi), S7 (46.2 - 3G, 58.8 - Wifi) and iPhone 5 (33.6 - 3G, 42 - Wifi) and 6s (42 - 3G, 46.2 - Wifi).

In overall, our results are promising but the current Twes+ implementation has some limitations that can be improved in the future. Firstly, in its current implementation of Twes+, the CSS properties for the image dimensions are not taken into account. This is mainly because processing external files can consume energy and delay the loading time of resources. Therefore, further work is required to investigate how to handle external CSS files without increasing the loading time of resources. Secondly, Twes+ only supports popular image formats and it does not consolidate animated GIF files because they can be the combination of several images with different dimensions. Finally, Twes+ also has some limitations with respect to dynamic web pages. When a web page modifies images with JavaScript, it creates several variations into the image tags. These cases are not currently handled by Twes+. However, even though Twes+ has these limitations, we believe by addressing these, Twes+ would have even more impact on the energy saving and battery life.

Finally, in this study, the evaluation of Twes+ is done with 16 web pages. Although our findings show the status with real websites, a further study can be conducted with more web pages. Due to space limitations, in this paper we only reported our measurements with a particular device but more measurements with other mobile devices can be found in [45]. In this paper, we also mainly looked into the energy saving on the client. However, with caching Twes+ can also contribute to green computing and especially to sustainability. In particular, if pages that are used commonly can be transcoded once by the proxy and can be accessed by many, this could potentially affect the sustainability and energy consumption by mass.

8   Conclusions

This paper presents our proxy-based system called Twes+. Twes+ transcodes and adapts the source code of web pages to save energy. Twes+ currently has two transcoding techniques: redirect transcoding and image transcoding. Redirect transcoding aims to improve temporary redirects, and image transcoding aims to reduce the number of image requests to the server by consolidating images. This paper also presents our systematic evaluation of these two techniques. The results of our studies show that Twes+ saves energy on the client side by reducing the number of redirect round trips and the number of image requests/responses without modifying the look&feel of the web pages. Twes+ was able to eliminate the temporary redirect round trips and in cumulative processor energy for infinite redirects 4.6% reduction is observed with Twes+. With image transcoding service, on average 7% reduction in cumulative processor energy is observed by reducing the number of request/response round trips between the client and the server. It is also observed that this percentage is more than 10% with pages that include many images.

We also investigated the battery improvements with Twes+ and our estimations shows that with the image transcoding service of Twes+, in average Twes+ provides 7.6% power reduction and if we consider specific pages, for example Ebay, we can say that Twes+ provides 40.27 min battery improvements when the cumulative processor energy is equal to the entire system energy consumption and 24.99 min improvement when the cumulative processor energy is equal to the 70% of the system energy consumption (assuming that 30% is used by other processes, and only Ebay is processed). This also gives us on average 40 to 60 min of battery life extension on typical mobile devices. With redirection service of Twes+, we observe the most improvement in case 2 where redirection module returns the new address to redirect transcoding service and the content of the web page is requested from this new address and forwarded to the client. Of course, these are our estimations on a specific configuration but we believe Twes+ is a promising system in saving energy while users are browsing the web and it can be further extended and improved by adding new transcoding services to Twes+.

References

1. Ahmadi, H., Kong, J.: Efficient web browsing on small screens. In: Proceedings of AVI 2008, pp. 23–30, USA (2008)
2. Ali, M., Alex, Y., von Jouanne, A.: Integration of thermoelectrics and photovoltaics as auxiliary power sources in mobile computing applications. J. Power Sources **177**, 239–246 (2008)
3. HTTP Archive. Trends and statistics, September 2015. http://httparchive.org/
4. Asakawa, C., Takagi, H.: Transcoding. In: Yesilada, Y., Harper, S. (eds.) Web Accessibility. Human-Computer Interaction Series, pp. 231–260. Springer, London (2008)
5. Barr, K.C., Asanović, K.: Energy-aware lossless data compression. ACM TOCS **24**(3), 250–291 (2006)
6. Ben Abdesslem, F., Phillips, A., Henderson, T.: Less is more: energy-efficient mobile sensing with senseless. In: Proceedings of MobiHeld, pp. 61–62. ACM (2009)
7. Chen, J., Zhou, B., Shi, J., Zhang, H., Fengwu, Q.: Function-based object model towards website adaptation. In: Proceedings of WWW 2001, USA, pp. 587–596 (2001)
8. Chi, C.-H., Deng, J., Lim, Y.-H.: Compression proxy server: design and implementation. In: USENIX Symposium on Internet Technologies and Systems (1999)
9. Cobbaut, P.: Introduction to squid, May 2015. http://linux-training.be/networking/ch09.html
10. Chrome DevTools overview, October 2014. https://developer.chrome.com/devtools
11. The chromium project - SPDY: An experimental protocol for faster web, September 2014. http://www.chromium.org/spdy/spdy-whitepaper
12. Pagespeed module - sprite images, August 2014. http://tinyurl.com/juc537v
13. Data compression proxy, January 2015. https://developer.chrome.com/multidevice/data-compression
14. Duane Wessels, R.C., Nordstrom, H., Jeffries, A.: Squid: optimising web delivery (2013). www.squid-cache.org
15. Etoh, M., Ohya, T., Nakayama, Y.: Energy consumption issues on mobile network systems. In: Proceedings of SAINT 2008, pp. 365–368. IEEE Computer Society, USA (2008)
16. Everts, T.: Rules for mobile performance optimization. Queue **11**(6), 40:40–40:51 (2013)

17. Fainberg, L., Ehrlich, O., Shai, G., Gadish, O., Dobo, A., Berger, O.: Systems and methods for acceleration and optimization of web pages access by changing the order of resource loading, 2 August (2010). US Patent Ap. 12/848,559

18. Fielding, R., Gettys, J., Mogul, J., Frystyk, H., Masinter, L., Leach, P., Berners-Lee, T.: Rfc2616 hypertext transfer protocol http/1.1 (1999)

19. Firtman, M.: Programming the Mobile Web, vol. 1, 2nd edn. OReilly Media Inc., March 2010

20. Frain, B.: Responsive Web Design with HTML5 and CSS3. Packt Publishing (2012)

21. Gochman, S., Mendelson, A., Naveh, A., Rotem, E.: Introduction to intel core duo processor architecture. Intel Technol. J. **10**(2), 109–122 (2006)

22. Google. Chrome accessibility, November 2015. http://tinyurl.com/pgyrgf2

23. Kothiyal, R., Tarasov, V., Sehgal, P., Zadok, E.: Energy and performance evaluation of lossless file data compression on server systems. In: Proceedings of SYSTOR 2009, USA, pp. 4:1–4:12 (2009)

24. Kumar, D.: Image sprites - how to merge multiple images, and how to split them, June 2012. http://tinyurl.com/j56g6h8

25. L3WS. Greasyspoon- open-source ICAP server factory for core network services (2015). http://greasyspoon.sourceforge.net/

26. Lai, P.P.Y.: Efficient and effective information finding on small screen devices. In: Proceedings of W4A 2013, USA, pp. 4:1–4:10 (2013)

27. Liu, Y., Guo, L.: An empirical study of video messaging services on smartphones. In: Proceedings NOSSDAV 2014, pp. 79:79–79:84, USA (2014)

28. Matsudaira, K.: Making the mobile web faster. Queue **11**(1), 40:40–40:48 (2013)

29. McLachlan, P.: On mobile, data URIs are 6x slower than source linking (new research), July 2013. http://www.mobify.com/blog/data-uris-are-slow-on-mobile/

30. Mobify. Image resizing with mobify.js (2013). https://www.mobify.com/mobifyjs/docs/image-resizing/

31. Narendran, T., Gaurav, A., Angela, N., Dan, B., Singh, J.P.: Who killed my battery? analyzing mobile browser energy consumption. In: Proceedings of WWW 2012, pp. 41–50. ACM (2012)

32. Opera. Faster browsing on slow networks with off-road mode, December 2015. http://help.opera.com/opera/Windows/1116/en/fasterBrowsing.html

33. Paul, K., Kundu, T.K.: Android on mobile devices: an energy perspective. In: Proceedings of CIT 2010, pp. 2421–2426. IEEE Computer Society, USA (2010)

34. Pentikousis, K.: In search of energy-efficient mobile networking. IEEE Commun. Mag. **48**, 95–103 (2010)

35. Song, R., Liu, H., Wen, J.-R., Ma, W.-Y.: Learning block importance models for web pages. In: Proceedings of WWW 2004, USA, pp. 203–211 (2004)

36. Takagi, H., Asakawa, C., Fukuda, K., Maeda, J.: Site-wide annotation: reconstructing existing pages to be accessible. In: Proceedings of ASSETS 2002, USA, pp. 81–88 (2002)

37. A.T.S. Team. Apache traffic server, January 2015. http://tinyurl.com/pv63dxj

38. M. team. Introducing jazzcat: A javascript and CSS concatenation service, August 2012. http://tinyurl.com/jf2v82u

39. Thoba, L., Mamello, T.: A transcoding proxy server for mobile web browsing. In: The Southern Africa Telecommunication Networks and Applications Conference (2011)

40. Ugo, E., Gennaro, I., Delfina, M., Vittorio, S.: Personalizable edge services for web accessibility. UAIS **6**, 285–306 (2007)

41. Yahoo. Best practices for speeding up your web site, January 2015. https://developer.yahoo.com/performance/rules.html

42. Yesilada, Y., Jay, C., Stevens, R., Harper, S.: Validating the use and role of visual elements of web pages in navigation with an eye-tracking study. In: Proceedings of WWW 2008, Beijing, China (2008)

43. Yin, X., Lee, W.S.: Using link analysis to improve layout on mobile devices. In: Proceedings of WWW 2004, USA, pp. 338–344 (2004)

44. Zakas, N.C.: The evolution of web development for mobile devices. Queue **11**(2), 30:30–30:39 (2013)

45. Köksal Ahmed, E.: Transcoding Web Pages For Energy Saving On the Client-Side. Middle East Technical University Northern Cyprus Campus (2016)

46. Apple: Iphone7. http://www.apple.com/sg/iphone-7/specs/

47. Ana, R., Mateos, C., Zunino, A.: Improving scientific application execution on android mobile devices via code refactorings. Softw. Pract. Exp. **47**(5), 763–796 (2016)

MateTee: A Semantic Similarity Metric Based on Translation Embeddings for Knowledge Graphs

Camilo Morales[1,2], Diego Collarana[1,2(✉)], Maria-Esther Vidal[2,3],
and Sören Auer[1,2]

[1] Enterprise Information Systems (EIS), University of Bonn, Bonn, Germany
{moralesc,collaran,auer}@cs.uni-bonn.de
[2] Fraunhofer Institute for Intelligent Analysis and Information Systems (IAIS),
Sankt Augustin, Germany
vidal@cs.uni-bonn.de
[3] Universidad Simón Bolívar, Caracas, Venezuela

Abstract. Large Knowledge Graphs (KGs), e.g., DBpedia or Wikidata, are created with the goal of providing structure to unstructured or semi-structured data. Having these special datasets constantly evolving, the challenge is to utilize them in a meaningful, accurate, and efficient way. Further, exploiting semantics encoded in KGs, e.g., class and property hierarchies, provides the basis for addressing this challenge and producing a more accurate analysis of KG data. Thus, we focus on the problem of determining relatedness among entities in KGs, which corresponds to a fundamental building block for any semantic data integration task. We devise MateTee, a semantic similarity measure that combines the *gradient descent* optimization method with semantics encoded in ontologies, to precisely compute values of similarity between entities in KGs. We empirically study the accuracy of MateTee with respect to state-of-the-art methods. The observed results show that MateTee is competitive in terms of accuracy with respect to existing methods, with the advantage that background domain knowledge is not required.

1 Introduction

We are living in the Big Data era where a large number of structured and semi-structured datasets are publicly available. Such datasets are collected from different social domains, e.g., government, scientific communities, or social media and social networks. The semantic representation of the data in RDF Graphs helps in the endeavor of automatically solving data-driven oriented tasks, providing as result, more useful and meaningful services from such big and heterogeneous data [2]. Particularly, the tasks affected by a good similarity metric between data entities are: semantic data integration of heterogeneous data, entity linking and clustering, as well as the generation of recommendations. The future of the Web of Data and the Web of Things brings even more heterogeneity and larger datasets. Streaming data coming at high rates need to be processed on-demand,

© Springer International Publishing AG 2017
J. Cabot et al. (Eds.): ICWE 2017, LNCS 10360, pp. 246–263, 2017.
DOI: 10.1007/978-3-319-60131-1_14

all of which only increases the need of automation in the process of creation and processing of semantics. Data management and Artificial Intelligence approaches play an important role on the task of KG data analysis. Machine Learning (ML), mostly in its supervised flavor, aims to give machines the capability to learn by examples, essentially, labeled data. ML field has achieved promising results with sophisticated techniques, such as Kernel methods or Deep Learning models. Furthermore, the Semantic Web, and in general, all the available Knowledge Graphs (KGs) such as DBpedia or Yago, have been built with a tremendous effort of the scientific community having the main objective of making the data understandable not only by humans but also by computers. Structured data facilitates the tasks of data integration, relations or associations discovery, as presented by Bordes et al. 2013 with TransE [3].

On one hand, we have an immense amount of available knowledge facts, encoded as structured data in knowledge graphs, and on the other, we have the Machine Learning boom and techniques able to have access to Big Data sets, for two main tasks: classification and link prediction. In the case of KGs, we are referring to classification of entities in a set of classes, and prediction (or discovery) of new relations between entities, i.e., RDF triples. In this paper, we focus on the problem of determining relatedness among entities in KGs. This problem serves as a building block for classification and link prediction, as well as for semantic data integration. Building on results from knowledge graph embedding, we devise MateTee a semantic similarity measure that combines *gradient descent* optimization method with semantics encoded in ontologies, to determine relatedness among entities in KGs.

We conduct an empirical evaluation to assess the quality of MateTee with respect to state-of-the-art similarity measures. Experiments are performed on two benchmarks: (a) The CESSM [15] KGs of proteins annotated with the Gene Ontology[1]; (b) A KG of people from DBpedia. Observed results suggest that MateTee is able to outperform existing similarity measures, while it does not require any background knowledge or domain expertise to be configured.

In summary, our main contributions are as follows:

- An *end-to-end* approach named MateTee able compute similarity value between entities in a KG. MateTee is based on TransE, which utilizes the gradient descent optimization method to learn the *features representation* of the entities automatically.
- An extensive empirical evaluation on existing benchmarks and state-of-the-art showing MateTee behavior. Results indicate the benefits of using embeddings for determining relatedness among entities in KGs. MateTee and experimental studies are publicly available[2].

The remainder of the paper is structured as follows. First, we motivate the problem of determining relatedness among entities in KGs using a practical example. Then, we briefly describe preliminaries and background concepts required to

---

[1] http://geneontology.org/
[2] https://github.com/RDF-Molecules/MateTee

understand the problem treated in the paper, as well as the proposed solution. Section 4 defines the problem, the proposed solution, and MateTee architecture. Section 5 reports on the empirical evaluation, while related work is summarized in Sect. 6. Conclusion and future work are outlined in Sect. 7.

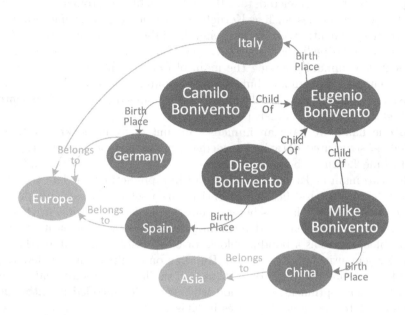

Fig. 1. Motivating Example. A portion of a Knowledge Graph (KG) describing relationships among persons and the places where they have been born. There exist different types of relations and multiple connectivity patterns among entities in KG

2   Motivating Example

Consider a knowledge graph in Fig. 1. Nodes of the same color indicate they share the same properties, while nodes of different colors differ in at least one property. Determining relatedness among same-colored nodes, e.g., Camilo with Diego, requires to compare, in a 1-1 fashion, values of each property of those entities and aggregate the results. This computation can be done as Camilo and Diego have the same set of properties, i.e., *Child_of* and *Birth_Place*. Contrary, if entities have different properties, i.e., they are on different colors, the problem is to measure their relatedness considering the complete set of properties of both nodes while is not possible to use the 1-1 approach, e.g., Germany and Camilo. Moreover, whenever entities are compared in terms of their *neighborhoods* and *reachable* nodes, Camilo should be more similar to Diego than to Mike, as Diego and Camilo are from Europe, while Mike is from China.

These difficulties come inherently with the multi-relational datasets. In relational data tables, all elements have the same properties, i.e., columns, and therefore, the similarity computation is performed aggregating a 1-1 similarity value between each pair of the properties. With multi-relational data, nodes need to be made *comparable*, which means that they all must have the same set of properties or features. This can be done manually, *handcrafting* the features, and creating a list of them for each node, based on previous knowledge of the specific field or domain of the data. These sets of features will be regarded as a new representation of the nodes in KG. Then, these sets of features can be compared, again, in a 1-1 fashion. The problem is that manual creation of the features requires deep domain knowledge, not to mention it is error-prone and time consuming. Thus, to solve these problems a similarity measurement approach that automatically creates a canonical entity representations is required. In this paper, we present MateTee, a similarity approach that relies on embedding the original KG into a vector space in order to make all entities comparable. Similarity values among embeddings are measured based on any distance metric defined for vector spaces, e.g., Euclidean distance.

3 Preliminaries

MateTee determines relatedness between entities in Knowledge Graphs based on Encoding Generation methods such as TransE [3]. MateTee combines the gradient descent optimization method (used in TransE) with the explicit knowledge encoded in the ontologies of a KG.

3.1 Translation Embeddings

MateTee is based on TransE [3], acronym of *Translation Embeddings*, presented by Bordes et al. 2013. TransE tackles the problem of embedding a Knowledge Graph (KG) into a low dimensional vector space (called embedding space) for subsequent prediction or classification objectives, e.g., predict missing edges. The core of TransE is to learn the embeddings of entities in a way that similar entities in the KG should be also close in the embedding space. Additionally, dissimilar entities in the KG should be also far in the embeddings space. Learning the embeddings is done by analysing the *connectivity patterns* between entities in a KG, and then encoding these patterns into their vector representation, i.e., their embeddings. The optimization technique *Stochastic Gradient Descent* is executed to compute this encoding.

Modeling RDF triples in the embedding space with relations as **translations** is the core contribution of TransE. The basic idea behind translation-based model is the following:

$$Subject + Translation \approx Object$$

TransE aims at minimizing the error when summing up the distance d between the embeddings of the *Subject + Translation* pair and the embedding of the *Object*. Stochastic Gradient Descent (SGD) meta-heuristic allows for learning entity embeddings by minimizing the error defined as the sum of the distances d of all the triples in the KG. A global minimum cannot be ensured because SGD depends on a *randomly* selected start position of the *descent*.

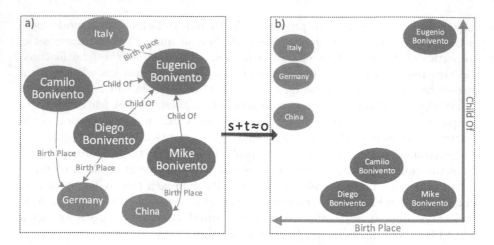

Fig. 2. TransE approach intuition. (a) An RDF Knowledge Graph where similar entities are in the same color; (b) Clusters of entities in the embedding space. Entities of the same color are close to each other in the identified cluster.

The random initialization procedure followed by TransE is presented in detail at [8]. Figure 2 illustrates the intuition of this approach.

4   MateTee: A Semantic Similarity Measure for RDF Knowledge Graphs

MateTee focuses on measuring the similarity between any pair of entities belonging to an input RDF Knowledge Graph. Measuring the similarity between entities is an important phase for any data integration problem, and for most machine learning tasks, e.g., clustering of nodes, or link prediction in KGs.

The main problem for computing the similarity of RDF knowledge graphs is that not all the nodes have the same properties, therefore, a 1-1 comparison at property level cannot be performed. State-of-the-art methods like GADES [20] perform a semantic analysis of the entities based on multiple aspects, i.e., 1-hop neighborhood, class hierarchy of the subjects/objects, class hierarchy of the properties, and mixtures of them. This analysis relies on domain knowledge and user expertise about the provenance of the data, e.g., GADES requires a good design of the hierarchy of classes and properties.

To overcome this problem, MateTee embeds an RDF knowledge graph into a vector space, once all the entities are represented as vectors with same dimensionality, it uses any common distance metric to calculate their similarity values. MateTee relies on finding a vector representation of graph entities to produce the similarity value. For this, MateTee utilizes TransE [3], a method based on Stochastic Gradient Descent that encodes the connectivity patterns of the entities into a low-dimensional embedding space. TransE ensures that similar nodes in the RDF graph are close in the embedding space, while dissimilar nodes in the graph are distant in the embedding space.

By using TransE approach, MateTee aims to calculate similarity values as close as possible to the *ground truth*: values accepted by the scientific community because they were calculated manually with deep domain expertise, e.g., Sequence Similarity in the Gene Ontology domain. Formally, MateTee can be defined as:

Definition 1 (MateTee Embedding). *Given a knowledge graph $G = (V, E)$ composed by a set T of RDF triples, where $V = \{s \mid (s,p,o) \in T\} \cup \{o \mid (s,p,o) \in T\}$ and $E = \{p \mid (s,p,o) \in T\}$, MateTee aims to find a set M of embeddings of each member of V, such that:*

$$\arg\min_{\mathbf{m_1},\mathbf{m_2} \in M} Error(M) = \arg\min_{\mathbf{m_1},\mathbf{m_2} \in M} \sum_{\mathbf{m_1},\mathbf{m_2} \in M} |S_1(\mathbf{m_1},\mathbf{m_2}) - S_2(\mathbf{m_1},\mathbf{m_2})|$$

where S_1 is a similarity metric computed using any distance measure defined for vector spaces, e.g., Euclidean distance, and S_2 is a similarity value given by the Gold Standards. The Gold Standards are the values considered as ground truth.

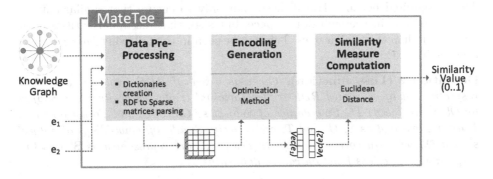

Fig. 3. The MateTee Architecture. MateTee receives as input an RDF Knowledge Graph (KG), and entities e_1 and e_2 from the KG. MateTee outputs a similarity value between e_1 and e_2 according to the connectivity patterns found in KG. A pre-processing step allows for the transformation of a KG into a matrix-based representation. Then, n-dimensional embeddings are generated. Finally, values of similarity are computed

4.1   The MateTee Architecture

Figure 3 depicts the *end-to-end* MateTee architecture. MateTee receives as input an RDF Knowledge Graph (KG), and entities e_1 and e_2 belonging to the KG. The objective of the complete process is to calculate the similarity value between e_1 and e_2. The first step is to **Pre-Process** the original data in order to transform it into the format required by the optimization method. As the optimization methods are numeric based, we need a numerical representation of the data.

In other words, the string-based triples coming as input must be translated into a numeric format, usually sparse matrices. The implementation of TransE employs three sparse matrices: one representing the Objects, another for the Subjects, and a third one for the Translations. The matrices have as many columns as RDF triples are in the original KG, and as many rows as entities, i.e., number of Subjects + number of Translations + number of Objects. Note that if a Subject appears also as Object in another RDF triple, it is considered as one. Moreover, in order to map the original entities to their respective *encodings*, i.e., embeddings, dictionaries need to be created. Dictionaries map the original URIs of the entities with the ID of their embeddings.

Once the numerical representation and dictionaries of the RDF triples are created, the embeddings of the entities can now be *learned*. Learning embeddings happens at the **Encoding Generation** phase. This numerical representation of the data is now fed to the optimization method. The method aims to update the value of the embeddings in order to minimize an overall error according to a proposed model. MateTee is based on TransE, this method aims at minimizing the distance (in MateTee Euclidean Distance is used) between the sum of the embeddings of the Subject and Translation to the embeddings of the Object. TransE also defines *corrupted triples*, which are triples with either the Subject or Object replaced by another randomly selected resource from the set of entities. This is required because TransE needs not only to ensure that similar entities should be close in the embedding space, but also, that dissimilar entities must be farther than the similar ones. This can be seen in the following *Loss Function* used by TransE:

Definition 2 (Loss function). *Given is a set of RDF triples T and their respective set of corrupted RDF triples (original triples with either the Subject or Object replaced) T'. Embeddings of Subject s, Object o, and Transitions t in T are represented as S, O, and T, respectively. Similarly, embeddings of Subject s' and Object o' in corrupted RDF triples in T' are represented as S' and O', respectively. The loss function can be defined as:*

$$Loss(T, T') = \sum_{(s,t,o) \in T} \sum_{(s',t,o') \in T'} [margin + d(\mathbf{S} + \mathbf{T}, \mathbf{O}) - d(\mathbf{S'} + \mathbf{T}, \mathbf{O'})]_+$$

The key is to notice that the loss function only considers the positive part of the difference of the distances, plus the margin; this is denoted by $[x]_+$ in the loss formula. Considering positive values is crucial because if the distance between entities of the original triple, i.e., $d(\mathbf{S} + \mathbf{T}, \mathbf{O})$, is greater than the distance between the entities of the corrupted triple, i.e., $d(\mathbf{S'} + \mathbf{T}, \mathbf{O'})$, then the difference between the two is positive (regardless of the margin) and this number will increase the overall error. This situation should not occur according to the model $S + T \approx O$ as we want this difference to be as close to zero as possible. On the other hand, if the opposite situation happens, the distance between the entities of the original RDF triple, i.e., $d(\mathbf{S} + \mathbf{T}, \mathbf{O})$, is smaller than the distance between the entities of the corrupted triple, i.e., $d(\mathbf{S'} + \mathbf{T}, \mathbf{O'})$. This state is exactly what the model looks for, and since the difference between both distances is negative, the overall error is not increased as only the positive part is considered. In the case when the entities of the original RDF triple is smaller than the distance between the entities of the corrupted triple, the margin tightens the model as the negative difference between both distances must be at least as big as the margin, otherwise the overall error will be increased.

TransE - Gradient Descent Algorithm. The core of TransE learning algorithm performs the following steps:

1. **Initialization:** The embedding of each entity (Subject/Object) is initialized uniformly and randomly between $\frac{-6}{\sqrt{k}}$ and $\frac{6}{\sqrt{k}}$ where k is the dimensionality of the embeddings. At this point only the relations are normalized, they will not be normalized again during the optimization. Entities will be normalized at the beginning of each iteration.
2. **Training (loop):**
   (a) **Entity embeddings normalization:** In each iteration, first current embeddings of the entities are normalized. This is important because it prevents the optimization to minimize the error by artificially increasing the length i.e., norm, of the embeddings.
   (b) **Creation of mini-batches:** Triples to be used as training examples for each iteration of the GD are selected. First, a random sample of set of triples from the input data set is chosen, and then, for each triple in the sample, a corrupted triple is created.
      – **Corrupted triples:** A corrupted triple is the same as the original but with either its Subject or Object replaced by another randomly selected entity from the data set, always just one, not both at the same time, as show in Fig. 4:
   (c) **Embeddings update:** Once the training set of examples, i.e., real triples ∪ corrupted triples is set, it proceeds with the actual optimization process:
      – For each one of the dimensions of each one of the embeddings in the data set, we calculate the derivative of the overall error with respect to this parameter. This derivative gives the direction on which the overall error is growing with respect to this parameter. Then, to know how to update this parameter so that the overall error decreases, it changes the direction to the opposite of the derivative, and moves one unit of the learning rate (which is also an input hyper-parameter). This process iterates until a maximum number of iterations is reached.

When the optimization reaches the termination condition, e.g., the maximum number of iterations in TransE, the embeddings of the entities have been already learned. Having the embeddings of all the entities in the input KG, including e_1 and e_2, MateTee can now proceed to the **Similarity Measure Computation** of both entities. Any distance metric for vector spaces can be used to calculate this value, e.g., any Minkowski distance, Euclidean for MateTee. It is important to notice that MateTee calculates the similarity and not the distance.

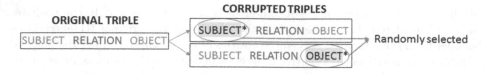

Fig. 4. Corrupted triples. An original RDF triple t and two corrupted versions of t are presented on the left and right hand of the figure, respectively. Corrupted triples have either the Subject or the Object replaced by another randomly selected entity from the set of entities from the input KG

Therefore, using the Euclidean distance MateTee finds a similarity value between 0 and 1, using the following formula:

$$similarity(A, B) = \frac{1}{1 + EuclideanDistance(A, B)}$$

5   Empirical Evaluation

We empirically study the effectiveness of MateTee on solving the problem of measuring the semantic similarity between entities in a KG. We assess the following research questions: **(RQ1)** Does the translations embeddings method used in MateTee improve the accuracy of determining relatedness between entities in a KG? **(RQ2)** Is MateTee able to perform as good as the state-of-the-art similarity measures? **(RQ3)** Does MateTee perform well in Knowledge Graphs from different domains? To answer our research questions, we evaluate MateTee in two different scenarios. In the first evaluation, we compare Proteins annotated with the Gene Ontology[3]. In the second evaluation, we compare people extracted from DBpedia, we prepare a dataset named *DBpedia People* [5].

Implementation. MateTee is implemented in Python 2.7.10. The experiments were executed on a Ubuntu 14.04 (64 bits) machine with CPU: Intel(R) Xeon(R) E5-2660 2.60GHz (20 physical cores) with 132 GB RAM, and GPU card GeForce GTX TITAN X. Source code and a Docker set up are available in Git[4].

5.1   Similarity Among Proteins Annotated with the GO Ontology

Datasets. This experiment is conducted on the collections of proteins published at the Collaborative Evaluation of GO-based Semantic Similarity Measures [16] (CESSM) websites 2008[5] and 2014[6]. The CESSM 2008 collection is composed of 13,430 pairs of proteins from UniProt with 1,039 distinct proteins, while

---

[3] http://geneontology.org/
[4] https://github.com/RDF-Molecules/MateTee
[5] http://xldb.di.fc.ul.pt/tools/cessm/
[6] http://xldb.di.fc.ul.pt/biotools/cessm2014/

the CESSM 2014 dataset includes 22,302 pairs of proteins also from UniProt with 1,559 distinct proteins. The sets of annotations of CESSM 2008 and 2014 comprise 1,908 and 3,909 distinct GO terms, respectively. The original CESSM collections are presented in a multi-file fashion, one file per protein. Technical details in Table 1 refer to the unified (single file) dataset, after data transformations are applied. CESSM computes the Pearson's correlation coefficients with respect to three similarity measures from the genomic domain[7]: *ECC similarity* [7], *Pfam* [16], and the *Sequence Similarity* (SeqSim) [19]. Furthermore, the CESSM evaluation framework makes the results of eleven semantic similarity measures available. These state-of-the-art semantic similarity measures are specific for the genomic domain and exploit the knowledge encoded in the Gene Ontology (GO) to determining relatedness among proteins in the CESSM collections. These semantic similarity measures are extensions of well-known similarity measures to consider GO annotations, Information Content (IC) of these annotations, and pair-wise combinations of common ancestors in GO hierarchy. The extended similarity measures are the following: Resnik (R) [17]; Lin (L) [11]; and Jiang and Conrath (J) [13]. Additionally, the CESSM evaluation framework considers the average of the ICs of pairs of common ancestors during the computation of these measures; this measure is denoted with the label A. Following the approach reported by Sevilla et al. [18], the maximum value of IC of pairs of common ancestors is computed; combined measures are distinguished with the label M. As proposed by Couto et al. [6], the best-match average of the ICs of pairs of disjunctive common ancestors (DCA) is also computed; measures labelled with B or G correspond to combinations with the best-match average of the ICs. Finally, the Jaccard index is applied to sets of annotations together with domain-specific information in the similarity measures simUI (UI) and simGIC (GI) [15].

Table 1. CESSM 2008 and 2014 - Dataset description. Shows dataset size in Megabytes, overall number of triples, number of left entities (Subjects), right entities (Objects), and shared entities (appearing as Subject and as Object), and number of relations, to present a comparison of size between datasets from 2008 and 2014

		CESSM 2008	CESSM 2014
Size (MBs)		1	1
Triples		8,359	20,153
Entities	Left	1,039	1,559
	Shared	0	0
	Right	1,908	3,909
Relations		1	1

---

[7] The area in molecular biology and genetics that studies the genetic material of an organism.

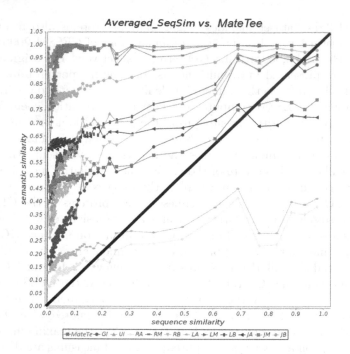

Fig. 5. Results from the CESSM evaluation framework for the CESSM 2008 collection. Results include: average values for MateTee with respect to SeqSim. The black diagonal line represents the values of SeqSim for the different pairs of proteins in the collection. The similarity measures are: simUI (UI), simGIC (GI), Resnik's Average (RA), Resnik's Maximum (RM), Resnik's Best-Match Average (RB/RG), Lin's Average (LA), Lin's Maximum (LM), Lin's Best-Match Average (LB), Jiang & Conrath's Average (JA), Jiang & Conrath's Maximum (JM), J. & C.'s Best-Match Average (JB). MateTee outperforms eleven measures and reaches a value of Pearson's correlation of **0.787**

Results. Figures 5 and 6 report on the comparison of MateTee and the rest of the eleven similarity measures with SeqSim; both plots were generated by the CESSM evaluation framework. The black diagonal lines represent the values assigned by SeqSim. The majority of the studied similarity measures assign high values of similarity to pairs of proteins that SeqSim considers as similar proteins, i.e., in pairs of proteins with high values of SeqSim, the majority of the curves of the similarity measures are close to the black line. Nevertheless, the same behavior is not observed for the pairs of proteins that are not similar according to SeqSim, i.e., the corresponding curves are far from the black line. Contrary to state-of-the-art similarity measures, MateTee is able to compute values of similarity that are more correlated to SeqSim, i.e., the curve of MateTee is close to the black line in both collections. MateTee is able to reach values of the Pearson's correlation of **0.787** and **0.817** in CESSM 2008 and 2014, respectively.

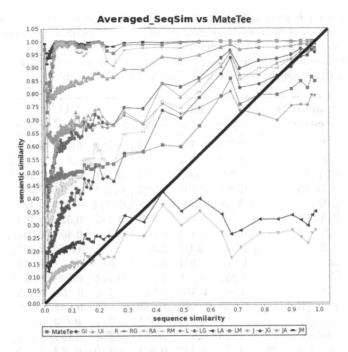

Fig. 6. Results from CESSM evaluation framework for the CESSM 2014 collection. Results include: average values for MateTee with respect to SeqSim. The black diagonal line represents the values of SeqSim for the different pairs of proteins in the collection. The similarity measures are: simUI (UI), simGIC (GI), Resnik's Average (RA), Resnik's Maximum (RM), Resnik's Best-Match Average (RB/RG), Lin's Average (LA), Lin's Maximum (LM), Lin's Best-Match Average (LB), Jiang & Conrath's Average (JA), Jiang & Conrath's Maximum (JM), J. & C.'s Best-Match Average (JB). MateTee outperforms eleven measures and reaches a value of Pearson's correlation of **0.817**

Additionally, we present the results of the comparison of MateTee and eleven similarity measures with respect to the gold standard similarity measures: ECC, Pfam, and SeqSim; Table 2 presents these results. Moreover, additional similarity measures are included in the study: d_{tax} [1], d_{ps} [14], OnSim [22], IC-OnSim [21], and GADES [20]. As before, values of the Pearson's correlation represent the quality of a measurement of similarity, the higher the correlation with the gold standards, the better the measurement. The top 5 similarity measures (before introducing MateTee) with higher quality are highlighted in gray, and the highest is highlighted in bold.

Discussion: From the results, the following insights can be concluded; MateTee already outperforms the quality of GADES for both collections 2008 and 2014, which is the best-performing measurement before our method, for the Sequence Similarity. In the 2008 collection, MateTee stands at the 5th position against the other two gold standards, only at 0.015 points to the GADES for ECC, and

Table 2. GO - CESSM 2008 and 2014 - Results. Quality in terms of Pearson's correlation coefficient between three gold standards, i.e., SeqSim (Sequence) similarity, Pfam (Protein Families) similarity and EC (Enzyme Commission) similarity, and thirteen in-house similarity measures of CESSM, plus OnSim, IC-OnSim and GADES. With gray background the best 6 correlations and the best in bold

Similarity measure	2008			2014		
	SeqSim	*ECC*	*Pfam*	*SeqSim*	*ECC*	*Pfam*
GI [16]	0.773	0.398	0.454	0.799	0.458	0.421
UI [16]	0.730	0.402	0.450	0.776	0.470	0.436
RA [19]	0.406	0.302	0.323	0.411	0.308	0.264
RM [22]	0.302	0.307	0.262	0.448	0.436	0.297
RB [7]	0.739	0.444	0.458	0.794	0.513	0.424
LA [13]	0.340	0.304	0.286	0.446	0.325	0.263
LM [22]	0.254	0.313	0.206	0.350	0.460	0.252
LB [7]	0.636	0.435	0.372	0.715	0.511	0.364
JA [11]	0.216	0.193	0.173	0.517	0.268	0.261
JM [22]	0.234	0.251	0.164	0.342	0.390	0.214
JB [7]	0.586	0.370	0.331	0.715	0.451	0.355
d_{tax} [1]	0.650	0.388	0.459	0.682	0.434	0.407
d_{ps} [15]	0.714	0.424	0.502	0.750	0.480	0.450
OnSim [26]	0.733	0.378	0.514	0.774	0.455	0.457
IC-OnSim [25]	0.779	0.443	**0.539**	0.810	0.513	0.489
GADES [24]	0.780	**0.446**	**0.539**	0.812	**0.515**	0.49
MateTee	**0.787**	0.431	0.496	**0.817**	0.501	0.461

0.043 for Pfam. While in the 2014 collection, MateTee stands at the 3th position against the Pfam gold standard, only at 0.029 points to the GADES (the best before MateTee), and at the 5th position against the ECC gold standard, only at 0.014 points to the GADES (the best before MateTee).

It can be observed that GADES [20] is the greatest competitor for MateTee. It performs better comparing with the ECC and Pfam gold standards, but it is outperformed against SeqSim. As the results of GADES and MateTee are rather close. For the three gold standards, the advantage of MateTee against GADES is that the former requires domain expertise to define its final similarity measure (GADES defines multiple measures based on: Class hierarchy, Neighborhood, Relation Hierarchy, Attributes, and mixtures of them). While the latter learns the embeddings in an automatic way (through an optimization process called Stochastic Gradient Descent), and then uses any common vector similarity measure, e.g., Euclidean or Cosine, to calculate their similarity.

5.2   Similarity Among People from DBpedia

Dataset: Table 3 shows technical details of the datasets used in the *DBpedia People* experiment. The **Gold Standard (GS):** the collection was extracted from the live version of DBpedia (July 2016); it contains 20,000 subjects of type

Table 3. Description of the data set *DBpedia People*. Shows the datasets size in Megabytes, overall number of triples, overall number of persons, number of left entities (Subjects), right entities (Objects), and shared entities (appearing as Subject and as Object), and number of relations, to present a comparison of size between three dumps

		Data set 1	Data set 2	Data set 3
Size (MBs)		80	80	80
Triples		552,355	553,232	552,527
Subjects (Persons)		20,000	20,000	20,000
Entities	Left	60,000		
	Shared	0		
	Right	247,465		
Relations		1,981		

Person[8], i.e., 20,000 subjects with all available properties and their values. The overall number of RDF triples is 829,184. The Gold Standard is used to compute Precision and Recall during the evaluation. The **Test Datasets (TS):** are created from the Gold Standard with their properties and values were randomly split among three test datasets. Each triple is randomly assigned to one or several test datasets. The selection process takes two steps: (1) a number of test datasets to copy a triple to is chosen randomly under a uniform distribution; (2) the chosen number is used as a sample size to randomly select particular test datasets to write a triple. URIs are generated specifically for each test data set. Eventually, each test dump contains a subset of the properties in the gold standard. Each subset of properties of each person is composed randomly using a uniform distribution.

Metrics: We measure the behavior of MateTee in terms of the following metrics: **(a) Precision** From all matched pairs (pairs with similarity greater than the threshold), percentage of correct matches.

$$Precision = \frac{\text{Number of correctly matched pairs}}{\text{Total number of matched pairs}}$$

(b) Recall From all expected matches (all, including below and above the threshold), percentage of correct matches.

$$Recall = \frac{\text{Number of correctly matched pairs}}{\text{Total number of expected matches}}$$

Results: We tested the quality of MateTee by comparing its results with two other similarity measurements: Jaccard [5] and GADES [20]. For each one we calculate the Precision and Recall, considering different values of **Threshold**. The Threshold is the minimum similarity value so that the pair of people is considered in the matched-pairs set. Table 4 show the results obtained using Jaccard, GADES, and MateTee similarity approaches.

---

[8] <http://dbpedia.org/ontology/Person>.

Table 4. DBpedia People Test Datasets. Results comparison of precision and recall using Jaccard, GADES and MateTee similarity measurements, obtained with different threshold values: 0.6, 0.7, 0.8 and 0.9. In bold the best value for each threshold

	T0.6		T0.7		T0.8		T0.9	
	Precision	Recall	Precision	Recall	Precision	Recall	Precision	Recall
Jaccard	0.36	0.01	0.30	0.01	0.30	0.01	0.30	0.01
GADES	0.87	0.73	0.83	0.43	0.80	**0.16**	**0.63**	**0.05**
MateTee	**0.93**	**0.79**	**0.99**	**0.59**	**1.00**	0.10	0.00	0.00

Discussions: From the results we extract the following insights. Regarding Precision, MateTee similarity measurement has the best quality among all the three measurements, and for all the considered thresholds. Regarding Recall, our method is the best up until a threshold of 0.7. For higher thresholds, e.g., 0.8, the recall rapidly goes down to 0.1, and to absolute 0 for 0.9. The explanation for this is that MateTee, being an optimization-based method, will always have an error as small as possible, so even if the neighborhoods of two entities are exactly the same, it is very unlikely to have similarities higher than 0.9 or 1.0, they will for sure be higher than between people which neighborhoods are absolutely different, but very unlikely be equal to 1.0. Then, using a threshold equal to 0.9, very few pairs of people will be considered, and with 1.0, absolutely no pairs are considered to count in the numerator of the Recall formula.

6   Related Work

Griver et al. [9] present **Node2Vec**, the latest method of the *everything-2-vec* saga. **Node2Vec** tackles the problem of giving a vector representation to nodes in graphs. Node2Vec focuses mainly on two common prediction tasks: *Multi-label classification of nodes*, where the objective is to classify new unknown nodes into one of the known classes, and *Link prediction* with the objective of predicting if a link i.e., relation, should be established (or re-established in case of incomplete datasets) between a pair of given nodes. Further, node2vec identifies the type of link. The main contribution and uniqueness of node2vec, compared with similar techniques, is the flexible notion it gives to the meaning of *neighbourhood*. It is based on the idea that nodes, and their connectivity patterns in the network, can be described based on two factors: First, on the *communities* to which they belong, i.e., *homophily* or essentially the set of their 1-hop neighbors, and second, on the role the nodes play in the network, i.e., *structural equivalence* or the type of node they are, e.g., border node, internal node, etc. Therefore, a node could have multiple neighbourhoods, and it can only be considered k of these neighbourhoods, the problem turns into how to sample them.

Based on *Breadth-first Sampling (BFS)* and *Depth-first Sampling (DFS)*, node2vec proposes a new sampling approach called *Random Walks*. It consists on explore the connectivity patterns based on both BFS and DFS manners, interpolating between both approaches based on a bias term. This idea comes from the fact that the connectivity patterns in real life graphs, e.g., Social Networks, are not exclusively based on structural equivalence and *homophily*, but commonly in a mixtures of both. The bias term aims to accommodate the random walks to the actual structure of the sub-graph being analyzed.

Another publication also aiming on finding vector representations of entities in KGs, and that in fact was the predecessor of TransE, is the **Structured Embeddings** (SE), presented in [4] also by Bordes et al. SE is a method on which the vector representation of the entities is established by a neural network acting like a bridge between the entities in the original data and their feature representation. The fundamental characteristics of their approach includes *Flexibility* and domain independence, meaning that it should work and be easily adaptable for most of available KBs, and *Compactness* in the sense that each entity is assign one low-dimensional vector in the feature space, and only one matrix to each relation. As for TransE, SE considers relations between the entities, i.e., subjects and objects, as operators. If certain operation is performed in the feature space between the subject vector and the relation matrix, the resulting vector must be the object vector (or a nearest in its neighbor). The main difference between SE and further approaches from the same authors, e.g., TransE [3], is that SE models relations as pair of matrices. For TransE the relations are normal embeddings with the special characteristic that they are not normalized after each iteration, as for Subjects and Objects.

7   Conclusions and Future Work

In this paper we presented MateTee, a method to compare entities in knowledge graphs, based on the vectorization of the nodes, and specially without any domain expertise. To test the accuracy of MateTee, we compared its results with the state-of-the-art methods like GADES or OnSim, as well as state-of-the-art similarity measures available in the CESSM evaluation framework. MateTee exhibited high accuracy and competitive results, even outperforming the results of GADES, one of the best-performing similarity metric. This behavior was observed in the collections of proteins for UniProt, and the collection of persons from DBpedia. Therefore, observed results suggest that representing knowledge encoded in KGs in the embedding space and using vector based similarity metrics to compare the embeddings of KG entities, provides an accurate method for determining relatedness among entities in the KG.

In the future work, we plan to extend MateTee, to consider the implicit knowledge encoded in the ontologies used to describe a KG. Furthermore, MateTee will be modified to be able to produce values of similarity on-demand, i.e., MateTee workflow will start to work on a pre-training of the original data set. Thus, the optimization process will converge more rapidly whenever any new entity are added to the KG because all other embeddings are positioned in the vector space in a way that the error is already minimized.

Acknowledgments. This work is supported in part by the European Union under the Horizon 2020 Framework Program for the project BigDataEurope (GA 644564) as well as by the German Ministry of Education and Research with grant no. 13N13627 for the project LiDaKrA. We thank Mikhail Galkin for creating the DBpedia collection used in our experiments, and Ignacio Traverso Ribón for his support on the experimental comparison with GADES.

References

1. Benik, J., Chang, C., Raschid, L., Vidal, M.-E., Palma, G., Thor, A.: Finding cross genome patterns in annotation graphs. In: Bodenreider, O., Rance, B. (eds.) DILS 2012. LNCS, vol. 7348, pp. 21–36. Springer, Heidelberg (2012). doi:10.1007/978-3-642-31040-9_3
2. Bernstein, A., Hendler, J.A., Noy, N.F.: A new look at the semantic web. Commun. ACM **59**(9), 35–37 (2016)
3. Bordes, A., Usunier, N., Garcia-Duran, A., Weston, J., Yakhnenko, O.: Translating embeddings for modeling multi-relational data. In: Burges, C.J.C., Bottou, L., Welling, M., Ghahramani, Z., Weinberger, K.Q. (eds.) Advances in Neural Information Processing Systems 26, pp. 2787–2795. Curran Associates Inc. (2013)
4. Bordes, A., Weston, J., Collobert, R., Bengio, Y.: Learning structured embeddings of knowledge bases (2011)
5. Collarana, D., Galkin, M., Lange, C., Grangel-González, I., Vidal, M., Auer, S.: Fuhsen: a federated hybrid search engine for building a knowledge graph on-demand (short paper). In: Debruyne, C., et al. (eds.) OTM Conferences - ODBASE. LNCS, vol. 10033, pp. 752–761. Springer, Heidelberg (2016)
6. Couto, F.M., Silva, M.J., Coutinho, P.: Measuring semantic similarity between gene ontology terms. Data Knowl. Eng. **61**(1), 137–152 (2007)
7. Devos, D., Valencia, A.: Practical limits of function prediction. Proteins: Struct., Funct., Bioinf. **41**(1), 98–107 (2000)
8. Glorot, X., Bengio, Y.: Understanding the difficulty of training deep feedforward neural networks. In: Proceedings of the International Conference on Artificial Intelligence and Statistics (AISTATS 2010). Society for Artificial Intelligence and Statistics (2010)
9. Grover, A., Leskovec,J.: node2vec: scalable feature learning for networks (2016). arXiv:1607.00653. In: Proceedings of the 22nd ACM SIGKDD International Conference on Knowledge Discovery and Data Mining (2016)
10. Jiang, J.J., Conrath, D.W.: Semantic similarity based on corpus statistics and lexical taxonomy. In: Proceedings of 10th International Conference on Research in Computational Linguistics, ROCLING 1997 (1997)
11. Lin, D.: An information-theoretic definition of similarity. In: ICML, vol. 98 (1998)
12. Lin, D.: An information-theoretic definition of similarity. In: Proceedings of the Fifteenth International Conference on Machine Learning, ICML 1998, San Francisco, CA, USA, pp. 296–304. Morgan Kaufmann Publishers Inc. (1998)
13. Lord, P., Stevens, R., Brass, A., Goble, C.: Investigating semantic similarity measures across the gene ontology: the relationship between sequence and annotation. Bioinformatics **19**, 1275–1283 (2003)
14. Pekar, V., Staab, S.: Taxonomy learning: factoring the structure of a taxonomy into a semantic classification decision. In: COLING 2002 Proceedings of the 19th International Conference on Computational Linguistics, vol. 2, pp. 1–7. Association for Computational Linguistics (2002)

15. Pesquita, C., Faria, D., Bastos, H., Falcão, A.O., Couto, F.M.: Evaluating go-based semantic similarity measures. In: Proceedings of the 10th Annual Bio-Ontologies Meeting (BIOONTOLOGIES), pp. 37–40 (2007)
16. Pesquita, C., Pessoa, D., Faria, D., Couto, F.: Cessm: collaborative evaluation of semantic similarity measures. JB2009: Challenges Bioinform. **157**, 190 (2009)
17. Resnik, P.: Semantic similarity in a taxonomy: an information-based measure and its application to problems of ambiguity in natural language. J. Artif. Intell. Res. **11**, 95–130 (1998)
18. Sevilla, J.L., Segura, V., Podhorski, A., Guruceaga, E., Mato, J.M., Martínez-Cruz, L.A., Corrales, F.J., Rubio, A.: Correlation between gene expression and go semantic similarity. IEEE/ACM Trans. Comput. Biol. Bioinform. **2**(4), 330–338 (2005)
19. Smith, T., Waterman, M.: Identification of common molecular subsequences. J. Mol. Biol. **147**(1), 195–197 (1981)
20. Traverso-Ribón, I., Vidal, M.-E., Kämpgen, B., Sure-Vetter, Y.: Exploiting relation and class taxonomy semantics to compute similarity in knowledge graphs. In: SEMANTICS (2016)
21. Traverso-Ribón, I., Vidal, M.-E.: Exploiting information content and semantics to accurately compute similarity of go-based annotated entities. In: IEEE Conference on Computational Intelligence in Bioinformatics and Computational Biology, CIBCB (2015)
22. Traverso-Ribón, I., Vidal, M.-E., Palma, G.: OnSim: a similarity measure for determining relatedness between ontology terms. In: Ashish, N., Ambite, J.-L. (eds.) DILS 2015. LNCS, vol. 9162, pp. 70–86. Springer, Cham (2015). doi:10.1007/978-3-319-21843-4_6

Improved Developer Support for the Detection of Cross-Browser Incompatibilities

Alfonso Murolo[✉], Fabian Stutz, Maria Husmann, and Moira C. Norrie

Department of Computer Science, ETH Zurich, 8092 Zurich, Switzerland
{amurolo,stutzf,husmannm,norrie}@ethz.ch

Abstract. Various tools are available to help developers detect cross-browser incompatibilities (XBIs) by testing the documents generated by their code. We propose an approach that enables XBIs to be detected earlier in the development cycle by providing support in the IDE as the code is being written. This has the additional advantage of making it clear to the developers where the sources of the problems are and how to fix them. We present wIDE which is an extension to an IDE designed specifically to support web developers. wIDE uses a compatibility knowledge base to scan the source code for XBIs. The knowledge base is extracted automatically from online resources and periodically updated to ensure that the compatibility information is always up-to-date. In addition, developers can query documentation from within the IDE to access descriptions and usage examples of code statements. We report on a qualitative user study where developers provided positive feedback about the approach, but raised some issues to address in future work.

Keywords: Testing · Compatibility · Documentation · Web applications · IDE

1 Introduction

An important part of web development is ensuring that a website will have the desired look and behaviour on different browsers and devices. This is non-trivial given the pace at which web technologies continue to evolve in terms of both their specifications and their implementations in different browsers. Taken together with the diversity of devices and the number of browsers and browser versions, with differences between implementations for specific operating systems, the task of ensuring compatibility across browsers is significant. Responsive design partly addresses the problem by ensuring that adaptation to different viewing contexts is central to the design and development processes [1], however developers still need to keep abreast of variations in browser support for specifications and test their code in a large number of viewing contexts.

Fortunately, there are a number of tools and services that developers can use to test their code and check for cross-browser incompatibilities (XBIs). For example, Browsershots[1] will perform testing of websites on a wide range of browsers

---

[1] http://browsershots.org.

© Springer International Publishing AG 2017
J. Cabot et al. (Eds.): ICWE 2017, LNCS 10360, pp. 264–281, 2017.
DOI: 10.1007/978-3-319-60131-1_15

in a distributed screenshot factory and, therefore, return feedback in the form of screenshots. However, while these services can detect XBIs, the developer still has to trace the source of any problems and find out how to correct them.

Online resources such as Can I Use[2] help developers by providing compatibility information about which code features are fully supported, partially supported or not supported in different browser versions. Depending on developer experience, these sites may be consulted frequently either during code development, or after testing, to track the source of XBIs. This requires the developer to switch from the IDE to the browser and studies have shown that web developers spend around 19% of their time in a browser [2].

To address these issues and provide improved developer support for the detection of XBIs, we propose an approach that integrates a tool for compatibility analysis into the IDE so that XBIs can be detected in the source code ahead of testing. This is achieved by performing a static analysis of the HTML, CSS and JavaScript source code, using a compatibility knowledge base extracted from online resources such as Can I Use. Developers can specify the set of browser versions that they want to support and they then receive a compatibility report with feedback on the level of support covered by the current implementation, together with links to the parts of the source code which cause compatibility issues. Although our approach could be applied to any IDE, we have implemented it as an extension to IntelliJ IDEA to specifically support web developers. The resulting environment, called wIDE, also provides direct access to documentation extracted from the Mozilla Development Network[3] for code features used so that developers can avoid having to switch between the IDE and the browser.

We discuss the background to this work and related research in Sect. 2 before going on to detail our approach in Sect. 3. The wIDE system that we developed to support our investigations is presented in Sect. 4. We then report on a qualitative user study that we carried out to evaluate the system in Sect. 5. Concluding remarks and a discussion of future work are given in Sect. 6.

2   Background

In the software engineering community, it has been recognised that developers spend a lot of their time browsing the web looking for solutions and documentation. This has led to various proposals for integrating support for browsing activities into the IDE. For example, HyperSource [3] is an IDE augmentation that links source code modifications to browsing histories, making use of a browser extension that logs pages visited, a facility that tracks use activity in the IDE and an interface that allows developers to interact with the browser histories. Since many developers actually prefer to view the full documentation pages in the browser rather than IDE pop-ups, Codetrail [4] instead integrated Eclipse with a browser extension for Firefox which automatically opens documentation pages

---

[2] http://caniuse.com.
[3] https://developer.mozilla.org.

when the developer browses them in the IDE. As well as creating links between the source code and online documentation based on an analysis of browsing history, Codetrail creates links when a user performs interactions such as copying code from an online resource, thereby guaranteeing traceability in the future.

Often developers browse collections of code examples to assist in the coding task, and the goal of Fishtail [5] is to deliver code examples and documentation related to the current task. The system is integrated with the task management system Mylyn[4] and it tries to determine relevant web resources by analysing the interaction history of the developer with the source code. Suggestions are based on keywords related to program elements such as methods or classes which last changed their *degree of interest*, a value that grows the more frequently the developers interact with them.

XSnippet [6] is another system that suggests relevant code examples to developers. In this case, the system mines Java source code to extract the types used, type hierarchies and method signatures. These are then used to construct search queries that deliver example code snippets from crawled web sources. Muse [7] also has the goal of sharing code snippets, in this case, by extracting them from existing software or developer discussions, pruning irrelevant statements and duplicates.

Rather than coordinating the browser and the IDE, another approach is to integrate online resources into the IDE directly. For example, amAssist [8] introduces a search process inside a Java IDE which makes use of queries to online resources augmented by the usage context as observed by the system. The context is also used to refine the search and rank the results. Similarly, Blueprint [9] is a system targeted at the Adobe Flex community that generates search queries augmented with code context from Adobe Flex builder and creates links between copied code and its source.

Some projects have integrated code examples from online Q&A communities such as StackOverflow into the IDE to help developers find solutions to errors that lead to exception stack traces [10,11]. Seahawk [12,13] integrates StackOverflow code examples directly in the IDE, allowing code snippets to be dragged-and-dropped into the IDE and using annotations for the traceability of snippets with the original crowdsourced solutions. Prompter [14] also goes in this direction, combining the relevance ranking from StackOverflow with an internal ranking based on the context of the code.

Clearly, the problem of switching between the browser and the IDE can also be addressed using the opposite approach of moving the IDE to the browser entirely. Arvue [15] is a browser-based tool that allows developers to create and publish web applications to the cloud. Adinda [16] also investigates the collaborative aspect, integrating web services into the IDE that not only allow traditional development tasks for Java to be performed, but also boost collaboration and communication tasks appropriate for development projects.

In our opinion, introducing support from online sources into existing IDEs (web-based or not) is a preferable strategy as opposed to navigating these resources manually, since modern IDE applications usually provide a baseline

---

[4] http://www.eclipse.org/mylyn/.

of functionalities which can be easily extended to achieve the desired integration with online resources. Further, it can be argued that such an approach is more likely to achieve wider acceptance since developers are already using these IDEs. We therefore chose to investigate how web developers could be better supported by developing an extension to an IDE that would check for XBIs and provide links to online documentation resources.

A number of tools and services already exist to help developers check for cross-browser compatibility and avoid having to manually test their applications in a wide variety of browser/device viewing contexts. One of the earliest research proposals addressing this problem was by Eaton and Memon [17], who used an inductive model based on HTML tags which end-users and developers were expected to keep up-to-date. Systems such as X-PERT [18], WebDiff [19], and its follow-up CrossCheck [20], rely on detecting XBIs by comparing various cues, such as the visual rendering or the DOM of the web pages. Mesbah and Prasad [21] present a fully automated solution for detecting XBIs under different browser environments based on building and comparing hierarchical screen models. XD-Testing [22] is a recent tool specifically designed for testing cross-device applications on simulated devices where developers write test cases that run in different browser environments, exploiting the paradigm of UI Testing.

A number of commercial tools and services for cross-browser testing exist. These include the previously mentioned BrowserShots as well as BrowserStack[5], BrowserSandbox[6], and Browsera[7]. Such services run the web application on either real or simulated devices, accessing the rendered page in various browsers and reporting layout discrepancies or, in some cases, functionality errors such as console logs.

Our goal was to improve developer support by checking for XBIs earlier in the development process, at the time of writing the code rather than later during testing. This can be done using a static analysis of the source code to anticipate any failures that could arise in services and tools like the ones mentioned above. In contrast to such services, it not only detects an XBI but also identifies the code that is the source of the problem and hence makes it much easier to resolve it. To the best of our knowledge, the only previous work that uses static code analysis to detect XBIs is a recent proposal by Xu and Zeng [23], which aims at finding HTML5 incompatibilities with a manually crafted database of XBIs. In contrast to this, we check for XBIs in CSS and JavaScript as well as HTML, and do so using a compatibility knowledge base generated automatically from online resources.

3   Approach

The static analysis in wIDE can provide documentation and compatibility reports about each specific element in the project using online resources. These elements differ between each supported language, and we refer to them as

---

[5] https://www.browserstack.com.
[6] https://turbo.net/browsers.
[7] http://www.browsera.com.

Elements of Interest (EOI). EOIs can be identified by each language handler in the system.

3.1 Elements of Interest

EOIs can be of three different types. A *Central EOI* is a crucial type of element in a language which is interesting as a target for documentation and/or compatibility support. A *Satellite EOI* is, instead, an element which is also relevant for compatibility and documentation lookup, but generally depends on a Central EOI. Finally, *Potentially Foreign EOIs* are elements that usually are not relevant for documentation or compatibility lookups. However, in certain cases and conditions, these can still produce results for a documentation lookup or a compatibility analysis.

The languages currently supported in wIDE are the three main client-side languages connected to cross-browser compatibility issues: HTML, CSS and JavaScript. We will now provide details about the different EOIs that we have defined in each language.

HTML. Tags are clearly central EOIs since they are the principal means of structuring and defining the content of an HTML document. Documentation sources will be centred around HTML tags and therefore will be able to provide information about their support in various browsers as well as providing documentation on their usage, together with some examples. These also usually include attributes of the various tags, which we therefore define as satellite EOIs, since they obviously depend on the tag to which they belong. Finally, we define two potentially foreign EOIs in HTML. Attributes may have limitations for their values, implying that, in these cases, only a subset of attribute values can be applied to an attribute, each with a specific effect. This of course means that these effects will be described in documentation sources, and can also be subject to cross-browser compatibility analysis. Plain-text elements are also categorised as potentially foreign EOIs since usually they have no result in documentation lookups. However, sometimes the content of HTML pages can be written in other languages, which need to be analysed by the appropriate language handler.

CSS. Provides four different elements: selectors, pseudo-selectors, properties and property values. While properties are clearly a central EOI, property values necessarily need to be a satellite EOI since the possible values depend on the property. Both are very relevant elements for documentation and compatibility lookup. Selectors and pseudo-selectors, such as: nth-child, are instead generally supported across every browser. Pseudo-selectors are also generally self-explanatory and, therefore, generally not very interesting in terms of documentation lookup. IDE support for automatic suggestions of these pseudo-selectors is, in our opinion, enough to provide support to the developers in terms of usage. For these reasons, we have decided not to include these in our EOIs.

JavaScript is more complicated. Functions are crucial elements in this language, because developers often use a lot of built-in functions that are dependent on the native browser implementation, making them the perfect central EOI. References to built-in objects and types are categorised as satellite EOIs and may be very interesting for documentation purposes. JavaScript also contains many other elements such as language keywords, blocks and other constructs which are common to programming languages in general. While these could return results for a documentation lookup, we argue that these should generally be supported across the various browsers and, more importantly, are not very interesting in terms of documentation lookup. Therefore, we discarded these as candidates for EOIs.

EOIs are the inputs for the two main functionalities offered by wIDE to developers: the documentation lookup and the compatibility scan. The former will be triggered on individual EOIs that appear in the source code, while the latter can be triggered on both single and multiple EOIs at the same time, for example a project-wide compatibility scan. We will now describe both functionalities in detail.

Fig. 1. An IDE suggestion from IntelliJ.

3.2 Documentation Lookup

The documentation lookup process queries online documentation pages, such as the Mozilla Developer Network (MDN). Usually, the content of these resources is rather exhaustive, including syntax of features, code examples, compatibility information and available attributes/parameters and so on. Currently, little support is provided in IDEs for describing web technologies in order to assist web developers in writing code, forcing them to switch to the browser to navigate to the documentation pages online.

However, IDEs usually provide at least suggestions on which features can be used once the developer starts to type, in an auto-completion fashion, as shown in Fig. 1. While the developer navigates the suggestions, for example by hovering on them or through arrow keys, wIDE will request documentation information to be shown in a sidebar panel, dividing it into separated extensible panes (see Fig. 2). As an alternative, a documentation lookup can be triggered from existing source code by highlighting the corresponding EOI, through mouse or keyboard selection, and using a hotkey. The extensible panes offer the advantage of reducing the size needed to display the various sections, still allowing them to be expanded if the developer wants to see more of any of them, thereby reducing the time needed to navigate across the sections and also avoiding information overload.

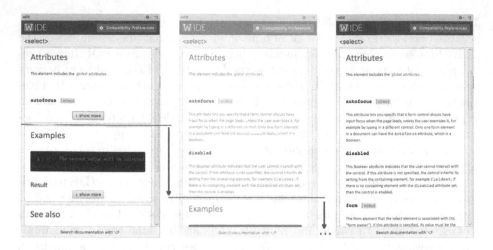

Fig. 2. Extending a documentation pane.

3.3   Compatibility Scan

The second functionality offered by wIDE is the compatibility scan, which can target anything from a single selected EOI to the whole project, scanning for all the EOIs contained in the entire code base. This functionality involves querying an online compatibility knowledge base such as Can I Use. Depending on the type of EOI, sometimes there can also be solutions or work-arounds suggested to improve compatibility for the browsers, for example the use of *polyfills* in the case of JavaScript.

Clearly there needs to be a specified *compatibility set*, which is a set of browsers that the developers want to support, in order to detect any incompatibilities which are relevant for the project. Developers using wIDE can define this set from a preferences panel (see Fig. 3), where they can specify ranges of versions that they want to support. They can also include preview releases, to address cross-browser incompatibilities before these reach a broader set of users, which can be quite useful in the maintenance stage. An additional flag can be selected to keep the project preferences up with new browser releases, as soon as they come out. In this way, when a new browser version is released, it will automatically be included in the compatibility set. By default, all browser versions, from the first release to the latest preview release, are included in the compatibility set.

The compatibility scan can be invoked at three different levels: on a single EOI, on an entire file, and on an entire project. The first level is triggered automatically together with the documentation lookup: an extensible pane about compatibility will be added to the functionality, as shown in Fig. 4. The compatibility scan will show colour-coded results for each browser. For each browser, a vertical bar is presented with the first and the latest versions shown as side labels. The most recent will be at the top, while the oldest will appear at the

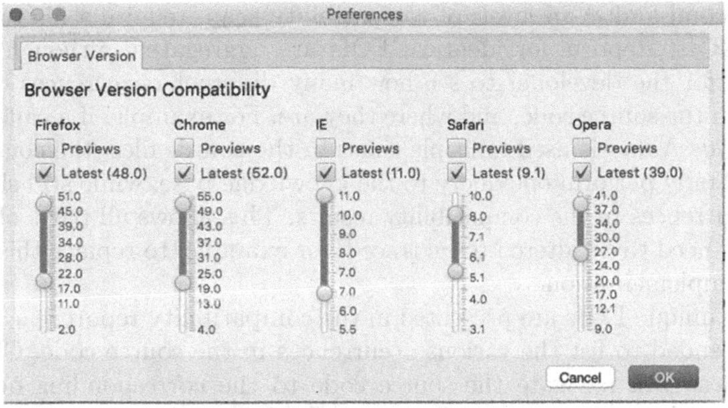

Fig. 3. The preferences panel in wIDE, to set compatibility goals.

bottom. Every transition in the support state, for example a browser starting to support a feature, shows the label of the version implementing the transition on the side. The bar itself is coloured in red if the versions shown do not support the feature, while they are shown in green if they do. Partial support, such as a prefixed, browser-specific implementation, is marked in yellow. Browser versions which are not of interest to the developer, based on the preferences set, are coloured in grey. Finally, the different browser versions are related to their corresponding user base as reported by the knowledge-base, showing the percentage of users that are left out because of the selected EOI. Figure 4 shows this in detail.

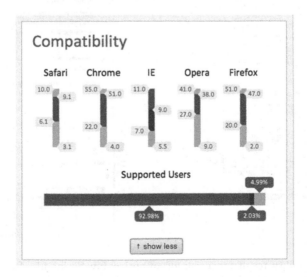

Fig. 4. Compatibility result for an EOI. (Color figure online)

The second and third levels of compatibility scan, require a file traversal to scan for EOIs. Reports for identical EOIs are aggregated, although it is still important for the developer to see how many different occurrences of an EOI there are in the source code, and where they are. For example, if a call to *getElementsByClassName* is used multiple times in the source files, the compatibility report will only perform one query to the knowledge base, while still showing all of the occurrences in the compatibility results. This allows all parts of the code which may need to be altered to be traced, for example, to replace this call with a polyfill implementation.

All the unique EOIs are presented in the compatibility report, each of which can be extended to list the various occurrences in the source code. Clicking on each of these will navigate the source code to the corresponding occurrence. Since the results are aggregated by unique EOIs, these receive a compatibility score which is used to rank the issues by compatibility, in ascending order, so that the most severe issues appear at the top. The compatibility score $C_s \in [0, 1]$ where 0 means that no browser supports the functionality, while 1 stands for full support in every browser version.

Let S_s be the number of browser versions that support the feature, P_s be the number of those that partially support it, Pf_s be the number of those that only

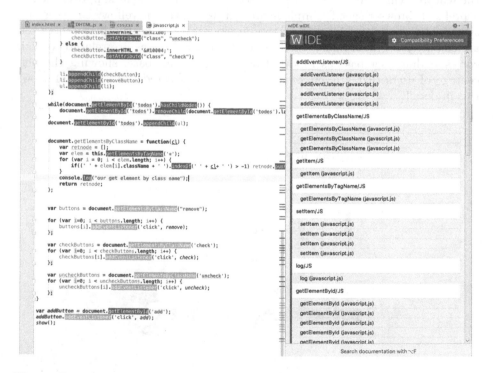

Fig. 5. Compatibility result for a scan of an entire JavaScript file. Issues with higher severity (i.e. lower compatibility score) are ranked at the top, with a colour marking on the side. The same colour is used in the source to locate the issues quickly. (Color figure online)

have the features in a prefixed implementation, N_s be the number of those that do not support the functionality and n be the total number of browser versions in the knowledge base. Then the compatibility score C_s is calculated as follows:

$$C_s = \begin{cases} 0.45 * \frac{S_s + P_s + Pf_s}{n}, & if \ N_s > 0 \\ 0.5 + 0.45 * \frac{S_s}{n}, & else \ if \ P_s + Pf_s > 0 \\ 1, & otherwise \end{cases}$$

An example of the results from a compatibility scan of the second level is shown in Fig. 5. The results are ordered by severity, with a yellow colour tone for the top EOIs, while the ones at the bottom appear green. In addition, the occurrences of the EOIs in the code are also highlighted with the corresponding colour, so the developers can locate the occurrences and the issues in the source code at a glance.

For the last two levels of compatibility scan, the colour coding is more complex than the first level. The compatibility score C_s is mapped to a colour in the range between red and green. The brightest tones of every colour are not used in order to improve readability.

4   Architecture and Implementation

The architecture of wIDE was conceived with the goal of having a shared server that can be deployed for a team or small development organisation. Therefore, it has a client-server infrastructure which is shown in Fig. 6. The client is implemented as a plugin for IntelliJ IDEA, which is responsible for handling the user interaction, parsing the project source code, and communicating with the server. In our implementation of the approach, the server is also responsible for fetching the content from the online sources for the relevant EOIs and parsing it, before caching it in a MySQL database. In this way, the lookup results can be stored in the knowledge base for the entire development team, rather than in each client, thereby minimising the number of queries performed to the sources. In addition, the privacy of the source code in the project is ensured since only the EOI is sent in a request to each online source, and nothing else.

As seen in Fig. 7, each client listens to the various actions that the developer performs in the IDE, such as triggering the lookup of an EOI through a keyboard shortcut. The client contains an appropriate code parser for each language in the language registry, which contains HTML, CSS and JavaScript in our current implementation. Note that other languages such as PHP and Ruby could be added to the registry and we provide more details on how this could be done later in this section.

The identification of EOIs is implemented through code parsing that builds up a Program Structure Interface tree (PSI tree), integrated with IntelliJ. An example of such a tree is shown in Fig. 8. Based on these trees, wIDE automatically detects the best suited EOI for a lookup by searching for pre-defined patterns. For example, in the case of JavaScript functions, the interesting nodes

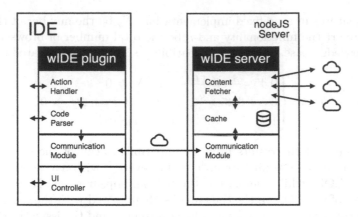

Fig. 6. The architecture of wIDE.

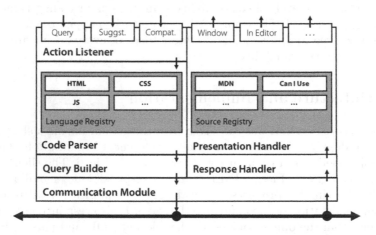

Fig. 7. The components of the wIDE client. The left side processes a query for the knowledge base, while the right side manages the response from the server once the queries have produced a result.

are mostly represented by a REFERENCE node as the first child of a CALL node. Clearly, different patterns need to be applied depending on the language. In addition, different parsing rules have to be considered when wIDE has to handle lookup requests on incomplete code.

Identifying which are the interesting EOIs in a selection of HTML or CSS is quite simple. In HTML, any code selection between the tag delimiters <> will lead to a lookup either of the tag or of the attribute. In the case of CSS, wIDE will only allow lookups on properties, meaning that a selection that contains either the property or its value, will lead in any case to a lookup of the property through the PSI tree.

In the case of JavaScript, wIDE needs to perform an additional step before performing a lookup in the knowledge base because of two main issues. First, the

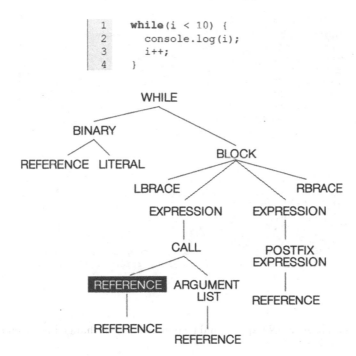

```
1    while(i < 10) {
2        console.log(i);
3        i++;
4    }
```

Fig. 8. A simple JavaScript code snippet with the corresponding PSI tree. The red box highlights the EOI which will be found as appropriate for the query, based on pattern matching. (Color figure online)

potential amount of defined functions is unlimited and, second, there could be name conflicts between functions, especially in the case where multiple libraries are included in the project. Therefore, wIDE traces back the function definition to distinguish whether it is natively defined, for example *getElementById*, or was defined in a library that has been included. We found this step quite challenging for two reasons. First of all, JavaScript is loosely typed and the receiver of a function call is not simply distinguishable before runtime. Second, it may be possible that some libraries are available in the project files, while others may be linked from a content delivery network. In our implementation, wIDE is only able to look for JavaScript definitions within the files of the project workspace and the native functions. Once possible definitions have been traced back, wIDE performs a lookup of all potential function definitions, ordered by the probability that each individual function may be the one called at runtime.

Once the EOI is identified through the appropriate language handler, the client communicates with the server by sending a JSON-based request that contains various information items such as the type of EOI and the content, and it even allows for composite requests, as shown in Fig. 9. The composite requests allow for a request to contain additional sub-requests, which can be useful in special cases, such as when a central EOI is sent, and additional EOIs need to be shipped with the parent request, for example satellite EOIs.

```
 1   'WideLookupRequest':    {
 2     'lang': 'HTML',       // required - The language (HTML, CSS, JS)
 3     'type': 'tag',        // required - The type of the element
 4     'key': 'form',        // required - The name of the element
 5     'value': '',          // required for some types of elements
 6     'children': [         // optional - requests may include subrequests
 7       {
 8         'lang': 'HTML',
 9         'type': 'attribute',
10         'key': 'method',
11         'value': 'POST'
12       }]
13   }
```

```
 1   'WideLookupResponse':   {
 2     'lang': 'HTML',
 3     'type': 'tag',
 4     'key': 'form',
 5     'value': '',
 6     'children': [...],       // subresponses, if there were subrequests
 7     'documentation': [       // one entry for every documentation source
 8       'mdn': {...},
 9       ...]
10   }
```

Fig. 9. The format of JSON requests and responses being exchanged between the wIDE client and the server.

The format of the server response is also shown in Fig. 9. In the server response, the lookup results are attached as a payload of individual objects for each source. Once the response is received, the client can present the received lookup result through the presentation handler of each source, as shown in Sect. 3.

On the server, requests get analysed and, if the response has already been stored from previous queries, it can be returned immediately. If not, the request gets decomposed by a Query handler and a Compatibility handler, which, respectively, will query resources from the corresponding source registry (see Fig. 10). Source handlers will manage the communication with the respective source. Note that while Can I Use offers an API to query for individual EOIs, MDN does not. Once the querying is complete, the result can be sent back to the client and saved in the cache, with an expiration time of 7 days.

As mentioned earlier, wIDE currently only supports JavaScript, HTML and CSS, but could be extended to support other languages commonly used in web applications such as PHP and Ruby, for example to compare support for functionalities across different runtime versions. The architecture is designed so that such extensions can be supported easily. To add support for a language, the developer must implement a language handler for the client and a source handler for both the client and the server. A language handler consists of three main components:

- an *abbreviation* of the language to distinguish on the server side which language is being queried,
- a *language parser*, which identifies the EOIs and builds up the lookup request to the server,

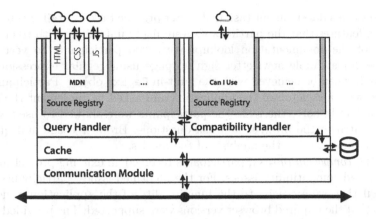

Fig. 10. The decomposition of the wIDE server.

- a *window factory*, that handles the presentation of the lookup result in the wIDE sidebar, allowing for language specific presentation styles of the results.

In addition to the language handler, it may be possible that the existing client source handlers are not able to present any information for the new language being added. Therefore, wIDE provides extensibility of the source handlers for result presentation in the client. Essentially, any implementation of a source handler needs to define three main methods:

- a constructor that takes and parses the server response in JSON format,
- a method that allows the construction and the population of the extensible panes with content from the new source,
- a method to compute compatibility support, if required, for example for different runtime versions in the case of PHP.

Clearly, the source handlers on the server need to implement a method to query and extract the appropriate content depending on each source's public interface. For example, the source handler for MDN will extract data and styling for the appropriate sections from the public website.

5 Evaluation

We evaluated wIDE in a qualitative user study with 9 developers to receive feedback on our approach of integrating documentation and compatibility information into the IDE.

5.1 Tasks and Procedure

We designed two tasks: one that focused on the documentation lookup and one for the compatibility scan. For the *documentation lookup* task, participants were asked to solve a development task in JavaScript. The task was tailored to include language features that the participants were likely unfamiliar with to encourage the use of the documentation lookup. We asked participants to verify email addresses for a simple newsletter sign up page using regular expressions. This functionality can be achieved with the built-in RegExp object. Participants were provided with a skeleton of the application and asked to implement the missing functionality. To solve the task, the participants were allowed to use the wIDE plugin, switch to the browser and ask questions. Browser usage and questions were noted and used in the analysis of the results.

For the *compatibility* task, participants received an existing project and were asked to find compatibility issues. For the identified issues, participants had to analyse if they were crucial to the functionality of the application and resolve them so that the required browser versions were supported. The provided project was a simple to-do application consisting of one HTML, one JavaScript and one CSS file and had the following CSS and JavaScript compatibility issues.

- **CSS border-radius** not supported in IE versions <9. Cosmetic.
- **CSS text-decoration** full support only in Firefox, partially supported by Safari and behind experimental flags in Chrome and Opera. Cosmetic.
- **JavaScript getElementsByClassName** not supported in IE versions <9. Crucial.
- **JavaScript addEventListener** not supported in IE versions <9. Crucial.

Before the first tasks, participants were introduced to the wIDE plugin and guided through an introductory task with the aim of familiarising them with the plugin's features and usage. After all tasks had been completed, participants were asked to fill in a questionnaire. In addition, we logged usage data from the plugin and took notes of the participants' behaviours such as when they struggled, when they used the browser and what they were looking for in that case.

5.2   Participants

We recruited nine participants from our university. Participants were between 22 and 41 years old, with an average age of 27 years. Two participants were female and seven were male. All but one had a background in computer science and the remaining participant was a student of electrical engineering. We required at least basic knowledge of HTML, CSS, and JavaScript to participate in the study and relied on self-assessment for these skills. All participants reported having encountered compatibility issues as developers before our study. When asked about documentation sources that participants were familiar with, the most common answer was *StackOverflow*, which was known to all of them, followed by *w3schools* (6 mentions) and the official *w3c* web standard documentation (5 mentions). Only three users were aware of *Can I Use* and *MDN*.

5.3   Results

For the documentation tasks, participants triggered 11.4 documentation lookups explicitly on average, while 71.5 lookups were trigged automatically as users were scrolling through the auto-complete suggestions. On average, a user opened 7.7 documentation panes to look at the content. We observed two participants who mainly accessed documentation through the browser and only opened one and two panes, respectively. For the compatibility, we measured 2.5 project scans and 3.2 file scans on average per participant. Analysing the questionnaires, we found that all participants stated that wIDE provides relevant or very relevant information. Some participants would have liked to have even more example-centric content that demonstrates the usage of a feature or provides alternative solutions in the case of compatibility issues. Based on our observations, users might benefit from a natural language access point to the documentation. Our approach based on the auto-complete suggestions requires the user to be aware of the name of an EOI, or at least the first few characters of it. In our study, many participants struggled to find the `RegExp` object and either switched to the browser or asked for help. Only three participants managed to access the documentation for this object using the auto-complete suggestions without help.

The compatibility scans received positive feedback[8]. Suggestions for improvement included automatic background scans for changed code that would highlight problematic elements as the user is writing code. Currently, wIDE will only highlight compatibility issues but developers have to figure out how to address them. Multiple participants requested more assistance for resolving issues. For example, the system could provide alternative solutions or polyfills. One participant suggested to have small tiles that visually display the output of an application loaded into different browsers. Another participant requested an extension for PHP.

6   Conclusion

We have presented wIDE, a system to support web developers obtain compatibility and documentation information directly in the development environment. To provide support for web technologies in the IDE, wIDE builds a knowledge base that centralises information extracted from various online resources and displays it to the user non-intrusively and in-place, thereby reducing the need to switch the context to the browser. The sources are queried through a context-aware parsing of the source code, to provide information about documentation for the various functionalities and their compatibility across various browsers. This helps in locating cross-browser compatibility issues, either before moving to a cross-browser testing stage, or after such stage has been completed and the issues in the source code have to be located and addressed.

Currently, our implementation of wIDE for IntelliJ IDEA supports the languages which are relevant for a cross-browser compatibility analysis, namely HTML, CSS and JavaScript. However, it could be extended to even perform compatibility analysis of server-side languages, with the goal of analysing the compatibility of functionalities across different runtime versions. In addition,

---

[8] Questionnaire and responses at https://github.com/fabwid/wIDE/blob/master/ user_study/userStudyResponses.xlsx - Accessed 21 March 2017.

together with the support for additional languages, wIDE also offers support for introducing new knowledge resources, which might be required if additional languages were added.

We designed and conducted a user study to evaluate the implementation of wIDE in terms of the feedback that it provides to developers. The study consisted of a development task and a compatibility analysis task. The data logs, observations and qualitative feedback provided by the participants have shown that wIDE provides relevant information based on the user's needs. However, they also expressed the desire for additional support for the automatic resolution of compatibility issues, and more assistance in looking for features beyond the current support which relies on the IDE auto-completion.

The problem of automatically resolving compatibility issues could be addressed in future work. For example, in some cases, MDN provides a *polyfill* implementation of missing JavaScript functionalities, which could be included automatically. We also plan to introduce more example-centric sources such as StackOverflow based on compatibility issues in order to improve support for resolving issues once these are located. We have opened the project sources[9] with the hope that this might encourage others to extend the set of languages supported, and even extend the scope of the compatibility analysis to consider server runtimes.

References

1. Mohorovičić, S.: Implementing responsive web design for enhanced web presence. In: 36th International Conference on Information & Communication Technology Electronics & Microelectronics (MIPRO), pp. 1206–1210. IEEE (2013)
2. Brandt, J., Guo, P.J., Lewenstein, J., Dontcheva, M., Klemmer, S.R.: Two studies of opportunistic programming: interleaving web foraging, learning, and writing code. In: Proceedings of the SIGCHI Conference on Human Factors in Computing Systems. ACM (2009)
3. Hartmann, B., Dhillon, M., Chan, M.K.: HyperSource: bridging the gap between source and code-related web sites. In: Proceedings of the SIGCHI Conference on Human Factors in Computing Systems. ACM (2011)
4. Goldman, M., Miller, R.C.: Codetrail: connecting source code and web resources. J. Vis. Lang. Comput. **20**(4), 223–235 (2009)
5. Sawadsky, N., Murphy, G.C.: Fishtail: from task context to source code examples. In: Proceedings of the 1st Workshop on Developing Tools as Plug-ins. ACM (2011)
6. Sahavechaphan, N., Claypool, K.: XSnippet: mining for Sample Code. ACM Sigplan Not. **41**(10), 413–430 (2006)
7. Moreno, L., Bavota, G., Penta, M.D., Oliveto, R., Marcus, A.: How can I use this method? In: Proceedings of 37th IEEE International Conference on Software Engineering, vol. 1. IEEE (2015)
8. Li, H., Zhao, X., Xing, Z., Bao, L., Peng, X., Gao, D., Zhao, W.: amAssist: In-IDE ambient search of online programming resources. In: Proceedings of IEEE 22nd International Conference on Software Analysis, Evolution, and Reengineering (SANER). IEEE (2015)

---

[9] https://github.com/fabwid/wIDE.

9. Brandt, J., Dontcheva, M., Weskamp, M., Klemmer, S.R.: Example-centric programming: integrating web search into the development environment. In: Proceedings of the SIGCHI Conference on Human Factors in Computing Systems. ACM (2010)
10. Cordeiro, J., Antunes, B., Gomes, P.: Context-based recommendation to support problem solving in software development. In: 3rd International Workshop on Recommendation Systems for Software Engineering (RSSE). IEEE (2012)
11. Rahman, M.M., Yeasmin, S., Roy, C.K.: Towards a context-aware IDE-based meta search engine for recommendation about programming errors and exceptions. In: IEEE Conference on Software Maintenance, Reengineering and Reverse Engineering (CSMR-WCRE). IEEE (2014)
12. Ponzanelli, L., Bacchelli, A., Lanza, M.: Seahawk: stack overflow in the IDE. In: Proceedings of the International Conference on Software Engineering. IEEE Press (2013)
13. Ponzanelli, L., Bacchelli, A., Lanza, M.: Leveraging crowd knowledge for software comprehension and development. In: 17th European Conference on Software Maintenance and Reengineering (CSMR). IEEE (2013)
14. Ponzanelli, L., Bavota, G., Penta, M.D., Oliveto, R., Lanza, M.: Mining StackOverflow to turn the IDE into a self-confident programming prompter. In: Proceedings of the 11th Working Conference on Mining Software Repositories. ACM (2014)
15. Aho, T., Ashraf, A., Englund, M., Katajamäki, J., Koskinen, J., Lautamäki, J., Nieminen, A., Porres, I., Turunen, I.: Designing IDE as a service. Commun. Cloud Softw. 1(1) (2011)
16. van Deursen, A., Mesbah, A., Cornelissen, B., Zaidman, A., Pinzger, M., Guzzi, A.: Adinda: a knowledgeable, browser-based IDE. In: Proceedings of the 32nd ACM/IEEE International Conference on Software Engineering, vol. 2. ACM (2010)
17. Eaton, C., Memon, A.M.: An empirical approach to evaluating web application compliance across diverse client platform configurations. Int. J. Web Eng. 3(3), 227–253 (2007)
18. Choudhary, S.R., Prasad, M.R., Orso, A.: X-PERT: accurate identification of cross-browser issues in web applications. In: Proceedings of the 2013 International Conference on Software Engineering. IEEE Press (2013)
19. Choudhary, S.R., Versee, H., Orso, A.: WEBDIFF: automated identification of cross-browser issues in web applications. In: International Conference on Software Maintenance (ICSM). IEEE (2010)
20. Choudhary, S.R., Prasad, M.R., Orso, A.: Crosscheck: combining crawling and differencing to better detect cross-browser incompatibilities in web applications. In: Proceedings of the 5th International Conference on Software Testing, Verification and Validation. IEEE (2012)
21. Mesbah, A., Prasad, M.R.: Automated cross-browser compatibility testing. In: Proceedings of the 33rd International Conference on Software Engineering. ACM (2011)
22. Husmann, M., Spiegel, M., Murolo, A., Norrie, M.C.: UI testing cross-device applications. In: Proceedings of the ACM Conference on Interactive Surfaces and Spaces. ACM (2016)
23. Xu, S., Zeng, H.: Static analysis technique of cross-browser compatibility detecting. In: Proceedings of the International Conference on Applied Computing and Information Technology/International Conference on Computational Science and Intelligence. IEEE (2015)

Proximity-Based Adaptation of Web Content on Public Displays

Amir E. Sarabadani Tafreshi$^{(\boxtimes)}$, Kim Marbach, and Moira C. Norrie

Department of Computer Science, ETH Zurich, 8092 Zurich, Switzerland
{tafreshi,norrie}@inf.ethz.ch, marbachk@student.ethz.ch

Abstract. Viewers of public displays perceive the content of a display at different sizes according to their distance from the display. While responsive design adapts web content to different viewing contexts, so far only the characteristics of the device and browser are taken into account. We show how these techniques could be extended to consider viewer proximity as part of the viewing context in the case of public displays. We propose a general model for proximity-based adaptation and present a JavaScript framework that we developed to support experimentation with variants of the model in both single and multi-viewer contexts. We also report on an initial user study based on single viewers which yielded promising results in terms of improved user perception and engagement.

Keywords: Responsive design · Proximity-based adaptation · Web development framework

1 Introduction

Responsive design is an approach to web design aimed at crafting sites to provide an optimal viewing and interaction experience. Currently, only the characteristics of devices and browsers are taken into account in adapting a website to a particular viewing context. However, it is interesting to note that earlier drafts proposing extensions to media queries in CSS4[1] did also consider the environmental factor of ambient light level. Since the distance of the viewer from the screen is highly variable in public display contexts with the effect that users perceive content in radically different sizes, we propose that responsive design techniques should be extended to take user proximity into account in such settings.

An open issue in the design of pervasive display systems for public and semi-public spaces is how to attract and retain user interest [1]. We wanted to investigate how web content could be adapted according to the proximity of the viewers and whether this would improve the user experience. In this paper, we present a model that uses the distance of the viewers to the screen in combination with the display characteristics to adapt the display content. While the basic model is based on a single viewer, public displays also differ from traditional web viewing

---

[1] https://www.w3.org/TR/2016/WD-mediaqueries-4-20160126/.

© Springer International Publishing AG 2017
J. Cabot et al. (Eds.): ICWE 2017, LNCS 10360, pp. 282–301, 2017.
DOI: 10.1007/978-3-319-60131-1_16

contexts in the fact that there are typically multiple viewers. We therefore propose a number of variants of the model for multi-viewer contexts.

We also introduce a development framework that implements the model and acts as an experimental platform for proximity-based interaction. To detect the number and distance of viewers, we use the camera-based Kinect sensing technology which is readily available as a commercial product. The framework, called ResponDis, is based on JavaScript and builds on the Kinect SDK. Similar to responsive design breakpoints specified using CSS3 media queries, the framework supports a zone concept which allows designers to specify more radical changes to the content and layout as viewers move between zones.

The ResponDis framework is characterised by four main aspects: (1) useful programming abstractions for the use of our proposed approach (2) lightweight support for cross-platform, multi-device development based on native web technologies with which many developers are already familiar, (3) a flexible client-server architecture enabling a variety of multi-device ecosystems around Kinect, (4) an extensible architecture which allows multiple Kinects to be used and connected to different distributed servers in order to track more people in different locations and at different viewing angles.

We begin with a discussion of related work in Sect. 2, before introducing our proximity-based adaptation model in Sect. 3. The features of the ResponDis framework are presented in Sect. 4 followed by a description of the architecture and implementation in Sect. 5. We then report on an initial user study carried out to evaluate our adaptation model in Sect. 6. The paper concludes with a discussion of our results in Sect. 7 followed by final remarks in Sect. 8.

2   Related Work

Responsive design is a recent trend in web development that caters for the wide diversity of devices now used to access websites [2]. There are three main parts to responsive design: a fluid rather than fixed layout that adapts to the viewport size, media queries that optimise the design for different viewing contexts, and means of selecting and sizing images according to viewing context [3]. *Fluid layout* relies on the use of relative units such as percentages and em instead of absolute units such as pixels and points, along with flexible grid layouts such as that supported by Flexbox[2]. *Media queries* are used to specify design breakpoints in terms of alternative CSS style rules to be applied in specific viewing contexts defined by values such as the viewport size, pixel density and device orientation [4].

In the research domain, the focus has been on support for desktop-to-mobile adaptation [2], while an increasing number of practitioners advocate a mobile-first strategy in conjunction with responsive design that could be categorised as mobile-to-desktop adaptation. However, relatively little attention has been paid in either research or practice to adaptation to very large screens, even though

---

[2] https://css-tricks.com/snippets/css/a-guide-to-flexbox/.

these are now common in public and semi-public spaces. An exception was work investigating adaptation to large screens, particularly in the case of text-centric websites such as online newspapers [5]. In this work, they proposed a set of design metrics, developed an adaptive layout template and carried out a user study to show the benefits of the adaptation.

While CSS3 introduced media queries to allow designs to be adapted to particular viewing contexts, they only take into account characteristics of the device and browser such as viewport size and pixel density. Earlier draft proposals for extensions to media queries in CSS4 included the property *luminosity* for the ambient light level and therefore considered contextual factors beyond the device and browser. We note however that this has been removed from the latest draft.

A factor that has not yet been taken into account in responsive design is the proximity of the user to the screen which is an issue in the case of public displays where the distance of the user to the screen can be hugely variable. In contrast, the variation in viewing distances for desktop and mobile displays is relatively small. For example, one study showed that the average smartphone viewing distance is about 12 in. [6], and it is easy for a user to adjust the distance to improve legibility if required. In the case of a public screen, it might require that they significantly adjust their position or motion path, which they would only do if they were already engaged with the display. Therefore, we advocate that, ideally, several other factors need to be taken into account when adapting web content to public displays such as the distance of the user, their visual acuity, and the resolution of content.

Screen manufacturers typically recommend either a fixed sitting distance or a range within which a viewer should be seated to have an optimal view. However, the recommendations differ depending on which manufacturer you ask and on the use of the display itself. We summarise some of the available recommendations below.

The most common fixed viewing distance recommendations are proposed by SMPTE[3] (Society of Motion Picture and Television Engineers) and the company THX[4], as well as manufacturers and retailers. A widescreen high-definition television (HDTV) normally has horizontal lines of vertical resolution (pr) and an aspect ratio of 16:9. One of the popular fixed viewing distance recommendations for a 1080pr resolution is 2.5 times the display's diagonal size (DS) which corresponds to a 20-degree viewing angle. However, SMTPE standards recommend 1.6264 times the DS for a 16:9 TV which is a very popular viewing distance recommendation in the home theatre enthusiast community [7]. This recommendation corresponds to a position where the display occupies a 30-degree field of view. In contrast, THX, which develops high fidelity audio/visual reproduction standards, recommends that the "best seat-to-screen distance" is one where the view angle approximates 40 degrees. To achieve this, they recommend multiplying the DS by 1.2.

---

[3] http://www.smpte.org.
[4] http://www.thx.xom.

In addition to the fixed viewing distance, different optimum viewing range recommendations based on the screen size also exist. For the minimum and maximum viewing distance, some manufacturers recommend a view angle of approximately 31 and 10 degrees, respectively. On the other hand, some retailers recommend the minimum viewing distance that allows a view angle of just a little over 32 degrees on average and the maximum viewing distance that provides a viewing angle of approximately 16 degrees. However, THX certified cinema screen placements offer a different range. THX still contends that minimum viewing distance is set to approximate a 40-degree view angle, and the maximum viewing distance is set to approximate 28 degrees. Their maximum horizontal view angle recommendation is based on the average human vision, and, in their opinion, a 40-degree view angle provides the most "immersive cinematic experience". Therefore, they consider their minimum viewing distance recommendation as the optimum viewing distance which provides the maximum viewing angle based on average human vision.

Regardless of inconsistencies in the recommendations, almost all of them are based on variable screen size but on a fixed display resolution with the same content resolution and what is considered to be normal vision. However, these factors certainly affect the calculation of optimum viewing distance taking into account the limitations of the human visual system [8].

Previous research [9] in information interaction at multiple distances and angles has shown that adaptive interfaces are useful for addressing the user's various attention states. Dostal et al. [10] implemented a multi-user interface in a wall-sized display that exploits people's movements and distance to the content display to enable collaborative navigation. This work also offers a toolkit called SpiderEyes for designing attention- and proximity-aware collaborative interfaces for wall-sized displays. However, the interface design in such works are still based on experiments for their specific setup. Consequently, migration of such designs to a new setup requires new design decisions. Therefore, a general model that enables responsive web design based on the limitations of the human visual system and influencing factors is a necessity.

3   The Model

As discussed in the previous section, the ideal viewing distance is based on visual acuity. The human eye with *normal* vision (20/20) can detect or resolve details down to as small as $1/60th$ of a degree of arc [11]. This distance represents the point beyond which some details in the image can no longer be resolved. Being closer to the screen than this results in the need for higher image resolution per degree of arc (or angular resolution) as well as increased pixel count of the display. This value should be lowered if visual acuity is worse than normal human vision, and raised if it is better.

An equation that considers these factors to calculate the minimum (optimum) viewing distance has been proposed [12]:

$$VD = \frac{DS}{\sqrt{((\frac{NHR}{NVR})^2 + 1)} \times CVR \times \tan(\frac{\frac{1}{60}}{eyeAccuracy} \times \frac{\pi}{180})} \tag{1}$$

where the meaning of the parameters is shown in Table 1.

Table 1. Explanation of the parameters of the equation.

VD	Viewing distance (in inches)
NHR	Display's native horizontal resolution (in pixels)
NVR	Display's native vertical resolution (in pixels)
DS	Display's diagonal size (in inches)
CVR	Vertical resolution of the content being displayed (in pixels)
eyeAccuracy	Visual acuity, 20/20 is considered as "normal" eye sight

Given these parameters, one can see that many factors play a role in computing the appropriate viewing distance. Not only is the distance of the user to the display an important factor, but also factors such as the size and resolution of the display, the resolution of the content being displayed and human visual acuity. Using this formula, one can see, for example, that the optimum viewing distance becomes closer to the screen, as the screen size decreases.

Since the goal of responsive web design is to provide an optimum web interface, we need to calculate the optimal content size corresponding to the viewing distance of users, rather than the optimal distance for a viewer given fixed content. Therefore, we transformed the formula given above to instead have the viewing distance (VD) as an input and the resolution of the content being displayed as an output. The resulting formula calculates the optimal content vertical resolution (CVR) as follows.

$$CVR = \frac{DS}{\sqrt{((\frac{NHR}{NVR})^2 + 1)} \times VD \times \tan(\frac{\frac{1}{60}}{eyeAccuracy} \times \frac{\pi}{180})} \tag{2}$$

Similar to the use of screen size in current responsive design methods, the calculated CVR could be used to define design breakpoints in the form of media queries as well as for fluid web layout.

This can also be done using the optimal content horizontal resolution (CHR) which is equivalent to CVR multiplied by the screen aspect ratio ($\frac{NHR}{NVR}$). Defining media queries using optimal content resolution rather than fixed user distances automatically results in the specification of distance ranges which define *zones* in front of a display.

For example, consider an application that displays a world information map on a public display. We could define three zones as shown in Fig. 1.

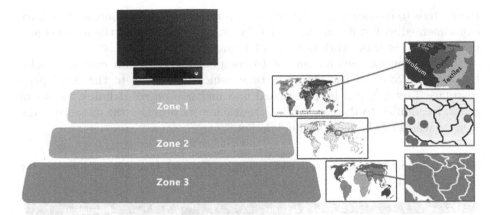

Fig. 1. Zone based UI adaptation of world information map

The furthest zone, Zone 3, could be designed to attract the attention of users by simply displaying a world map showing the continents. This could be defined as the range $0pr < \text{CHR} < 360pr$. Curious users might move closer into Zone 2 defined by the range $360pr < \text{CHR} < 720pr$. At this point, the display content would be adapted to show more detailed information such as a world population cluster visualisation. Moving even closer to the screen, they enter Zone 1 defined by the range $720pr < \text{CHR}$. Here, even more details could be provided such as a commodity word cloud which would be easily read by viewers at this distance.

Using the optimal content resolution to define the zones makes the layout adaptation independent of fixed distance ranges which also means that it is easy to cater for different setups just as responsive design caters for multiple devices. Our model can therefore support pervasive display systems (PDS) which manage content for heterogeneous display networks [1].

4 The ResponDis Framework

As a first solution to test the model and enable rapid prototyping, we designed and developed ResponDis–a JavaScript framework for proximity-based adaptive display user interfaces. The framework provides developers with crucial information such as the proximity of viewers, the current zones of the viewers, individually optimal content resolution (in pixels) of all tracked viewers, and the recommended activated zone for the case of multiple viewers based on the number of viewers in each zone as well as the total number of viewers.

As already mentioned in the previous section, our intention was to use the optimal content resolution as design breakpoints, similar to how viewport size is often used in CSS3 media queries to define layout breakpoints. In the case of our adaptation model, these breakpoints correspond to a zone-scheme so that content is adapted depending on the zone in which a user is currently located. As will be discussed later, in the case of multiple users located in different zones, there needs to be a strategy that determines which zone to activate and we have experimented with different strategies. We start by describing the features and operation of the framework in terms of a single user.

The framework uses our model to compute the optimum content resolution "*contentSize*" at each point in time, which, according to the developer's customised setting, can be partitioned into multiple ranges defining the set of zones. In addition to the *contentSize*, the framework provides the distance of the viewers which can be used for fluid design.

Table 2. ResponDis features with code examples. Callback functions are based on the *settings* object.

Features	Examples	Description
settings	ResponDis({trackedBodyPart: 'neck', eyeAccuracy: 20/20, NHR : 1366, NVR : 768, groupTreatment: 'medianZone', DS: 32, zones: [[0, 480],[480, 1080]]] });	Set the formula variables, body part to calculate the distance, treatment method to serve multiple viewers, content resolution mode, and zones breakpoints
proximity	ResponDis(ourSetting, function(proximity) { console.log("viewer"+ z +"distance: " + proximity[z]); });	Register callback for proximity of viewers
zone	ResponDis(ourSetting, function(zone) { console.log("viewer"+ z +" is in zone: "+zone[z]); });	Register callback for current zone of each viewer
contentSize	ResponDis(ourSetting, function(contentSize) { console.log("viewer"+ z +" content optimum size:"+ contentSize[z]); });	Register callback for appropriate content resolution (pixels) of each viewer
multiViewerZone	ResponDis(ourSetting, function(multiViewerZone) { console.log("recommended zone for multiple viewers:" + multiViewerZone});	Register callback for a recommended zone to serve all the viewers
numViewersPerZone	ResponDis(ourSetting, function(numViewersPerZone) { console.log("number of viewers in"+ z + numViewersPerZone[z]); });	Register callback for the number of viewers in each zone
totalViewers	ResponDis(ourSetting, function(totalViewers) { console.log(totalViewers + "viewer(s) are detected"); });	Register callback for the number of viewers in each zone

Table 2 presents the key features that are encapsulated in the ResponDis framework. *Proximity*(s) provides an array containing the proximity of all current viewers of the display. *Zone*(s) gives an array containing the zone of all current viewers. *contentSize*(s) makes available an array containing the optimal content size (in pixels) of all current viewers. *multiViewerZone*(s) gives a number representing the recommended zone considering all current viewers. *totalViewers*(s) provides a number that shows the number of simultaneous viewers.

Settings is one of the key features of ResponDis and allows the framework parameters and computation methods to be configured. Table 3 shows the parameters together with their possible and default values along with the corresponding description.

trackedBodyPart defines the proximity of a viewer by measuring the distance between the screen and the viewer's body part. The proximity of the head

Table 3. ResponDis' configurable setting options

Option	Descriprion	Default Value	Possible Values
trackedBodyPart	the body part used to measure the distance to the display	head	{'head', 'spine base', 'spine middle point', 'neck'}
eyeAccuracy	vision acuity	20/20	some number representing the average eye sight, e.g. 20/20 is considered as "normal" eye- sight
groupTreatment	computation method to serve multiple viewers	averageProximity	{'averageProximity', 'averageZone', averageContentSize', 'medianProximity', 'medianZone', 'medianContentSize', 'mostCrowdedZone', 'closestViewer', 'firstDetectedViewer'}
horizontalResolution	whether the zone should stand for the content horizontal resolution (horizontal or vertical)	true	{'true', 'false'}
DS	display's diagonal - should be provided by the developer to ensure better results	getDS() – approximates the display's diagonal size; can be off by a couple of inches in some cases	number in inches
NHR	display's native horizontal resolution (in pixels)	window.outerWidth	number in pixels
NVR	display's native vertical resolution (in pixels)	window.outerHeight,	number in pixels
zones	Zone breakpoints based on the appropriate content size (lowerLimit ≤ contentSize < upperLimit)	[[0,360], [360,480], [480,720], [720,1080]]	two dimensional array of limits [[lowerLimitZone1, upperLimitZone1], ..., [lowerLimitZoneN, upperLimitZonenN]]

indicates at what distance the passer-by perceives the content of the display. *eyeAccuracy* corresponds to the user's vision acuity. As a default, we assume the common viewer has normal or corrected-to-normal eyesight (20/20). *groupTreatment* offers some possible group treatment methods to be able to serve groups of users an optimal view. The integration of this feature into the framework is because public displays are considered to have multiple simultaneous viewers. We set the *averageProximity* as a group Treatment because it is a simple model and should provide all of them with a relatively good view. *horizontalResolution* specifies which resolution mode should be used to get the optimal content resolution. This resolution mode is important when deciding which zone is currently active. We set the default mode to the horizontal resolution to make it similar to the common use of screen width in responsive design. *DS* stands for the diagonal size of the display in inches. However, using JavaScript and HTML, it is impossible to calculate the exact diagonal size since it is not feasible to determine how many pixels correspond to exactly one inch. Therefore, as a default value, we approximate the *DS* using a "div" element of size 1 cm. By relating it to its pixel width and height, we compute the size of the display and its diagonal. But since a "div" of width 1 cm always gets assigned a fixed number of pixels independent of the display size, the result might be off by a couple of inches. Due to this, we strongly depend on the *DS* setting of the developer to ensure precise zone computations.

NHR is the native horizontal resolution of the screen and we set the default value to the width of the entire visible section of the screen. *NVR* is the native vertical resolution of the screen and the default value is set to the height of the full screen. *zones* defines media queries for optimal content resolution ranges. We assigned the default zones in the breakpoints at the standard resolutions used in practice i.e. 360pr, 480pr, 720pr, and 1080pr.

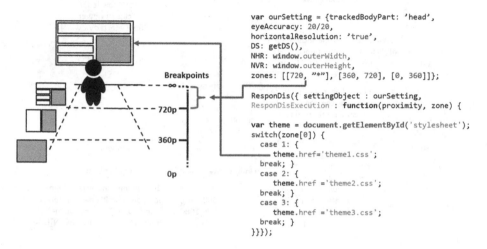

Fig. 2. A sample use of ResponDis framework for one viewer

As shown in Fig. 2, to use the framework, the developer has to construct a *settingObject*, in which they can redefine all factors of their choosing. They then have to define the *RespondDis* object, which takes as arguments the constructed *settingObject* and the function *ResponDisExecution*. The *ResponDisExecution* function has to be defined by the developer. Depending on the given parameters (*proximity, zone, contentSize, multiViewerZone, numViewersPerZone, totalViewers*), the content, layout or design of the display can be responsive. The *ResponDisExecution* function will be re-executed automatically by the ResponDis framework every time new data arrives from the Kinect. To simplify the definition of media queries based on the viewers' optimal content resolution ranges, these breakpoints can be defined in the *zones* parameter of the *settingObject*. Then, as represented in Fig. 2, the framework based on the configuration in the *settingObject* provides which zone each user is standing in. Likewise, the designer can choose what CSS style should be loaded for each optimal content resolution range (zone). When considering a single viewer, the developer can either extract the zone information of that one viewer from the parameter *zone* or use the zone value of *multiViewerZone*, as it returns the same value for a single viewer. Using the calculated parameters, the designer would also be able to make the adaptation fluid. To do so, they can use the *contentSize* as well as *proximity* of viewers.

As mentioned before, the framework also offers different methods for dealing with the case of multiple simultaneous users. These methods include: (1) average proximities of all current viewers *"averageProximity"*, (2) the average zone number of individual viewers *"averageZone"*, (3) the average appropriate content size of all viewers *"averageContentSize"*, (4) the median proximity of all viewers *"medianProximity"*, (5) the median of the zones of individual viewers *"medianZone"*, (6) the median of the appropriate content size for all viewers *"medianContentSize"*, (7) the zone with the most viewers *"mostCrowdedZone"*, (8) the zone of the closest viewer *"closestViewer"*, (9) the zone of the first detected viewer *"firstDetectedViewer"*

These are used to calculate which zone should be activated in order to determine what and how content should be displayed. The developer can specify the preferred method as part of the framework setting. If different elements of the UI should be adapted based on different *groupTreatments*, several setting objects can be constructed, and multiple functions can be defined. Since the framework, in addition to the appropriate content size, provides other functionalities such as the total number and the proximity of the viewers, the zone of each viewer, and the number of viewers in each zone, developers can easily define their own group treatment method. As shown in Fig. 3, the designer using the *multiViewerZone* feature of the framework can decide what CSS style should be be loaded for each optimal content resolution range (zone).

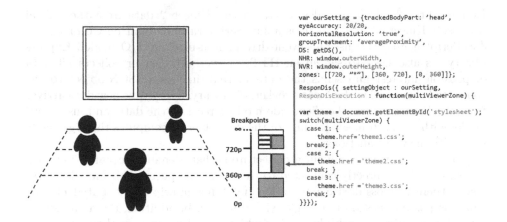

Fig. 3. A sample use of ResponDis framework for multiple viewers

5   Architecture and Implementation

ResponDis is based on a client-server architecture (see Fig. 4). The architecture consists of a *Kinect 2* device connected to a Node.js server through the kinect2

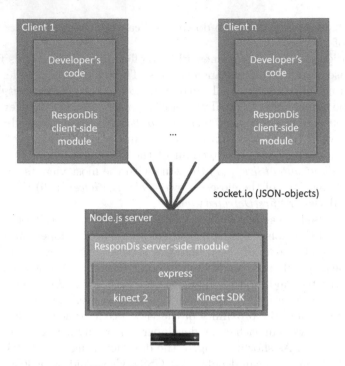

Fig. 4. ResponDis architecture

Nodejs library[5] which provides access to the Kinect 2 data from the official Microsoft Kinect SDK[6]. Node.js is a JavaScript runtime built on Chrome's V8 JavaScript engine and uses an event-driven, non-blocking I/O model. Express library[7] is also used to set up an HTTP server. The server collects all data of possible interest and sends the data in real-time as a JSON-object to the connected clients using the Nodejs socket.io[8] library. Every time new data arrives from the Kinect device, the client-side module receives the data, and, using our model with the customised settings of the developer, it computes the arguments for the framework functions.

One of the advantages of our architecture is that it enables scenarios in which the Kinect is not directly connected to the client computer. This includes cases where there is cross-device interaction involving multiple distributed clients, which is possible based on a single Kinect server. In addition, the architecture could be extended to multiple Kinects which would allow more than six people to be tracked given that Kinect can track a maximum of six people. It could also be used to handle proximities in different locations and/or at different angles. To do this, we could run multiple servers, each of which could be connected to

---

[5] https://www.npmjs.com/package/kinect2.

[6] https://developer.microsoft.com/en-us/windows/kinect.

[7] https://www.npmjs.com/package/express.

[8] https://www.npmjs.com/package/socket.io.

multiple Kinects and clients. Additionally, some slave servers (S) could not only send their data to their clients, but also to a master server (M). In this case, all the received data from the n slaves is combined into a single bulk object: $D_M = \{D_{S1}, D_{S2}, ..., D_{Sn}\}$ which is then be sent to all clients of (M).

6   User Study

To evaluate the proposed adaptation model, we ran a brief user study in a controlled lab setting. Our experiment had two primary goals. First, we aimed to examine whether our model improves the perception of viewers as well as usability and engagement. Second, we wanted to evaluate how our model compares to current characteristics-based static UIs. Our investigation of user engagement was due to the fact that the means of attracting and engaging viewers are considered major factors in PDS [13]. It has been shown that current user interfaces for large displays often cause difficulties in information perception with the result that user engagement is relatively low [14].

6.1   Participants

The study had 13 participants (6 females; age range (median): 19–40 (25) years). Participants were recruited at our university and were mostly ($n = 10$) from the Department of Computer Science. All the participants had normal or corrected-to-normal eyesight.

6.2   Methodology and Procedure

Before participants started performing the tasks, we introduced the system, the purpose of the study, and asked for their consent to record the experiment using a video camera.

The main task in the user study was to find a specific character in a wimmel-book [15] picture. The characters and pictures were chosen so that the character was relatively well known to most people and fitted well into the pictures. The size of the character was carefully adjusted so that it integrated well with the other characters in the picture. The presentation size of the picture for the static UI was adjusted to cover 50% of the entire display so that the view of the content for both the closet and furthest viewers was fairly good. For the adaptive UI, we defined four zones and resized the presentation of the picture for each so that a user standing in a zone had a close to optimal view for that distance.

We kept the type of adaptation simple on purpose, as our goal was to compare the adaptive and static approaches, and we thus wanted to avoid as far as possible that the user focused on the adaptation method itself.

Each participant performed the tasks using these two different interfaces. The design of the studies was cross-over, i.e. the starting order of both approaches was randomised in such a way that any user was equally likely to start with one or other UI and then use the other in the second half of the study. Furthermore, the content orders in each case were randomised.

In the study, the participants had to perform two tasks: (1) Each participant was first shown an image of a character and then was asked to enter the room from the farthest distance to the display and walk around freely to locate the character on the display. The participants were instructed to stop moving and inform the experimenter as soon as they found the character. In this way, we were able to record the time that each participant took to find the character. (2) In the second task, we wanted to measure the effect of each method for different distances. Therefore, each participant was asked to stay within a particular zone. For each zone, a different wimmelbook picture was displayed and the corresponding character shown. Similar to the first task, the participants had to look for a specific character and the time to find it was noted. This experiment was repeated for each of the zones in a random order. After finding the character in each particular zone, the participants were asked which system they preferred for that zone.

After performing the tasks, the participants were free to move around to experience and examine both the static and adaptive approach. Afterwards, we asked participants to fill out a questionnaire and answer several semi-structured questions about their experience. The questionnaire first asked participants to provide demographic information about themselves and their visual acuity before prompting questions from the Software Usability Scale (SUS) questionnaire [16] as well as the questions focusing on different aspects of user engagement. In addition, the participants were asked to give an overall rating on a 10-point Likert-scale to each of the methods, separately.

The SUS consisted of 10 questions each with a 5 point Likert scale, resulting in a single measure of usability that is between 0 and 100. We used the SUS score as the main measure of the usability of both the static and adaptive UI. Above 68%, 74%, and 80.3% usability scores are considered as average (grade C), good (grade B), and excellent (grade A) usability performances, respectively [16,17].

To measure user engagement, multiple scales are required. Previous research has proposed several user engagement scales, such as exploratory information search [18], mobile user engagement scales [19] and video game-based [20] each focusing on different aspects of user engagement. Since our study is concerned with both physical and virtual content navigation, we used O'Brien and Tom's [18] user engagement scales (UES), which combine a wide range of user engagement attributes from previous studies and consists of six dimensions (see Table 4). This wide range of dimensions enabled us to evaluate our scheme from many different points of view. While one design might perform better in some dimensions, it is possible that no difference might be found in other dimensions. By differentiating between these dimensions, we are capable of learning from where the differences come [18].

Using the O'Brien and Toms's guideline [18], we designed eleven questions relevant to the UES six-dimensions on a 5-point Likert-scale. The corresponding questions for each dimension are shown in Table 5. To evaluate each dimension,

Table 4. Factors of User Engagement (six dimensions) and their definitions

Aesthetic appeal	The visual presentation of the interface
Endurability	Overall success, whether they would recommend it to others
Felt involvement	User's feeling of being involved, having fun while interacting
Focused attention	Users's concentration during the interaction, whether they are absorbed in their task
Novelty	Whether the system is new to the user
Perceived usability	Users's cognitive and emotional feedback to the system

Table 5. User Engagement dimensions and their corresponding questions

Aesthetic appeal:	I like the kind of interaction of the adaptive/static display.
Endurability:	I would recommend the adaptive/static display to my friends and family.
	Searching using the adaptive/static display was worthwhile.
Felt involvement:	It was fun interacting with the adaptive/static display.
	I felt involved in the searching task.
Focused attention:	I was absorbed in my searching task.
	The time I spent searching just slipped away.
Novelty:	This kind of interaction with the adaptive/static display is new to me.
	I have not seen anything alike before.
Perceived usability:	Using the adaptive/static display confused me.
	I felt annoyed with the adaptive/static display.

we combined the results of the questions related to each dimension by adding up the received scores.

Related-samples Wilcoxon signed rank tests were used to compare the characteristics of the static and adaptive approaches. Furthermore, when appropriate (i.e. no violation of the normality assumption $p > 0.05$ by the Shapiro-Wilk test etc.), repeated measure ANOVA was deployed. To perform the statistical analysis, we used IBM SPSS Statistics (version 22.0, Armonk, NY: IBM Corp.) and set the minimum significance level at $p = 0.05$.

For our experiments, we used a 27" LED display, operating in landscape mode. We implemented the tasks for the user studies using our ResponDis framework. We configured the framework with the corresponding information about the display, namely DS = 27, NHR = 1920, and NVR = 1080. The other configurable options of the framework remained untouched to use the default settings. As shown in Fig. 5, we indicated all four zones on the ground using tape. A camera was positioned next to the display to film the participants while they performed the tasks.

Fig. 5. Setup

6.3   Results

Viewer Perception. For the first task, the difference in walk-in time measurements for the adaptive (Median = 5.2 s) and static approaches (Median = 6.9 s) were not statistically significant ($Z = -1.503, p = 0.133$). However, the difference in the zone where the participants ended up finding the character was statistically significant for the two approaches ($Z = -2.226, p = 0.026$). Using the adaptive approach (Median = Zone 4), participants moved less distance towards the display in comparison to the static method (Median = Zone 3). Only the difference in time measurement for "Zone 4" was statistically significant (see Table 6).

To have an overall evaluation, we averaged the measured times from all four zones. Then we ran a one-way repeated measure ANOVA ($F(1, 12) = 8.191, p = 0.014$) which revealed a statistically significant difference between the overall measured times of the adaptive and static approaches. The mean difference between the two approaches was 1.867 s, in favour of the adaptive approach, showing that it had required less time. These results show that the adaptive approach (Mean ± standard deviation = 3.5 ± 1.35 s) in comparison to the static approach (Mean ± standard deviation = 5.37 ± 2.39 s) improves the content perception of viewers by 24.35%.

User Engagement. Table 7 summarises the results of the UES for both approaches where related-samples Wilcoxon signed rank tests were used. The comparison column expresses which approach received a statistically significant higher value, where a mark(?) indicates that there was no statistically significant difference between them. The conclusion column contains a check mark(✓) if the

Table 6. The outcome of the statistical analyses comparing the time measurements of the approaches for each individual zone

	Z	p	Adaptive time (median) [s]	Static time (median) [s]
zone 1	$Z = -1.363$	$p = 0.173$	2.7	2.2
zone 2	$Z = -0.629$	$p = 0.529$	2.2	2.5
zone 3	$Z = -1.155$	$p = 0.248$	2.5	3.6
zone 4	$Z = -3.180$	$p = 0.001$	2.7	7.2

Table 7. The results and comparison of the single-viewer user study on user engagement dimensions for adaptive and static approaches. ✓: the adaptive approach performs better, ?: no difference between the two approaches was found.

User engagement dimension	Median	Z-value	p-Value	Comparison	Conclusion
Aesthetic appeal	Adaptive: 4 Static: 3	$Z = -1.997$	$p = 0.046$	Adaptive > Static	✓
Endurability	Adaptive: 8 Static: 7	$Z = -1.930$	$p = 0.054$	Adaptive ? Static	?
Felt involvement	Adaptive: 8 Static: 6	$Z = -2.381$	$p = 0.017$	Adaptive > Static	✓
Focused attention	Adaptive: 6 Static: 7	$Z = -0.187$	$p = 0.852$	Adaptive ? Static	?
Novelty	Adaptive: 9 Static: 3	$Z = -3.099$	$p = 0.002$	Adaptive > Static	✓
Perceived usability	Adaptive: 4 Static: 4	$Z = -0.486$	$p = 0.627$	Adaptive ? Static	?

adaptive approach performed better than the static model. There was no case where the static approach was considered better than the adaptive one.

Usability and Overall Rating. The overall rating on a 10-point Likert scale for the adaptive approach (Median = 8) was statistically significantly higher than for the static approach (Median = 5), $Z = -2.641, p = 0.008$. The adaptive approach (Median = 4) also achieved a statistically significant higher score compared to the static approach (Median = 2), $Z = -3.088, p = 0.002$, when participants scored the statement "I did not feel the urge to step out of the assigned zone, I felt comfortable in my zone" statement. However, there was no statistically significant difference between the usability score of adaptive (Median = 77.5) and static (Median = 82.5) approaches ($Z = 1.652, p = 0.099$).

Qualitative Feedback. The feedback provided as comments gave us a better insight into the opinions of the participants.

The static approach was considered as the current state of the art: *"The static display is everywhere, so I've already got used to it and it was not that difficult to find something in a picture. ..."*(P11). *"This is the normal situation. So I go closer to see picture better. ..."*(P4).

The behaviour of the adaptive UI confused some viewers: *"I first was puzzled when the picture got smaller. I was approaching the screen to see it better and the picture got smaller. So I was thinking about going back to get the bigger picture.*"(P4). (P2) highlighted one of the advantages of providing the optimal view when resizing the content "*... to display additional content or hints.*" on the display's empty spaces. Another participant suggested that the content should have a fluid design: "*It would be nice if the image scaling would be smooth between the distance states.*"(P5).

The adaptive approach was generally well-received: "*Adaptive approach was superior to the static approach, as I could always have a larger view when necessary. ...*" (P11) and "*I understand the approach and find it very good. ...*" (P4). At the same time, participants suggested using the approach for other purposes: "*... I was just thinking what about enlarging the picture if you step closer? This would be useful in cases I am interested in some details. Comfortable would be that I don't need to go very close but the system detects my goal and enlarges the picture for me.*" (P4). Another participant added: "*I think the zones are interesting, but I'm not sure that making the content smaller is the way to go. Maybe showing different content or more content could be more interesting.*" (P2). These are good inputs and could be considered in future work. A few participants commented on the design decision for Zone 1: "*At least in this setting I would not go to zone 1. It was too close for me since I had to look at the whole image to solve the task. Maybe for searching some particular spot on a map I would go as close as zone 1.*" (P3). Another participant added: "*... in zone one (1) there was no significant difference between the two approaches, which was logical as I was so close to the display that I didn't need any enhancement/picture enlargement by the display.*" (P11).

7   Discussion

As reported in the previous section, we found no statistically significant difference in the time it took participants to find the character when walking into the room. Nevertheless, the medians differ by 1.7 s in favour of the adaptive design. We did, however, find statistical evidence that users did not have to walk as close to the screen with the adaptive approach compared to the static one.

We noticed a vast difference in the manner in which participants walked into the room. Some walked in quickly, eager to find the character, while others expressed uncertainty and walked slowly. This could be one of the reasons why there is no significant difference in the time measurements.

When taking the average of the zone-by-zone time measurements, one can see a statistically significant difference between the two approaches, showing that the seek-time on the adaptive display was less. This could be because participants did not have to walk and adapt their view to the content. Since participants were asked to stay within their assigned zone, the difference in the seek-time measurements could not differ due to their slow or quick movement and, thus, the observed significant difference can be associated with the corresponding approaches.

Although the adaptive approach generally performed better as participants were faster in finding the character in the pictures, the static approach performed slightly better for "Zone 1" (see Table 6). However, the difference was not statistically significant. We believe that the better performance of the static approach for "Zone 1" could be a result of the framework's default setting for "Zone 1" which forces too small a distance to the screen. This closeness sometimes caused Kinect user detection failure and, thus, inappropriate adaptation, which might be the reason why the participants could not quickly find the characters. In addition, we also received feedback that *"At least in this setting I would not go to zone 1"* (P3). This suggests that careful design decisions for each zone is required.

Studying Table 7, it can be seen that the adaptive approach was always rated better or the same as the static approach. The adaptive approach was systematically better in the *aesthetic appeal, felt involvement and novelty* of the user engagement dimensions. The participants' rating of the interaction of the adaptive design was significantly better, which could be due to the fact that it adapts itself to the viewer.

The participants also had fun interacting using the adaptive approach and rated it as more novel. This is not surprising since the static approach is the one used in most existing systems, while adaptive UIs are still a topic of research.

While no systematic difference between the approaches was observed in the perceived usability dimension, there was clear feedback that the adaptive UI confused some viewers. Reviewing the comments, we learned that some participants walked closer to try to perceive more detail, and were confused that the content got smaller instead of larger. Others wished that the intentional blank space around the content used for the study had been used to zoom in or provide additional hints when the content size decreased as they approached the screen. While this highlights a potential advantage of our approach over the static approach, we had decided not to use the free space to display more content during the study to avoid potential conflicts in content. This is something that we now feel would be an important addition in future studies.

Previous research has shown that low user engagement is a result of poor system usability [21]. The statistical analysis of the usability scores for both approaches showed that the scores are close, with both above the usability average, and that there is not a systematic difference between the approaches. This means that the low user engagement rating of the static approach is not due to the low usability.

The adaptive approach was considered to be significantly better when looking at the overall rating compared to the static approach. Participants also did not feel the urge to step out of their assigned zone when using the adaptive model, whereas they did while using the static design. We noticed however that when participants were asked to stay within a particular zone while using the static approach, some leant forward as far as their balance permitted in order to get a better view.

8   Conclusion and Future Work

We have proposed a model which could be used to integrate the proximity of viewers to a public display as an additional dimension of the viewing context considered in responsive design. In order to experiment with the model, we developed a JavaScript framework that supported the rapid prototyping of applications. This was used to carry out a basic user study which compared a conventional static UI typical of current public displays with an adaptive UI based on our model. The results of the study with single viewers showed that the adaptive approach not only provides a better view, but also improves the user engagement in terms of aesthetic appeal, felt involvement and novelty. In the future, we aim to investigate the effect of the model on multiple viewers and in more realistic settings such as the deployment of an information service within our department.

References

1. Sarabadani Tafreshi, A.E., Norrie, M.C.: ScreenPress: a powerful and flexible platform for networked pervasive display systems. In: Proceedings of the 6th ACM International Symposium on Pervasive Displays (PerDis) (2017)
2. Nebeling, M., Norrie, M.C.: Responsive design and development: methods, technologies and current issues. In: Daniel, F., Dolog, P., Li, Q. (eds.) ICWE 2013. LNCS, vol. 7977, pp. 510–513. Springer, Heidelberg (2013). doi:10.1007/978-3-642-39200-9_47
3. Gardner, B.S.: The spark of innovation begins with collaboration. Inside the Digital Ecosyst. **11**(1) (2011)
4. Frain, B.: Responsive Web Design with HTML5 and CSS3. Packt Publishing Ltd. (2015)
5. Nebeling, M., Matulic, F., Streit, L., Norrie, M.C.: Adaptive layout template for effective web content presentation in large-screen contexts. In: Proceedings of the 11th ACM Symposium on Document Engineering (DocEng) (2011)
6. Bababekova, Y., Rosenfield, M., Hue, J.E., Huang, R.R.: Font size and viewing distance of handheld smart phones. Optom. Vis. Sci. **88**(7), 795–797 (2011)
7. Rushing, K.: Home Theater Design. Rockport Publishers (2004)
8. Sugawara, M., Masaoka, K., Emoto, M., Matsuo, Y., Nojiri, Y.: Research on human factors in Ultrahigh-Definition Television (UHDTV) to determine its specifications. SMPTE Motion Imaging J. **117**(3), 23–29 (2008)
9. Dostal, J., Kristensson, P.O., Quigley, A.: Multi-view proxemics: distance and position sensitive interaction. In: Proceedings of the 2nd ACM International Symposium on Pervasive Displays (PerDis) (2013)
10. Dostal, J., Hinrichs, U., Kristensson, P.O., Quigley, A.: SpiderEyes: designing attention-and proximity-aware collaborative interfaces for wall-sized displays. In: Proceedings of the 19th International Conference on Intelligent User Interfaces (IUI) (2014)
11. Snellen, H.: Probebuchstaben zur bestimmung der sehschärfe. H. Peters (1873)
12. Jiang, F., Zhou, L., Liu, K., He, X., Liu, X., Li, Z.: P-35: the prospect assessment of 65+ Inch TVs based on the size of mainstream living rooms in china. In: SID Symposium Digest of Technical Papers, vol. 46. Wiley Online Library (2015)

13. Müller, J., Wilmsmann, D., Exeler, J., Buzeck, M., Schmidt, A., Jay, T., Krüger, A.: Display blindness: the effect of expectations on attention towards digital signage. In: Tokuda, H., Beigl, M., Friday, A., Brush, A.J.B., Tobe, Y. (eds.) Pervasive 2009. LNCS, vol. 5538, pp. 1–8. Springer, Heidelberg (2009). doi:10.1007/978-3-642-01516-8_1

14. Li, A.X., Lou, X., Hansen, P., Peng, R.: Improving the user engagement in large display using distance-driven adaptive interface. Interact. Comput. **28**, 462–478 (2015)

15. Rémi, C.: Reading as Playing. Emergent Literacy: Children's Books from 0 to 3 13 (2011)

16. Sauro, J.: A Practical Guide to the System Usability Scale: Background, Benchmarks & Best Practices. Measuring Usability LLC (2011)

17. Sauro, J.: Measuring Usability with the System Usability Scale (SUS) (2011)

18. O'Brien, H.L., Toms, E.G.: Examining the generalizability of the user engagement scale (UES) in exploratory search. Inf. Process. Manage. **49**(5), 1092–1107 (2013)

19. Kim, Y.H., Kim, D.J., Wachter, K.: A study of mobile user engagement (MoEN): engagement motivations, perceived value, satisfaction, and continued engagement intention. Decis. Support Syst. **56**, 361–370 (2013)

20. Wiebe, E.N., Lamb, A., Hardy, M., Sharek, D.: Measuring engagement in video game-based environments: investigation of the user engagement scale. Comput. Hum. Behav. **32**, 123–132 (2014)

21. Robertson, G., Czerwinski, M., Baudisch, P., Meyers, B., Robbins, D., Smith, G., Tan, D.: The large-display user experience. IEEE Comput. Graph. Appl. **25**(4), 44–51 (2005)

Ontology-Enhanced Aspect-Based Sentiment Analysis

Kim Schouten[✉], Flavius Frasincar, and Franciska de Jong

Erasmus University Rotterdam,
P.O. Box 1738, 3000 DR Rotterdam, The Netherlands
{schouten,frasincar}@ese.eur.nl, f.m.g.dejong@eshcc.eur.nl

Abstract. With many people freely expressing their opinions and feelings on the Web, much research has gone into modeling and monetizing opinionated, and usually unstructured and textual, Web-based content. Aspect-based sentiment analysis aims to extract the fine-grained topics, or aspects, that people are talking about, together with the sentiment expressed on those aspects. This allows for a detailed analysis of the sentiment expressed in, for instance, product and service reviews. In this work we focus on knowledge-driven solutions that aim to complement standard machine learning methods. By encoding common domain knowledge into a knowledge repository, or ontology, we are able to exploit this information to improve classification performance for both aspect detection and aspect sentiment analysis. For aspect detection, the ontology-enhanced method needs only 20% of the training data to achieve results comparable with a standard bag-of-words approach that uses all training data.

1 Introduction

With many people freely expressing their opinions and feelings on the Web, much research has gone into modeling and monetizing opinionated, and usually unstructured and textual, Web-based content [14]. A popular option to extract information from Web texts is to perform sentiment analysis. Given a certain unit of text, for instance a document or a sentence, the task of sentiment analysis is to compute the overall sentiment expressed by the author of the text. Text can be tagged for sentiment by using labels for emotions, or by assigning a polarity value to the processed unit of text, which is a more commonly adopted method. Aspect-based sentiment analysis goes a step deeper. Rather than labeling a document or a sentence, it aims to extract and tag semantic units. It captures the topics, or aspects, that are being talked about, together with the sentiment expressed about those aspects [20]. Relating the expressed sentiment directly to certain topics enables the extraction of opinions expressed in product and service reviews in a much more focused way. Instead of an overall score in which both positive and negative aspects are combined, a breakdown can now be provided, showing the aspects for which the reviewer said positive things and the aspects he or she was less enthusiastic about.

J. Cabot et al. (Eds.): ICWE 2017, LNCS 10360, pp. 302–320, 2017.
DOI: 10.1007/978-3-319-60131-1_17

Generally speaking, we can define two sub-problems that together comprise aspect-based sentiment analysis: aspect detection and aspect sentiment classification. We define aspect detection as capturing the topics, or aspects, that are being talked about. This can be done within the textual unit of choice, for instance per sentence or per document. Most of the aspects will be explicitly mentioned in the textual unit, and the exact phrase that mentions the aspect is defined as the target expression of the aspect. Aspects without a target expression are not explicitly mentioned but rather are implied by a, usually larger, portion of text. Since target expressions are often very specific, it can be informative to group aspects together in aspect categories. Implicit aspects, even though lacking a specific target expression, can be categorized in the same manner. Even within sentences, multiple aspects, both explicit and implicit, can occur, as shown in Example 1. This sentence contains two explicit aspects: "chow fun" and "pork shu mai", both belonging to the broader 'food' category, as well as an implicit aspect about sharing the table with a loud and rude family, which can be categorized under 'ambiance'.

Example 1. *"Chow fun was dry; pork shu mai was more than usually greasy and had to share a table with loud and rude family."*

Aspect sentiment analysis, the second sub-problem of aspect-based sentiment analysis, can be defined as determining the sentiment expressed on that sentiment in the text where the aspect is mentioned. For explicit aspects, the target expression indicates where in the text the aspect is mentioned and this information can be useful in determining the relevance of each piece of sentiment carrying text as textual units, such as sentences, can contain multiple aspects that have differing sentiment values. This is illustrated in Example 2 in which one sentence contains two aspects, one about the food and one about the service, but the expressed sentiment on these two aspects is completely different. For implicit aspects, where target expressions are not available, the complete textual unit can be relevant, but the aspect category usually provides some information on which part of the sentence might be relevant.

Example 2. *"The food was great, if you're not put off by the rude staff."*

Current approaches for aspect-based sentiment analysis rely heavily on machine learning methods because this yields top performance [17]. Deep learning approaches are especially popular and include techniques such as word embeddings [16], convolutional neural networks [11], recursive neural tensor networks [21], and long short-term memory networks [22]. While the above methods have shown very promising results in various natural language processing tasks, including sentiment analysis, there are some downsides as well. For example, while the methods learn their own features, often better than what an individual researcher could come up with, they do so at the cost of requiring much training data. While this may not be a problem for resource-rich languages, such as English, and resource-rich domains such as reviews or tweets, it is a real issue for many other languages and domains where training data are not abundant or simply unavailable.

With the previous argumentation in mind, we propose a knowledge-driven approach to complement traditional machine learning techniques. By encoding some common domain knowledge into a knowledge repository, or ontology, we limit our dependence on training data [3]. The idea is that, compared to using only information from the text itself, relating the text to the concepts described in a knowledge repository will lead to stronger signals for the detection of aspects as well as the prediction of sentiment. Consequently, having stronger signals limits the amount of necessary training data, as the relation between input and desired output is easier to discover.

While knowledge repositories, such as ontologies, have to be adjusted for every domain and language, this is a less resource-intensive task than manually annotating a big enough corpus for the training of a deep neural network. Furthermore, since ontologies are designed to be reused, for example within the Linked Open Data cloud, it is easy to share knowledge. Last, with ontologies, being logically sound structures, it is possible to reason with the available data, and to arrive at facts not directly encoded in the ontology. For example, if meat is a type of food, and food is edible, we can state that meat is edible, even without directly specifying this. Furthermore, inferencing functionality can help to disambiguate sentiment carrying phrases given the context they appear in, for example by taking into account that "cold" is positive for "beer", but negative for "pizza". This opens up some exciting possibilities when performing sentiment analysis.

This paper is structured as follows. In the next section, some of the related work is presented, followed by a discussion of the problem and the used data set in Sect. 3. In Sect. 4, an overview of the proposed method is given, and in Sect. 5, its performance is compared and evaluated. This paper concludes with Sect. 6, providing both conclusions and possible avenues for future work.

2   Related Work

In [3], a short overview is given of the field of affective computing and sentiment analysis, and the author makes a case for the further development of hybrid approaches that combine statistical methods with knowledge-based approaches. The leading idea in that field is that the intuitive nature and explanatory power of knowledge-based systems should be combined with the high performance of machine learning methods, which also forms the research hypothesis of our current work.

Sentic computing is presented in [1], which combines statistical methods with a set of linguistic patterns based on SenticNet [2]. Each sentence is processed in order to find the concepts expressed in it. The discovered concepts are linked to the SenticNet knowledge repository, which enables the inference of the sentiment value associated to the sentence. If there are no concepts expressed in this sentence or if the found concepts are not in the knowledge base, then a deep learning method that only uses the bag of words is employed to determine the sentiment for that sentence. Note that this is a sentence level approach and not an aspect based approach.

In [5], a multi-domain approach to sentence-level sentiment analysis is presented, where the task is to assign sentiment to each sentence, but where the sentences could come from a variety of domains. The proposed method is designed in such a way that sentiment words can be disambiguated based on the recognized domain the sentence originates from. Similar to what is typical of our approach, the used knowledge graph is split into two main parts: a semantic part in which the targets are modeled, and a sentiment part in which the links between concepts and sentiment are described. A big difference with our approach is that while we opt for a focused domain ontology, in [5], due to the multi-domain nature of the problem, a very broad knowledge graph is created that combines several resources such as WordNet [7] and SenticNet [2]. Another difference is the use of fuzzy membership functions to describe the relations between concepts and domains, as well as between concepts and sentiment. This gives more flexibility in terms of modeling, but makes it harder to reason over the knowledge graph.

In [18], an extended version of the Sentilo framework is presented that is able to extract opinions, and for each opinion its holder, expressed sentiment, topic, and subtopic. The framework converts natural language to a logical form which in turn is translated to RDF that is compliant with Semantic Web and Linked Data design principles. Then, concepts with relations that signal sentiment are identified. The sentiment words receive a sentiment score based on SentiWordNet [6] and SenticNet [2].

A typical problem for which external knowledge can be very useful is the issue of sentiment ambiguity: a certain expression can be positive in one context and negative in the other (e.g., the cold pizza and cold beer example from the previous section). This problem is tackled in [24] by means of a Bayesian model that uses the context around a sentiment carrying word, in particular the words that denote the aspect, to determine the polarity of the sentiment word. When there is not enough information in the context to make the decision, a backup method is to retrieve inter-opinion data, meaning that if the previous opinion was positive and there is a conjunction between that one and the current one, it is very likely that the current opinion is positive too.

In contrast to the previous approaches, high performing SemEval submissions on the aspect-based sentiment analysis task [17], are typically limited in their use of external knowledge or reasoning. For instance, in the top performing system for aspect category classification [23], a set of binary classifiers is trained, one for each aspect category, using a sigmoidal feedforward network. It uses words, bigrams of words, lists of opinion target words extracted from the training data, syntactic head words, and Brown word clusters as well as k-mean clusters from word2vec [16]. The highest performing system in the sentiment classification task [19] also exclusively focuses on using lexical features. Besides the information that can be directly extracted from the text, a number of lexical resources such as sentiment lexicons were used to detect the presence of negation and sentiment words. While lexical resources can be seen as being external knowledge, they are limited in functionality and do not, for example, support reasoning.

3 Specification of Data and Tasks

The data set used in this research is the widely known set of restaurant reviews from SemEval [17], with each review containing one or more sentences: it contains 254 reviews with in total 1315 sentences. Each sentence is annotated with zero, one, or multiple aspects, and each aspect is put into a predefined aspect category and is labeled as either positive, neutral, or negative. For explicit aspects, the target expression is also provided. Some statistics related to aspects and sentiment can be found in Fig. 1. In Fig. 1a, the number of times each category label appears is presented and in Fig. 1b, the proportion of sentences with multiple aspects is shown in which not all aspects have the same sentiment label. This is related to Fig. 1c, which shows the distribution of aspects per sentence. Figure 1d presents the distribution of sentiment values over aspects, showing that this data set, especially the training data set, is unbalanced with respect to sentiment.

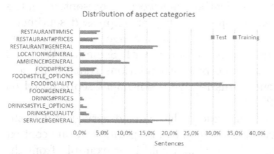

(a) Relative frequencies of each aspect category label

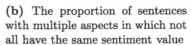
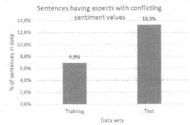

(b) The proportion of sentences with multiple aspects in which not all have the same sentiment value

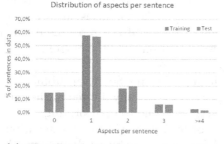

(c) The distribution of aspects per sentence

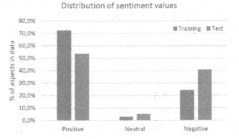

(d) Relative frequencies of each sentiment label

Fig. 1. Some statistics related to the used data set

A snippet of the used data set is shown in Fig. 2. The data set is already organized by review and by sentence, and each sentence is annotated with zero or more opinions, which represent the combination of an aspect and the sentiment expressed on that aspect.

```
<sentence id="1032695:1">
<text>Everything is always cooked to perfection, the
    service is excellent, the decor cool and understated.</
    text>
<Opinions>
<Opinion target="NULL" category="FOOD#QUALITY" polarity="
    positive" from="0" to="0"/>
<Opinion target="service" category="SERVICE#GENERAL"
    polarity="positive" from="47" to="54"/>
<Opinion target="decor" category="AMBIENCE#GENERAL"
    polarity="positive" from="73" to="78"/>
</Opinions>
</sentence>
```

Fig. 2. A snippet from the used dataset showing an annotated sentence from a restaurant review.

For the aspect detection task, only the sentence is given. The task is to annotate the sentence with aspects but the `polarity` field can be left empty. While some variations of the task exist, we limit ourselves to predicting only the `category` field of each aspect. Hence, the `target` field and corresponding `to` and `from` fields are ignored. The category labels themselves consist of an entity and an attribute that are separated by a hash symbol. In this work, however, we regard the category as just a single label. In the evaluation, every category that is in the data and is also predicted is a true positive, every category that is predicted but is not in the data is a false positive and every category that is not predicted, even though it is in the data, is a false negative. From the number of positives and negatives, the standard evaluation metrics of precision, recall, and F_1 score can be computed.

For the sentiment classification task, the sentence with the aspects are given. Thus, we get everything in Fig. 2, except for the values of the `polarity` fields. Every correctly predicted polarity is a true positive and every incorrectly predicted polarity is both a false positive and a false negative, so precision, recall, and F_1 have the same value for this task.

4   Method

Since both detecting aspects and determining their sentiment can be seen as a classification task, we choose an existing classifier to work with. In this case, we choose to use a linear Support Vector Machine (SVM), since it has shown good performance in the text classification domain, with relatively many input features compared to the number of training examples [4]. For aspect detection, we train an independent binary classifier for each different aspect category. In this way, per sentence, each classifier will determine whether that aspect is present or not, enabling us to find zero, one, or more aspects per sentence. For sentiment classification, we train only one (multiclass) model, that is able to predict one of three outcomes: positive, neutral, or negative. We use the lib-svm [4] implementation of the SVM classifier.

Using natural language processing (NLP) techniques, we gather information from the review texts that will comprise the input vector for the SVM. In Fig. 3, the components of the NLP pipeline are shown. First, an automated spelling correction is performed, based on the JLanguageTool library [10]. Given the nature of consumer-written reviews, this is a very helpful step. Next, the text is split into tokens, which are usually words, but also includes punctuation. Then, these tokens are combined into sentences. With sentence boundaries defined, the words in each sentence can be tagged with Part-of-Speech tags, which denote the word types (e.g., noun, verb, adjective, etc.). Once that is known, words can be lemmatized, which means extracting the dictionary form of a word (e.g., reducing plurals to singulars). The syntactic analysis then finds the grammatical relations that exist between the words in a sentence (e.g., subject, object, adjective modifier, etc.). All these components are provided by the Stanford CoreNLP package [15]. The last step connects each word with a specific meaning called a synset (i.e., set of synonyms), given the context in which it appears, using a Lesk [13] variant and WordNet semantic lexicon [7]. This particular version of the Lesk algorithm is provided by DTU [9].

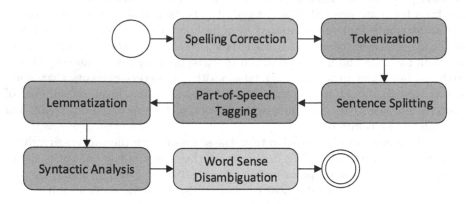

Fig. 3. The NLP pipeline used at the basis of the methods

4.1 Ontology Design

The ontology consists of three main parts, modeled as top-level classes. The first is a *Sentiment* class, which has individuals representing the various values of sentiment. In our case, that is only positive, neutral, and negative, but one can imagine a more fine grained classification using basic emotion classes like anger, joy, etc. The second major class is *Target*, which is the representation of an aspect. The higher level concepts correspond to aspect categories, while the sub-classes are often target expressions of an aspect. Subclasses of *Target* are domain specific, and for our restaurant domain we use *Ambience, Sustenance, Service,*

Restaurant, Price, Persons, and *Quality*. Some of these have only a handful of individuals, such as *Quality* since quality is more expressed in evaluative words and not in concepts, while *Sustenance*, unsurprisingly, has many subclasses and individuals. Because we want to use object relations, all subclasses of Target are modeled as having both the class role and the individual role, much like classes in OWL FULL. For every subclass of *Target*, there is an individual with the same (resource) identifier that is of that same type. Hence, there is a subclass *Beer*, and an individual *Beer* that is of type *Beer*. This duality allows us to use the powerful subclass relation and corresponding reasoning, as well as descriptive object relations. The latter are mainly used for the third part of the ontology, which is the *SentimentExpression* class.

The *SentimentExpression* class only has individuals describing various expressions of sentiment that can be encountered. Each sentiment expression is linked to a *Sentiment* value by means of an object relation called *hasSentiment*, and to a *Target* with an object relation called *hasTarget*. In most cases, the *hasTarget* relation points to the top-level concept *Target*, since the word "good" is positive regardless of the target. However, the word "cold", when linked to the concept *Pizza* has the *negative* sentiment value, while it has the *positive* value when linked to *Beer*.

The ontology is lexicalized by means of a data property that is added to each concept. The targets have a *targetLexicalization* property, and the sentiment expressions have a *sentimentLexicalization* property. By means of these lexicalizations, which can be one or more words, the concepts in the ontology can be linked to words or phrases in the text.

In Fig. 4, the sentiment expression for "cold beer" is shown with its related concepts. Note that the ellipse around the *Beer* class and the *Beer* individual denotes the fact that those are two roles of the same concept.

This ontology design allows us to perform two types of reasoning, one for aspect detection and one for sentiment classification. The first is that if we encounter a sentiment word, we know that its target is also in this sentence. For example, when we find the word "delicious", we will find the SentimentExpression with the same *sentimentLexicalization*. This concept has a target, namely the *Sustenance* concept. Because of this, we know that the sentence where we find "delicious", the target aspect is something related to food or drinks. The second type of reasoning is that when we encounter a sentiment word in a sentence and that word is linked to a *SentimentExpression* in the ontology, the aspect for which we want to determine the sentiment has to be of the same type as the target of that *SentimentExpression* in order for its sentiment value to be relevant. For example, we again find the word "delicious" in the text, but we want to determine the sentiment for the aspect FOOD#PRICE, we should not take the positive value of "delicious" into account, since it is not relevant to the current aspect we are classifying. This is especially useful when a sentence has more than one aspect.

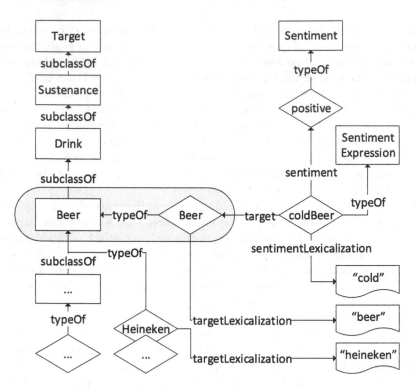

Fig. 4. Snippet of the ontology showing a sentiment expression and its related concepts

The ontology is created manually, using the OntoClean methodology [8], so it is guaranteed to fit with the restaurant domain of the reviews. To keep the ontology manageable, we have deliberately opted for a relatively small, but focused, ontology. As such, it contains 56 sentiment expressions, 185 target concepts, and two sentiment concepts: positive and negative. The maximum depth of its class hierarchy is 5.

4.2 Features

Since the aspect detection task is defined as predicting zero or more aspect category labels per sentence, we extract the following features from each sentence: the presence or absence of lemmatized words, the presence or absence of Word-Net synsets, and the presence or absence of ontology concepts. For the latter, we use words and phrases in the sentence to find individuals of top-level class *Target* that have a matching *targetLexicalization*. When a concept is found, we include all its types as features. For example, when we find the concept *Steak*, we also include the concepts *Meat*, *Food*, *Sustenance*, and *Target*. Furthermore, when a word or phrase matches with a *sentimentLexicalization*, we include the *target* of that SentimentExpression as being present as well. All these features are binary, so the input vector for the SVM will contain 1 for features that are present, and 0 otherwise. The same features are used for each binary aspect classifier

This process of gathering features can be formalized as follows.

$$L_W = \{l | l = lemma(w), w \in W\} \tag{1}$$

$$Z_W = \{z | z = synset(w), w \in W\} \tag{2}$$

$$C_W = \{c | k : c, (k, lemma(w)) : targetLexicalization, w \in W\} \cup$$
$$\{c | (k, c) : target, (k, lemma(w)) : sentimentLexicalization, w \in W\} \tag{3}$$

where W represents a set of words, given as a parameter, $i : c$ represents an individual k of type c, and $(k, c) : target$ represents that k is related to c through relation type $target$. Then, let W' be the set of all words in the data set. Every word has its own unique representation in this set, so the same word appearing in three different places will have three entries in this set. And let S be the indexed set of all sentences in the data set, with functions $g : I \rightarrow S$, and $g' : S \rightarrow I$, so that $i \rightarrow s_i$, $s \rightarrow i_s$ and $I = \{i \in \mathbb{N} | i \geq 0, i < |S|\}$, resulting in a unique one-to-one mapping between I and S. Then W_i is defined as the set of words in sentence s_i.

Using W', we gather all possible features from the full data set into set $F_{W'}$.

$$F_{W'} = L_{W'} \cup Z_{W'} \cup C_{W'} \tag{4}$$

Similar to S, set $F_{W'}$ is indexed with a one-to-one mapping: $h : F_{W'} \rightarrow J$ and $h' : J \rightarrow F_{W'}$, so that $j \rightarrow f_j$ and $f \rightarrow j_f$ with $J = \{j \in \mathbb{N} | j \geq 0, j < |F_{W'}|\}$.

Given this mapping between J and F_{W_i}, the index numbers of only the features that are present in a given sentence i are retrieved through $h(F_{W_i})$. This leads to defining the input matrix $\mathbf{X}$ as having

$$x_{ij} = \begin{cases} 1, \text{if } j \in h(F_{W_i}) \\ 0, \text{otherwise} \end{cases} \tag{5}$$

where i specifies the row in the matrix, representing the current sentence and j specifies the column in the matrix, representing the current feature.

For sentiment classification, the process of gathering features is similar, so due to the page limit, the formalization is omitted here. The difference is that the scope here is a single aspect for which we want to determine the sentiment. An aspect already has the category information given, together with its position in the sentence, if applicable. Besides the features for lemmas, synsets, and ontology concepts, we also include the aspect category information of an aspect as a feature (e.g., FOOD#QUALITY or FOOD#PRICE. In addition, we include some sentiment information, using existing sentiment tools or dictionaries together with our own ontology. Utilizing the Stanford Sentiment tool [21], which assigns sentiment scores (decimals between -1 and 1) to every phrase in the parse tree of a sentence, we add a feature that represents the sentiment of the whole sentence, as well as a feature that represents the sentiment of the smallest phrase containing the whole aspect. The latter is only available for explicit aspects, whereas the former is always available. Since the sentence sentiment score is additional information that can be useful, for instance when the aspect sentiment, as determined by this tool, is incorrect, we chose to always add this feature.

Since the lowest level of the parse tree is comprised of the words in a sentence, we use the same tool to get sentiment values for each word. A special review sentiment dictionary [12] is used to retrieve sentiment values for some of the words as well, and as a third source of sentiment information we use the ontology to find sentiment information for any word that can be linked to a *SentimentExpression* in the ontology. As explained in the previous section, we only take the latter into account when the aspect for which we want to determine the sentiment can be linked to a concept in the ontology that matches with the target concept of the sentiment expression. The *positive* concept is translated to a value of 1, and the *negative* concept is translated to a value of -1. All these sentiment values are averaged to arrive at a single sentiment value for a given word or phrase. However, when we do find an applicable sentiment expression in the ontology, preliminary experiments suggest to use double the weight for this value in the average computation. Assigning a higher weight is an intuitive course of action, since we are sure this is relevant information.

Some of the aspects have location information provided, so we know which words in the sentence are describing that particular aspect. When this information is available, we construct a scope around the aspect so words that are closer to the aspect are more valuable than words further away. The distance is measured in terms of grammatical steps between words. In this way, words that are grammatically related are close by, even though there might be many words in between them in the actual sentence. Based on some preliminary experiments, we compute the distance correction value as:

$$distanceCorrection = \max\left(0.1, 3 - grammaticalSteps\right) \qquad (6)$$

Instead of having binary features (cf. Eq. 5), denoting just the presence of features, we multiply the distance correction value with the average sentiment value and use this as the weight for each word-based feature. The aspect category features, as well as the sentence sentiment and aspect sentiment features are not affected by this because distance and sentiment are irrelevant seeing that these features are not directly linked to specific words.

5   Evaluation

In this section, we will evaluate the proposed method and discuss the results. First, the used data sets are described, followed by a comparative overview of the performance of the method. Then, to test our hypothesis that less data is needed for knowledge-driven approaches, a series of experiments is performed with varying amounts of training data. This is followed by a feature analysis, showing which features are useful and demonstrating that the output of the algorithm can be explained.

5.1   Performance

To test the performance of the proposed method, we compare the full ontology-enhanced method (+SO) against more basic versions of the same algorithm, having the exact same setup except for some missing features: a version without synsets, but with ontology features (+O); a version with synsets but without ontology features (+S); and a version without both synsets and ontology features (base). The two tasks of the algorithm are tested separately, so the aspect sentiment classification is performed with the gold input from the aspect detection task. This prevents errors in aspect detection from influencing the performance analysis of the sentiment classification. The two performances on the two tasks are given in Tables 1 and 2, respectively. The reported F_1 scores are averages over 10 runs where each run is using 10-fold cross-validation with randomly assigned folds. The standard deviation is also reported, together with the p-values of the two-sided t-test.

Table 1. The performance on the aspect detection task

	avg. F_1	st.dev.	p-values of two-sided t-test			in-sample F_1	out-of-sample F_1
			base	+S	+O	(training data)	(test data)
base	0.5749	0.0057	-	-	-	0.803	0.5392
+S	0.6317	0.0039	<0.0001	-	-	0.896	0.5728
+O	0.6870	0.0026	<0.0001	<0.0001	-	0.858	0.6125
+SO	0.6981	0.004	<0.0001	<0.0001	<0.0001	0.920	0.6281

Table 2. The performance on the aspect sentiment classification task

	avg. F_1	st.dev.	p-values of two-sided t-test			in-sample F_1	out-of-sample F_1
			base	+S	+O	(training data)	(test data)
base	0.7823	0.0079	-	-	-	0.831	0.7372
+S	0.7862	0.0049	0.0294	-	-	0.847	0.7349
+O	0.7958	0.0069	0.0002	0.0008	-	0.863	0.7479
+SO	0.7995	0.0063	<0.0001	<0.0001	0.0029	0.884	0.7527

For aspect detection the picture is most clear, showing that every step towards including more semantic features, starting with synsets and going to ontology concepts, is a significant improvement over not including those features. Note that most of the improvement with respect to the base algorithm comes from the ontology. The synsets, while showing a solid improvement on their own, are able to increase the performance with much less when the ontology is also used (i.e., going from +O to +SO).

For aspect sentiment classification, the results are less pronounced than for aspect detection, but it still shows the same overall picture. Adding more semantic features improves the performance, and while the improvement is less, it is statistically significant. A key observation here is that for sentiment analysis, we already employ a number of features that convey sentiment values, and as such, it is a strong signal that in spite of all that information, the ontology is still able to boost the performance.

This work is mainly focused on showing how ontologies have an added value for aspect-based sentiment analysis. Nevertheless, our methods show a competitive performance. In Table 3, an overview of the top performances of SemEval submissions on the same task are given [17]. These methods have been tested on the exact same test data, so their reported F_1 scores are directly comparable. For ease of reference, our +SO method is shown in bold, together with the top 6 out of 15 submissions for both tasks.

Table 3. Ranks of the proposed methods in top of SemEval-2015 ranking

Aspect detection		Aspect sentiment classification	
Team	Performance	Team	Performance
+SO	0.628	Sentiue	0.787
NLANGP	0.627	ECNU	0.781
NLANGP	0.619	Isislif	0.755
UMDuluth-CS8761-12	0.572	**+SO**	0.752
UMDuluthTeamGrean	0.572	LT3	0.750
SIEL	0.571	UFRGS	0.717
Sentiue	0.541	wnlp	0.714

5.2 Data Size Sensitivity

Since we hypothesize that an ontology-enhanced algorithm needs less data to operate than a traditional machine learning method, we perform the following experiment. Taking a fixed portion of the data as test data, we train on an ever decreasing part of the total available training data. In this way, the test data remains the same, so results can be easily compared, while the size of the training data varies. This maps the sensitivity of the algorithms to training data size and the results are shown in Fig. 5 for the aspect detection task and in Fig. 6 for the aspect sentiment classification task.

When looking at the sensitivity of the aspect detection method, we can see that the base algorithm is quite sensitive to the size of the data set, dropping the fastest in performance when there are fewer training data. With synsets, the performance is more stable, but with less than 40% of the original training data, performance drops significantly. The two versions that include ontology features are clearly the most robust when dealing with limited training data. Even at 20% of the original training data, the performance drop is less than 10%.

For sentiment classification, it shows that all methods are quite robust with respect to the amount of training data. This might be because all variants are

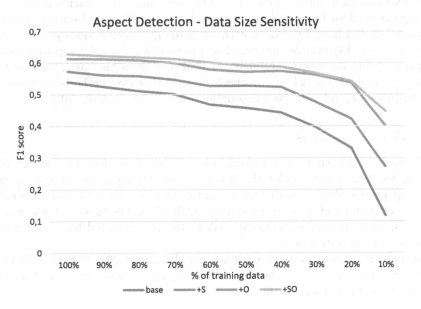

Fig. 5. The data size sensitivity for the aspect detection task

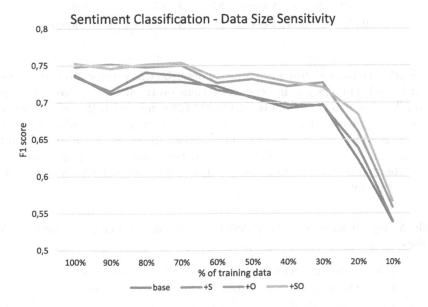

Fig. 6. The data size sensitivity for the aspect sentiment classification task (note that the y-axis does not start at 0 to improve readability)

using external information in the form of the sentiment dictionaries, already alleviating the need for training data to a certain extent. The gap between ontology-enhanced methods and the ones without ontology features does not widen, so contrary to our hypothesis, ontology features do not reduce the required number of training data for aspect sentiment analysis. On the other hand, the ontology-enhanced methods consistently outperform the other two methods, so the ontology features stay relevant even at smaller numbers of training data.

5.3   Feature Analysis

To investigate whether the ontology concepts are indeed useful for the SVM model, we take a look under the hood by investigating the internal weights assigned by the SVM to each feature. Since we use a binary classifier for each aspect category label, there are too many trained models to show these weights for all of them. Hence, we will only look at the top-weighted features for some of the more illustrative ones.

A very nice example is the trained SVM for the DRINKS#STYLE_OPTIONS category, which deals with the variety of the available drinks. The four most important features are listed in Table 4.

Table 4. Top 4 features for DRINKS#STYLE_OPTIONS according to weight assigned by SVM classifier

1	0.369 - Ontology Concept: Menu	3	0.307 - Synset: list
2	0.356 - Ontology Concept: Drink	4	0.265 - Synset: enough

Clearly, the top two features perfectly describe this category, but the synsets for "list" and "enough", as in "enough options on the list", are also very appropriate. Another good example is DRINKS#PRICES, for which the top 10 features are shown in Table 5. Here, again, the top two concepts are typical of this aspect category. However, we see that certain lemmas and synsets are also high on the list, but with a lower weight. Note the word "at" which is often used to denote a price, but which, being a function word, is not associated with a synset or an ontology concept.

Table 5. Top 10 features for DRINKS#PRICES according to weight assigned by SVM classifier

1	0.428 - Ontology Concept: Price	6	0.180 - Lemma: price
2	0.303 - Ontology Concept: Drink	7	0.176 - Synset: wine
3	0.232 - Synset: drink	8	0.175 - Lemma: wine
4	0.204 - Lemma: drink	9	0.165 - Ontology Concept: Wine
5	0.184 - Lemma: at	10	0.157 - Synset: value

An example where the ontology could make less of a difference is the category RESTAURANT#MISC, where, due to the miscellaneous nature of this aspect, no ontology concepts were applicable. Another category with interesting weights is FOOD#QUALITY, where next to the obvious ontology concept *Food*, a lot of adjectives were useful for the SVM since they convey the quality aspect. The fact that people write about quality is demonstrated by the strong use of sentiment words, such as "amazing" and "authentic". Hence, it is rather difficult to define an ontology concept *Quality*, and while this concept is present in the ontology, it is not very useful, being ranked the 14th most useful feature here.

Looking at the RESTAURANT#GENERAL category, we find that some concepts are really missing from the ontology, namely the idea of someone liking a restaurant. This is often expressed as wanting to go back to that place, recommending it to others, or that it is worth a visit or worth the money. The top 10 features for this category are listed in Table 6 below to illustrate this.

Table 6. Top 10 features for RESTAURANT#GENERAL according to weight assigned by SVM classifier

1	0.636 - Lemma: worth	6	0.402 - Lemma: up
2	0.505 - Lemma: love	7	0.399 - Lemma: again
3	0.487 - Lemma: back	8	0.368 - Lemma: overall
4	0.419 - Lemma: wrong	9	0.360 - Lemma: favorite
5	0.406 - Lemma: return	10	0.355 - Lemma: recommend

At a first glance, the word "wrong" looks a bit out of place here, but at closer inspection of the data, it seems this word is often used in the phrase "you can't go wrong ...", which is indeed a positive remark about a restaurant in general.

For sentiment classification, a clear feature analysis is not very feasible, since the features are not binary, but are weighted according to distance and sentiment dictionary values, when applicable. Furthermore, the SVM model is trained on three sentiment values, which means that internally, a 1-vs-all strategy is performed, so the weights would be a combination of three trained models and not so descriptive anymore.

6 Conclusion

In this paper we presented a method for aspect-based sentiment analysis that utilizes domain ontology information, both for aspect detection and for sentiment analysis of these aspects. The performance of the SVM classifier is improved with this additional knowledge, although the improvement for aspect detection is more pronounced.

For aspect detection, it can indeed be concluded that fewer training data are needed, because performance drops much less when fewer training data are available compared to the same method without ontology features. This is not the case for the sentiment analysis method applied, where due to the fact that all methods use external resources in the form of sentiment dictionaries, the added value of the ontology remains limited. However, the ontology-enhanced method keeps outperforming the basic methods on the sentiment analysis task with about the same difference, even at smaller amounts of available training data.

When interpreting the internal weights assigned by the SVM, we see that for aspect detection the ontology features are in most cases the most informative. Only when the aspect categories themselves are not clearly defined (e.g., RESTAURANT#MISC), or when that category is not described in the ontology (e.g., RESTAURANT#GENERAL), do we see the SVM using mostly word-based features.

This leads us to conclude that ontology concepts are useful, especially for aspect detection, but also for sentiment analysis, and that ontology-enhanced aspect-based sentiment analysis is a promising direction for future research.

In terms of future work, we would suggest expanding the ontology by including more domain-specific sentiment expressions. Given the fact that there are more than three times as many target concepts in the current ontology than sentiment expressions, focusing as much on sentiment expressions could lead to a more pronounced increase in performance on the sentiment analysis task as well. Furthermore, this process could be automated, scraping the information from the Web, where this type of data is readily available. Linking the ontology to others in the Linked Open Data cloud is also a direction that could be further explored.

While we propose methods for both aspect detection and sentiment analysis, there is still a subtask that is not yet covered, which is determining the target of an aspect within the sentence. Even though we use the target location information for the sentiment analysis task, we currently do not determine this information, predicting only the aspect category label. To create a complete method that can be deployed on real-life data, this missing link will need to be dealt with.

Acknowledgments. The authors of this paper are supported by the Dutch national program COMMIT.

References

1. Cambria, E., Hussain, A.: Sentic Computing: A Common-Sense-Based Framework for Concept-Level Sentiment Analysis. Springer, Cham (2015)
2. Cambria, E., Olsher, D., Rajagopal, D.: SenticNet 3: a common and common-sense knowledge base for cognition-driven sentiment analysis. In: Proceedings of the 28th AAAI Conference on Artificial Intelligence (AAAI 2014), pp. 1515–1521. AAAI (2014)
3. Cambria, E.: Affective computing and sentiment analysis. IEEE Intell. Syst. **31**(2), 102–107 (2016)

4. Chang, C.C., Lin, C.J.: LIBSVM: a library for support vector machines. ACM Trans. Intell. Syst. Technol. **2**, 27:1–27:27 (2011). Software, http://www.csie.ntu.edu.tw/~cjlin/libsvm
5. Dragoni, M., Tettamanzi, A.G.B., Costa Pereira, C.: A fuzzy system for concept-level sentiment analysis. In: Presutti, V., Stankovic, M., Cambria, E., Cantador, I., Iorio, A., Noia, T., Lange, C., Reforgiato Recupero, D., Tordai, A. (eds.) SemWebEval 2014. CCIS, vol. 475, pp. 21–27. Springer, Cham (2014). doi:10.1007/978-3-319-12024-9_2
6. Esuli, A., Sebastiani, F.: SentiWordNet: a publicly available lexical resource for opinion mining. In: Proceedings of the 5th International Conference on Language Resources and Evaluation (LREC 2006), vol. 6, pp. 417–422. European Language Resources Association (ELRA) (2006)
7. Fellbaum, C.: WordNet: an Electronic Lexical Database. MIT Press (1998)
8. Guarino, N., Welty, C.: Evaluating ontological decisions with OntoClean. Commun. ACM **45**(2), 61–65 (2002)
9. Jensen, A., Boss, N.: Textual Similarity: Comparing Texts in Order to Discover How Closely They Discuss the Same Topics. Bsc. thesis, Technical University of Denmark (2008). http://etd.dtu.dk/thesis/220969/bac08_15.pdf
10. JLanguageTool (2016). http://wiki.languagetool.org/java-api
11. Kim, Y.: Convolutional neural networks for sentence classification. In: Proceedings of the 2014 Conference on Empirical Methods in Natural Language Processing (EMNLP 2014), pp. 1746–1751. The Association for Computer Linguistics (2014)
12. Kiritchenko, S., Zhu, X., Cherry, C., Mohammad, S.: NRC-Canada-2014: detecting aspects and sentiment in customer reviews. In: Proceedings of the 8th International Workshop on Semantic Evaluation (SemEval 2014), pp. 437–442. Association for Computational Linguistics and Dublin City University (2014)
13. Lesk, M.: Automatic sense disambiguation using machine readable dictionaries: how to tell a pine cone from an ice cream cone. In: Proceedings of the 5th Annual International Conference on Systems Documentation (SIGDOC 1986), pp. 24–26. ACM (1986)
14. Liu, B.: Sentiment Analysis and Opinion Mining, Synthesis Lectures on Human Language Technologies, vol. 16. Morgan & Claypool (2012)
15. Manning, C.D., Surdeanu, M., Bauer, J., Finkel, J., Bethard, S.J., McClosky, D.: The Stanford CoreNLP natural language processing toolkit. In: Proceedings of 52nd Annual Meeting of the Association for Computational Linguistics: System Demonstrations, pp. 55–60. Association for Computational Linguistics (2014)
16. Mikolov, T., Sutskever, I., Chen, K., Corrado, G.S., Dean, J.: Distributed representations of words and phrases and their compositionality. In: Advances in Neural Information Processing Systems, pp. 3111–3119 (2013)
17. Pontiki, M., Galanis, D., Papageorgiou, H., Manandhar, S., Androutsopoulos, I.: SemEval-2015 task 12: aspect based sentiment analysis. In: Proceedings of the Ninth International Workshop on Semantic Evaluation (SemEval 2015), pp. 486–495. Association for Computational Linguistics (2015)
18. Recupero Reforgiate, D., Presutti, V., Consoli, S., Gangemi, A., Nuzzolese, A.G.: Sentilo: frame-based sentiment analysis. Cogn. Comput. **7**(2), 211–225 (2015)
19. Saias, J.: Sentiue: target and aspect based sentiment analysis in SemEval-2015 task 12. In: Proceedings of the 9th International Workshop on Semantic Evaluation (SemEval 2015), pp. 767–771. Association for Computational Linguistics (2015)
20. Schouten, K., Frasincar, F.: Survey on aspect-level sentiment analysis. IEEE Trans. Knowl. Data Eng. **28**(3), 813–830 (2016)

21. Socher, R., Perelygin, A., Wu, J.Y., Chuang, J., Manning, C.D., Ng, A.Y., Potts, C.P.: Recursive deep models for semantic compositionality over a sentiment tree-bank. In: Proceedings of the 2013 Conference on Empirical Methods on Natural Language Processing (EMNLP 2013), pp. 1631–1642. Association for Computational Linguistics (2013)
22. Tai, K.S., Socher, R., Manning, C.D.: Improved semantic representations from tree-structured long short-term memory networks. In: Proceedings of the 53rd Annual Meeting of the Association for Computational Linguistics and the 7th International Joint Conference on Natural Language Processing of the Asian Federation of Natural Language Processing (ACL 2015), pp. 1556–1566. The Association for Computer Linguistics (2015)
23. Toh, Z., Su, J.: NLANGP: supervised machine learning system for aspect category classification and opinion target extraction. In: Proceedings of the 9th International Workshop on Semantic Evaluation (SemEval 2015), pp. 496–501. Association for Computational Linguistics (2015)
24. Xia, Y., Cambria, E., Hussain, A., Zhao, H.: Word Polarity Disambiguation Using Bayesian Model and Opinion-Level Features. Cogn. Comput. **7**(3), 369–380 (2015)

Vision Papers

Inter-parameter Constraints
in Contemporary Web APIs

Nathalie Oostvogels$^{(\boxtimes)}$, Joeri De Koster, and Wolfgang De Meuter

Vrije Universiteit Brussel, Brussels, Belgium
{noostvog,jdekoste,wdmeuter}@vub.ac.be

Abstract. Today's web applications often rely on a myriad of external
web APIs, communicating with them through various HTTP requests
spread throughout the application. These APIs are often textually
described by constraints on the inputs and outputs of their entry points.
In this paper we discuss constraints in web APIs that span multiple
parameters. We show that these constraints are common in web APIs,
but cannot be expressed in existing machine-readable API specification
languages. We envision the emergence of *constraint-centric* specification
languages which focus on expressing constraints and describe a prototypical language that supports constraints over multiple parameters.

Keywords: Web application · Web API · API specifications · Inter-
parameter constraints

1 Introduction

Today it is hard to imagine a web site without cross-website functionality such
as a "like" button from Facebook, a video from YouTube or a Twitter feed. Such
cross-website functionality is typically provided by a web service that exposes
its functionality through an Application Programming Interface (API), which is
primarily accessed by means of requests over HTTP(S).

A web API is comprised of a number of entry points (also referred to as endpoints, methods or routes) that are usually described in a publicly accessible textual documentation. For every entry point, it lists inputs (accepted parameters)
and outputs (which are returned to the client). Additionally, the documentation
often lists constraints on the input parameters of each entry point. Examples of
such constraints are the type of a parameter, whether a parameter is required
or optional, allowed values, etc.

Satisfying the constraints set by the API providers is essential for a request
to succeed. Modern web applications contain many requests to many different
APIs. Manually verifying each request in a web application is a difficult and
time-consuming task. Fortunately, machine-readable API specifications aid the
automatic verification of requests and generation of documentation.

N. Oostvogels—Funded by a PhD Fellowship of the Research Foundation - Flanders (FWO).

© Springer International Publishing AG 2017
J. Cabot et al. (Eds.): ICWE 2017, LNCS 10360, pp. 323–335, 2017.
DOI: 10.1007/978-3-319-60131-1_18

These specifications are written using machine-readable API specification languages [6], which describe the same information found in textual API documentations: general information (location and authentication requirements) and a list of every entry point of the API. The input and output data is described for every entry point, together with constraints for the input data.

As an API evolves, so do the constraints on parameters. This has led to the emergence of constraints over multiple parameters, which are currently not supported by existing specification languages. Table 1 shows an example of such an *inter-parameter* constraint in the Twitter API. The direct_messages/new entry point expects up to three parameters: user_id, screen_name and text. As for constraints, Twitter only indicates whether a parameter is required (text) or optional (user_id, screen_name). These *single-parameter* constraints are mostly well-documented and supported by existing machine-readable specification languages.

Table 1. Excerpt from Twitter API documentation (https://dev.twitter.com/rest/reference/post/direct_messages/new)

Field name	Optional?	Description
user_id	Optional	The ID of the user who should receive the direct message.
screen_name	Optional	The screen name of the user who should receive the direct message.
text	Required	The text of your direct message.

Note: One of user_id or screen_name are required. (At the time of conducting our survey, the note below the table was explicitly mentioned in the API. Recently, the description has changed — omitting the note — but the constraint still holds.)

However, as indicated below the table, either the screen name or the user ID should be provided as well. This is an example of a constraint that spans multiple parameters. In contrast to single-parameter constraints, inter-parameter constraints are not captured by existing machine-readable API specifications and can only be verified manually. This limits their usability for expressing more complex scenarios that are commonly found in today's web APIs.

In this paper, we reflect on inter-parameter constraints such as the one above, and we make three contributions:

1. We identify three categories of inter-parameter constraints, and show that they are commonly found in existing web APIs (Sect. 2);
2. We show where current machine-readable API specification languages fall short (Sect. 3);
3. We introduce a new *constraint-centric* API specification language, an extension of OpenAPI, that addresses these shortcomings (Sect. 4).

2   Inter-parameter Constraints

Section 1 already showed an example of an inter-parameter constraint. In this section, we discuss several instances of inter-parameter constraints found in existing web APIs, and classify them into three categories: exclusive constraints, dependent constraints and group constraints. Inter-parameter constraints are as much part of an API as single-parameter constraints, but they are often not properly represented or enforced.

2.1   Exclusive Constraints

We call a constraint an *exclusive constraint* when **exactly one of a set of parameters is required**. Table 1 illustrates an exclusive constraint: Twitter expects either a `user_id` or a `screen_name` to indicate the recipient of the private message. In this case the textual representation does not truly reflect the correct requirements for the parameters of the request. Both parameters are tagged as optional in the API documentation, which is contradicted by the requirement that one of them must be supplied for the request to succeed.

Our study of existing API documentation (in Sect. 2.5) reveals that many of them contain exclusive constraints. Examples include entry points in

- Facebook for publishing a status update[1] where *"either link, place or message must be supplied"*;
- Stripe[2], where *"either source or customer is required"* when creating a charge;
- YouTube[3], where you may only provide one filter when retrieving a playlist (*"specify exactly one of the following parameters"*).

2.2   Dependent Constraints

The second category of inter-parameter constraints are *dependent* constraints, where **constraints on a parameter depend on a property of another parameter** (which we call the *base parameter*). This dependency can be on either the presence of a parameter or its value. There are three sub-categories of dependent constraints[4]:

- Present-Present (PP) dependent constraint: the presence of a parameter depends on the presence of the base parameter;
- Present-Value (PV) dependent constraint: the presence of a parameter depends on the value of the base parameter;
- Value-Value (VV) dependent constraint: the accepted set of values for a parameter depends on the value of the base parameter.

---

[1] https://developers.facebook.com/docs/graph-api/reference/v2.8/user/feed.

[2] https://stripe.com/docs/api/node#create_charge.

[3] https://developers.google.com/youtube/v3/docs/playlists/list.

[4] Value-Present dependent constraints are omitted, because no example was found in this survey.

Table 2 shows an example of a PP-dependent constraint in the Facebook API. When posting a `link` on someone's wall, you can specify a `picture`, `name`, `caption` and `description` for that link. These four parameters may only be included when `link` (the base parameter) itself is also present.

Table 2. Dependent constraints in the Facebook API (see Footnote 1)

Field name	Optional?	Description
`link`	Optional	The URL of a link to attach to the post. Additional fields associated with `link` are shown below.
`picture`	Optional	Determines the preview image associated with the link.
`name`	Optional	Overwrites the title of the link preview.
`caption`	Optional	Overwrites the caption under the title in the link preview.
`description`	Optional	Overwrites the description in the link preview.

Our study of existing API documentation (in Sect. 2.5) also revealed dependent constraints in other web APIs. For example, in the "add a member to a list" entry point in the Twitter API[5], the parameters `owner_screen_name` or `owner_id` are only taken into account if the `slug` parameter is also present. This is actually a combination of two inter-parameter constraints: next to the PP-dependent constraint, there is also an exclusive constraint on the parameters `owner_id` and `owner_screen_name`. The Google Maps API has a PV-dependent constraint for rendering the directions[6]: the property `infoWindow` is ignored when `suppressInfoWindows` is true. Finally, the Amazon API for product advertisement contains an example of a VV-dependency: when searching for an item[7], `condition` cannot be set to "`new`" when the `availability` parameter is set to "`available`".

2.3 Group Constraints

We classify inter-parameter constraints as *group* constraints when **a set of parameters should be either all excluded or all included in the request.** Table 3 shows a group constraint found in the Twitter API: when creating a new tweet, the user's current location can be provided via the `lat` and `long` parameters. However, it is an error to pass along only `lat` *or* only `long`: both parameters must be included to specify the location of the resulting tweet.

In Sect. 2.5, we show that group constraints are found in many APIs. In Flickr[8], for example, all coordinates of a person in a picture (`x`, `y`, `width` and

---

[5] https://dev.twitter.com/rest/reference/post/lists/members/create.

[6] https://developers.google.com/maps/documentation/javascript/reference#DirectionsRenderer.

[7] http://docs.aws.amazon.com/AWSECommerceService/latest/DG/ItemSearch.html.

[8] https://www.flickr.com/services/api/flickr.photos.people.add.html.

Table 3. A group constraint in the Twitter API (https://dev.twitter.com/rest/reference/post/statuses/update)

Field name	Optional?	Description
lat	Optional	The latitude of the location this Tweet refers to. This parameter will be ignored unless it is inside the range −90.0 to +90.0 (North is positive) inclusive. It will also be ignored if there isn't a corresponding long parameter.
long	Optional	The longitude of the location this Tweet refers to. The valid ranges for longitude is −180.0 to +180.0 (East is positive) inclusive. This parameter will be ignored if outside that range, if it is not a number, or if there is not a corresponding lat parameter.

height) must be provided. The YouTube API[9] requires that when creating a playlist, the parameter onBehalfOfContentOwnerChannel may only be present when there is a value for the onBehalfOfContentOwner parameter. In addition, onBehalfOfContentOwnerChannel can only be used in conjunction with onBehalfOfContentOwner.

2.4 Identifying Unsatisfied Constraints in API Requests

When only a textual representation of an API documentation is available, an IDE is unable to automatically verify whether requests comply with a specification. Developers are then forced to rely on the API provider to respond with a meaningful error message in case of a malformed request, or are forced to manually verify each request in the application. The problem with the former is that this means that bugs can only be identified after deployment of the application. Additionally, this approach requires full coverage of every API request by the application's test suite. Furthermore, every API provider responds differently — and not always with an error message — to requests that do not satisfy its constraints. We can classify the responses to unsatisfied inter-parameter constraints in three categories:

The API provider returns an *error message*: in the best-case scenario the API provider returns a meaningful error message whenever inter-parameter constraints are not satisfied. Unfortunately, this is not always the case. For example, when the exclusive constraint from the YouTube API is not met by supplying more than one filter for a playlist, the following error message is returned: *"Incompatible parameters specified in the request"*. Twitter returns a more detailed error message when a dependent constraint is not satisfied: *"You must specify either a list ID or a slug and owner"*. For unsatisfied group constraints, Flickr returns as error message: *"Some co-ordinate parameters were blank"*.

The API provider makes a *silent choice*: API providers can opt to tolerate certain malformed requests in order to be compatible with a wider variety

---

[9] https://developers.google.com/youtube/v3/docs/playlists/insert.

of clients. For example, Twitter does not complain when both the screen name and user ID are passed along when sending a direct message. However, when the screen name and the user ID belong to different users, Twitter chooses the screen name and silently ignores the user ID instead of raising an error. The same applies for group constraints present in the Twitter API: if not all group parameters are present, all incomplete groups are ignored. Similarly, Facebook just silently ignores all the dependent parameters when the *base* parameter is not provided. These kinds of errors are very difficult to debug, because the developer does not receive any feedback about the incorrect requests.

The API documentation is incorrect: in the case of Facebook, where their API documentation mentions the exclusive constraint *"either link, place or message must be supplied"* for publishing a status update, supplying all parameters results in a sensible status update, where all provided values are combined.

2.5   Inter-parameter Constraints in the Wild

To investigate how frequently the three categories of inter-parameter constraints occur in web APIs in the wild, we analysed the six most popular APIs of ProgrammableWeb[10] (based on usage in mashups). Other catalogs and metrics exist, e.g. API Harmony[11] checks usage on GitHub. Table 4 summarises our results.

In every API documentation, we looked for keywords that indicate an inter-parameter constraint. Exclusive constraints are often indicated with *either* or *one of*, dependent constraints with keywords such as *additional* and *providing*, and keywords for group constraints include *corresponding* and *providing*.

Table 4. Inter-parameter constraints in web APIs

	Exclusive	Dependent	Group	# entry points
Google Maps JavaScript API	10	3	3	117
Twitter REST API	32	14	6	97
YouTube Data API	11	3	5	50
Flickr API	12	0	1	206
Facebook Graph API	11	4	1	209
Amazon Product Advertising API	2	5	2	9

Table 4 shows that every inter-parameter constraint occurs in every API, except for Flickr. Exclusive constraints are the most common inter-parameter constraint in web APIs with a total of 78 occurrences. Although less frequent, group constraints occur in all the APIs we investigated. Apart from Flickr, every API has dependent constraints in their API documentation.

---

[10] http://www.programmableweb.com/apis/directory.
[11] https://apiharmony-open.mybluemix.net/.

In summary, inter-parameter constraints are present in modern web APIs and the way services respond to requests that do not satisfy constraints is not always well-defined. These diverse ways of responding to invalid requests stem from a disconnect between the documentation of an API and its implementation. Ideally, a specification is available that defines every constraint in a machine-readable manner, including inter-parameter constraints. This specification enables automatic verification of every constraint on the client side and reject invalid requests before they are sent. However, inter-parameter constraints are currently second-class concepts in specification languages and can therefore only be described textually.

3   Machine-Readable Specification Languages for Web APIs

Machine-readable specification languages for web APIs have been around since 2000, with the introduction of WSDL (Web Services Description Language) [2]. Since then, many new languages have emerged, such as WADL [4] (Web Application Description Language), OpenAPI specification (formerly known as Swagger), MSON (Markdown Syntax for Object Notation) and RAML (RESTful API Modeling Language). These languages primarily form the input for tools that generate human-readable documentation, but also enable automated testing of APIs and code analysis. If a machine-readable API is not available or cannot fully represent the constraints of an API, developers have to resort to manual verification of each API call.

Existing specification languages already support single-parameter constraints such as types, minimum and maximum values and whether parameters are required or optional. However, we argue that inter-parameter constraints are not supported by specification languages for web APIs. Table 5 shows four specification languages, but there exist many more specification languages for web APIs such as WSDL, WifL [3], Web IDL [1] and hRESTS [5]. However, to the best of our knowledge, none of these deal with inter-parameter constraints.

Table 5. Constraints in web API specifications

	Exclusive	Dependent	Group
OpenAPI	×	×	×
MSON	×	×	×
RAML	×	×	×
WADL	×	×	×
JSON Schema	×	✓	✓

We also include JSON Schema in our discussion — even though it only validates one object at a time — because it supports some inter-parameter constraints.[12] However, it is not a good fit for describing web APIs, as there are several mismatches between JSON Schema and web API specifications. First, JSON Schema by default allows fields that were not described in the schema. This is undesired behaviour while validating web APIs: the list of parameters should be exhaustive and extra parameters should be rejected. To ensure that unmentioned fields are rejected, the field `additionalProperties` must be added and set to `false` in *every* JSON Schema object.

Note that parameters in OpenAPI are described using a *subset* of JSON Schema, which excludes the features below.

JSON Schema does not directly support the specification of inter-parameter constraints, but programmers are able to express group constraints and Present-Present dependent constraints. The following snippet shows an encoding in JSON Schema of the PP-dependent constraint specified in Table 2. In this constraint we use `dependencies` to specify that the preview picture and the title name can only be present if a link was also provided with the request.

```
{type: 'object', properties: {link:    {type:'string'},
                              picture: {type:'string'},
                              name:    {type:'string'}},
              dependencies: {picture:'link', name:'link'}}
```

The snippet below shows an encoding in JSON Schema of the group constraint specified in Table 3 as a mutual dependency between both fields.

```
{type:'object', properties: {long: {type:'number'}, lat:  {type:'number'}},
              dependencies: {long:'lat', lat:'long'}}
```

For humans creating or interpreting the specification, it can be difficult to see which dependency maps to which logical constraint. Likewise, it is difficult to combine multiple constraints on parameters in JSON Schema. Readability and maintainability could be improved by separating constraints from the structure of the object, eg. by having language constructs for defining custom constraints.

Finally, Table 5 shows that JSON Schema does not support exclusive constraints. Ostensibly, the *oneOf* construct appears suitable, but we show this is not the case with a counterexample in Listing 1.1. This example attempts to encode the constraint given in Sect. 2.1, where an object may only contain a `screen_name` or `user_id` field. Nonetheless, an object {`screen_name:42`, `user_id:42`} would be accepted as well: the `screen_name` parameter is not a string, therefore the first schema is not considered valid, and therefore the `oneOf` constraint passes! This is not a good fit for the exclusive constraints found in web APIs: we want to ensure that exactly one of the fields is present.

Listing 1.1. Attempt at using `oneOf` for exclusive constraints

```
1  {oneOf: [
2    {type: "object",
3     properties: {screen_name: {type: "string", required: true}}},
4    {type: "object",
5     properties: {user_id: {type: "number", required: true}}}]}
```

We can conclude that the current API specification languages have very minimal support for inter-parameter constraints. There is a need for a specification

---

[12] JSON Hyper-Schema is an extension of JSON Schema for describing APIs. However, it does not add additional expressiveness for describing web APIs.

language that enables the specification of inter-parameter constraints (next to single-parameter constraints). Moreover, this language needs to be future-proof such that new kinds of inter-parameter constraints can be easily supported in the language as well. In the next section, we present a specification language that embeds support for inter-parameter constraints in its core.

4   OAS-IP: A Constraint-Centric Specification Language

In this section we introduce OAS-IP, a new specification language for web APIs, focused on defining and imposing constraints on parameters of entry points. By defining constraints using propositional logic, writers of API specifications can factor out patterns in constraints and impose constraints on single or multiple parameters. This enables the discovery and implementation of novel inter-parameter constraints.

OAS-IP is an extension of the OpenAPI specification language, which aims to be a vendor-neutral specification language for web services, and is supported by many companies such as Google and Microsoft. OAS-IP offers two extensions to the specification language, both described below. The first extension enables developers to define predicates for common constraints, and the second introduces a new way to impose constraints on query and payload parameters. Contraints on path and header parameters are not supported, as they were not found in our survey.

4.1   Constraint Definitions

Figure 1 shows the syntax of constraints in OAS-IP: a constraint is a logical formula that consists of operations over parameters, joined together with logical connectives. As usual, precedence can be indicated by parentheses. Parameters target regular and nested fields, as well as "array" fields — constraints on which apply to every element of the targeted array. Operations test properties on these parameters, such as whether it is present, its type, and restrictions on its value or its length. Finally, to promote reusability, constraint definitions enable the abstraction of common constraint patterns.

$$
\begin{array}{rl}
\mathsf{v} \in \text{Values} & ::= \text{Number, String, Boolean or Parameter} \\
\mathsf{f} \in \text{Parameters} & ::= \mathsf{s} \mid \mathsf{f.s} \mid \mathsf{f.[]} \\
\mathsf{t} \in \text{Types} & ::= \mathsf{string} \mid \mathsf{number} \mid \mathsf{boolean} \mid \mathsf{object} \mid \mathsf{t[]} \mid \mathsf{null} \\
\mathsf{cd} \in \text{Constraint definitions} & ::= \mathsf{s(s_1, ..., s_n)} = \mathsf{c} \\
\mathsf{c} \in \text{Constraint} & ::= \mathsf{o} \mid \mathsf{lc} \mid \mathsf{s(v_1, ..., v_n)} \\
\mathsf{o} \in \text{Operations} & ::= \mathsf{present(f)} \mid \mathsf{type(f)}{=}\mathsf{t} \mid \mathsf{length(f)} \oplus \mathsf{v} \mid \mathsf{value(f)} \oplus \mathsf{v} \\
\mathsf{lc} \in \text{Logical connectives} & ::= \mathsf{and(c, c)} \mid \mathsf{or(c, c)} \mid \mathsf{not(c)} \mid \mathsf{implic(c, c)} \mid \mathsf{iff(c, c)} \\
\oplus \in \text{Math operators} & ::= \mathsf{=, \mathsf{!=}, <, >, <=, >=}
\end{array}
$$

Fig. 1. Syntax definition for constraints

Listing 1.2 shows the definition of several single-parameter constraints (lines 2–5) as well as the three categories of inter-parameter constraints we identified in Sect. 2 (lines 6–10). Expressing the exclusive constraint requires that either present(f1) or present(f2) is true. Dependent parameters are expressed using an implication. Finally, group constraints are expressed with a double implication: f1 must be present when f2 is present, and vice versa.

Listing 1.2. Sample constraint definitions in the YAML syntax

```
1  x-constraint-definitions:
2     - minimum(f, v)       := value(f) >= v
3     - required(f)         := present(f)
4     - string?(f)          := type(f) = string
5     - number?(f)          := type(f) = number
6     - xor(f1, f2) := or(and(present(f1), not(present(f2))), and(present(f2), not(present(f1))))
7     - pp-dependent(f1, f2)           := implic(present(f1), present(f2))
8     - pv-dependent(f1, f2, v)        := implic(present(f1), value(f2) = v)
9     - vv-dependent(f1, f2, v1, v2) := implic(value(f1) = v1, value(f2) = v2)
10    - group(f1, f2)       := iff(present(f1), present(f2))
```

4.2 Constraints

Listing 1.3 shows how (inter-parameter) constraints are imposed on entry points in OAS-IP. It imposes the exclusive constraint already discussed in Sect. 2.1.

Listing 1.3. Expressing constraints for an API operation in OAS-IP

```
1  /direct_messages/show:
2    post:
3      parameters:
4        - { name: screen_name, in: query, type: string }
5        - { name: user_id, in: query, type: string }
6        - { name: text, in: query, type: string}
7      x-constraints:
8        - xor(screen_name, user_id)
```

In OpenAPI, and thus in OAS-IP as well, entry points are grouped under the paths key, with the different HTTP methods they support nested under the entry point. Lines 4–6 list the parameters for the POST method of this entry point, including single-parameter constraints to impose a type on the parameters. The inter-parameter constraint on these parameters is listed on line 8, indicating that exactly one of screen name and user ID must be supplied.

In OpenAPI, constraints are only specified per parameter, thus limiting it to supporting single-parameter constraints. Such constraints can be trivially translated to the *x-constraints* section, using the single-parameter constraint definitions in Listing 1.2. In Sect. 4.4 we discuss in detail how OAS-IP supports the constraints that can be expressed in OpenAPI.

4.3 Composing Inter-parameter Constraints

Section 2 showed that inter-parameter constraints are common in the documentation of web APIs. Sometimes these inter-parameter constraints are even nested. We will discuss the composability of inter-parameter constraints by means of an example from the Twitter API:

"You can identify a list by its slug *instead of its* list_id. *If you decide to do so, note that you will also have to specify the list owner using the* owner_id *or* owner_screen_name *parameters."*

This sentence denotes a dependent constraint between slug and *two* fields (owner_id and owner_screen_name), which have an exclusive constraint imposed on them in turn. There is also an exclusive constraint between these three fields and the list_id field. Using the constraint definitions in Listing 1.2, we would like to write this down as:

```
and(xor(list_id, slug), iff(present(slug), xor(owner_screen_name, owner_id)))
```

However, this is subtly wrong: the constraint is also valid if every field except slug is present.[13] Instead, this constraint may be written down as a set of smaller, non-nested constraints:

```
xor(slug, list_id)
implic(present(slug), xor(owner_screen_name, owner_id))
pp-dependent(owner_screen_name, slug)
pp-dependent(owner_id, slug)
```

This set of constraints is not as concise, but it is correct. During the course of developing OAS-IP (and accompanying examples), we found it beneficial to work with sets of singular constraints rather than nesting constraints.

4.4   Comparison with Other Web API Specification Languages

Section 3 discussed several languages for web API specifications. Our main concern with existing languages is the lack of support for inter-parameter constraints: only JSON Schema has limited support for expressing constraints over multiple parameters. More specifically, JSON Schema expresses dependent constraints and group constraints using its dependencies construct.

This section introduced OAS-IP, a specification language for web APIs with constructs for defining constraints by means of predicates over parameters. Writers of web API specifications do not have a limited set of constraints to choose from: they can use any combination of the provided operations. Moreover, complicated predicate combinations are abstracted to a custom constraint definition. This is an advantage over JSON Schema, where lack of abstractions gives rise to readability and maintainability issues. For example: a group constraint results in two dependencies expressions, which is not always intuitive to the reader.

To gauge the expressiveness of OAS-IP for modeling existing web APIs, we examined the constraint keywords of OpenAPI and JSON Schema and attempted to replicate them with constraint definitions in OAS-IP. Apart from the type and whether the parameter is required, the majority of keywords are single-parameter constraints on either numeric values of parameters or the size of arrays or objects, and thus supported with the existing operations in OAS-IP. Others, such as pattern, uniqueItems andmultipleOf can be supported by adding new

---

[13] It is possible to come up with alternative formulations, but the reader needs to construct a truth table in order to convince him or herself.

operations. OAS-IP currently does not support the JSON Schema `items` and `additionalItems` constraints which provide a different schema for each item of an array.

A final difference is the handling of unspecified fields. The `patternProperties` keyword allows constraining parameters whose name matches a regular expression, while the `additionalProperties` keyword either validates or forbids unknown parameters. As we mentioned before, OAS-IP defaults to rejecting requests with unknown fields. Both can be supported with the addition of *patterned fields*: for example, `string?(metadata./^x-/)` would require that unspecified fields of the `metadata` object starting with `x-` must be strings.

With these extensions, OAS-IP can express the same single-parameter constraints as OpenAPI and JSON Schema. In addition, it is capable of describing inter-parameter constraints, which other specification languages cannot.

5   Conclusion and Future Work

Today, APIs of modern web applications are described in a public API documentation. This documentation is often generated using a machine-readable API specification language. Such specifications can also enable automatic verification of constraints in the API. Traditionally such specification languages have focused on *single-parameter* constraints.

We surveyed current API documentation and identified three kinds of *inter-parameter* constraints. These constraints are currently only described textually and are not covered by the existing specification languages. This limits the power of existing tooling around API specification languages. Our survey indicates that inter-parameter constraints are essential to achieving comprehensive, machine-readable web API specifications. Therefore, we designed OAS-IP, an API specification language based on the OpenAPI specification, extended with new language constructs to define and impose both single-parameter and inter-parameter constraints. With this added flexibility, developers can describe *all* constraints in their APIs, which in turn makes tooling more powerful.

As a proof of concept, we have developed a preprocessor which, given an OAS-IP specification, produces a list of constraints for every entry point in the API.[14] We envision that this preprocessor will form the basis for various tools that help developers interact with web APIs. As a first step, existing documentation generation tools can use the preprocessor to help web developers understand the constraints on particular entry points. Going further, we envision that constraint-centric languages will form the basis of new tools that statically analyse interactions with web APIs in web applications.

---

[14] https://github.com/noostvog/Verify-Request.

References

1. Bae, S., Cho, H., Lim, I., Ryu, S.: SAFEWAPI: Web API misuse detector for web applications. In: Foundations of Software Engineering, pp. 507–517 (2014)
2. Christensen, E., Curbera, F., Meredith, G., Weerawarana, S., et al.: Web services description language (WSDL) 1.1 (2001)
3. Danielsen, P.J., Jeffrey, A.: Validation and interactivity of Web API documentation. In: International Conference on Web Services (ICWS), pp. 523–530 (2013)
4. Hadley, M.J.: Web application description language (WADL) (2006)
5. Kopecky, J., Gomadam, K., Vitvar, T.: hRESTS: an HTML microformat for describing RESTful web services. In: Web Intelligence and Intelligent Agent Technology (WI-IAT), vol. 1, pp. 619–625 (2008)
6. Verborgh, R., Harth, A., Maleshkova, M., Stadtmüller, S., Steiner, T., Taheriyan, M., Van de Walle, R.: Survey of semantic description of REST APIs. In: Pautasso, C., Wilde, E., Alarcon, R. (eds.) REST: Advanced Research Topics and Practical Applications, pp. 69–89. Springer, New York (2014). doi:10.1007/978-1-4614-9299-3_5

Collaborative Item Embedding Model for Implicit Feedback Data

ThaiBinh Nguyen[1]([⊠]), Kenro Aihara[2], and Atsuhiro Takasu[2]

[1] Department of Informatics,
SOKENDAI (The Graduate University for Advanced Studies),
Shonan Village, Hayama, Kanagawa 240-0193, Japan
`binh@nii.ac.jp`
[2] National Institute of Informatics, 2-1-2 Hitotsubashi,
Chiyoda, Tokyo 101-8430, Japan
`{kenro.aihara,takasu}@nii.ac.jp`

Abstract. Collaborative filtering is the most popular approach for recommender systems. One way to perform collaborative filtering is matrix factorization, which characterizes user preferences and item attributes using latent vectors. These latent vectors are good at capturing global features of users and items but are not strong in capturing local relationships between users or between items. In this work, we propose a method to extract the relationships between items and embed them into the latent vectors of the factorization model. This combines two worlds: matrix factorization for collaborative filtering and item embedding, a similar concept to word embedding in language processing. Our experiments on three real-world datasets show that our proposed method outperforms competing methods on top-n recommendation tasks.

Keywords: Recommender system · Collaborative filtering · Matrix factorization · Item embedding

1 Introduction

Modern recommender systems (RSs) are a core component of many online services. An RS analyzes users' behavior and provides them with personalized recommendations for products or services that meet their needs. For example, Amazon recommends products to users based on their shopping histories; an online newspaper recommends articles to users based on what they have read.

Generally, an RS can be classified into two categories: Content-based approach and collaborative filtering-based (CF-based) approach. The content-based approach creates a description for each item and builds a profile for each user's preferences. In other words, the content-based approach recommends items that are similar to items for which the user has expressed interest in the past. In contrast, the CF-based approach relies on the past behavior of each user, without requiring any information about the items that the users have consumed.

© Springer International Publishing AG 2017
J. Cabot et al. (Eds.): ICWE 2017, LNCS 10360, pp. 336–348, 2017.
DOI: 10.1007/978-3-319-60131-1_19

An advantage of the CF-based approach is that it does not require collection of item contents or analysis. In this work, we focus on the CF-based approach.

Input data for CF-based methods are the user-item interaction matrix, in which each entry is the feedback of a user to an item. The feedback can be explicit (e.g., rating scores/stars, like/dislike) or implicit (e.g., click, view, purchase). Early work mainly focused on explicit feedback data such as SVD++ [1], timeSVD [2], or probabilistic matrix factorization [3]. One advantage of explicit feedback is that it is easy to interpret because it directly expresses the preferences of users for items. However, explicit feedback is not always available and is extremely scarce, as few users provide explicit feedback.

Implicit feedback, in contrast, is generated in abundance while users interact with the system. However, interpreting the implicit feedback is difficult, because it does not directly express users' opinions about items. For example, a user's click on an item does not mean that he or she likes it; rather, the user may click and then find that he or she does not like the item. On the other hand, even though a user does not interact with an item, this does not imply that the user dislikes it; it may be because the user does not know that the item exists.

Hu et al. proposed the weighted matrix factorization (WMF) [4], a special case of the matrix factorization technique targeted to implicit datasets. The model maps each user and item into a low-dimensional vector in a shared latent space, which encodes all the information that describes the user's preference or the item's characteristics. Locations of users and items in the space show their relationships. If two items are close together in the space, they are considered to be similar. On the other hand, if a user and an item are close in the space, that user is considered to like that item.

Detecting the relationships between items is crucial to the performance of the RS. We consider two kinds of relationships, the global relationship and a local one. The former indicates the global structure that relates simultaneously to most or all items, and is captured from the overall information encompassed in all user–item interactions. The latter, in contrast, indicates the relationships between a small set of closely related items [1,5]. Detecting the local relationship will benefit the RS in recommending correlated items. One example of correlated items in the movie domain is the three volumes of the film series "Lord of the Rings." Usually, a user who watches one of them will watch the others. The detection of local relationships gives the system the ability to capture such correlations and recommend one of these volumes when it knows that the user has watched the others. However, while WMF as well as other MF-based algorithms are strong at capturing the global relationships, they are poor at detecting the local relationships [1,5].

In this work, we propose a model that can capture both global and local relationships between items. The idea is to extract the relationships between items that frequently occur in the context of each other, and embed these relationships into the factorization model of WMF [4,6]. The "context" can be the items in a user's interaction list (i.e., the items that the user interacts with), or the items in a transaction. Two items are assumed to be similar if they often appear in a context with each other, and their representations should be located close to each other in the space. The proposed model identifies such relationships and reflects them into WMF. This was inspired by word-embedding techniques in natural language processing that represent words by vectors that can capture the relationships between each word and its surrounding words [7–10].

In detail, we build an item–item matrix containing the context information and embed information from this matrix into the factorization model. The embedding is performed by factorizing the user–item matrix and the item–item matrix simultaneously. In the model, the role of the item–item matrix factorization is to adjust the latent vectors of items to reflect item–item relationships.

The rest of this paper is organized as follows. In Sect. 2, we present the background knowledge related to this work. Section 3 presents the details of our idea and how we add item embedding to the original factorization model. In Sect. 4, we explain our empirical study and the experimental results. After reviewing some related work in Sect. 5, we discuss the results of this work and show some directions for future work in Sect. 6.

2   Preliminary

2.1   Weighted Matrix Factorization

Suppose we have N users and M items. For each user u and item i, we denote by r_{ui} the number of times user u has interacted with item i. We assume that user u likes item i if he or she has interacted with item i at least once. For user u and item i, we define a reference value p_{ui} indicating whether user u likes item i (i.e., $p_{ui} = 1$ if $r_{ui} > 0$ and $p_{ui} = 0$ otherwise), and a confidence level c_{ui} to represent how confident we are about the value of p_{ui}. Following [4], we define c_{ui} as:

$$c_{ui} = 1 + \alpha r_{ui}, \tag{1}$$

where α is a positive number.

Weighted matrix factorization (WMF) [4,6], is a factorization model to learn the latent representations of all users and items in the dataset. The objective function of the model is:

$$\mathcal{L}(X,Y) = \sum_{u,i} c_{ui}(p_{ui} - \mathbf{x}_u^\top \mathbf{y}_i)^2 + \lambda \left(\sum_u ||\mathbf{x}_u||_F^2 + \sum_i ||\mathbf{y}_i||_F^2 \right), \tag{2}$$

where $X \in \mathbb{R}^{d \times N}$ and $Y \in \mathbb{R}^{d \times M}$ are matrices with columns $\mathbf{x}_u$ and $\mathbf{y}_i$ that are the latent vectors of users and items, respectively; $||.||_F$ is the Frobenius norm of a vector. This optimization problem can be efficiently solved using the Alternating Least Square (ALS) method as described in [4].

2.2 Word Embedding

Word embedding models [7,8,10,11] have gained success in many natural language processing tasks. Their goal is to find vector representations of words that can capture their relationship with their context words (i.e., the surrounding words in a sentence or paragraph).

Given a corpus and a word w, a context word c of w is a word that occurs within a specific-size window around w (context window) in the corpus. Let $\mathcal{D}$ denote the set of all word–context pairs, i.e., $\mathcal{D} = \{(w,c) | w \in V_W, c \in V_C\}$, where V_W and V_C are the set of words and set of context words, respectively. Word embedding models represent a word $w \in V_W$ and a context word $c \in V_C$ by vectors $\mathbf{w} \in \mathbb{R}^d$ and $\mathbf{c} \in \mathbb{R}^d$, respectively, where d is the embedding's dimensionality.

Mikolov et al. proposed an efficient model for learning word vectors [7], which is performed by maximizing the log-likelihood function for every word-context pair $(w,c) \in \mathcal{D}$:

$$\log \sigma(\mathbf{w}^\top \mathbf{c}) + k \mathbb{E}_{c_N \propto P_D} \sigma(-\mathbf{w}^\top \mathbf{c}_N), \tag{3}$$

where $\sigma(.)$ is the sigmoid function: $\sigma(x) = 1/(1 + \exp(-x))$, P_D is a distribution for sampling false context words (hence, negative sampling) and k is a hyperparameter specifying the number of negative samples. This model is called Skip-gram negative sampling (SGNS) [7]. Based on this model, Mikolov et al. released a well-known open source package named word2vec[1].

Levy et al. [9] showed that the optimal solutions $\mathbf{w}^*, \mathbf{c}^*$ of Eq. (3) satisfy:

$$\mathbf{w}^{*\top} \mathbf{c}^* = \mathrm{PMI}(w,c) - \log k, \tag{4}$$

where $\mathrm{PMI}(w,c)$ is the *pointwise mutual information* between word w and context word c. The symbol k, again, is the number of negative samples.

The PMI [12] of a word-context pair (w,c) is a measure that quantifies the association between a word w and a context word c. It is defined as:

$$\mathrm{PMI}(w,c) = \log \frac{P(w,c)}{P(w)P(c)}, \tag{5}$$

where $P(w,c)$ is the probability that c appears in the context of w; $P(w)$ and $P(c)$ are the probabilities that word w and context word c appear in the corpus, respectively. Empirically, PMI can be estimated using the actual number of observations in a corpus:

$$\mathrm{PMI}(w,c) = \log \left(\frac{\#(w,c)|\mathcal{D}|}{\#(w)\#(c)} \right), \tag{6}$$

where $|\mathcal{D}|$ is the size of $\mathcal{D}$; $\#(w,c)$ is the number of times the pair (w,c) appears in $\mathcal{D}$; and $\#(w) = \sum_c \#(w,c)$ and $\#(c) = \sum_w \#(w,c)$ are the numbers of times w and c appear in $\mathcal{D}$, respectively.

---

[1] https://code.google.com/archive/p/word2vec/.

Levy et al. [9] then proposed a word embedding model by factorizing the matrix S, which has elements S_{wc} that are defined in Eq. (7). This matrix is called the shifted positive pointwise mutual information matrix (SPPMI matrix).

$$S_{wc} = \max\{\text{PMI}(w, c) - \log k, 0\}. \tag{7}$$

In other words, the SPPMI matrix S is obtained by shifting the PMI matrix by $\log k$ and then replacing all negative values with zeroes (hence, shifted positive pointwise mutual information).

3   Co-occurrence-based Item Embedding for Collaborative Filtering

3.1   Co-occurrence-based Item Embedding

By considering each item as a word, we aim to extract the relationships between items in the same way as word embedding techniques do. Our motivation is that the representation of an item is governed not only by the users who interact with it but also by the other items that appear in its context. In this work, we define "context" as the items occurring in the interaction list of a user (i.e., the items that the user interacts with). However, other definitions of context can also be used without any problems. We argue that if items co-occur frequently in the interaction lists of some users, they are similar, and their latent vectors should be close in the latent space.

Inspired by the work of Levy et al. [9], which we present in Sect. 2.2, we construct an SPPMI matrix of items based on co-occurrences and embed it into the factorization model.

Constructing the SPPMI Matrix for Items. We now show how to construct the SPPMI matrix for items according to their co-occurrences.

Let $\mathcal{D} = \{(i, j) | i, j \in I_u, i \neq j, u \in U\}$, where I_u is the set of items with which user u has interacted. We use $\#(i, j)$ to denote the number of times the item pair (i, j) appears in $\mathcal{D}$ and $\#(i) = \sum_j \#(i, j)$ to denote the number of times item i appears in $\mathcal{D}$.

For example, if we have three users u_1, u_2, and u_3 whose interaction lists are $I_1 = \{1, 2, 4\}$, $I_2 = \{2, 3\}$, and $I_3 = \{1, 2, 3\}$, respectively, we will have:

- $\mathcal{D} = \{(1, 2), (1, 4), (2, 4), (2, 3), (1, 2), (1, 3), (2, 3)\}$
- $\#(1, 2) = 2, \#(1, 3) = 1, \#(1, 4) = 1, \#(2, 3) = 2, \#(2, 4) = 1$
- $\#(1) = 4, \#(2) = 5, \#(3) = 3, \#(4) = 2.$

The item–item matrix S has elements:

$$s_{ij} = \log\left(\frac{\#(i, j)|\mathcal{D}|}{\#(i)\#(j)}\right) - \log k, \tag{8}$$

where $\log\left(\frac{\#(i,j)|\mathcal{D}|}{\#(i)\#(j)}\right)$ is the pointwise mutual information of pair (i,j), as mentioned above, and k is a positive integer corresponding to the number of negative samples in the SGNS model [7]. In our experiments, we set $k = 1$.

Because S defined above is symmetric, instead of factorizing S into two different matrices as in [9], we factorize it into two equivalent matrices. In more detail, we factorize S to the latent vectors of items:

$$S = Y^\top Y \tag{9}$$

In this way, S can also be viewed as a similarity matrix between items, where element s_{ij} indicates the similarity between item i and item j.

3.2 Co-occurrence-based Item Embedded Matrix Factorization (CEMF)

We can now show how to incorporate the co-occurrence information of items into the factorization model. The SPPMI matrix will be factorized to obtain the latent vectors of items. The learned latent factor vectors of items should minimize the objective function:

$$\sum_{i,j:s_{ij}>0} \left(s_{ij} - \mathbf{y}_i^\top \mathbf{y}_j\right)^2. \tag{10}$$

Combining with the original objective function in Eq. (2), we obtain the overall objective function:

$$\mathcal{L}(X,Y) = \sum_{u,i} c_{ui}\left(p_{ui} - \mathbf{x}_u^\top \mathbf{y}_i\right)^2 + \sum_{\substack{i \\ j>i \\ s_{i,j}>0}} \left(s_{ij} - \mathbf{y}_i^\top \mathbf{y}_j\right)^2$$

$$+ \lambda\left(\sum_u ||\mathbf{x}_u||_F^2 + \sum_i ||\mathbf{y}_i||_F^2\right). \tag{11}$$

Learning Method. This function is not convex with respect to $\mathbf{x}_u$ and $\mathbf{y}_i$, but it is convex if we keep one of these fixed. Therefore, it can be solved using the Alternating Least Square method, similar to the method described in [4].

For each user u, at each iteration, we calculate the partial derivative of $\mathcal{L}$ with respect to $\mathbf{x}_u$ while fixing other entries. By setting this derivative to zero, $\frac{\partial \mathcal{L}}{\partial \mathbf{x}_u} = 0$, we obtain the update rule for $\mathbf{x}_u$:

$$\mathbf{x}_u = \left(\sum_i c_{ui}\mathbf{y}_i\mathbf{y}_i^\top + \lambda\mathbf{I}_d\right)^{-1}\left(\sum_i c_{ui}\mathbf{y}_i p_{ui}\right). \tag{12}$$

Similarly, for each item i, we calculate the partial derivative of $\mathcal{L}$ with respect to $\mathbf{y}_i$ while fixing other entries, and set the derivative to zero. We obtain the update rule for $\mathbf{y}_i$:

$$
\mathbf{y}_i = \left(\sum_u c_{ui} \mathbf{x}_u \mathbf{x}_u^\top + \sum_{j:s_{i,j}>0} \mathbf{y}_j \mathbf{y}_j^\top + \lambda \mathbf{I}_d \right)^{-1}
$$
$$
\left(\sum_u c_{ui} p_{ui} \mathbf{x}_u + \sum_{j:s_{ij}>0} s_{ij} \mathbf{y}_j \right),
$$

$$(13)$$

where $\mathbf{I}_d \in \mathbb{R}^{d \times d}$ is the identity matrix (i.e., the matrix with ones on the main diagonal and zeros elsewhere).

Computational Complexity. For user vectors, as analyzed in [4], the complexity for updating N users in an iteration is $\mathcal{O}(d^2|\mathcal{R}| + d^3 N)$, where $|\mathcal{R}|$ is the number of nonzero entries of the preference matrix P. Since $|\mathcal{R}| >> N$, if d is small, this complexity is linear in the size of the input matrix. For item vector updating, we can easily show that the running time for updating M items in an iteration is $\mathcal{O}(d^2(|\mathcal{R}| + M|\mathcal{S}|) + d^3 M)$, where $|\mathcal{S}|$ is the number of nonzero entries of matrix S. For systems in which the number of items is not very large, this complexity is not a big problem. However, the computations become significantly expensive for systems with very large numbers of items. Improving the computational complexity of updating item vectors will be part of our future work.

4   Empirical Study

In this section, we study the performance of CEMF. We compare CEMF with two competing methods for implicit feedback data: WMF [4,6] and CoFactor [13]. Across three real-world datasets, CEMF outperformed these competing methods for almost all metrics.

4.1   Datasets, Metrics, Competing Methods, and Parameter Setting

Datasets. We studied datasets from different domains: movies, music, and location, with varying sizes from small to large. The datasets are:

- *MovieLens-20M (ML-20M)* [14]: a dataset of users' movie ratings collected from MovieLens, an online film service. It contains 20 million ratings in the range 1–5 of 27,000 movies by 138,000 users. We binarized the ratings thresholding at 4 or above. The dataset is available at GroupLens[2].

---

[2] https://grouplens.org/datasets/movielens/20m/.

- *TasteProfile*: a dataset of counts of song plays by users collected by Echo Nest[3]. After removing songs that were listened to by less than 50 users, and users who listened to less than 20 songs, we binarized play counts and used them as implicit feedback data.
- *Online Retail Dataset (OnlineRetail)* [15]: a dataset of online retail transactions provided at the UCI Machine Learning Repository[4]. It contains all the transactions from December 1, 2010 to December 9, 2011 for a UK-based online retailer.

For each user, we selected 20% of interactions as ground truth for testing. The remaining portions from each user were divided in two parts: 90% for a training set and 10% for validation. The statistical information of the training set of each dataset is summarized in Table 1.

Table 1. Statistical information of the datasets after post-preprocessing

	ML-20M	TasteProfile	OnlineRetail
# of users	138,493	629,113	3,704
# of items	26,308	98,486	3,643
# of interactions	18M	35.5M	235K
Sparsity (%)	99.5	99.94	98.25
Sparsity of SPPMI matrix (%)	75.42	76.34	66.24

Evaluation Metrics. The performance of the learned model was assessed by comparing the recommendation list with the ground-truth items of each user. We used Recall@n and Precision@n as the measures for evaluating the performance.

Recall@n and Precision@n are usually used as metrics in information retrieval. The former metric indicates the percentage of relevant items that are recommended to the users, while the latter indicates the percentage of relevant items in the recommendation lists. They are formulated as:

$$\text{Recall@}n = \frac{1}{N} \sum_{u=1}^{N} \frac{|S_u(n) \cap V_u|}{|V_u|}$$

$$\text{Precision@}n = \frac{1}{N} \sum_{u=1}^{N} \frac{|S_u(n) \cap V_u|}{n} \tag{14}$$

where $S_u(n)$ is the list of top-n items recommended to user u by the system and V_u is the list of ground-truth items of user u.

---

[3] http://the.echonest.com/.
[4] https://archive.ics.uci.edu/ml/datasets/Online+Retail.

Competing Methods. We compared CEMF with the following competing methods.

– *CoFactor* [13]: factorizes user–item and item–item matrices simultaneously as we do, where the item–item co-occurrence matrix is factorized into two matrices.
– *WMF* [4]: a weighted matrix factorization matrix for the implicit feedback dataset.

Parameters.

– *Number of factors d*: we learn the model with the number of factors running from small to large values: $d = \{10, 20, 30, 40, 50, 60, 70, 80, 90, 100\}$.
– *Regularization term*: we set the regularization parameter for the Frobenius norm of user and item vectors as $\lambda = 0.01$.
– *Confidence matrix*: we set $c_{ui} = 1 + \alpha r_{ui}$ ($\alpha > 0$). We changed the value of α and chose the one that gave the best performance.

4.2 Results

We evaluated CEMF by considering its overall performance and its performance for different groups of users. Results for Precision@n and Recall@n show that our method outperformed the competing methods.

Overall Performance. Overall prediction performance with respect to Precision and Recall are shown in Tables 2 and 3 respectively. These are the results for $d = 30$; larger values of d produce higher accuracy but the differences in performance between the methods do not change much. The results show that CEMF improves the performances for the three datasets over almost all metrics, except for some metrics with $n > 20$ for the *TasteProfile*. If we use only small values of n, say $n = 5$ or $n = 10$, CEMF outperforms all competing methods over the three datasets.

Performance for Different Groups of Users. We divided the users into groups based on the number of items they had interacted with so far, and evaluated the performance for each group. There were three groups in our experiments:

– *low*: users who had interacted with less than 20 items
– *medium*: users who had interacted with 20–100 items
– *high*: users who had interacted with more than 100 items.

The Precision@n and Recall@n for these groups are presented in Fig. 1. The results show that CEMF outperforms the competing methods for almost all groups of users. For users with small numbers of interactions, CEMF is slightly better than WMF and much better than CoFactor. For users with many items

Table 2. Precision@n of WMF, CoFactor, and CEMF over three datasets

Dataset	Model	Pre@5	Pre@10	Pre@20	Pre@50	Pre@100
ML-20M	WMF	0.2176	0.1818	0.1443	0.0974	0.0677
	CoFactor	0.2249	0.1835	0.1416	0.0926	0.0635
	CEMF	**0.2369**	**0.1952**	**0.1523**	**0.1007**	**0.0690**
TasteProfile	WMF	0.1152	0.0950	0.0755	**0.0525**	**0.0378**
	CoFactor	0.1076	0.0886	0.0701	0.0487	0.0353
	CEMF	**0.1181**	**0.0966**	**0.0760**	0.0523	0.0373
OnlineRetail	WMF	0.0870	0.0713	0.0582	0.0406	0.0294
	CoFactor	0.0927	0.0728	0.0552	0.0381	0.0273
	CEMF	**0.0959**	**0.0779**	**0.0619**	**0.0425**	**0.0302**

Table 3. Recall@n of WMF, CoFactor, and CEMF over three datasets

Dataset	Model	Recall@5	Recall@10	Recall@20	Recall@50	Recall@100
ML-20M	WMF	0.2366	0.2601	0.3233	0.4553	0.5788
	CoFactor	0.2420	0.2550	0.3022	0.4101	0.5194
	CEMF	**0.2563**	**0.2750**	**0.3331**	**0.4605**	**0.5806**
TasteProfile	WMF	0.11869	0.1148	**0.1377**	**0.2129**	**0.2960**
	CoFactor	0.1106	0.1060	0.1256	0.1947	0.2741
	CEMF	**0.1215**	**0.1159**	0.1369	0.2092	0.2891
OnlineRetail	WMF	0.1142	0.1463	0.2136	0.3428	0.4638
	CoFactor	0.1160	0.1384	0.1891	0.3020	0.4159
	CEMF	**0.1232**	**0.1550**	**0.2191**	**0.3466**	**0.4676**

in their interaction lists, CEMF shows much better performance than WMF and better than CoFactor.

In a system, we usually have users with few interactions and users with many interactions; therefore, using CEMF is more efficient than either WMF or CoFactor.

5   Related Work

Standard techniques for implicit feedback data include weighted matrix factorization [4,6], which is a special case of the matrix factorization technique that is targeted to implicit feedback data, where the weights are defined from the interaction counts, reflecting how confident we are about the preference of a user for an item. Gopalan et al. [16] introduced a Poisson distribution-based factorization model that factorizes the user–item matrix. The common point of these methods for matrix factorization is that they assume that the user–item interactions

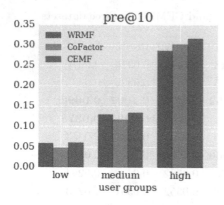 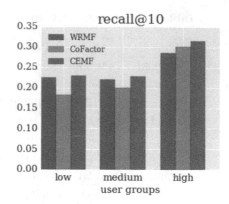

Fig. 1. Precision@10 and Recall@10 for different groups of users with the ML-20M dataset

are independent; thus, they cannot capture the relationships between strongly related items in the latent representations.

Collective matrix factorization (CMF) [17] proposes a framework for factorizing multiple related matrices simultaneously, to exploit information from multiple sources. This approach can incorporate the side information (e.g., genre information of items) into the latent factor model.

In [18], the authors present a factorization-based method that uses item–item similarity to predict drug–target interactions. While this model uses the item–item similarity from additional sources as side information, we do not require side information in this work. Instead, we exploit the co-occurrence information that is drawn from the interaction matrix.

The CoFactor [13] model is based on CMF [17]. It factorizes the user–item and item–item matrices at the same time in a shared latent space. The main difference between our method and CoFactor is how we factorize the item–item co-occurrence matrix. Instead of representing each item by two latent vectors as in [13], where it is difficult to interpret the second one, we represent each item by a single latent vector.

6   Discussion and Future Work

We have examined the effect of co-occurrence on the performance of recommendation systems. We proposed a method that combines the power of two worlds: collaborative filtering by MF and item embedding with item context for items in the interaction lists of users. Our goal is a latent factor model that reflects the strong associations of closed related items in their latent vectors. Our proposed method improved the recommendation performance on top-n recommendation for three real-world datasets.

We plan to explore several ways of extending or improving this work. The first direction is to consider different definitions of "context items". One approach is to define context items as items that co-occur in the same transactions as the given items. In this way, we can extract relationships between items that frequently appear together in transactions and can recommend the next item given the current one, or recommend a set of items.

The second direction we are planning to pursue is to reduce the computational complexity of the current algorithm. As we mentioned in Sect. 3, the computational complexity for updating item vectors is $\mathcal{O}(d^2(|\mathcal{R}| + M|\mathcal{S}|) + d^3M)$, which becomes significantly expensive for systems with large numbers of items. We hope to develop a new algorithm that can improve this complexity. An online learning algorithm, which updates user and item vectors when new data are collected without retraining the model from the beginning, is also in our plan to improve this work.

Acknowledgments. This work was supported by a JSPS Grant-in-Aid for Scientific Research (B) (15H02789).

References

1. Koren, Y.: Factorization meets the neighborhood: a multifaceted collaborative filtering model. In: 14th ACM SIGKDD International Conference on Knowledge Discovery and Data Mining, pp. 426–434. ACM (2008)
2. Koren, Y.: Collaborative filtering with temporal dynamics. Commun. ACM **53**(4), 89–97 (2010)
3. Salakhutdinov, R., Mnih, A.: Probabilistic matrix factorization. In: Advances in Neural Information Processing Systems, vol. 20 (2008)
4. Hu, Y., Koren, Y., Volinsky, C.: Collaborative filtering for implicit feedback datasets. In: Eighth IEEE International Conference on Data Mining 2008, ICDM 2008, pp. 263–272. IEEE (2008)
5. Bell, R.M., Koren, Y.: Lessons from the netflix prize challenge. SIGKDD Explor. **9**(2), 75–79 (2007)
6. Pan, R., Zhou, Y., Cao, B., Liu, N.N., Lukose, R.M., Scholz, M., Yang, Q.: Oneclass collaborative filtering. In: IEEE International Conference on Data Mining (ICDM 2008), pp. 502–511 (2008)
7. Mikolov, T., Sutskever, I., Chen, K., Corrado, G.S., Dean, J.: Distributed representations of words and phrases and their compositionality. In: NIPS, pp. 3111–3119 (2013)
8. Mikolov, T., Chen, K., Corrado, G., Dean, J.: Efficient estimation of word representations in vector space. arXiv preprint arXiv:1301.3781 (2013)
9. Levy, O., Goldberg, Y.: Neural word embedding as implicit matrix factorization. In: Advances in Neural Information Processing Systems 27, pp. 2177–2185. Curran Associates Inc. (2014)
10. Le, Q.V., Mikolov, T.: Distributed representations of sentences and documents. ICML **14**, 1188–1196 (2014)
11. Bengio, Y., Ducharme, R., Vincent, P., Jauvin, C.: A neural probabilistic language model. J. Mach. Learn. Res. **3**(Feb), 1137–1155 (2003)
12. Church, K.W., Hanks, P.: Word association norms, mutual information, and lexicography. Comput. Linguist. **1**(16), 22–29 (1990)

13. Liang, D., Altosaar, J., Charlin, L., Blei, D.M.: Factorization meets the item embedding: regularizing matrix factorization with item co-occurrence. In: RecSys, pp. 59–66. ACM (2016)
14. Maxwell Harper, F., Konstan, J.A.: The movielens datasets: history and context. TiiS **5**(4), 19 (2016)
15. Chen, D., Sain, S.L., Guo, K.: Data mining for the online retail industry: a case study of rfm model-based customer segmentation using data mining. J. Database Mark. Customer Strategy Manage. **19**(3), 197–208 (2012)
16. Gopalan, P., Hofman, J.M., Blei, D.M.: Scalable recommendation with poisson factorization. CoRR, abs/1311.1704 (2013)
17. Singh, A.P., Gordon, G.J.: Relational learning via collective matrix factorization. In: KDD, pp. 650–658. ACM (2008)
18. Zheng, X., Ding, H., Mamitsuka, H., Zhu, S.: Collaborative matrix factorization with multiple similarities for predicting drug-target interactions. In: KDD, pp. 1025–1033. ACM (2013)

Short Papers

Impact of Referral Incentives
on Mobile App Reviews

Noor Abu-El-Rub$^{(\boxtimes)}$, Amanda Minnich, and Abdullah Mueen

University of New Mexico, Albuquerque, NM 87131, USA
{nabuelrub,aminnich,mueen}@unm.edu

Abstract. Product owners occasionally provide referral incentives to the customers (e.g. coupons, bonus points, referral rewards). However, clever customers can write their referral codes in online review pages to maximize incentives. While these reviews are beneficial for both writers and product owners, the core motivation behind such reviews is monetary as opposed to helping potential customers. In this paper, we analyze referral reviews in the Google Play store and identify groups of users that have been consistently taking part in writing such abusive reviews. We further explore how such referral reviews indeed help the mobile apps in gaining popularity when compared to apps that do not provide incentives. We also find an increasing trend in the number of apps being targeted by abusers, which, if continued, will render review systems as crowd advertising platforms rather than an unbiased source of helpful information.

Keywords: Online reviews · Advertisement · Referrals · Google play · Abuse · Cliques · Incentives

1 Introduction

Providing incentives is a common strategy in modern marketing. In Google Play store, users are promised that if new users apply their referral codes, both new and old users will get reward points to spend in that app or to redeem for cash or gift cards [1]. As a result, users broadcast their referral code by posting referral reviews to increase their earning through incentives, destroying the purpose of a review system. Figure 1 shows a set of referral reviews in Google Play. These reviews are advertising a referral code in the ChampCash app with identical text and different referral codes. Most often, There is a notable difference in the average rating between referral and non-referral reviews. For example, the app com.tapgen.featurepoints has 6,037 reviews at the time of writing, and 2,147 (35.6%) of them are referral reviews. The average rating of the referral reviews (4.73) and the remaining reviews (4.08) suggests that the referral reviews and ratings are creating an undesirable bias in the review system.

Existing work focus on detecting spam and collaborative frauds. Referral reviews are different; while fake and paid reviews are written by abusers for

© Springer International Publishing AG 2017
J. Cabot et al. (Eds.): ICWE 2017, LNCS 10360, pp. 351–359, 2017.
DOI: 10.1007/978-3-319-60131-1_20

Fig. 1. Referral reviews in Google Play found in an app called ChampCash.

untraceable incentives, the incentive from spreading referral code via reviews is obtained from the product (i.e. app) owner, thus it is traceable. All referral reviews are untrustworthy for their monetary motivation. However, some referral reviews are worse for their spamming and adverse nature. Our goal, in this paper, is to understand how abusers are going beyond few random reviews to manipulating referral reviews and impacting the trustworthiness of the review system.

In this work, we collect and analyze referral reviews from the Google Play store. We develop a system to detect and extract referral reviews. We design a parsing pipeline that extracts app names and code words mentioned in reviews with high precision. The key challenge in the extraction process is that app names contain variable number of words of many forms (e.g. abbreviations, languages) and parts of speech. For example, it is hard to differentiate the app name "Uninstall" from the phrase "uninstall". Moreover, apps are added, edited and deleted frequently in the Play store. We use a dictionary based technique to extract app names and code words in a highly precise manner. To estimate the impact of referral reviews, we tracked the apps that provide incentives continuously for six months (October, 2015 – March, 2016). We discover that the apps that employ a rewarding mechanism gain significantly high star-ratings and downloads compared to those that do not.

2   Background and Related Work

A review is *referral* if the writer of the review gains any benefit in writing the review. For example, the reviews in Fig. 1 are broadcasting referral codes, which, if used by some new users, can earn reward points for their writers. Note that both the Android and iTunes platforms provide the functionality for incorporating a reward system in apps. In January 2016, Google began providing app developers with lists of alphanumeric codes that can be used as promotional codes [2], and Apple has supported referral codes for several years. This service encourages developers to use promotional codes. The apps we identified as rewarding apps were using these codes before Google provided the service; thus while we expect more apps to start using this reward system, we have a snapshot of them in our dataset.

Current works focus on identifying fraud reviews and reviewers, while we focus on understanding the (potentially abusive) impact of incentives on online reviews. Existing work can be categorized based on the methodologies they adopt to detect frauds. Fraud detection using graphical/network structure is studied in [3–5] where authors exploit network effects and clique structures among reviewers and products to identify fraud. Text-based detection of fraud is studied to spot a fake review without having the context of the reviewer and reviewed product [6–8]. Temporal patterns, such as bursts, have been identified as a fraudulent behavior of businesses [9,10]. In contrast, our work looks at specific textual features of referral reviews such as referral codes, app mentions, and keywords related to a reward system. Our method also utilizes unique contextual features such as the number of downloads, the number of reviews, and the average rating, which help us gauge the impact of referral reviews.

The closest work to ours is finding fraud and malware apps in Google Play, FairPlay [11]. The article discusses a method to automatically find such apps using review-based features related to apps and their users. We focus more specifically on referral reviews and consider finding abusive users and apps which are taking part in this segment.

2.1 Data Collection

We have implemented a **two-stage** data collection process. In the *first stage*, we searched in Google play store for apps that could potentially use incentives using specific keywords. We have collected a set of 10,355 apps. For each app, we collect up to 4480[1] of the most recent reviews. In the *second stage*, we develop an algorithm to detect referral reviews and apps. We have identified 4,029 apps that have some referral reviews. To understand how these apps benefit in gaining downloads and positive ratings, we have monitored the apps continuously from October, 2015 to March, 2016. For each app, we collect its metadata (e.g. app size, app description, and rating) and developer information. The total number of reviews we have collected is 14,555,502. Each review contains title, body, date, rating, and author. The total number of unique users in our dataset is 10,327,089 users. In this stage, we have collected 74,013 referral reviews with codes.

2.2 Codes and AppNames Extraction

We develop an algorithm to detect referral reviews by identifying *codes* from the reviews. Obviously other kinds of referral reviews may exist; however, referral incentives are almost always implemented through promo codes, which gives us a significant coverage on referral reviews. We first manually generate a *blackList* and *whiteList* based on extensive examination of the dataset. The *blackList* is used to identify reviews that likely contain an app code. Some example terms from the *blackList* are: `points`, `referral`, and `code`. A *whiteList* is used to identify reviews that may contain a string that could be confused for a referral code. Some example terms from the *whiteList* include `barcode`, `PayPal`, and `zip code`.

---

[1] The limit is set by Google Play.

In most development platforms, codes are a random sequence of numbers, alphabets, or a combination of alphabets and numbers. We retain English reviews that contain at least one keyword from the *blackList*, then perform different checks to test if any mentioned word is a code. We use *whiteList* to exclude cases where a word can be confused for a referral code such as game scores, reward points, or abbreviations (e.g. `mp3`, `Car4you`, and `galaxy4`).

In addition to codes, abusive reviews also contain references to other apps. Knowing the app that a review is referring to will help us to measure the impact of the reward system in that referred app. Our system generates a list of app identifiers from the metadata (i.e. app title). Usually, app names are long and app developers tend to put keywords in the title describing functionality (e.g. AppCoins (How to make money)). When users refer to an app in reviews they usually use the first couple of words without mentioning the whole title. To cover all possible cases, we generate different app identifiers from each app's title. We have generated 20,682 app identifiers from 10,355 app titles. Each identifier is tagged with an appID that connects the exact app with its identifiers. Next Step takes the app identifiers and a review as input, and outputs the app name that appears in the review. We find that 2.4% of the referral reviews reference to other apps.

We detect promotional reviews with 91% precision and extract codes with 93% precision. We detect and extract the app names with 95% precision. The precision values are calculated over an unbiased sample of one hundred reviews evaluated by two judges. Note that calculating the recall rate is impossible because there is no ground truth. We also argue that our analysis does not depend on the recall rate as we have thousands of users, apps, and reviews, which are precise and large enough for accurate statistical analysis. We have provided all of our code, data, and spreadsheet of results in our supporting page [12].

3   Comparative Study Among App Groups

In this section, we categorize the apps into five groups and perform a comparative study to understand them better. The groups are: *non-promoting, promoting, source, non-promoting target*, and *promoting target* apps. Below we formally define them.

Sources: Source apps are apps that have been mentioned in promotional or referral reviews written on other apps' review pages at least once. Source apps can have some promotional reviews in their own pages. We find 25 such apps. In Fig. 2a we show the distribution of the source apps over various app categories in Google Play. The most frequent source-type is entertainment, while source apps exist in six other categories.

Promoting Targets: An app is *"targeted"* by a source app when users write reviews in the target app about promotions in the source app. If a target app has a rewarding system implemented, we call them promoting targets. A promoting target app can also be a source app in some reviews. We use a threshold of minimum five targeted reviews to separate a source from a promoting target. We find 126 apps in this category.

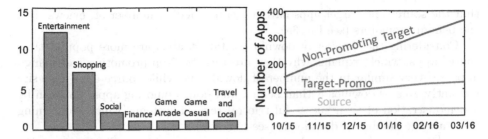

Fig. 2. (a) Source Apps Categories. (b) Growth trends of three groups of apps related to referral reviews.

Non-promoting Targets: Non-promoting targets are apps whose review page has been abused by some reviewers and have not implemented a rewarding system. We find 361 apps that are being targeted "by" source apps. These apps are mostly popular apps from top developers including Skype, Facebook, Twitter, Google, Amazon, and eBay.

Benign-Promoting Apps: Benign promoting apps have a rewarding system implemented and have reviews with promotional or referral codes. However, they are not sources or targets. We have 1150 apps in this category. We call the apps benign to distinguish them from the sources and targets. In reality, they are also abused by the reviewers.

Non-Promoting Apps: A set of randomly selected 6,328 apps that have no referral or promotional codes in reviews. We have collected the reviews for non-promoting apps during the period between October 2015 to March 2016.

Table 1 shows the summary statistics for the five app groups. Source apps have the highest average rating with the lowest variance.

Table 1. Statistics for app groups

	Benign promoters	Sources	Targets-Promo	Targets-NonPromo	Non-promoter
Number of apps	3,408	25	126	361	6,328
Average rating	3.98	4.17	3.99	4.02	3.99
ccStandard deviation rating	1.46	1.39	1.48	1.45	1.45
Number of promotional reviews	23,643	13,353	32,926	1,724	-

3.1 Feature Comparison

We compare the five groups of apps based on their total number of reviews, the number of downloads, burstiness and the number of promotional reviews. We show the results in box-plots in Fig. 3.

Based on the number of reviews, we see that the apps participating in reward systems have more reviews than random non-promoting apps. We also observe that the source and target apps have a greater median number of reviews than the benign promoters (see Fig. 3a).

Considering the number of downloads, target apps are more popular than source apps, which explains why they are targets. Non-promoting and benign apps are very similar in the number of downloads, while source apps have significantly greater downloads than benign and non-promoting apps. This can be a demonstration of their successful referral reward systems, which are earning them a large number of downloads (see Fig. 3b).

In [10,13], authors have shown that bursts of reviews indicate spamming activities. We measure the maximum and average number of reviews an app has received in a day and take their ratio as a measure of "burstiness." We see a relatively high burstiness in source apps compared to the benign promoters. Non-promoting targets, which are also popular apps, show similar burstiness as source apps (See Fig. 3c). If we only consider the number of promotional reviews, we identify that benign apps have very few promotional reviews while source apps have a large number of such reviews. This is a significant difference that motivates further analysis on the source apps. Target apps, although having a large number of total reviews, show much less promotional reviews compared to the source apps because target apps mostly do not have their own referral systems (Fig. 3d).

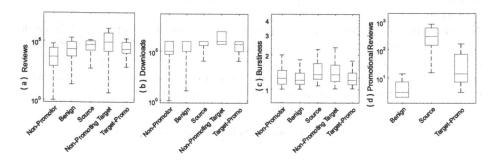

Fig. 3. Comparison among the five app groups based on four features.

Although we categorize source and target apps separately based on the reviews they have received, we have no evidence to say that the app owners have initiated such reviews.

3.2 Trend Comparison

As we demonstrate the significant difference between the app groups, we need to understand if the number of apps in the source and target groups is increasing. We show in Fig. 2b the trends for each group over the six months period of data collection. We observe that source apps are growing at a much smaller rate than the target apps. The most alarming fact is that the non-promoting target apps have almost doubled in six months. This suggests that we need to save non-promoting apps from abusive reviewers of promoting apps.

4 Discovering Abusive User Groups

In this section, we analyze the *users* who participate in writing referral reviews. We apply graph mining techniques to discover user groups who are involved in collaborative abuse. We also perform temporal analysis to understand trends in the users who are writing referral reviews. We create three different graphs connecting the review writers: *app-graph*, *text-graph* and *code-graph*.

App-graph: Two reviewers writing reviews for the same source app are connected by an edge. We expect a random graph to be formed in an unbiased review system where the reviewers mention other apps randomly without any bias.

Text-graph: We use Levenshtein distance [14] as a metric to measure text similarity. We set an error threshold of 6 for the distance function to allow an approximately 10% difference in a review as the average review length is 75 characters. An edge is added between two users if at least one pair of promotional reviews between the users has a distance less than or equal to the threshold. As shown in the Introduction, there are near duplicate reviews in the app reviews. The major reason for near duplicates is that writing an identical review to the most "helpful" review increases the chance of being ranked highly on the review page. If a group of users is posting similar text, we investigate further to identify if they are copying from each other. We find 6237 reviews using 401 unique templates by just changing the code part. The templates range from 6 to 497 characters in length, not including the code.

Code-graph: We add an edge between two users if they promote the same code. Codes are generated at random in an unbiased system, such edges, therefore, should not exist. However, we find many users who post the same code. Thus, code-graph creates an opportunity to spot groups of abusive users.

4.1 Clique Discovery

We consider a clique only if it has at least 3 users and set a minimum edge weight of 1 to find the cliques. We describe the largest cliques we have found in the three graphs defined above. In the app-graph, the largest clique was of size 346 users, which means all these users were referring to some common apps (not necessarily the same). In the text-graph, 36 users form the largest clique, and in the code-graph, the largest clique contains 65. we observe that the app-clique is disjoint to the code-clique and text-clique, while text-clique is a subset of the code-clique.

Table 2. 12 User Names from code-clique

Shreya Gupta	Shreya Gupta	chetan sahu	John Smith	Bhavna Sharma	Faqat Khan
Nancy Gupta	samita sah	chetan sahu	Harvey Dend	Sagar Sharma	Ankita Diwan

We perform a qualitative check on the code cliques. A random subset of 12 users is shown in Table 2. 100% of the reviews these users have ever written are promotional reviews, 82% have exactly 2 distinct codes and the remaining have one code and they all share the same codes (123900 and 201470). Three users have the same name and profile picture and 72% users have changed their profile names at least once.

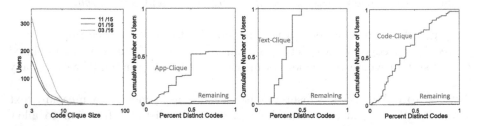

Fig. 4. (a): Size distribution of code-cliques over time, Remaining: Empirical CDFs for the three graphs.

4.2 Clique Properties

To perform more principled analysis on the cliques, we select a stricter edge weight of 10 and find the extreme incentivized users. We find 317 cliques in the app-graph, 37 cliques in the code-graph, and 6 cliques in the text-graph. For each graph, we compare the users participating in any clique against the remaining users who did not participate in cliques. We use the percentage of distinct codes over the number of promotional reviews. If the percentage is 100%, it means the reviewers are not reusing codes in their reviews. If the percentage is 20%, it means the reviewers are, on average, writing five reviews per referral code. We show the CDF (cumulative distribution function) of this metric over all the users who are in some clique (see Fig. 4). We also show the CDFs for users who are not in any clique. There is a significant difference between the CDFs for all of the three cliques, demonstrating that the users forming cliques are reusing promotional codes in multiple reviews.

We show the size distribution of the cliques from the code-graph in Fig. 4a. Naturally, we have a large number of small cliques and a few large cliques. We show three distributions for three datasets accumulated at two months interval. We identify a trend in both number and size of the cliques. This is an alarming indication that the number of abusers is growing rapidly.

5 Conclusion

This paper identifies a new type of abuse in review systems. Mobile apps support referral rewards, which create opportunities for users to write incorrect, untruthful, and abusive reviews. In this paper, we identify referral reviews, and analyze them to understand the following: ⋆ the number of referral reviews is rapidly increasing, ⋆ and apps are indirectly benefited from referral reviews in terms of the number of reviews and downloads.

References

1. Champcash: champcash (2015). https://play.google.com/store/apps/details? id=com.cash.champ
2. Google: developer console promotional code terms of service (2016). https://play. google.com/about/promo-code-developer-terms.html
3. Akoglu, L., Chandy, R., Faloutsos, C.: Opinion fraud detection in online reviews by network effects. In: Proceedings of ICWSM, pp. 2–11 (2013)
4. Chau, D.H., Pandit, S., Faloutsos, C.: Detecting fraudulent personalities in networks of online auctioneers. In: Fürnkranz, J., Scheffer, T., Spiliopoulou, M. (eds.) PKDD 2006. LNCS (LNAI), vol. 4213, pp. 103–114. Springer, Heidelberg (2006). doi:10.1007/11871637_14
5. Wang, G., Xie, S., Liu, B., Yu, P.S.: Review graph based online store review spammer detection. In: Proceedings - IEEE International Conference on Data Mining, ICDM 2011, pp. 1242–1247 (2011)
6. Ott, M., Choi, Y., Cardie, C., Hancock, J.T.: Finding deceptive opinion spam by any stretch of the imagination. In: Proceedings of HLT 2011, pp. 309–319 (2011)
7. Jindal, N., Liu, B.: Opinion spam and analysis. In: Proceedings of WSDM 2008, pp. 219–230. ACM, New York (2008)
8. Sun, H., Morales, A., Yan, X.: Synthetic review spamming and defense. In: KDD 2013, p. 1088 (2013)
9. Xie, S., Wang, G., Lin, S., Yu, P.S.: Review spam detection via temporal pattern discovery. In: Proceedings of KDD 2012, p. 823 (2012)
10. Fei, G., Mukherjee, A., Liu, B., Hsu, M., Castellanos, M., Ghosh, R.: Exploiting burstiness in reviews for review spammer detection. In: Proceedings of ICWSM, pp. 175–184 (2013)
11. Rahman, M., Rahman, M., Carbunar, B., Duen, H., Chau, G., Tech: fairplay: fraud and malware detection in google play. In: Proceedings of the 2016 SIAM International Conference on Data Mining, pp. 99–107. SIAM (2016)
12. Supporting Page - supporting webpage containing experimental result, data and code. http://www.cs.unm.edu/~nabuelrub/referral_reviews/
13. Minnich, A.J., Chavoshi, N., Mueen, A., Luan, S., Faloutsos, M.: TrueView: harnessing the power of multiple review sites. In: Proceedings of WWW 2015, pp. 787–797 (2015)
14. Wagner, R.A., Fischer, M.J.: The string-to-string correction problem. J. ACM (JACM) 21(1), 168–173 (1974)

Towards Stochastic Performance Models
for Web 2.0 Applications

Johannes Artner, Alexandra Mazak$^{(\boxtimes)}$, and Manuel Wimmer

Business Informatics Group (BIG), TU Wien, Vienna, Austria
{artner,mazak,wimmer}@big.tuwien.ac.at

Abstract. System performance is one of the most critical quality characteristics of Web applications which is typically expressed in response time, throughput, and utilization. These performance indicators, as well as the workload of a system, may be evaluated and analyzed by (*i*) model-based or (*ii*) measurement-based techniques. Given the complementary benefits offered by both techniques, it seems beneficial to combine them. For this purpose we introduce a combined performance engineering approach by presenting a concise way of describing user behavior by Markov models and derive from them workloads on resources. By means of an empirical user test, we evaluate the Markov assumption for a given Web 2.0 application which is an important prerequisite for our approach.

Keywords: Web application performance engineering · Markov models · Queueing theory

1 Introduction

Performance quality describes the degree to which a Web application meets its performance requirements. Typically, performance is expressed in *response time*, *throughput*, and *utilization* of the application [10]. In this paper, we concentrate on performance quality with respect to *response time*, since this is a central characteristic for users [10]. It is quantified by measuring the time between sending a request and receiving the response (e.g., response time of components/operations).

There are approaches for performance evaluation that are typically used "off-line" like the traditional performance analysis cycle [8]. In practice, performance problems mostly occur after the development of an application which means when the application is deployed and running. In addition to off-line techniques *performance management* combines measuring, analyzing, and improving measures as well as automates their interplay [11]. However, performance management is often done ad-hoc by trial & error, rather than based on engineering

---

A. Mazak and M. Wimmer are affiliated with the CD-Labor MINT at TU Wien funded by the Austrian Federal Ministry of Science, Research and Economy (BMWFW) and the National Foundation for Research, Technology and Development (CDG).

© Springer International Publishing AG 2017
J. Cabot et al. (Eds.): ICWE 2017, LNCS 10360, pp. 360–369, 2017.
DOI: 10.1007/978-3-319-60131-1_21

principles [6]. Performance indicators as well as the workload of a system may be evaluated and analyzed by using (predictive) model-based or (descriptive) measurement-based techniques [5]. Both approaches have different limitations that may be mitigated by each other, e.g., to integrate empirical measures in predictive models over time to re-estimate these models and to use predictive models to quickly analyse performance aspects already in the design phase instead of running costly and time-consuming tests on the implementation level.

The essential and most basic elements for describing the performance characteristics of Web applications are (i) resources that offer computing capabilities, (ii) the workload that describes how the resources are being used, and (iii) the workload intensity in terms of inter-arrival times. Based on these elements, we introduce an stochastic-based approach for evaluating the performance of Web Applications by systematically combining model-based and measurement-based performance techniques. Thereby, we measure, predict, and evaluate performance indicators by timely responses of components. We outline the approach by means of a real example for a typical Web 2.0 application. Furthermore, we present a new concept for describing user behavior by Markov models to derive workloads on resources.

The remainder of this paper is structured as follows. The next section presents an overview of relevant concepts underlying our approach. In Sect. 3, we introduce the main concepts of the approach. In Sect. 4, we present a Web 2.0 application designed for travellers and an empirical user test based on this application. In Sect. 5, we discuss related work, and finally, we conclude with an outlook on future work in Sect. 6.

2   Background

In this section, we briefly describe the theoretic background and main building blocks necessary for the context of this paper.

Performance Engineering. PE is an engineering discipline within systems engineering. It represents the whole collection of engineering activities and related analysis used throughout the development cycle to meet performance engineering requirements. In this context, it is necessary to distinguish between two types of models, *design models* and *performance models* [21]. Design models primarily focus on system topologies, workflows and interactions between entities in a prescriptive way. On the other hand, performance models focus on the evaluation of performance by using simulations, or other analytical means (i.e., queueing networks). As mentioned before there are two different approaches for performance evaluation, the *model-based* and the *measurement-based* ones. Typically, the model-based approach is carried out in an early project phase to predict the performance quality of the system, while the measurement-based approach is performed after the system is deployed [21].

Markov model. A Markov model is a stochastic model that fulfills the Markov property which implies that the future is independent of the past, given the present [7]. This means that all necessary information is encoded in the present state, and therefore, information about the past is not needed. This is why the Markov property is also known as "memoryless". A more formal definition of the Markov property is given here:

$$\Pr(X_{t_k}|X_{t_{k-1}}, X_{t_{k-2}}, ..., X_{t_1}) = \Pr(X_{t_k}|X_{t_{k-1}}) \tag{1}$$

where t_k denotes a set of times $t_k > t_{k-1} > ... > t_1$.

Many theories are built upon this simplified Markov model of first order. However, for modeling real-world systems it is important to check the Markov assumption, whether the Markov property holds or not. The simplest form of a Markov model is a *Markov chain*. Another special form of a Markov model is a *Hidden Markov Model (HMM)*, where states cannot be directly observed, and therefore, they are "hidden".

Queueing theory. The queueing theory provides a formalism to describe systems in which "waiting" plays a key role. Waiting lines occur whenever the demand exceeds the service availability. The main goals of queueing theory are prediction as well as proposing design improvements of systems. There are two basic elements in a queueing system: (*i*) a number of (limited) resources that are capable of executing tasks, called *servers*, and (*ii*) *customers* that request tasks handled by servers. In terms of software systems, a server might be, e.g., an image server or a CPU. Examples for customers in the context of Web applications are users browsing a website.

3   Stochastic Performance Models for Web Applications

In this section, we present our approach based on two metamodels to systematically combine model-based with measurement-based techniques. This hybrid approach bases on the assumption that the software development process is done in an agile manner. In this context "agile" means that parts of the software system can be continuously delivered to a test- or production environment in which the running software can be profiled. This means that as much information as possible is gathered from continuously observing parts of the running system. Based on these observations a mix of predictive models and measured data can be derived for computing performance indicators.

CETO (Components Emission and Timely Observations). CETO represents the central data format. It is enriched over time with static information like the functionality and topology of the system as well as predictive and measured data of user behavior. After every development iteration the system is deployed and profiled. In particular, CETO tracks the observations of use case and operation calls and the duration time of operations. There are existing models such as the *Core Scenario Model* [14], or the *Performance Model Interchange Format* [19] that also provide capabilities to model static and predictive data, but no measurement data during development.

MUPOM (Markov Usage Process and Operation Measurements). MUPOM is designed with the goal to model stochastic processes of user behavior. We use this model to describe user behavior by applying the Markov formalism. A single state in this model represents a single use case. The user behavior combined with the workload intensity represents the workload of the system. This workload is described by arrival rates of resources, which we compute by using *queueing networks (QN)*. The arrival rate of resources is determined by external and internal arrivals. There are three available distributions to model these arrivals: Poisson, Exponential, and Normal distribution.

Transfer CETO to MUPOM. In order to combine the CETO and MUPOM metamodels, we apply *model-to-model transformations* as introduced in Berardinelli et al. [2]. We apply these transformations to transform: (*i*) the use cases and operation topology to CETO, (*ii*) CETO to MUPOM, and (*iii*) MUPOM to QN. This means that all necessary information is kept in a CETO model and solved by transforming it to a MUPOM model. To utilize these transformations the following ten steps are needed: (*i*) to (optional) classify user types, (*ii*) to (optional) unveil the underlying Markov chain, (*iii*) to canonicalize operation durations, (*iv*) to calculate probability distributions for think times and operation durations, (*v*) to construct the transformation matrix P_{ij} of the Markov model, (*vi*) to check whether the Markov property holds, (*vii*) to calculate the steady state distribution, (*viii*) to (optional) derive the causal orderings and parallelisms of operations, (*viv*) to check whether the system satisfies the product form assumption of QN, and finally, (*vv*) to construct a performance model.

Transferring Markov Models to QN. Resource utilization in software systems is mainly a result of user behavior. The MUPOM model, besides the Markov model itself, also contains the solution for a steady state if it exists. A Markov chain has a limiting distribution if it is irreducible and all its states are positive recurrent [17]. If such a limiting distribution exists, it can be referred to as *steady state* that describes the long-run behavior of a system independent of the starting state. For calculating the steady state in our approach we use approximations.

In order to transfer a Markov models to QN, we have to achieve the characteristics of an *ergodic* system. Such a system requires the following assumptions that must hold: (*i*) *irreducibility*, if it is possible to reach each state from any other state; (*ii*) *aperiodicity*, if the system state is not systematically connected to time; and (*iii*) *recurrence*, if all states are recurrent. The reasons therefore are defined in *Little's law*, which is a very prominent operational law applied to queueing theory. It states that the long-term average number of customers in a stable system L is equal to the long-term average effective arrival rate λ multiplied by the average time a customer spends in the system W; or expressed formally: $L = \lambda W$.

Given these findings, we are able to translate MUPOM models to QN. In a first step, we calculate the average number of users in a system by applying Little's law:

$E(N) = \lambda * E(T)$. In a second step, we calculate the average number of users in each state by the approximated steady state distribution (π): $N_i = \pi_i * E(N)$.

The Markov chains in the MUPOM model are described in discrete time. The time step is in steps of seconds. Since, we know how many users are in each stage on average, the transition rates between each state can be simply calculated by multiplying the average number of users with the corresponding transition probability: $\lambda_j = p_{ij} * N_i$. In a final step, we calculate the total number of new enterings on average to a state per second and sum up all of these incoming transitions: $\Lambda_J = \sum_{j=1}^{k} \lambda_j$. The final outcome is the sum of all arrival rates to a certain state. By the inverse, we get the inter arrival rate, which forms together with the service rate, the basis for analyzing queuing networks for the purpose of performance evaluation.

4   Evaluating the Markov Assumption for an Example Application

In order to combine the model-based with the measurement-based approaches of performance engineering, we defined a transformation chain based on ten steps (cf. Sect. 3). Step (*vi*) of this chain checks whether the Markov property holds when transferring CETO to MUPOM. Due to page limitations, we primarily focus on this step in our case study. The evaluation of the validity of the Markov assumption is an essential pre-condition that must hold before we can check whether the Web application satisfies the product form assumption of QN to finally construct a stochastic performance model. For more insights, we refer the interested reader to our project website[1] where all artifacts of the case study, including the Travelistr software, are available.

Research question. *Is it sufficient to describe the transition probability between pages/features of a Web 2.0 application with Markov models and does the transition probability fulfill the Markov assumption?* Thereby, we assume that Web 2.0 applications are ergodic system. And if not, they can be transferred to one.

Case Study Design. As an essential part of our case study, we evaluate the Markov assumption for an example Web 2.0 application. For this purpose we use the approximative approach introduced in Li et al. [13] as a basis. In particular, we compare transition probabilities of a Markov model of first order to transition probabilities of a Markov model of second order. If the probabilities of order one and two are not diverging more than a certain threshold, the Markov assumption is fulfilled.

Selected Web 2.0 Application. *Travelistr* is a Web 2.0 application designed for travellers. It enables people to share pictures of their journey with other travellers that are nearby. Pictures can be enriched with information, e.g., of the geographic location of the photograph. Interested users get three pictures and

---

[1] http://www.johannes-artner.at/#ASPE.

are able to like one of them. Travelistr is a JavaEE application based on Spring Web MVC. For database access, Hibernate is used together with the c3p0 connection pool. The frontend is created with static HTML pages and dynamic JSP pages. Log4j is used as logging framework and OperationsAndTraceMonitor instructions are added at the language level. Travelistr is hosted at an Apache Tomcat server. The data is stored in a PostgreSQL database and the images are stored using the external vendor Cloudinary.

Tool Setting for Data Collection. Based on the CETO and MUPOM metamodels, we implement two Java-based tools to track and analyze observations, the *OperationsAndTraceMonitor* tool and the *UserTrace2Markov* tool. The observational data is gathered by using the *UserTrace* component of the *OperationsAndTraceMonitor* tool. It writes these observations to a CSV-file in a thread-safe manner in order to provide an appropriate format to use it as input to other tools. The *UserTrace2Markov* tool is a semi-automatic tool to transform user traces, tracked with the *OperationsAndTraceMonitor*, from the CSV-format to a Markov model of first and second order.

Results. We represent the results of applying our approach for the given set up. The user test took place for one week and in total 32 users participated. The participants of our evaluation study were students from our institute. They uploaded 146 pictures, and 839 *Likes* were received. 4520 transitions between states were observed within 173 distinct user interactions. Two consecutive transactions from a user are considered to be in the same user interaction if their timely difference is equal or less than 120 s. In its first version, the action of the *UserTrace2Markov* tool can only be of type *GET*. The results of the analysis are written to a txt-file (i.e., path to the output file). The tool calculates transition rates between states for a first and second ordered Markov model. Furthermore, think times in every state are calculated and described by mean values on basis of a standard normal distribution. Additionally, the total aggregated time spent in each state by every user is summed up. This is useful for analyzing the steady state assumption. The figure showing the results of the empirical user test is available for downloading under http://www.sysml4industry.org/?download=811.

To distinguish users, a randomly generated ID was assigned to each session. A user session was never invalidated during the whole test. So, if a user hasn't logged out on purpose, the user could continue where she left previously without logging in again. Table 1 shows the results of this approximative evaluation of the Markov assumption. A sample trace of every state and transition logged during the user test is available at http://www.sysml4industry.org/?download=822.

Interpretation of Results. We conclude that Markov models of order one are good approximations for describing the user behavior of typical Web 2.0 applications. The Markov assumption is an important prerequisite and holds for the Travelistr-App. Thus, it can be assumed that it also holds for other Web 2.0 applications.

Table 1. Results of comparing the first-order with second-order Markov model solutions.

State	Avg. information loss from 2nd to 1st order model	Observed transitions (n)
Nearby	4.36%	1067
Dashboard	11.45%	506
Profile	6.72%	193
Publish	2.87%	190
Published	1.77%	153
Start	7.58%	128
Login	3.14%	106
Register	12.48%	35

Threads to Validity. The empirical user test for evaluating the Markov assumption was successful for the given example. However, only 32 users took part in the test and not every state/transition was visited as often as desirable. We are aware that this sample is not large enough, so that any bias may change the conclusion. To reduce the risk of evaluating insufficient data, only states and their respective transitions with more than 20 observed transitions should be taken into consideration for evaluating the Markov assumption. Furthermore, there is a risk that the users did not use Travelistr as they would, if Travelistr would be an application they would use in their daily life. An additional risk is the time period of the conducted user test. It can be assumed that the behavior of users over a longer period accordingly will change, e.g., due to a better understanding of the system. The fractions of some states (e.g., *Register*) is therefore assumed to be different in the long run. And last but not least the results may differ if a Web application with more functionality had been considered.

5   Related Work

There are measurement methods that need performance probes of different levels (e.g., resource level, application level), or request-based techniques to estimate specific parameters of performance models. Kraft et al. [12] present a high-level request-based measurement technique to estimate service resource consumptions based on two approaches, a linear regression method and a maximum likelihood function. Their approach has several advantages compared to lower-level approaches, e.g., request measurements are easy to obtain. Barham et al. [1] present a tool chain, named "Magpie", which extracts the workload of a software system based on hardware, middleware, or application level traces. The resulting traces are correlated to requests. The computing demand for the requests is calculated and dependencies are ordered based on causality. The results of this procedure are used as basis for computing performance models. *Kieker* [16] is a framework to continuously monitor and analyze a system's runtime behavior. It has two main components, *Kieker.Tpmon* and *Kieker.Tpan*. Kieker.Tpmon monitors and logs data, while Kieker.Tpan is an analysis component which logs the runtime behavior to reverse engineer it (e.g., by sequence diagrams, class diagrams).

Markov models and QN are two formalisms widely used for performance engineering of systems as, e.g., introduced in [3,6,7]. Also, Hidden Markov models (HMM) have been used in many different fields, predominantly in the field of speech recognition [15]. To use HMMs for performance evaluation of software systems is a relatively new application field [18]. Hoorn et al. [20] introduce approaches based on statistical modeling for generating probabilistic and intensity-varying workloads for Web-based software systems. They present three models: (*i*) an *application model* to specify possible sequences of usage by a hierarchical finite state machine, (*ii*) a *user behaviour mix-model* to define which user behaviors are used for computing a workload, and (*iii*) a *workload intensity model* to specify the varying numbers of users over time. Based on these models, the authors extend JMeter to a probabilistic-based workload generator. Jespersen et al. [9] evaluate the Markov assumption for mining Web usage. The authors examine the quality of rules derived from two example websites by using the *Hypertext Probabilistic Grammar (HPG)* model introduced in [4], which relies on the Markov assumption and is mostly used for website analysis. The authors define *similarity* and *accuracy* for quality evaluation. On the one hand, similarity compares the amount of rules that are derived by using HPG and that are equal to the true usage patterns. On the other hand, accuracy compares the derived probabilities of the rules with the true usage patterns. As a result, the authors suggest that Markov-based approaches are better suited for tasks which require less accuracy. In addition, Li et al. [13] present a simple approach for evaluating the Markov property of Markov chains to model user behavior. In their approximative approach, they compare transition probabilities only considering the current state to transition probabilities considering the current and one previous state. They demonstrate their approach for Web usage profiling.

6   Conclusion and Future Work

Using Markov models to express user behavior is not new. However, to express the usage profile by Markov models in a model-driven way by model-to-model transformations is to the best of our knowledge a novel idea. We systematically combine the model-based and the measurement-based approach in an agile manner. We use Markov models to define the user behavior of a system in a predictive way. This implies that the underlying stochastic processes of user behavior fulfill the Markov assumption. Therefore, we conducted an empirical user study to show that this assumption holds for our implemented Web 2.0 application.

We outlined how to retrieve QN from Markov models based on user behavior. For this, necessary conditions have to hold. In addition to the Markov assumption, this is the ergodicity of the system. For Travelistr, both conditions hold.

Other software system types may be of the same nature. In this context, it is required to understand the characteristics of software systems that are ergodic. Furthermore, it is of interest to extrapolate observed user interactions of a limited set of features to a user behavior model of an entire system, and how the measurements can be validated. The practical benefit would be that usage patterns can be quantified already during development. Besides enhancing the accuracy of operation durations, also the user behavior model may be adapted and enriched with findings from measurements. A possible solution might be to mock the behavior of missing features.

References

1. Barham, P., Donnelly, A., Isaacs, R., Mortier, R.: Using Magpie for request extraction and workload modelling. In: OSDI (2004)
2. Berardinelli, L., Maetzler, E., Mayerhofer, T., Wimmer, M.: Integrating performance modeling in industrial automation through AutomationML and PMIF. In: INDIN (2016)
3. Bolch, G., Greiner, S., de Meer, H., Trivedi, K.S.: Queueing Networks and Markov Chains: Modeling and Performance Evaluation with Computer Science Applications. Wiley, Hoboken (2006)
4. Borges, J., Levene, M.: Data mining of user navigation patterns. In: Masand, B., Spiliopoulou, M. (eds.) WebKDD 1999. LNCS, vol. 1836, pp. 92–112. Springer, Heidelberg (2000). doi:10.1007/3-540-44934-5_6
5. Franks, G., Al-Omari, T., Woodside, M., Das, O., Derisavi, S.: Enhanced modeling and solution of layered queueing networks. IEEE TSE **35**, 148–161 (2009)
6. Harchol-Balter, M.: Performance Modeling and Design of Computer Systems: Queueing Theory in Action, 1st edn. Cambridge University Press, New York (2013)
7. Hevizi, G., Biczó, M., Poczos, B., Szabo, Z., Takics, B., Lorincz, A.: Hidden Markov model finds behavioral patterns of users working with a headmouse driven writing tool. In: IJCNN (2004)
8. Jain, R.: The Art of Computer Systems Performance Analysis: Techniques for Experimental Design, Measurement, Simulation, and Modeling. Wiley, New York (1991)
9. Jespersen, S., Pedersen, T.B., Thorhauge, J.: Evaluating the markov assumption for web usage mining. In: WIDM (2003)
10. Kappel, G., Pröll, B., Reich, S., Retschitzegger, W.: Web Engineering. Wiley, New York (2006)
11. Kotsis, G., Pinzger, M.: AWPS – an architecture for pro-active web performance management. In: Hummel, K.A., Hlavacs, H., Gansterer, W. (eds.) PERFORM 2010. LNCS, vol. 6821, pp. 215–226. Springer, Heidelberg (2011). doi:10.1007/978-3-642-25575-5_18
12. Kraft, S., Pacheco-Sanchez, S., Casale, G., Dawson, S.: Estimating service resource consumption from response time measurements. In: VALUETOOLS (2009)
13. Li, Z., Tian, J.: Testing the suitability of Markov chains as Web usage models. In: COMPSAC (2003)
14. Petriu, D.B., Woodside, M.: An intermediate metamodel with scenarios and resources for generating performance models from uml designs. SoSyM **6**(2), 163–184 (2007)

15. Rabiner, L.: A tutorial on hidden Markov models and selected applications in speech recognition. Proc. IEEE **77**(2), 257–286 (1989)
16. Rohr, M., Van Hoorn, A., Matevska, J., Sommer, N., Stoever, L., Giesecke, S., Hasselbring, W.: Kieker: continuous monitoring and on demand visualization of Java software behavior. In: IASTED-SE (2008)
17. Serfozo, R.: Basics of Applied Stochastic Processes. Springer, Heidelberg (2009)
18. Souza e Silva, E., Leão, R.M.M., Muntz, R.R.: Performance evaluation with hidden Markov models. In: Hummel, K.A., Hlavacs, H., Gansterer, W. (eds.) PERFORM 2010. LNCS, vol. 6821, pp. 112–128. Springer, Heidelberg (2011). doi:10.1007/978-3-642-25575-5_10
19. Smith, C.U., Lladó, C.M., Puigjaner, R.: Performance Model Interchange Format (PMIF 2): a comprehensive approach to queueing network model interoperability. Perform. Eval. **67**(7), 548–568 (2010)
20. Hoorn, A., Rohr, M., Hasselbring, W.: Generating probabilistic and intensity-varying workload for web-based software systems. In: Kounev, S., Gorton, I., Sachs, K. (eds.) SIPEW 2008. LNCS, vol. 5119, pp. 124–143. Springer, Heidelberg (2008). doi:10.1007/978-3-540-69814-2_9
21. Woodside, M., Petriu, D.C., Merseguer, J., Petriu, D.B., Alhaj, M.: Transformation challenges: from software models to performance models. SoSyM **13**(4), 1529–1552 (2014)

Web Intelligence Linked Open Data
for Website Design Reuse

Maxim Bakaev[1(⊠)], Vladimir Khvorostov[1], Sebastian Heil[2],
and Martin Gaedke[2]

[1] Novosibirsk State Technical University, Novosibirsk, Russia
{bakaev,xvorostov}@corp.nstu.ru
[2] Technische Universität Chemnitz, Chemnitz, Germany
{sebastian.heil,martin.gaedke}@informatik.
tu-chemnitz.de

Abstract. Code and design reuse are as old as software engineering industry itself, but it's also always a new trend, as more and more software products and websites are being created. Domain-specific design reuse on the web has especially high potential, saving work effort for thousands of developers and encouraging better interaction quality for millions of Internet users. In our paper we perform pilot feature engineering for finding similar solutions (website designs) within Domain, Task, and User UI models supplemented by Quality aspects. To obtain the feature values, we propose extraction of website-relevant data from online global services (DMOZ, Alexa, SimilarWeb, etc.) considered as linked open data sources, using specially developed web intelligence data miner. The preliminary investigation with 21 websites and 82 human annotators showed reasonable accuracy of the data sources and suggests potential feasibility of the approach.

Keywords: Linked data quality · Software reuse · Web design patterns · Data mining · Model-driven development

1 Introduction

Nowadays, web engineering (WE) has become an established, multi-billion dollar industry, and the number of websites worldwide exceeded 1 billion, although only a quarter of them are believed to be truly active. Given the multitude of existing websites and the amount of work effort put into producing and debugging the respective code up to date, their reuse would seem to be an attractive opportunity. Reuse is consistently named by Software Engineering (SE) experts among the advances and techniques that increased programmers' productivity the most, but its applicability on a large scale is domain-dependent [1]. On the web, the current stage of the industry development implies "mass-production" of functionality (code) and design, while content and usability need to be "hand-crafted" and their reuse seems problematic. Simple reuse of code is enabled via development environments and content management systems/ frameworks, while more advanced methods involve self-organizing component-based WE (e.g. [2]), evolutionary programming, etc. Reuse of design, which is considered to

© Springer International Publishing AG 2017
J. Cabot et al. (Eds.): ICWE 2017, LNCS 10360, pp. 370–377, 2017.
DOI: 10.1007/978-3-319-60131-1_22

be even more promising than reuse of code [1] and that is in our focus in the current research work, started to attract special interest in the 1990s and at the time was popularly shaped as design guidelines or patterns.

Recently, after data mining and content mining, the term *design mining* came to denote automated extraction of design patterns and trends from large collections of design examples. In case of the potent *Webzeigeist* tool that implements a kind of *design search engine*, designs are structured form web pages that are conveniently collected from the WWW, and then direct, query-based or stream-based access to design elements can be performed effectively [3]. The Webzeigeist authors rightfully claim that "in a database of ten million pages, the likelihood (that a designer will find a useful example) increases", but one shouldn't fall into the same pit as the early Internet search engines that valued results' quantity over relevance. Populating the database with web designs or design patterns shouldn't be a problem, given the currently existing billion of websites, but selecting the ones appropriate to the project context doesn't seem to be resolved.

In design example repositories, like Webzeigeist, search can be carried out on rather technical design parameters, like page aspect ratio or element styles. Extensive libraries of website templates, which have been named the "killers" of web design for more than a decade now, encompass many advanced tools (e.g. [4]), but suffer from the same organizational issues, as virtually none of them can adequately perform search based on problem description or design context. Thus, feature engineering is generally not performed in such collections, and moreover, there seems to be no agreed set of features for website reuse. In addition, there's a problem with identifying values for these features, especially for a website you don't own – in this case project specifications and website use logs are not available for data mining.

So, our current paper is a study in progress dedicated to identifying a set of features important for reuse of website design and finding the ways to obtain their concrete values. Particularly, we explore the feasibility of "web intelligence" (WI) approach, where mining of the website code is supplemented with extraction of website-related data (we rather not call them "metadata", since it denotes a different thing in HTML) from external sources. In Sect. 2, we overview feature engineering process for websites and propose model-based UI development approach as the appropriate framework. Then, we describe the architecture and capabilities of the dedicated web intelligence linked open data miner that we developed. In Sect. 3, we test the formulated hypotheses on some WI data to make inferences regarding the data accuracy and choosing data sources of higher quality. Finally, we make the conclusions and outline directions for further research work in the field.

2 Method

2.1 Feature Engineering for Website Design Reuse

There's a general consensus that feature engineering (FE) is crucial in applied machine learning, building recommender systems, case-based reasoning, etc. [5]. The major stages of the conventional FE process can be identified as: forming the excessive list of potential features (e.g. through brainstorming session), implementing all or some of them in a prototype, and selecting relevant features by optimizing the considered subset. Then, the corresponding similarity (distance) calculation approaches may be used to retain, usually via AI methods, the website designs that are most relevant for the current web project and offer the best chance of reuse.

A fair amount of research works deal with feature selection for web pages, particularly for automated classification purposes [6, 7]. Indeed a web page is a technically opportune object for analysis, as it is represented in easily processable code (HTML, CSS, etc.), but it's not self-contained, either goal-wise or in terms of design resolutions. So, we believe that FE for reuse should be performed for a whole website (web project) and that model-based (MB) approach to web UI development provides a good starting point for assembling the potential features (since for a conventional website user, web interface basically equals website). The MB paradigm identifies three groups of models: (1) per se interface models – Abstract UI, Concrete UI, and Final UI, (2) functionality-oriented models – Tasks and Domain, and (3) context of use models – User, Platform, and Environment. Of these, we consider the Domain, Tasks, and User of higher relevance to website design reuse and will apply them in the FE, while Platform and Environment models rather relate to website's back-office. Also, not all existing website designs are equally good (in contrast to e.g. re-usable programming code), so quality aspects must be reflected in the feature set. Let us further consider the selection of features and the corresponding similarity calculation approaches in more detail.

Domain: theoretically, the domain of a website may be inferred from its content, but this is quite complex and computationally expensive problem. Alternatively, website classifications in major web catalogues may be used, with the distance (similarity measure) defined as the minimal number of steps to get from one category item to another via hierarchical relations, which can be then divided by the "depth" of the item, to reduce potential bias for less specifically classified websites.

Tasks: extraction of tasks from website code is probably the best developed one in reverse WE (e.g. [8]), and the resulting model is generally specified in UML. In the simplest case, given that the domain is known, conventional tasks can be represented with the website chapter labels extracted from the code and arranged as tag cloud, with the subsequent employment of well-developed semantic similarity/distance methods [9].

User: stereotype modeling (with FOAF, WebML, etc.) implies identifying user groups and developing the corresponding user profiles or "personas", where features commonly include gender, age, experience, education level, etc. The methods for assessing similarity between profiles of website or social network users are reasonably well developed [10], but concrete demographics of the target users for someone's website aren't easy to obtain. It's naturally available in web project's specifications, while real user behavior patterns can be mined from access/interaction logs [11], but without access to either, a popular approach is employment of human annotators.

Quality: for the purposes of reuse, two dimensions of quality may be identified: (1) website's intrinsic quality of implementation – how well it was made from technical perspective and (2) quality-in-use – how well the website performs in online environment, satisfying target users in their tasks. The features for the former are reasonably well developed and quantifiable: website code correctness, accessibility, size of web pages, response times, etc. The latter closely matches the notion of usability whose concrete value is hard to auto-assess, but which is reflected in visitor behavior factors collected by web analytics services: bounce rates, average page views, conversion and completion rates, etc., though these data are generally not made openly available.

2.2   Linked Data Sources and Web Intelligence

Already more than a decade ago, the concept of publishing data in a semantic-aware, machine-readable form, ready for use by remotely accessing software, was shaped as Linked Data, and nowadays many web services, mashups, etc. rely on freely obtainable Linked Open Data (LOD). Finding an appropriate LOD source and estimating its quality is highly important in such web projects, but there's lack of research on the topic [12]. Some specific dimensions of LOD quality are: *Amount of Data, Conciseness, Completeness, Navigability* and *Interlinking* [12], but undoubtedly fitness for use and data accuracy are of foremost consideration for data users [13, 14].

As we mentioned before, the values for many of the features potentially significant for web design reuse are hard to determine in the absence of the website specifications and use statistics. However, virtually any operational website is regularly explored by crawlers, robots, spiders, etc. of numerous global web services, and is presented in web catalogues and search/indexing systems. For example, DMOZ catalogue claims to contain more than 1 million hierarchically-organized categories, and the number of included websites is about 4 million, which implies reasonably detailed classification – far more thorough than most website content analysis approaches could provide. Further, global web "aggregating" services, such as *Alexa* or *SimilarWeb*, are capable of indirectly estimating certain quality-related parameters even for websites with closed web statistics. Since the "fingerprints" of most websites are all around the web, the term *Web Intelligence* may be loosely applied to the process of website-related data gathering from LOD sources and their accuracy cross-checking.

To automate data collection, we developed a prototype WI miner capable of extracting data from specified locations, structuring them, and keeping them in the database. The current version (at http://webmining.khvorostov.ru) receives a website URL as the input, collects and structures the data (presented in Table 1), then outputs and stores them in the database. The main classes of the prototype, which in general correspond to the model-view-controller (MVC) pattern, are:

- AbstractController – abstract class for application controllers;
- SiteController – controller that displays the results;
- SiteAjaxController – controller responsible for processing AJAX queries;
- DBData – the model component class that interacts with the DB;
- IMiner – interface for implementation by all the classes related to mining (AlexaMiner, SimilarMiner, SectionsMiner, DMOZMiner, etc.);
- MinerFabric – factory class that returns the object of the necessary class for mining;
- AbstractHtmlParserMiner – abstract class for the miners that parse HTML.

Table 1. Web Intelligence data collected by the WI miner

Model/Realm	Features	WI sources
Domain	1. website category	DMOZ, SimilarWeb
Tasks	1. website chapter names, 2. number of website chapters	The website code (main navigation only)
User	1. demographics (Male, Female, No College, Some College, Graduate School, College), 2. Flesch-Kincaid Grade Level ($\sim$age)	Alexa (1), readability-score.com (2 – homepage only)
Quality	1. number of errors and warnings, 2. page load time, 3. bounce rate, 4. popularity rank (global), 5. number of visits	validator.w3.org (1 – homepage only), Alexa (2, 3, 4), SimilarWeb (3, 5)

2.3 The WI LOD Accuracy Investigation

The goal of our investigation was to perform preliminary analysis of the LOD sources accuracy, by testing them against some "common sense" from the WE field. To this end, we decided to supplement the data collected by the WI miner with website usability evaluations provided by human annotators, considered to be representative of the quality-in-use. Specifically, we employed official websites of 11 German and 10 Russian universities (all English versions) and 82 annotators representing the target user group (more detailed description of the experimental setup can be found in [15]). The reason we decided not to vary the Domain was that accuracy of website category data is obvious (the validity of the similarity measure based on these data is a different issue). So, the following hypotheses (H_i) were formulated for the WI LOD:

Cross-checking: analogous values provided by Alexa and SimilarWeb (the two *bounce rates* and *popularity rank* vs. *number of visits*) should correspond to each other (H_1).

Domain: none – the factor was fixed as *Career and Education* (SimilarWeb).

Tasks: more straightforward *website chapter names* should result in lower *bounce rates* (H_2) and higher *usability evaluation* (H_3).

User: since web content is important in user subjective impression of a website, *Flesch-Kincaid Grade Level* should affect *usability evaluation* (H_4). Given the target audience of university websites and presumably their dedicated usability engineering, higher share of *College*-level users should result in higher *usability evaluation* (H_5).

Quality: the technical quality (*number of errors and warnings, page load time*) and the quality-in-use factors (*bounce rates, popularity rank, number of visits, usability evaluation*) should be positively correlated (H_6).

3 Results

3.1 The Data Validity and Cross-Checking

Our preliminary analysis of the data validity found one outlier – a website for which the extracted *number of visits* was at 28 visitors per month and *bounce rate* (SimilarWeb) was at 100%. So, the 20 websites (95.2% of the data) were valid for the analysis. Also, user education-related data extracted from Alexa was incomplete, as only *College* and *Graduate School* were available for all the websites.

H_1: correlation between *bounce rate* values extracted from Alexa and from SimilarWeb was r = 0.582 (p = 0.007), while correlation between the *popularity rank* and the *number of visits* was r = −0.600 (p = 0.005).

3.2 The Tasks Model

To pinpoint the tasks that correspond to the *Career and Education* domain (SimilarWeb), we identified 8 most typical chapter labels, which were found on 6 or more of the websites: *University/About us* (present on 21 websites), *Research/Science* (19), *International* * (12), *Study* (10), *Faculties* (8), *Prospective students/Admissions* (7), *Contacts* (6), *News/Media/Press* (6). Then for each website we divided the number of the typical chapters it has by the total number of chapter in its main navigation, thus receiving the website "conventionality" value (ranging from 0.4 to 1, average 0.622, SD = 0.159).

H_2: negative correlation between the website "conventionality" and the SimilarWeb *bounce rate* was significant (p = 0.002, r = −0.587), unlike for Alexa *bounce rate*. Also, we found no significant correlation with the *usability evaluation* (H_3).

3.3 The User Model

H_4: correlation between the *Flesch-Kincaid Grade Level* and the *usability evaluations* was significant (p = 0.01; r = 0.561), which may imply positive effect of sophisticated texts on website evaluation by the target group.

H_5: correlation between the *College* share and the *usability evaluations* was significant at $\alpha = 0.06$ (p = 0.056; r = 0.433),

3.4 Quality

H_6: somehow unexpectedly, we found significant negative correlation between the *number of errors and warnings* (summarized) and the *bounce rate* extracted from Alexa (p = 0.033; r = −0.479). Also, correlation between *popularity rank* and *page load time* was significant at $\alpha = 0.08$ (p = 0.078; r = 0.403).

Further exploring whether usability evaluations provided by annotators can be predicted by the mined WI LOD, we constructed the regression model with comprehensive list of factors, using the *Backwards* inclusion method. We selected the model that had the highest adjusted R^2, and it included 4 factors: the *Flesch-Kincaid Grade Level* (FK), the Alexa *College* share (C), the *number of errors and warnings* (E), and the *number of visits* (V, in millions). The model was significant (p = 0.01), but had moderate $R^2 = 0.559$:

$$U_{eval} = 2.94 + 0.09 * FK + 0.79 * C - 0.01 * E - 0.19 * V \tag{1}$$

4 Discussion and Conclusions

The general idea of design reuse seems to be repeatedly re-invented at different stages of SE industry development under different names and in various sub-fields. Currently, existing website design repositories focus on technical, structural or stylistic aspects, but not on problem- or user-oriented ones; neither they contain a quality metric. Since code analysis alone can't provide values for most features important for retaining appropriate solutions for reuse, and employment of human annotators is restricted in scale and budget, our proposal is to gather website-related data (metadata, in non-technical sense) from linked open data sources. In this, our first step in the current pilot research work was to investigate the accuracy of the data sources and general feasibility of the approach.

To this end, we developed prototype "web intelligence" data miner capable of extracting data (see Table 1) for any given website from Alexa, SimilarWeb and certain other online services. To test the accuracy of the data, we performed automated data collection for 21 German and Russian university websites (*Career and Education* domain) and asked 82 human annotators who represented a target user category – students – to provide subjective evaluations of the websites' usability. More information on the experimental setup can be found in [15], while the extracted data and some supplementary materials are available at http://webmining.khvorostov.ru/docs.zip. In regard to the 6 hypotheses formulated for accurate data, the results of the analysis showed the following:

H_1 **(cross-checking):** effect found, at reasonably high statistical significance.

H_2 **and** H_3 **(Tasks):** effect found only for *bounce rate* provided by SimilarWeb (in line with usability guidelines about not making users think any more than necessary), which should be considered the preferred data source.

H_4 **and** H_5 **(User):** effects found (arguably appropriate for the target user group).

H_6 **(Quality):** few effects found, and the effect for Alexa was the opposite of what was expected. The regression model for the *usability evaluation* with the quality factors was significant (p = 0.01, $R^2 = 0.559$). *Usability evaluation* was negatively affected by the *number of errors and warnings* as well as by higher *number of visitors* (we can only speculate that websites of smaller universities may be better tailored to the needs of their target users).

So, we suggest that WI data mined from web analytic services, search engines, even advertisement networks, can indeed be a useful supplement to analysis of the actual website code and manual annotations, when website design similarity is evaluated for the purposes of reuse. Limitations of the current pilot research include quite a small number of employed websites (since we were basically limited by the effort of human annotators) and rather informal approach to data accuracy analysis. In our further research we also plan to focus on auto-engineering features (deep learning) and finding out the values for the User model that seems to be under-explored in modern literature, for which we plan to extend the capabilities of our prototype WI data miner to gather data from web analytic services. Moreover, the analysis of the Domain model should include both similarity calculation in hierarchical web catalogues and, possibly, natural language processing of titles and descriptions submitted by website owners.

Acknowledgement. The reported study was funded by RFBR according to the research project No. 16-37-60060 mol_a_dk. The authors also thank S. Firmenich and J.M. Rivero from LIFIA (Argentina) who contributed to the discussion of the paper topics.

References

1. Glass, R.L.: Facts and Fallacies of Software Engineering. Addison-Wesley Professional, Boston (2002)
2. Gaedke, M., Rehse, J.: Supporting compositional reuse in component-based Web engineering. ACM Symp. Appl. Comput. **2**, 927–933 (2000)
3. Kumar, R., et al.: Webzeitgeist: design mining the web. In: SIGCHI Conference on Human Factors in Computing Systems, pp. 3083–3092 (2013)
4. Norrie, M.C., Nebeling, M., Geronimo, L., Murolo, A.: X-Themes: supporting design-by-example. In: International Conference on Web Engineering (ICWE 2014), pp. 480–489 (2014)
5. Anderson, M.R., et al.: Brainwash: a data system for feature engineering. In: 6th Biennial Conference on Innovative Data Systems Research (2013)
6. Mangai, J.A., Kumar, V.S., Balamurugan, S.A.: A novel feature selection framework for automatic web page classification. Int. J. Autom. Comput. **9**(4), 442–448 (2012)
7. Saraç, E., Özel, S.A.: An ant colony optimization based feature selection for web page classification. Sci. World J., **2014**, 1–16 (2014). doi:10.1155/2014/649260, Article ID: 649260
8. Paganelli, L., Paterno, F.: A tool for creating design models from web site code. Int. J. Softw. Eng. KEng. **13**(02), 169–189 (2003)
9. Park, J., Choi, B.C., Kim, K.: A vector space approach to tag cloud similarity ranking. Inf. Process. Lett. **110**(12), 489–496 (2010)
10. Kosinski, M., et al.: Manifestations of user personality in website choice and behaviour on online social networks. Mach. Learn. **95**(3), 357–380 (2014)
11. Varnagar, C.R., et al.: Web usage mining: a review on process, methods and techniques. In: IEEE Information Communication and Embedded Systems (ICICES), pp. 40–46 (2013)
12. Cappiello, C., Di Noia, T., Marcu, B.A., Matera, M.: A quality model for linked data exploration. In: International Conference on Web Engineering (ICWE), pp. 397–404 (2016)
13. Zaveri, A., et al.: Quality assessment for linked data: a survey. Seman. Web **7**(1), 63–93 (2016)
14. Wang, R.Y., Strong, D.M.: Beyond accuracy: what data quality means to data consumers. J. Manage. Inf. Syst. **12**(4), 5–33 (1996)
15. Bakaev, M., Gaedke, M., Heil, S.: Kansei Engineering experimental research with University websites. TU Chemnitz Technical Report, CSR-16-01 (2016)

Exploratory Search of Web Data Services Based on Collective Intelligence

Devis Bianchini[✉], Valeria De Antonellis, and Michele Melchiori

Department of Information Engineering, University of Brescia,
Via Branze, 38, 25123 Brescia, Italy
{devis.bianchini,valeria.deantonellis,michele.melchiori}@unibs.it

Abstract. Developers of data-intensive web applications benefit from the integration of data sourced from the web. Web data services are solutions off-the-shelf, provided by third parties, that enable access to web data sources. Web data services are usually discovered according to different features, related to lightweight descriptions. Recent approaches in literature convey on new research challenges, considering also collective intelligence in developers' networks, containing information about service co-usage in existing applications and ratings on services given by developers who used them in their own development experiences. Following this direction, in this paper, we contribute with a distinguishing viewpoint, by proposing an *explorative approach*, that enables web applications developers to iteratively discover services of interest by also relying on collective intelligence, in a Web 2.0 context.

Keywords: Web data service model · Exploratory search · Collective intelligence · Web-oriented architecture

1 Introduction

Exploratory search techniques and tools, that enable users to browse and discover information shared over the web, facing the increasing volume and heterogeneity of available data, are attracting more and more interest from the research and industrial community [1]. Building data-intensive web applications more and more requires frameworks to support the discovery of web data services, that enable access to huge repositories of data and must not be developed from scratch (e.g., Google Maps), according to the Web-Oriented Architecture (WOA) style. In this context, data exploration is possible only through exploration of services used to access data.

Nevertheless, service exploration presents distinguishing features compared to data exploration. Firstly, applications are designed through a sequence of service selections, where a selection step starts from other services previously considered for the application that is being developed. Secondly, who is searching for services expects them to be provided in a specific and subjective way. This paves the way to approaches that, beyond functional and non functional

© Springer International Publishing AG 2017
J. Cabot et al. (Eds.): ICWE 2017, LNCS 10360, pp. 378–385, 2017.
DOI: 10.1007/978-3-319-60131-1_23

requirements, use collective intelligence to suggest services based on the service usage experiences of other developers [5–7]. Our contribution in this paper is the proposal of an explorative approach, that is modeled as a sequence of exploration steps between the system, used to select services, and the developer. The exploration takes care of the collective intelligence in developers' networks, containing information about past service usage experiences in terms of service co-occurrence and rating on services by other developers. The explorative approach is based on a multi-perspective model that includes: (i) a *service-base perspective*, focused on descriptions of services to be explored; (ii) a *service-experience perspective*, focused on the relationships between services and developers who used them to build their own applications and rated services according to their experience; (iii) a *service-abstraction perspective*, focused on features used to describe services (e.g., tags, categories); (iv) a *service-collective-intelligence perspective*, that gives an overview on the network of services as used by the community (developers' service usage experience to be exploited for enabling the exploration). The elements considered in the first two perspectives have been already defined in [2]. We add here the other perspectives as a basis to guide data service exploration. Developers are the main beneficiary of the approach, since they are provided with easy data service exploration to build new applications with minimal effort.

The paper is organized as follows. Section 2 provides the background of the approach and its motivations. Section 3 describes the multi-perspective model. In Sect. 4 we present the data service exploration procedure based on the model. Section 5 presents a proposal for an interactive interface for service exploration. Finally, Sect. 6 closes the paper.

2   Background and Motivations

The approach will be focused on web data services. It deals with available service descriptions (as published in public repositories, like `Mashape.com` and `ProgrammableWeb.com`), that are lightweight characterizations through textual descriptions, categories and technical features (such as protocols and data formats to use them). These lightweight descriptions, which relieve service providers from the burden of providing complex descriptions of supplied services, is one of the success factors for RESTful services. On the other hand, they made challenging the study of effective search facilities.

As an example, let's consider a developer who is developing a new application for hotel booking. Let's suppose that the developer is using `Programmable-Web.com` and starts by specifying `hotel booking` keywords to search for services. The repository returns 80 services[1], as shown in Fig. 1. The popularity of `Cleartrip Hotel` service (followed by 164 users) and applications (18 mashups) where hotel booking services have been used (highlighted in figure) are not exploited by the system and it is up to the developer to improve his/her search by considering them.

---

[1] https://www.programmableweb.com/search/hotel%20booking.

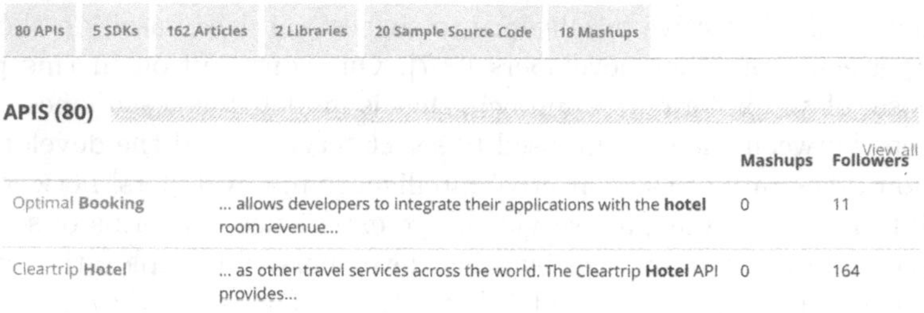
80 APIs
5 SDKs
162 Articles
2 Libraries
20 Sample Source Code
18 Mashups
APIS (80)
View all
Mashups
Followers
Optimal Booking
... allows developers to integrate their applications with the hotel room revenue...
0
11
Cleartrip Hotel
... as other travel services across the world. The Cleartrip Hotel API provides...
0
164

Fig. 1. An example of `hotel booking` service search using `ProgrammableWeb`.

Recent web service recommendation approaches rely on lightweight descriptions of services to overcome the complexity of traditional state-of-the-art approaches on data service discovery (e.g., [8]), that are hampered by the availability of complex, structured service descriptions (e.g., WSDL, WADL and semantic web service formalisms). They exploit categories, tags or semantic tags to search for services, with the application of advanced IR techniques to enhance topic-based service recommendation [3], natural language API description [4], latent factors (e.g., related to the perceived QoS) that affect users to make service selection, identified mainly using matrix factorisation techniques [6]. In this context, several approaches like leverage factors to estimate past experiences of service usage are considered, such as votes/ratings assigned by users to services [7] and the number of times a service has been used in the past and the co-occurrence of services in existing applications [5].

However, these approaches provide the developer with look-up search results, without taking into account decisions made by the developer during the development process by interacting with the search tool. Developers should be progressively supported in refining their requirements. They might not have still taken any decision about the use of additional criteria to search for hotels, such as the number of stars of the hotel (that may have an impact on the final price) as well as the proximity to a specific location. The system may help developers in refining the request, choosing among hotel booking services that accept further constraints such as the number of stars or the location, taking into account the popularity of different solutions and the kind of application that is being developed compared to existing applications where services have been used in the past.

3 Multi-perspective Model for Service Exploration

Starting from the motivations as presented in the previous section, we propose here a multi-perspective model to enable service exploration based on their lightweight descriptions and collective intelligence about their use. The model has inter-perspective relationships as shown in Fig. 2 and the explorative process presented in the next section is based on it.

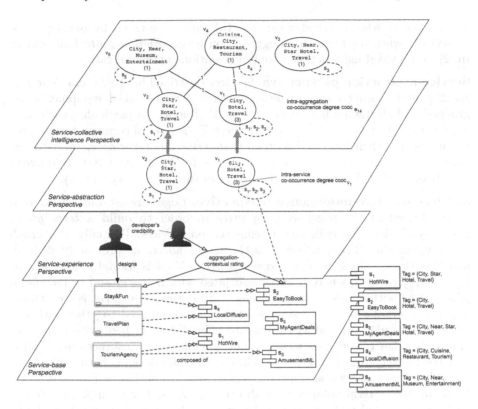

Fig. 2. Overview of the multi-perspective model for exploration purposes.

Service-base perspective. This perspective collects and describes services, aggregations and composition relationships between them. For the purpose of data service exploration, we describe a data service s_i as a set of tags $\mathcal{T}_{s_i}$ used to provide a terminological characterisation of the service (*terminological equipment*), as extracted from its lightweight description. For other service descriptive features, not explicitly mentioned here (e.g., categories, technical features like data formats and protocols), we refer to [2]. An *aggregation g* represents a set of services that will be integrated to deploy a web application. Concerning application development, according to the WOA style, developer has to explore the set of available services, select the most suitable ones, mashup them in the final application. We focus here on service exploration for selection purposes, and we talk about service aggregations, instead of web applications, where the latter ones are the final product of the development process. Figure 2 shows three aggregations, composed of five data services. Service terminological equipments are shown as well.

Service-experience perspective. This perspective is focused on the set $\mathcal{D}$ of developers who designed service aggregations. In this perspective we focus on collective knowledge related to aggregation design experience, modelled around two concepts: (a) the relationship between a service s, an aggregation g and a developer d, who designed g using s; (b) ratings assigned by developers to a service s when used within an aggregation g (*aggregation-contextual rating*). In [2] we provided more details about aggregation-contextual rating.

Service-abstraction perspective. Services described in the *Service-base perspective* and sharing common terms in their terminological equipments are grouped and abstracted here through a set $\mathcal{V}$ of nodes. In particular, each node $v_i \in \mathcal{V}$ is described as $v_i = \langle \mathcal{T}_{v_i}, cooc_{v_i} \rangle$, where $\mathcal{T}_{v_i}$ is a set of common terminological items (i.e., their intersection) used to describe a number $cooc_{v_i}$ of data services (*intra-service term co-occurrence degree*). In Fig. 2 {City, Hotel, Travel} have been used to describe three data services (namely, s_1, s_2 and s_3).

Service-collective-intelligence perspective. Edges are established between nodes defined in the previous perspective in order to build a *term graph*, which synthesizes the collective intelligence on services. Formally, the graph is represented as $\langle \mathcal{V}, \mathcal{E} \rangle$, where $\mathcal{V}$ is the set of nodes, as previously defined, and $\mathcal{E}$ is the set of edges. Each edge $e_{ij} \in \mathcal{E} \subseteq \mathcal{V} \times \mathcal{V} \times \mathbb{N}$ is formally described as $e_{ij} = \langle v_i, v_j, cooc_{e_{ij}} \rangle$, where services associated with nodes v_i and v_j have been jointly used within $cooc_{e_{ij}}$ aggregations (*intra-aggregation term co-occurrence degree*). In Fig. 2 $cooc_{e_{14}} = 2$ since service s_4 has been used together with services s_1 (TravelPlan) and s_2 (Stay&Fun). The difference here with respect to approaches that consider co-occurrence of specific service instances (e.g., [5]) is that the intra-aggregation term co-occurrence enables developers to explore services that have not been aggregated yet, but can be considered for aggregation based on their "term similarity" with other services. For example, service s_3 could be suggested to be used together with s_4 because of its term similarity with s_1 and s_2. This will enable a greater coverage of proposed solutions, at the cost of a lower precision, that can be acceptable in an explorative process. The term graph can be built and maintained in a fully automatic way, without the need of human intervention.

4   Data Service Exploration

We envision the service exploration process as a sequence of exploration steps between the developer and the system, used to search for services. The developer starts the exploration by specifying the set $\mathcal{T}^r$ of terms used within the search request, that provide some initial hints about developer's interests. The system suggests services by computing similarity, filtering and ranking techniques such as the ones introduced in [2]. Let's denote with $\mathcal{S}^e$ the set of search results, recommended by the system. The developer can modify the set $\mathcal{T}^r$ to look for new services or can choose some services from the search results $\mathcal{S}^e$ to include them in the work-in-progress aggregation g^r. The system reacts to developer's actions by supporting exploration according to three modalities.

– **Exploration by simple search.** The system also looks for nodes $v_i \in \mathcal{V}$ such that $\mathcal{T}^r \subseteq \mathcal{T}_{v_i}$. If multiple nodes are found, for each $v_i \in \mathcal{V}$ the system will suggest to the developer additional terms to be included within the set $\mathcal{T}^r$ considering the set $\mathcal{T}_{v_i} \setminus \mathcal{T}^r$. A suggestion is given for each $v_i \in \mathcal{V}$, ranked in

$\mathcal{T}^r$	$\mathcal{S}^e$	$\mathcal{G}^e$
{City, Hotel, Travel}	s_1 s_2 s_3	s_1

$$\Downarrow \qquad\qquad \Downarrow$$

$\mathcal{T}^r$	$\mathcal{S}^e$	$\mathcal{G}^e$		$\mathcal{T}^r$	$\mathcal{S}^e$	$\mathcal{G}^e$
{}	s_4	s_1		{}	s_5	s_1
$(cooc_{e_{14}} + cooc_{e_{24}} = 3)$				$(cooc_{e_{14}} + cooc_{e_{24}} = 2)$		

Fig. 3. Example of exploration by proactive completion.

decreasing order with respect to the $cooc_{v_i}$ value. The developer can explore these suggestions in order to consider services alternative to $\mathcal{S}^e$ and to formulate a different request. For instance, if $\mathcal{T}^r = \{\text{City, Hotel, Travel}\}$, the system might also suggests as additional terminological item the term $\{\text{Star}\}$ first $(cooc_{v_2} = 2)$, and $\{\text{Near}\}$ as second option $(cooc_{v_3} = 1)$. In this way, the developer might realize that hotels can be searched either based on the number of stars or based on the proximity to a given location and he/she might refine the request by choosing one of the two options.

- **Exploration by proactive completion.** The developer selects a subset $\overline{\mathcal{S}^e} \subseteq \mathcal{S}^e$ of services he/she is interested in. The system suggests services that could be used together with services in g^r, by updating the set $\mathcal{S}^e$, according to the *intra-aggregation co-occurrence*. Let's consider the example shown in Fig. 3. After performing a search based on $\mathcal{T}^r = \{\text{City, Hotel, Travel}\}$, thus obtaining $\mathcal{S}^e = \{s_1, s_2, s_3\}$ as results, the developer chooses s_1 to be included in g^r. With reference to Fig. 2, s_1 is associated with v_1 and v_2 nodes. Considering node v_1, other nodes connected to v_1 by graph edges are v_4 (associated with s_4, $cooc_{e_{14}} = 2$) and v_5 (associated with s_5, $cooc_{e_{15}} = 1$). Similarly, considering node v_2, $cooc_{e_{24}} = 1$ and $cooc_{e_{25}} = 1$. Therefore, the system ranks better the service s_4 than s_5, since $cooc_{e_{14}} + cooc_{e_{24}} > cooc_{e_{15}} + cooc_{e_{25}}$. The developer can accept one of these results. If more than one service is included in g^r, the step of retrieving services is repeated for each service in g^r.

- **Exploration by hybrid completion.** This explorative modality is a combination of proactive completion and simple search. After $\mathcal{S}^e$ has been updated, the developer selects a subset $\overline{\mathcal{S}^e} \subseteq \mathcal{S}^e$ of services he/she is interested in, as well as he/she specifies a new set $\mathcal{T}^r$ of terms. The system suggests services that could be used together with services in g^r, by updating the set $\mathcal{S}^e$. In order to obtain this set, a proactive completion step on g^r retrieves some services as explained before.

5   Considerations About the Exploration Interface

The interface of a tool aimed to support explorative search should better serve result set examination and item comparison, being conceived as a combination of

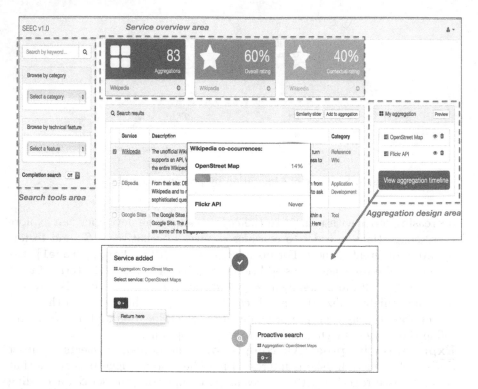

Fig. 4. Data service exploration web interface (with example of exploration timeline).

browsing and analytical strategies. Figure 4 depicts the web interface for explorative search of services. Faceted metadata (e.g., categories, technical features) provide an effective entry point for exploration and selection. Therefore, search and browsing facilities are provided to enable developers to search services by keywords, and browse available services by category or technical features (`Search tools area`). A flag also enables to activate completion search (either proactive or hybrid). When the developer moves the mouse on one of the search results, the number of uses of service and its rates are shown within the `Service overview area`, as well as co-occurrences of pointed service with other services in the aggregation that is being developed are shown in a popup window. This gives an immediate view of the search results suitability; the developer can move among search results according to overview information and perform multiple lookup searches, then he/she can select a service and include it into the aggregation that is being developed using the `Add to aggregation` button. By clicking on the `View aggregation timeline` button (on the right), the developer can browse in a timeline the history of his/her exploration steps, where for each step the status of the developed aggregation in that moment, the kind of search (simple, proactive or hybrid) and search criteria used in that step are shown. It is possible for the developer to return to any exploration step by selecting the `Return here` option on the step.

6   Concluding Remarks

In this paper, we proposed an approach for data service explorative search, based on collective intelligence in developers' networks, containing information about past service usage experiences in terms of service co-occurrence and ratings on services by other users. The approach is built on a sequence of steps, where the user receives suggestions from the system based on the past interactions. Future work will be devoted to the study of techniques for including latent factors (e.g., related to the perceived QoS) in the exploration process. Further open research challenges concern integration within the system, which implements the approach, of a query engine for multi-source data access and usability experiments on the visualization interface to further enhance the exploration experience of developers. This will move the approach towards non-expert users, who will be able to access the web of data without specific web application development skill.

References

1. Idreos, S., Papaemmanouil, O., Chaudhuri, S.: Overview of data exploration techniques. In: ACM Conference on Management of Data (SIGMOD) (2015)
2. Bianchini, D., Antonellis, V., Melchiori, M.: Capitalizing the designers' experience for improving web API selection. In: Meersman, R., Panetto, H., Dillon, T., Missikoff, M., Liu, L., Pastor, O., Cuzzocrea, A., Sellis, T. (eds.) OTM 2014. LNCS, vol. 8841, pp. 364–381. Springer, Heidelberg (2014). doi:10.1007/978-3-662-45563-0_21
3. Cao, B., Liu, X., Li, B., Liu, J., Tang, M., Zhang, T.: Mashup service clustering based on an integration of service content and network via exploiting a two-level topic model. In: Proceedings of the 23rd International Conference on Web Services (ICWS 2016) (2016)
4. Xiong, W., Wu, Z., Li, B., Gu, Q., Yuan, L., Hang, B.: Inferring service recommendation from natural language api description. In: Proceedings of the 23rd International Conference on Web Services (ICWS 2016) (2016)
5. Gao, W., Chen, L., Wu, J., Bouguettaya, A.: Joint modeling users, services, mashups and topics for service recommendation. In: Proceedings of the 23rd International Conference on Web Services (ICWS 2016) (2016)
6. Liu, X., Fulia, I.: Incorporating user, topic, and service related latent factors into web service recommendation. In: IEEE International Conference on Web Services (2015)
7. Balakrishnan, R., Kambhampati, S., Manishkumar, J.: Assessing relevance and trust of the deep web sources and results based on inter-source agreement. ACM Trans. Web 7(2), 32 (2013)
8. Vaculin, R., Neruda, C., Sycara, K.: Modeling and discovery of data providing services. In: Proceedings of the 2008 IEEE International Conference on Web Services, pp. 1032–1039 (2008)

Tweetchain: An Alternative to Blockchain for Crowd-Based Applications

Francesco Buccafurri[✉], Gianluca Lax, Serena Nicolazzo,
and Antonino Nocera

DIIES, University Mediterranea of Reggio Calabria,
Via Graziella, Località Feo di Vito, 89122 Reggio Calabria, Italy
{bucca,lax,s.nicolazzo,a.nocera}@unirc.it

Abstract. The assurance of information in the crowdsourcing domain cannot be committed to a single party, but should be distributed over the crowd. Blockchain is an infrastructure allowing this, because transactions are broadcast to the entire community and verified by *miners*. A node (or a coalition of nodes) with high computational power can play the role of miner to verify and approve transactions by computing the *proof of work*. Miners follows a highest-fee-first-served policy, so that a provider of a Blockchain-based application has to pay a non-negligible fee per transaction, to increase the likelihood that the application proceeds. This makes Blockchain not suitable for small-value transactions often occurring in the crowdsourcing paradigm. To overcome this drawback, in this paper we propose an alternative to Blockchain, leveraging an online social network (we choose Twitter to provide a proof of concept). Our protocol works by building a meshed chain of public posts to ensure transaction security instead of proof of work, and no trustworthiness assumption is required for the social network provider.

Keywords: Crowdsourcing · Blockchain · Twitter · Public ledger

1 Introduction

Coordination, record keeping, and irrevocability of transactions are features that make the Blockchain technology exploitable not just for cryptocurrencies. Indeed, a variety of applications can be built on top of this technology, making Blockchain a registry and inventory system for the recording, tracking, monitoring, and transacting of all assets [8]. Blockchain is considered a robust platform: even though a number of possible attacks are known [1,6,7,9,10], they are mostly not relevant in practice.

Blockchain is recently attracting the interest of both companies and researchers due to its power and the possibility to implement innovative solutions in those cases in which many users are spread over large spaces and may belong to different organizations. Despite this, there are a lot of situations in which the use of Blockchain appears not suitable. Indeed, in Blockchain, the consensus on

© Springer International Publishing AG 2017
J. Cabot et al. (Eds.): ICWE 2017, LNCS 10360, pp. 386–393, 2017.
DOI: 10.1007/978-3-319-60131-1_24

the ledger is reached thanks to the computation of the *proof of work* done by *miners*. A node (or a coalition of nodes) with high computational power can play the role of miner and follows (in general) a highest-fee-first-served policy to select transactions being verified. This implies that a provider of a Blockchain-based application has to pay a non-negligible fee per transaction, to increase the likelihood that its application proceeds. Obviously, in the case of small-value transactions often occurring in the crowdsourcing paradigm, this model appears little plausible.

There is another reason for which Blockchain seems not suitable for our setting. We expect that crowd-based applications are mostly relying on the participation of a large number of people through their mobile devices. But, any entity participating in the Blockchain system, even a simple user, must join a P2P network, and must exchange data in upload and download continuously. Moreover, a significant storage space is required, to store enough amount of past transactions to limit the need of contacting other nodes to retrieve them (how to do this in a secure way is not a trivial task). Both the above aspects are little compliant with the mobile and ubiquitous setting we refer to. Moreover, joining a P2P network, could be not well-accepted from many people, even for the perception of less security.

In this paper, we propose an alternative to Blockchain aimed at overcoming the above drawbacks and thus designed to make public-ledger-based solutions suitable to the crowdsourcing domain. Besides this domain, also Web-of-Things (WoT) applications could have benefits from our solution, because of similar motivations (battery consumption, computational and storage requirement, need of robust updates, need of mining on IoT devices, and so on).

This paper is organized as follows. In the next section, we provide a quick overview of the proposal. Then, in Sect. 3, we present our approach. We provide a preliminary security analysis in Sect. 4, and finally, we draw our conclusions in Sect. 5.

2   Overview of the Proposal

Our solution exploits the high capability of online social networks to share information among people. Besides communication media, online social networks can become part of people workflows. The advantages are reciprocal. People are used to operate on social networks profiles, there is no extra cost, availability of the social network services is very high. For social network providers, the role in the society would become more central, and this would increase business. Social networks may thus help us to overcome the limits of Blockchain in the crowd-sourcing domain.

To be effective, we focus our attention on a real-life social network, namely Twitter, but in principle, we could use any social network implementing a posting mechanism and a key-based search for posts. The idea is not to implement Blockchain over Twitter, but to reinvent a consensus protocol using tweets to encode transactions and meshed replications to substitute the proof of work and thus the need of fees for miners.

We call this protocol Tweetchain. Transaction tweets, together with a sufficient number of confirmation tweets posted by randomly selected members of the community, result in a meshed chain of replicated transactions. Under a proper non-collusion assumption, it is impossible for any party, including Twitter itself, to delete, alter, or forge valid transactions. From a conceptual point of view the protocol is truly decentralized. Indeed, there is no a central role of some node. However, no P2P tool must be enabled on the client machine. A user of the system only must run an application working on his Twitter profile. Importantly, Twitter does not play the role of trusted third party. Moreover, it is not the provider of some part of the protocol. The role of Twitter respect the protocol is transparent. As a consequence, Twitter cannot succeed as an adversary. This conclusion is based on the presence of the meshed network of transactions, for which no selective omission in publishing transaction tweets can be done. Twitter could, in principle, only stop the entire service, that is the entire Tweetchain community. Even though we can argue that this situation is very little plausible (like the fact that Blockchain is stopped by bodies providing the Internet), our protocol does not require a single social network. It could operate over multiple social networks having the basic required features. There are also models and technologies supporting multiple-social-network approaches [2,3]. This would further reduce the chance of success of a denial of service attack.

We stress that our goal is not that of replacing Blockchain, which obviously remains preferable in all the cases, as for cryptocurrencies, in which nobody should not have the possibility to stop the service. Tweetchain can be viewed as a lightweight public ledger for non-critical services, oriented at least to the domains of crowdsourcing and WoT.

This paper is at a preliminary stage. Therefore, the security of the protocol is only argued and intuitively understood. Moreover, a more abstract model can be designed to better present the idea and support its theoretical analysis. Anyway, we think that the direction of our research is promising, for the possible application impact.

3   The Tweetchain Model

In this section, we describe our approach and the structure of the messages exchanged by the involved entities.

The main actors of our model are:

- The Twitter social network, and, in particular, the followings features:
    1. the posting of *tweets* for registered users;
    2. the notification on the *follows* activity;
    3. the searching for information by *hashtags*.
- a welcome profile W used to implement a sort of yellow page support.
- the Tweetchain community, namely C, of users who join the Tweetchain protocol.

As a prerequisite, to participate in the `Tweetchain` community, a user must be able to build, by starting from a secret, a (SHA-256 based) hash chain [5] of a given size, say k, which will be used to maintain all his timeline activities linked together. This way, as will be clearer in the following, no modification can be done on older messages without compromising the remaining part of the user timeline. The value k, representing the length of the hash chain, is a system parameter which also limits the maximum size of the chunk of the user timeline (intended as number of tweets) that must be considered to verify the validity of a given message.

All the detail on the usage of this hash chain will be clarified in the following. As a further observation, we will consider our system in a steady-state, meaning that there are always at least $s = \frac{2t}{1-m}$ members in the `Tweetchain` community, where t and m are system parameters discussed in Sect. 4.

Now, we are ready to describe our proposal. The `Tweetchain` paradigm is composed of the following protocols.

Registration. It is executed by each user, say x, who wants to become member of the `Tweetchain` community C. Clearly, a prerequisite of this protocol is the sign up to Twitter in order to create a profile on it.

The first step he performs is to follow the welcome profile W and to publish a `hello` tweet with the following structure: $\langle \#HC_x^1 \#HC_W^1$ Hello @$W \rangle$, where $\#HC_x^1$ and $\#HC_W^1$ are Twitter hashtags with the base64 encoding of the first element of the hash chain of x and W as text, respectively, and @W is a Twitter reference to the welcome page W. After that, W verifies the tweet of x and sends a confirmation tweet as a `welcome` message with the reference to this user and a link to his `hello` tweet. Suppose that W has already posted $i-1$ tweets, then the `welcome` message for x will have the following structure: $\langle \#HC_W^i$ Welcome @$x \#HC_x^1 \#TID_x^1 \rangle$ where $\#HC_W^i$ and $\#HC_x^1$ are Twitter hashtags of the base64 encoding of the ith element of the hash chain of W and the first element of the hash chain of x, respectively, @x is a Twitter reference to the user x, and $\#TID_x^1$ is a Twitter hashtag with the ID of the first tweet (`hello`) of x as text. Observe that, the ID of a tweet (or status ID) is always unique inside Twitter.

As a consequence, W contains at least one tweet for each member of the community in join-chronological order. After this, x generates at random the set F_x of followings who will validate his transactions in the future. This set is built as follow:

- x retrieves his Twitter identifier. (Recall that each Twitter user has a unique 64-bit numeric identifier.)
- Then, this identifier is used as seed of a community-known PRNG to extract s random numbers and for each number, say n, computes $n \bmod w$, where w is the total number of tweets posted by W (i.e., the size of the `Tweetchain` community).
- At this point, the numbers computed above are used as indexes to select s distinct screen names from the welcome profile W.
- x sends a *private* message to each of the s profiles, whose screen names have been derived in the previous step, to ask them to follow him.

– After verifying the legitimacy of the request of x by using the community PRNG, each of the profiles contacted by x adds a follow link towards x and duplicates the welcome tweet of W by replacing $\#HC_W^i$ with their current hash chain element.

Transaction generation. This protocol allows the generation of a new transaction. Similarly to what happens in Blockchain, each transaction carries different information, such as: *(i) the timestamp of the generation,* (ii) a content, i.e., the transaction payload, *(iii) an input transaction, and* (iv) a target profile acting as transaction recipient.

In our protocol, the generation of a new transaction is associated with that of a new *tweet* by the user, in the following referred as *t-tweet*. According to the requirements described above, the i-th *t-tweet* of the user x, will have the following structure: $\langle \#HC_x^i \#TID_y^p$ content @$r\rangle$, where $\#HC_x^i$ is a hashtag with the base64 encoding of the ith element of the hash chain of x, $\#TID_y^p$ is a hashtag of the ID of the p-th tweet posted by the user y and used as input for this transaction, and @r is a Twitter reference to the recipient r of this transaction.

As soon as x posts a new *t-tweet*, the s users of F_x following him will be notified by the Twitter platform automatically. They will proceed by verifying the legitimacy of this new transaction by using the verification protocol described below. After running the verification procedure, they will publish a confirmation tweet on their timeline. Now, let v be one of the users of F_x and let $j-1$ be the number of tweets generated by v till now, his confirmation tweet for the transaction of x will be: $\langle \#HC_v^j @x \#TD_x^i$ status $\#TID_y^p$ content @$r\langle$, where $\#HC_v^j$ is the hashtag of the j-th element of the hash chain of the verifier v, @x is the reference to the user x, $\#TD_x^i$ is the hashtag of the ID of the *t-tweet* generated by x, and *status* can either be 1 for *success* or 0 for *failure* on the basis of the verification result. The remaining of this tweet is the essential part of the body of the *t-tweet* of x necessary to reconstruct the original tweet in case of deletion done by x.

Verification. This protocol is used to check the validity of a transaction. Now, suppose a verifier, say v, wants to verify the i-th transaction of the user x having the p-th transaction of a user y as input and the user r as target. The transaction of x will have the following structure: $\langle \#HC_x^i \#TID_y^p$ content @$r\langle$.

The protocol works by verifying each part of this *t-tweet*. Observe that, concerning the verification of the content, this is not considered here as it is strictly related to the objective of the transaction, which is not specified in this paper and, therefore, it is fully application dependent. As for the verification of $\#HC_x^i$, first v checks whether this hash chain element has been already used, in this case the verification will fail, otherwise the verifier has to compute the SHA-256 hash of $\#HC_x^i$. Due to the hash chain property, the results of this computation should be $\#HC_x^{i-1}$. Therefore, a search on Twitter for $\#HC_x^{i-1}$ should return the previous tweet posted by x. The goal of the verifier is to find the previous *t-tweet* of x or the initial hello message. Now, because x will also post confirmation tweets for other users' transactions, in the case $\#HC_x^{i-1}$ refers to a confirmation tweet, the procedure above is repeated until either a *t-tweet* or the `hello` tweet is found. Let $\#TID_x^{i-1}$ be the ID of such a tweet, the verifier has now to check whether he has confirmed this tweet in the past (i.e., he has posted a confirmation tweet with status 1 corresponding to it). If this is not the case, then the verifier will not confirm (*status* 0) the new transaction, otherwise he will proceed with the verification of the second part of the new *t-tweet* of x. The verification of $\#TID_y^p$ implies the verification of the validity of the input transaction. As said above, the input is the p-th *t-tweet* of the user y. The validity of this tweet is related to the presence of at least t confirmation tweets among the s generated by the verifiers associated with y. Because each confirmation tweet contains the ID of the corresponding *t-tweet*, a search in Twitter for $\#TID_y^p$ will return all the confirmation tweets for the p-th *t-tweet* of p. Now, the verifier has to check both the presence of t confirmations with status 1 and that they have been posted by the legitimate verifiers for y. Moreover, v checks whether the target of $\#TID_y^p$ is x and that it has not been already used as input in any other transaction. The last verification done is the check of the existence of r (i.e., the target user of this new transaction) in the `Tweetchain` community. This is obtained by searching for @r in Twitter and by verifying the presence of the `welcome` message in W along with t confirmations of the verifiers of r. At end of these steps, if all the verifications succeeded, then the verifier will post a confirmation tweet with status 1 for the new *t-tweet* of x, otherwise the status of this confirmation will be set to 0.

4   Security Analysis

In this section, we describe a security analysis of our approach showing that it accomplishes its objectives also in presence of attacks. Therefore, in the following, we describe the security model and sketch the security properties of our proposal.

To analyze the security properties, first our threat model includes the following assumptions:

A1 The adversary cannot change information shown on the social network page of any user of the `Tweetchain` community.

A2 The Twitter service is up and running as expressed in its use conditions and does not intend to block the `Tweetchain` community (i.e., we assume that Twitter only could block single individuals, not the whole community).

A3 The cryptographic hash function is robust, in the sense that collision, preimage and second preimage attacks are infeasible.

A4 The adversary cannot access secrets information of users from which hash chains are computed.

A5 Among x verifiers, at most $m \cdot x$ out of them (with $0 \le m < 1$) may not collaborate by executing correctly the *verification* protocol.

A6 At most t users can maliciously collude to break the security properties of the protocol.

Concerning the last assumption we have to recall that confirmations are generated collaboratively by multiple users playing the role of verifier. Some of these users might be corrupted by an attacker, but, as commonly assumed in the distributed domain [4,11], the majority of users at all times is considered honest. Therefore, our technique is parametric with respect to t. This value is chosen in such a way that the likelihood that at most t randomly selected users misbehave is negligible. Similar considerations can be done for Assumption A5, in which the parameter to set is m.

Now, we identify the security properties that our system has to assure and discuss the possible attacks.

SP1 - Transaction Authenticity. A transaction and the user generating it can always be verified by the `Tweetchain` community.

Attack AA1. *An attacker tries to impersonate the welcome profile W to tamper the list of verifiers for a user.*

Attack AA12. *An attacker creates multiple accounts and tries to use them as verifiers of his own transactions.*

SP2 - Transaction Integrity. The whole message (t-tweet) representing a transaction cannot be tampered once posted on the system.

Attack AI1. *Some of the verifiers do not execute the verification protocol invalidating a transaction.*

Attack AI2. *Some of the verifiers collude to compromise the integrity of a transaction.*

Attack AI3. *Twitter, playing as an attacker, tries to compromise the integrity of a transaction by deleting a portion of the chain (even the whole) this transaction depends on.*

SP3 - No repudiation of Transactions. The user generating a transaction cannot repudiate it.

Attack AR1. *An adversary tries to make ambiguous a transaction by forging another one with the same input transaction.*

Attack AR2. *The user, acting as an adversary, tries to repudiate a transaction by deleting the corresponding t-tweet.*

Being this paper a work-in-progress paper, we do not provide here the proof of the resistance of our protocol to the considered attacks.

5   Conclusion

Blockchain technology allows mutually distrustful parties to transact safely without trusted third parties and avoiding high legal and transactional costs. Despite the rapid grow of interest of both researchers and companies in Blockchain for all possible applications, Blockchain appears not suitable for small-value-transaction applications, typical of domains like crowdsourcing and WoT, due to the payment of fees related to the proof of work. In this paper, we propose a lightweight public ledger that, instead of the P2P network and the protocol of Blockchain, leverages the popular social network Twitter, by building a meshed chain of tweets to ensure transaction security. Importantly, Twitter does not play neither the role of trusted third party nor the role of ledger provider. Moreover, no proof of work or fees are needed. Incidentally, the elimination of the proof of work results in another relevant advantage. Indeed, many miners (sometimes unsuspecting victims of an attack) compete to reach the goal but only one succeeds every time a block is approved. In other words, the protocol produces an extraordinary quantity of wasted work, and thus lost of energy. Therefore, the protocol is really little sustainable from a global energy consumption point of view.

As a future work we plan to formalize our protocol in a more abstract fashion, to deeply analyze it in terms of security, and to highlight its relevance in a real-life specific application setting.

Acknowledgements. This work has been partially supported by the Program "Programma Operativo Nazionale Ricerca e Competitività" 2007–2013, Distretto Tecnologico CyberSecurity funded by the Italian Ministry of Education, University and Research.

References

1. Bahack, L.: Theoretical bitcoin attacks with less than half of the computational power (draft) (2013). arXiv preprint: arXiv:1312.7013
2. Buccafurri, F., Lax, G., Nicolazzo, S., Nocera, A.: A model to support design and development of multiple-social-network applications. Inf. Sci. **331**, 99–119 (2016)
3. Buccafurri, F., Lax, G., Nocera, A., Ursino, D.: A system for extracting structural information from social network accounts. Softw. Pract. Exp. **45**(9), 1251–1275 (2015)
4. Cramer, R., Gennaro, R., Schoenmakers, B.: A secure and optimally efficient multi-authority election scheme. Eur. Trans. Telecommun. **8**(5), 481–490 (1997)
5. Faye, Y., Niang, I., Noel, T.: A survey of access control schemes in wireless sensor networks. Proc. World Acad. Sci. Eng. Tech. **59**, 814–823 (2011)
6. Ron, D., Shamir, A.: Quantitative analysis of the full bitcoin transaction graph. In: Sadeghi, A.-R. (ed.) FC 2013. LNCS, vol. 7859, pp. 6–24. Springer, Heidelberg (2013). doi:10.1007/978-3-642-39884-1_2
7. Shultz, B.L., Bayer, D.: Certification of witness: mitigating blockchain fork attacks (2015)
8. Swan, M.: Blockchain: Blueprint for a New Economy. O'Reilly Media, Inc., Sebastopol (2015)
9. Swanson, T.: Consensus-as-a-service: a brief report on the emergence of permissioned, distributed ledger systems (2015)
10. Vasek, M., Thornton, M., Moore, T.: Empirical analysis of denial-of-service attacks in the bitcoin ecosystem. In: Böhme, R., Brenner, M., Moore, T., Smith, M. (eds.) FC 2014. LNCS, vol. 8438, pp. 57–71. Springer, Heidelberg (2014). doi:10.1007/978-3-662-44774-1_5
11. Zwierko, A., Kotulski, Z.: A light-weight e-voting system with distributed trust. Electron. Notes Theor. Comput. Sci. **168**, 109–126 (2007)

The Dimensions of Crowdsourcing Task Design

Ilio Catallo and Davide Martinenghi[✉]

DEIB - Politecnico di Milano, Piazza Leonardo da Vinci 32, 20133 Milano, Italy
{ilio.catallo,davide.martinenghi}@polimi.it

Abstract. Crowdsourcing, i.e., the provision of micro-tasks to be executed by a large pool of possibly anonymous workers, is attracting an increasing research attention, because it promises to help solving many scientific and practical problems where the harmonic cooperation of humans and machines delivers superior results. This paper proposes a systematic view of the crowdsourcing task design space and categorizes the dimensions that qualify the design decisions in crowdsourcing applications. For each dimension, we discuss the main open research problems and the most significant contributions, thereby offering guidelines for a principled understanding of current crowdsourcing marketplaces.

1 Introduction

The idea behind crowdsourcing is that, under the right circumstances, groups can be remarkably efficient. Indeed, groups do not need to be coordinated by exceptionally intelligent people in order to be smart, and group decisions are usually better than the decisions of the brightest group member. In this regard, crowdsourcing proved successful in business and marketing [23], medicine [25], and many other fields.

In order to harness the potential of crowdsourcing, requesters need to make a variety of decisions regarding the design of the crowdsourcing task they want to submit, such as how much to pay workers, or how to design an effective user interface for the task. Task requesters should not take these decisions lightly, as they affect both quality and quantity of the outcome [2,3]. Nevertheless, in many cases the design of crowdsourcing tasks is guided by the sole intuition, or, worse, is determined by imitating other successful designs, without properly understanding why such designs work well in their original application context. Task requesters may try to alleviate this issue by experimenting with different designs and select the one that better performs in practice. Unfortunately, since resources are usually limited, task requesters can rarely afford to pursue such a trial-and-error approach, which quickly becomes impracticable due to the very high number of possible designs. Moreover, complexity is further increased by the fact that some design choices can be used in combination. For instance, it is reasonable to imagine task designs in which different motivational techniques are used at the same time in order to engage workers.

In this paper, we propose a characterization of the task design space in terms of four main dimensions, organized in a taxonomy shown in Table 1, with the

© Springer International Publishing AG 2017
J. Cabot et al. (Eds.): ICWE 2017, LNCS 10360, pp. 394–402, 2017.
DOI: 10.1007/978-3-319-60131-1_25

Table 1. The dimensions of crowdsourcing design

What	Task type	sentiment analysis, content moderation & creation, classification, relevance feedback, survey, knowledge extraction, data enrichment, promotion, data cleaning
	Task features	skill variety, task identity, task significance, required effort
	Task output	determinism, evaluability
Who	Diversity	demographics, expertise
	Anonymity	requester's anonymity, workers' anonymity
	Hierarchy	flat, multi-level
Why	Intrinsic	fun, enjoyment, killing time
	Extrinsic	external regulation, introjection, identification, integration
How	Motivation	gamification, financial, framing, feedback, UI
	A-priory quality	qualification restrictions, qualification test, gold standard, redundancy, multi-review, peer consistency
	A-post. quality	statistical methods, expert review, automatic checks

objective of providing both researchers and practitioners with a systematic way of approaching the design of crowdsourcing tasks. The main goal of our approach, which we will address in the rest of the paper, is both to assist the conscious design of crowdsourcing tasks (and, ultimately, help to understand how task design affects the problem solving process) and to provide a classification tool for the improvement of existing platforms and the development of future designs. Another immediate outcome of the proposed taxonomy is that it directly enables an effective classification of the commercially-available platforms; we will develop this latter aspect in a forthcoming work, with the aim of providing an accurate analysis of the current state of the industry.

2   A Taxonomy for Crowdsourcing Tasks

Many taxonomies have been recently proposed for the fields of human computation [18], collective intelligence [16], and crowdsourcing [8–10]. Similarly to what was done by [16] for collective intelligence, we construct the task design space of crowdsourcing applications around four top-level design questions, namely, *what* is the task to be performed, *who* is going to perform it, *why* should the crowd work on the task, and *how* the task will be executed (see Table 1 for an overview). We will then detail the sub-dimensions of these high-level concerns, specifically for crowdsourcing application design.

2.1   The "What" Dimension

The *"what"* dimension regards the properties defining crowdsourcing tasks.

Task type. When humans are employed as computational components, the most relevant kinds of tasks, commonly used in current commercial platforms, include *sentiment analysis, content moderation and creation, classification, relevance feedback, survey, knowledge extraction, data enrichment, promotion, data cleaning.* All these involve interpretation, synthesis, or evaluation, which makes fully automated solutions not completely effective.

Task features. In the job characteristics model of [11], the outcome of a task is expressed as a function of different dimensions, three of which directly related to the task being executed: *skill variety, task identity,* and *task significance.* We refer to *skill variety* as to the set of skills that are needed for the resolution of a task. Workers' interest for the task increases as the number and variety of skills needed to solve the task increases. By *task identity* we mean the perception of completeness of a task, i.e., a worker considers as more meaningful those tasks that result in a tangible outcome. *Task significance* refers to the extent to which a task has a substantial impact on the lives or work of other people. Predictably, workers' interest increases as the task significance increases. Finally, for some tasks it is possible to define the associated *required effort*. For instance, in a task requiring workers to translate book pages, the task requester can dimension the required effort by varying the number of pages assigned to each worker.

Task output. Tasks can also be discriminated on the characteristics of their output. In this respect, some tasks exhibit *determinism*, i.e., there exists a unique correct answer. Conversely, tasks that admit more than one correct or admissible answer are nondeterministic. For example, different answers can be considered acceptable when working on content creation tasks. *Evaluability* refers instead to the simplicity of assessing correctness of an answer. For example, with binary classification tasks, verifying correctness has the same cost as executing the task. In other words, the evaluability of such tasks is really low, as carrying out an exhaustive assessment would cancel out any advantage related to the employment of crowd workers. Conversely, some classes of problems (e.g., combinatorial problems) feature a high degree of evaluability. This is because, even though the automatic generation of a solution may be extremely hard, the verification of such a solution can be automatically carried out at ease.

2.2   The "Who" Dimension

The "*who*" dimension pertains to the features of crowd workers that are going to undertake the task. Unlike traditional organizations, in which roles and responsibilities are usually clearly defined, in the case of crowdsourcing it may be difficult to assign beforehand responsibilities among crowd workers. Specifically, the characteristics of the crowd determine how much decision-making power is at disposal of the task requester. We identify such characteristics as *diversity, anonymity,* and *hierarchy.*

Diversity. Diversity is recognized as one of the most important conditions required to elicit the "wisdom of crowds" [21]. In order to ensure a high diversity, a system should keep the potential crowd of contributors as large and open as possible. In this regard, we distinguish between *demographic diversity* and *expertise diversity*. By the former, we refer to the property of the crowd of being composed of individuals differing in place of origin, gender, and age; whereas by the latter we refer to the degree to which the crowd differs in expertise and competence. Although intuitively one would say that a higher degree of expertise across the crowd is always advisable, a system should also allow for the serendipity of receiving high-quality contributions by amateurs or experts from different domains [10].

Anonymity. In the context of crowdsourcing, there could exist two forms of *anonymity*. Namely, the task requester could decide to keep her identity undisclosed to the crowd workers (*requester's anonymity*). Conversely, the crowd workers' identity could be kept hidden from the task requester (*workers' anonymity*). Depending on the specific crowdsourcing platform, workers' anonymity could also extend to past records about their performance, possibly impairing the capability of the task requester of selecting the most suitable workers.

Hierarchy. Most crowdsourcing systems adopt a *flat* hierarchy, i.e., the crowd is composed of peer contributors. However, some crowdsourcing platforms allow crowd workers to organize in *multi-level* hierarchies. For instance, in a two-level hierarchy, crowd workers are organized in groups, each guided by a group leader, whose increased responsibilities are associated with a larger compensation (or recognition by peers) w.r.t. the other members of the group.

2.3   The "Why" Dimension

The *"why"* dimension relates to the reasons why crowd workers participate in crowdsourcing initiatives. When subject to survey campaigns, many workers indicate money as the most important reason behind participation. Nevertheless, crowd workers tend to over-report money as the main motivation as consequence of a social desirability bias [1]. Such a response effect causes workers to avoid reporting other reasons, e.g., the desire of killing time, because perceived as socially less favorable. In light of this, we could expect the correct design of any crowdsourcing system to strictly depend on the accurate identification of workers' motivations, and in fact many works (see, e.g., [1], [16], [8]) come to the conclusion that misunderstanding workers' motivations is the single greatest factor behind failed crowdsourcing initiatives.

According to self-determination theory [7], individuals are motivated by a combination of *intrinsic* and *extrinsic* motivations; the former refer to doing something because it is inherently interesting or enjoyable, while the latter to doing something because it leads to a separable outcome (e.g., a monetary compensation). Hence, intrinsic motivation exists within the individual, and occurs only for those tasks the worker finds engaging, challenging, or worth doing in their own right. Conversely, extrinsic motivation comes from outside of the individual, who considers the task at hand just as an instrument for achieving a certain desired outcome. Based on this, we propose a taxonomy of workers' motivations, on which we comment on in greater detail next.

Intrinsic Motivations. Intrinsic motivations factors include *fun, enjoyment* and the *desire to kill time*. As mentioned above, it is worth noting that intrinsic motivation is highly subjective, as some individuals are intrinsically motivated for some activities and not others, and not everyone is intrinsically motivated for any particular task [7].

Extrinsic Motivations. Extrinsic motivations may be classified according to how much they succeed in entailing individuals' personal endorsement and a feeling of choice [19].

External regulation indicates that workers perform the task solely because of the presence of some external obligation. Predictably, monetary reward is the most common way of exerting external regulation on crowd workers. Social obligations is another form of external regulation. An example of social obligation is the need for an individual to provide support for her family. Finally, crowd workers can even be forced to execute a given task. This is indeed the case with the reCAPTCHA project[1], where the user is employed as forced labour (even though just for a few seconds) in exchange for a given service.

Introjection occurs when individuals perform actions in order to either avoid guiltiness and anxiety or to attain ego-enhancements and pride. In particular, individuals extrinsically motivated by a glory factor perform an action in order to increase or maintain their self-esteem, whereas peer recognition specifically refers to the desire of individuals to be recognized by peers for their contribution.

Identification implies that the individual recognizes the importance of the task, effectively making it her own. Workers experience such a motivation when, e.g., they perceive the task as a chance of learning something new, or because they think that the task can be valuable for their career.

Integration occurs when the task fully meets the worker's values and needs. Specifically, workers' acting becomes more and more self-determined as they assimilate the reasons for an action. In this respect, workers can be motivated by the desire of contributing to a higher cause (ideology factor). A second integration factor that can guide workers' actions is the need of a sense of belonging to a specific community, with which to share common values and objectives.

2.4   The "How" Dimension

The *"how"* dimension encompasses all those issues related to the actual execution of tasks on a crowdsourcing platform. These include decisions related to how the task requester can improve the completion rate of the submitted tasks, or ensure good quality of the outcome. Designing the "how" dimension correctly is crucial for the successful execution of the crowdsourcing tasks. For instance, identifying correct rewards for workers may affect both quality and quantity of results.

Motivational Techniques. Motivational techniques relate to the problem of properly engaging workers. While increasing monetary rewards will naturally

---

[1] https://www.google.com/recaptcha/intro/index.html.

tend to improve worker performance [17], there exist many crowdsourcing initiatives where high-quality contributions are collected independently of any monetary compensation. The effect of financial incentives is not straightforwardly predictable: for instance, paying more could even reduce quality after a certain point because it encourages speed [12]. The motivational aspect cannot be simplified into a mere linear relationship between money and performance, and, in fact, the financial dimension is just one out of several possible ways of motivating the crowd.

Gamification is the use of game thinking and game mechanics in non-game contexts in order to engage users in solving problems. Gamification approaches rely on the use of game elements to make tasks more enjoyable, in hopes of boosting workers' motivation. To this end, gamified systems typically make use of individual and social achievements, so that workers that succeed in completing the assigned tasks receive some form of reward, such as badges. Games with a purpose (GWAP) and ephemeral games can be used when the task can be completely hidden beyond a game-based interaction [22].

Financial incentives arguably represent the most common motivational technique. Different payment schemes are possible, namely non-performance-contingent (i.e., paying workers independently of their performance, like in hourly or piecework payment) and performance-contingent (i.e., paying by quality of the produced work, e.g., with bonus/malus strategies) [4].

Framing is a form of cognitive bias that causes individuals to react differently to the same information depending on its presentation (i.e., framing). To this end, the task requester may, e.g., emphasize the meaningfulness of the context [5] and improve clarity of instructions [14].

The task requester may try to motivate crowd workers by providing real-time *feedback* on the work being executed in order to keep the worker engaged, e.g., an explanation on the adopted payment scheme and the currently earned amount.

Ideally, the task requester should provide *user interfaces* that facilitate the efficient collection of input data with low effort. Even well-motivated workers may produce wrong data when interacting with a poorly designed user interface.

A-Priori Quality Assurance. Because of the open nature of crowdsourcing initiatives, quality assurance mechanisms are needed. A-priori mechanisms identify and reject poor quality contributions while the tasks are still under execution.

Qualification restrictions limit the approved worker pool by using qualitative or quantitative metrics [20] based, e.g., on worker's history and demographic information. Sometimes, workers must pass a *qualification test* before working on the actual tasks. Qualification tests provide instructions, background information and examples, and can therefore be used to train workers, but may not be enough to avoid spammers, who may deliberately start answering randomly after qualifying for the task. Workers' quality can also be estimated by means of a *gold standard* data set of tasks, for which the correct answers are known a priori. Tasks belonging to the gold standard data set are randomly mixed into the sequence of actual tasks. The estimated worker quality is updated at each gold standard checkpoint, and workers that repeatedly miss the correct answer are excluded from further tasks. *Redundancy* is the most common form of quality assurance. The idea is to collect several contributions for the same task, and apply, at the end of the collection phase, a statistical method in order to filter out poor quality results, at the price of an increased cost of the crowdsourcing campaign [6]. Other techniques include *multi-review*, which consists in subdividing workers into two groups mutually assessing the respective work, and *peer consistency*, where each worker's answer is compared with the answers of her peers on the same task.

A-Posteriori Quality Assurance. A-posteriori techniques identify and filter out poor quality contributions after workers' contributions have been collected.

When tasks expose some degree of determinism (see Sect. 2.1), *statistical methods* (e.g., majority voting or more sophisticated techniques that weigh workers according to their quality [24]) can be applied to aggregate the data and remove the effects of poor quality work. For tasks that are not deterministic, contributions can be assessed only by means of *expert reviewers*, who manually check the quality and relevance of the work. Many problems (e.g., combinatorial problems) are such that the automatic verification of a solution is really easy, while the automatic generation of a solution is instead extremely hard. For this class of problems, it is possible to carry out an *automatic assessment* (e.g., [13]).

3 Discussion

In this paper, we have presented a classification of crowdsourcing design activities developed around four classical dimensions ("what", "who", "why", and "how") and several sub-dimensions thereof. Some features are poorly supported by existing crowdsourcing marketplaces. Among these, hierarchies and gamification still play a little role today. Hierarchies would be a significant feature for both task requesters, who could then entrust higher-level workers with responsibilities for the work of others, and for workers, who could gain trust and thus achieve better rewards. We expect hierarchies to become more central in the near future. Although not available off the shelf, gamification can be implemented via external components built by the requester or by means of performance-based payment methods. Since gamification may highly motivate the completion of micro tasks, we expect it to be natively supported by crowdsourcing marketplaces.

Crowdsourcing marketplaces are the meeting point of two kinds of users, requesters and workers, often aiming to different goals. A discrepancy between the needs of requesters, who want to be able to choose specific workers, and platforms, in which worker anonymity and indistinguishability are commonplace, is apparent. While the observable trend is towards refined kinds of qualification restrictions (e.g., quality level and temporal deadlines), it is difficult to translate such quality constraints into the right (and appealing) price tags. This is perhaps the single aspect where the various marketplaces differ the most and where we expect future practice to bring some sort of standardization into play.

Another notable conflict in the requester-worker interaction lies in the noise introduced by spammers or unreliable workers. An important challenge is to design validation tasks that look similar to the other tasks, while minimizing their number, since test tasks must, too, be paid by the requester.

We expect new sub-dimensions to gain ground, e.g., hybrid engagement approaches involving a tighter interaction between crowdsourcing marketplaces, social networks, and GWAPs. Hybridization will entail mixtures of algorithmic execution and crowd validation, as testified to by recent projects (e.g., [15]).

Acknowledgements. The authors acknowledge support from the H2020-EU.3.3.1 "ENCOMPASS" project (ID: 723059).

References

1. Antin, J., Shaw, A.: Social desirability bias and self-reports of motivation: a study of amazon mechanical turk in the US and india. In: SIGCHI, pp. 2925–2934. ACM (2012)
2. Bernaschina, C., et al.: A web tool for the execution of crowdsourcing campaigns. In: WWW - Companion Volume, pp. 171–174 (2015)
3. Bernaschina, C., et al.: On the role of task design in crowdsourcing campaigns. In: HCOMP, pp. 4–5 (2015)
4. Catallo, I., et al.: A workload-dependent task assignment policy for crowdsourcing. WWW J. (2017). doi:10.1007/s11280-016-0428-7
5. Chandler, D., Kapelner, A.: Breaking monotony with meaning: motivation in crowdsourcing markets. J. Econ. Behav. Organiz. **90**, 123–133 (2013)
6. Ciceri, E., et al.: Crowdsourcing for top-K query processing over uncertain data. IEEE TKDE **28**(1), 41–53 (2016)
7. Deci, E.L., Ryan, R.M.: Self-Determination. Wiley Online Library (2010)
8. Doan, A., Ramakrishnan, R., Halevy, A.Y.: Crowdsourcing systems on the world-wide web. Commun. ACM **54**(4), 86–96 (2011)
9. Gadiraju, U., Kawase, R., Dietze, S.: A taxonomy of microtasks on the web. In HYPERTEXT, pp. 218–223. ACM (2014)
10. Geiger, D., et al.: Crowdsourcing information systems - definition, typology, and design. In: ICIS (2012)
11. Hackman, J.R., Oldham, G.R.: Motivation through the design of work: test of a theory. Organ. Behav. Hum. Perform. **16**(2), 250–279 (1976)
12. Heer, J., Bostock, M.: Crowdsourcing graphical perception: using mechanical turk to assess visualization design. In: SIGCHI, pp. 203–212. ACM (2010)
13. Khatib, F., et al.: Algorithm discovery by protein folding game players. Proc. Nat. Acad. Sci. **108**(47), 18949–18953 (2011)
14. Kittur, A., et al.: The future of crowd work. In: CSWC, pp. 1301–1318 (2013)
15. Loni, B., et al.: Fashion-focused creative commons social dataset. In: MMSys, pp. 72–77 (2013)
16. Malone, T.W., Laubacher, R., Dellarocas, C.: Harnessing crowds: mapping the genome of collective intelligence. MIT Sloan School of Management (2009)
17. Prendergast, C.: The provision of incentives in firms. J. Econ. Lit. **37**, 7–63 (1999)
18. Quinn, A.J., Bederson, B.B.: Human computation: a survey and taxonomy of a growing field. In: SIGCHI, pp. 1403–1412. ACM (2011)

19. Ryan, R.M., Deci, E.L.: Intrinsic and extrinsic motivations: classic definitions and new directions. Contemp. Educ. Psychol. **25**(1), 54–67 (2000)
20. Schulze, T., Krug, S., Schader, M.: Workers' task choice in crowdsourcing and human computation markets. In: ICIS (2012)
21. Surowiecki, J.: The wisdom of crowds. Random House LLC (2005)
22. Von Ahn, L.: Human computation. In: DAC 2009 46th ACM/IEEE Design Automation Conference, pp. 418–419. IEEE (2009)
23. Whitla, P.: Crowdsourcing and its application in marketing activities. Contemp. Manag. Res. **5**(1), 15–28 (2009)
24. Yan, Y., Rosales, R., Fung, G., Dy, J.G.: Active learning from crowds. In: ICML, pp. 1161–1168 (2011)
25. Yu, B., Willis, M., Sun, P., Wang, J.: Crowdsourcing participatory evaluation of medical pictograms using amazon mechanical turk. J. Med. Int. Res. **15**(6), e108 (2012)

Public Transit Route Planning Through Lightweight Linked Data Interfaces

Pieter Colpaert$^{(\boxtimes)}$, Ruben Verborgh, and Erik Mannens

Ghent University – imec – Internet and Data Lab, Ghent, Belgium
{pieter.colpaert,ruben.verborgh,erik.mannens}@ugent.be

Abstract. While some public transit data publishers only provide a data dump – which only few reusers can afford to integrate within their applications – others provide a use case limiting origin-destination route planning API. The Linked Connections framework instead introduces a hypermedia API, over which the extendable base route planning algorithm "Connections Scan Algorithm" can be implemented. We compare the CPU usage and query execution time of a traditional server-side route planner with the CPU time and query execution time of a Linked Connections interface by evaluating query mixes with increasing load. We found that, at the expense of a higher bandwidth consumption, more queries can be answered using the same hardware with the Linked Connections server interface than with an origin-destination API, thanks to an average cache hit rate of 78%. The findings from this research show a cost-efficient way of publishing transport data that can bring federated public transit route planning at the fingertips of anyone.

Keywords: Linked data · Public transport · Route planning · Open data

1 Introduction

The way travelers want route planning advice is diverse: from finding journeys that are accessible with a certain disability [2], to taking into account whether the traveler owns a (foldable) bike, a car or a public transit subscription, or even calculating journeys with the nicest pictures on social network sites [8]. However, when presented with a traditional route planning HTTP API taking origin-destination queries, developers of e.g. traveling tools are left with no flexibility to calculate journeys other than the functions provided by the server. As a consequence, developers that can afford a larger server infrastructure, integrate data dumps of the timetables (and their real-time updates), into their own system. This way, they are in full control of the algorithm, allowing them to calculate journeys in their own manner, across data sources from multiple authorities.

When publishing departure-arrival pairs (*connections*) in chronologically ordered pages – as demonstrated in earlier work [4] – route planning can be

© Springer International Publishing AG 2017
J. Cabot et al. (Eds.): ICWE 2017, LNCS 10360, pp. 403–411, 2017.
DOI: 10.1007/978-3-319-60131-1_26

executed at data integration time by the user agent rather than by the data publishing infrastructure. This way, the *Linked Connections (LC) framework*[1] allows for a richer web publishing and querying ecosystem within public transit route planners. It lowers the cost for reusers to start prototyping – also federated and multimodal – route planners over multiple data sources. Furthermore, other sources may give more information about a certain specific vehicle that may be of interest to this user agent. It is not exceptional that route planners also take into account properties such as fares [5], wheelchair accessibility [2] or criminality statistics. In this paper, we test our hypothesis that this way of publishing is more lightweight: how does our information system scale under more data reuse compared to origin-destination APIs?

The paper is structured in a traditional way, first describing state of the art, then introducing more details on the Linked Connections framework, to then describe the evaluation's design, made entirely reproducible, and discuss our open-source implementation. Finally, we elaborate on the results and conclusion.

2   State of the Art

The General Transit Feed Specification (GTFS)[2] is a framework for exchanging data from a public transit agency to third parties. It is – at the time of writing – the de-facto standard for describing and exchanging transit schedules. It describes the headers of several CSV files combined in a ZIP-file. Using a calendar and calendar exceptions, both periodic as aperiodic transit schedules can be described. GTFS also describes the geographic shape of trips, fare zones, accessibility, information about an agency, and so forth. We provided URIs to the terms in the GTFS specification through the Linked GTFS[3] vocabulary. Reusing these GTFS files, route planners exist in software as a service platforms such as Navitia.io, in end-user applications such as CityMapper, Ally or Google Maps, or as open-source software, such as Open Trip Planner or Bliksemlabs RRRR. It is also possible a that an agency, such as the Dutch railway company, the SNCB or the Deutsche Bahn, expose a route planner over HTTP themselves.

Data dumps and query APIs can be seen as two extremes on the *Linked Data Fragments* (LDF)[4] axis. Triple Pattern Fragments – similarly to Linked Connections [4] – already defined a way to publish an RDF dataset only through its triple patterns [10]. This moves the evaluation of all Basic Graph Patterns – and by extension SPARQL queries – to the client, reducing CPU load on the server.

In route planners, not SPARQL queries but *Earliest Arrival Time queries* (EAT) [9] need to be able to be solved. This is a question involving a time of departure, a departure stop and an arrival stop, expecting the earliest possible arrival time at the destination. More complex route planning questions exist,

---

[1] http://linkedconnections.org.
[2] https://developers.google.com/transit/gtfs/reference.
[3] Base URI: http://vocab.gtfs.org/terms#.
[4] http://linkeddatafragments.org.

such as the *Minimum Expected Arrival Time* (MEAT) [6] or *multi criteria profile* queries [1,5,11]. The *Connection Scan Algorithm* (CSA) is an approach for planning that models the timetable data as a directed acyclic graph [6,9]. By topologically sorting the graph by departure time, the shortest path algorithm only needs to scan through connections within a limited time window to solve an EAT query. CSA can be extended to solving problems where it also keeps the number of transfers limited, as well as calculating routes with uncertainty [6]. The ideas behind CSA can scale up to large networks by using multi-overlay networks [9].

3 Linked Connections

In this section, we design our information system for public transport data, called the Linked Connections framework. As our datasets need to be interoperable with other data published on the Web, we choose HTTP as our *uniform interface*. We allow cross origin resource sharing and enable cache header to indicate when we expect a document to change. In order to document our identifiers using this uniform interface as well, RDF is chosen as a knowledge representation model. We further only define the hypermedia controls needed in each document as well as the vocabularies to be used.

To solve the EAT problem using CSA, timetable information within a certain time window needs to be retrievable. Other route planning questions need to select data within the same time window, and solving these is expected to have similar results[5] Instead of exposing an origin-destination API or a data dump, a Linked Connections server paginates the list of connections in departure time intervals and publishes these pages over HTTP. Each page contains a link to the next and previous one. In order to find the first page a route planning needs, the document of the entry point given to the client contains a hypermedia description on how to discover a certain departure time. Both hypermedia controls are expressed using the Hydra vocabulary[6].

The base entities that we need to describe are connections, which we documented using the LC Linked Data vocabulary[7]. Each connection entity – the smallest building block of time schedules – provides links to an arrival stop and a departure stop, and optionally to a trip. It contains two literals: a departure time and an arrival time. Linked GTFS can be used to extend a connection with public transit specific properties such as a headsign, drop off type or fare information. Furthermore, it also contains concepts to describe transfers and their minimum change time. For instance, when transferring from one railway platform to another, Linked GTFS can indicate that the minimum change time from one stop to another is a certain number of seconds.

---

[5] Also preprocessing algorithms [11] scan through an ordered list of connections to, for example, find transfer patterns [1].

[6] http://www.hydra-cg.com/spec/latest/core/.

[7] http://semweb.mmlab.be/linkedconnections#.

In the proof of concept built for this paper available at http://linkedcon nections.org, we implemented this hypermedia control as a redirect from the entry point to the page containing connections for the current time. Then, on every page, the description can be found of how to get to a page describing another time range. In order to limit the amount of possible documents, we only enable pages for each X^8 minutes, and do not allow pages describing overlapping time intervals. When a time interval is requested for which a page does not exist, the user-agent will be redirected to a page containing connections departing at the requested time. The same page also describes how to get to the next or previous page. This way, the client can be certain about which page to ask next, instead of constructing a new query for a new time interval.

A naive implementation of federated route planning on top of this interface is also provided, as a client can be configured with more than one entry point. The client then performs the same procedure multiple times in parallel. The connections streams are merge sorted as they are downloaded. A Linked Data solution is to ensure that the client knows how to link a stop from one agency to a stop from another.

4   Evaluation Design

We implemented different components in JavaScript for the Node.js platform. We chose JavaScript as it allows us to use both components both on a command-line environment as well as in the browser.

CSA.js A library that calculates a minimum spanning tree and a journey, given a stream of input connections and a query[9]
Server.js Publishes streams of connections in JSON-LD, using a MongoDB to retrieve the connections itself[10]
Client.js Downloads connections from a configurable set of servers and executes queries[11]
gtfs2lc A tool to convert existing timetables as open data to the Linked Connections vocabulary[12]

We set up 2 different servers, each connected to the same *MongoDB* database which stores all connections of the Belgian railway system for October 2015: 1. a *Linked Connections server* with an NGINX proxy cache in front, which adds caching headers configured to cache each resource for one minute[13] and compresses the body of the response using gzip; 2. a *route planning server* which

---

[8] X is a configurable amount.
[9] https://github.com/linkedconnections/csa.js.
[10] https://github.com/linkedconnections/server.js.
[11] https://github.com/linkedconnections/client.js.
[12] https://github.com/linkedconnections/gtfs2lc.
[13] The Belgian railway company estimates delays on its network each minute.

exposes an origin-destination route planning interface instead using the csa.js library[14].

These tools are combined into different set-ups:

Client-side route planning. The first experiment executes the query mixes by using the Linked Connections client. Client caching is disabled, making this simulate the LC *without cache* set-up, where every request could originate from an end-user's device.

Client-side route planning with client-side cache. The second experiment does the same as the first experiment, except that it uses a client side cache, and simulates the LC *with cache* set-up.

Server-side route planning. The third experiment launches the same queries against the full route planning server. The query server code is used which relies on the same CSA.js library as the client used in the previous two experiments.

In order to have an upper and lower bound of a real world scenario, the first approach assumes every request comes from a unique user-agent which cannot share a cache, and has caching disabled, while the second approach assumes one user-agent does all requests for all end-users and has caching enabled. Different real world caching opportunities – such as user agents for multiple end-users in a software as a service model, or shared peer to peer neighborhood caching [7] – will result in a scalability in-between these two scenarios.

We also need a good set of queries to test our three set-ups with. The iRail project[15] provides a route planning API to apps with over 100 k installations on Android and iPhone [3]. The API receives up to 400 k route planning queries per month. As the query logs of the iRail project are published as open data, we are able to create real query mixes from them with different loads. The first mix contains half the query load of iRail, during 15 min on a peak hour on the first of October. The mix is generated by taking the normal query load of iRail during the same 15 min, randomly ordering them, and taking the half of all lines. Our second mix is the normal query load of iRail, which we selected out of the query logs, and for each query, we calculated on which second after the benchmark script starts, the query should be executed. The third mix is double the load of the second mix, by taking the next 15 min, and subtracting 15 min from the requested departure time, and merging it with the second mix. The same approach can be applied with limited changes for the subsequent query mixes, which are taken from the next days' rush hours. Our last query mix is 16 times the query load of iRail on the 1st of October 2015. The resulting query mixes can be found at https://github.com/linkedconnections/benchmark-belgianrail.

These query mixes are then used to reenact a real-world query load for three different experiments on the three different architectures. We ran the experiments on a quad core Intel(R) Core(TM) i5-3340M CPU @ 2.70 GHz with 8 GB

---

[14] The route planning server uses the CSA.js library to expose a route planning API on top of data stored in a MongoDB. The code is available at https://github.com/linkedconnections/query-server.

[15] https://hello.irail.be.

of RAM. We launched the components in a single thread as our goal is to see how fast CPU usage increases when trying to answer more queries. The results will thus not reflect the full capacity of this machine, as in production, one would run the applications on different worker threads.

Our experiments can be reproduced using the code at https://github.com/linkedconnections/benchmark-belgianrail. There are three metrics we gather with these scripts: the CPU time used of the server instance (HTTP caches excluded), the bandwidth used per connection, and the query execution time per LC connection. For the latter two, we use a per-connection result to remove the influence of route complexity, as the time complexity of our algorithm is $O(n)$ with n the total number of connections. We thus study the average bandwidth and query execution time needed to process one connection per route planning task.

5   Results

Figure 1 depicts the percentage of the time over 15 min the server thread was active on a CPU. The server CPU time was measured with the command pidstat from the sysstat package and it indicates the server load. The faster this load increases, the quicker extra CPUs would be needed to answer more queries.

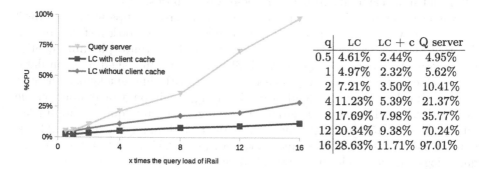

q	LC	LC + c	Q server
0.5	4.61%	2.44%	4.95%
1	4.97%	2.32%	5.62%
2	7.21%	3.50%	10.41%
4	11.23%	5.39%	21.37%
8	17.69%	7.98%	35.77%
12	20.34%	9.38%	70.24%
16	28.63%	11.71%	97.01%

Fig. 1. Server CPU usage under increasing query load shows that the Linked Connections server is more cost-efficient: more queries can be answered on one core.

When the query load is half of the real iRail load on October 1st 2015, we can see the lowest server load is the LC set-up with client cache. About double of the load is needed by the LC without client cache set-up, and even more is needed by the query server. When doubling the query load, we can notice the load lower slightly for LC with cache, and the other two raise slightly. Continuing this until 16 times the iRail query load, we can see that the load of the query server raises until almost 100%, while LC without cache raises until 30%, while with client cache, 12% is the measured server load.

In Fig. 2, we can see the query response time per connection of 90% of the queries. When the query load is half of the real iRail load on October 1st 2015,

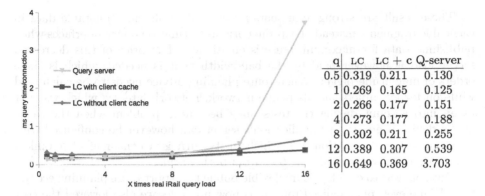

q	LC	LC + c	Q-server
0.5	0.319	0.211	0.130
1	0.269	0.165	0.125
2	0.266	0.177	0.151
4	0.273	0.177	0.188
8	0.302	0.211	0.255
12	0.389	0.307	0.539
16	0.649	0.369	3.703

Fig. 2. 90% of the journeys will be found with the given time in ms per connection under increasing query load.

we notice that the fastest solution is the query server, followed by the LC with cache. When doubling the query load, the average query execution time is lower in all cases, resulting in the same ranking. When doubling the query load once more, we see that LC with client cache is now the fastest solution. When doubling the query load 12 times, also the LC without client cache becomes faster. The trend continues until the query server takes remarkably longer to answer 90% of the queries than the Linked Connections solutions at 16 times the query load.

The average bandwidth consumption per connection in bytes shows the price of the decreased server load as the bandwidth consumption of the LC solutions are three orders of magnitude bigger: LC is 270B, LC with cache is 64B and query-server is 0.8B. The query server only gives one response per route planning question that is small. LC without client cache has a bandwidth that is three orders of magnitude bigger than the query server. The LC with a client cache has an average bandwidth consumption per connection that is remarkably lower. On the basis of these numbers we may conclude the average cache hit-rate is about 78%.

6   Conclusion

This paper introduced the Linked Connections framework for publishing queryable public transit data. We measured and compared the raise in query execution time and CPU usage between the traditional approach and Linked Connections. We could have hypothesized that we would find the origin-destination API approach to give faster response times, with a higher server CPU load that will increase even faster when the number of queries increase. We indeed achieved a better *cost-efficiency*: when the query-interface becomes saturated under an increasing query load, the lightweight LC interface only reached 1/4th of its capacity, meaning that the same load can be served with a smaller machine, or that a larger amount of queries can be solved using the same server. As the server load increases, the LC solution even give faster query results.

These result are strong arguments in favor of publishing timetable data in cacheable fragments instead of exposing origin-destination query interfaces when publishing data for maximum reuse is envisioned. The price of this decreased server load is however paid by the bandwidth that is needed, which is three orders of magnitude bigger. When route planning advice needs to be calculated while on for example a mobile phone network, network latency, which was not taken into account during the tests, may become a problem when the cache of the device is empty. An application's server can however be configured with a private origin-destination API, which in its turn is a consumer of a Linked Connections dataset, taking the best from both worlds.

Our goal was to enable a more flexible public transport route planning ecosystem. While even personalized routing is now possible, we also lowered the cost of hosting the data, and enabled in-browser scripts to execute the public transit routing algorithm. Furthermore, the query execution times of queries solved by the Linked Connections framework are competitive. Until now, public transit route planning was a specialized domain where all processing happened in memory on one machine. We hope that this is a start for a new ecosystem of public transit route planners.

References

1. Bast, H., Carlsson, E., Eigenwillig, A., Geisberger, R., Harrelson, C., Raychev, V., Viger, F.: Fast routing in very large public transportation networks using transfer patterns. In: Berg, M., Meyer, U. (eds.) ESA 2010. LNCS, vol. 6346, pp. 290–301. Springer, Heidelberg (2010). doi:10.1007/978-3-642-15775-2_25
2. Colpaert, P., Ballieu, S., Verborgh, R., Mannens, E.: The impact of an extra feature on the scalability of linked connections. In: COLD Workshop at ISWC2016 (2016)
3. Colpaert, P., Chua, A., Verborgh, R., Mannens, E., Van de Walle, R., Vande Moere, A.: What public transit API logs tell us about travel flows. In: WWW 2016: 25th International World Wide Web Conference (2016)
4. Colpaert, P., Llaves, A., Verborgh, R., Corcho, O., Mannens, E., Van de Walle, R.: Intermodal public transit routing using linked connections. In: Proceedings of the 14th International Semantic Web Conference: Posters and Demos, October 2015
5. Delling, D., Pajor, T., Werneck, R.F.F.: Round-based public transit routing. In: Proceedings of the 14th Meeting on Algorithm Engineering and Experiments (ALENEX 2012) (2012)
6. Dibbelt, J., Pajor, T., Strasser, B., Wagner, D.: Intriguingly simple and fast transit routing. In: Bonifaci, V., Demetrescu, C., Marchetti-Spaccamela, A. (eds.) SEA 2013. LNCS, vol. 7933, pp. 43–54. Springer, Heidelberg (2013). doi:10.1007/978-3-642-38527-8_6
7. Folz, P., Skaf-Molli, H., Molli, P.: CyCLaDEs: a decentralized cache for triple pattern fragments. In: Sack, H., Blomqvist, E., d'Aquin, M., Ghidini, C., Ponzetto, S.P., Lange, C. (eds.) ESWC 2016. LNCS, vol. 9678, pp. 455–469. Springer, Cham (2016). doi:10.1007/978-3-319-34129-3_28
8. Quercia, D., Schifanella, R., Aiello, L.M.: The shortest path to happiness: recommending beautiful, quiet, and happy routes in the city. In: Proceedings of the 25th ACM Conference on Hypertext and social Media, pp. 116–125. ACM (2014)

9. Strasser, B., Wagner, D.: Connection scan accelerated. In: Proceedings of the Meeting on Algorithm Engineering & Expermiments, pp. 125–137. Society for Industrial and Applied Mathematics (2014)
10. Verborgh, R., Vander Sande, M., Hartig, O., Van Herwegen, J., De Vocht, L., De Meester, B., Haesendonck, G., Colpaert, P.: Triple pattern fragments: a low-cost knowledge graph interface for the Web. J. Web Semant. **37**, 184–206 (2016)
11. Witt, S.: Trip-based public transit routing. In: Bansal, N., Finocchi, I. (eds.) ESA 2015. LNCS, vol. 9294, pp. 1025–1036. Springer, Heidelberg (2015). doi:10.1007/978-3-662-48350-3_85

A WebRTC Extension to Allow Identity Negotiation at Runtime

Kevin Corre[1,3]([⊠]), Simon Bécot[1], Olivier Barais[2], and Gerson Sunyé[2]

[1] Orange Labs, Cesson-Sevigne, France
{kevin1.corre,simon.becot}@orange.com
[2] INRIA, Rennes, France
{olivier.barais,gerson.sunye}@inria.fr
[3] IRISA, Rennes, France
kevin.corre@irisa.fr

Abstract. In this paper we describe our implementation of the WebRTC identity architecture. We adapt OpenID Connect servers to support WebRTC peer to peer authentication and detail the issues and solutions found in the process. We observe that although WebRTC allows for the exchange of identity assertion between peers, users lack feedback and control over the other party authentication. To allow identity negotiation during a WebRTC communication setup, we propose an extension to the Session Description Protocol. Our implementation demonstrates current limitations with respect to the current WebRTC specification.

1 Introduction

In business communications, especially with previously unknown party, when dealing with high value information, or when having privacy concerns, ensuring the identity of the other party and the confidentiality of the communication is of the greatest importance. Usually Communication Services (CS) require and manage user authentication, secure call signaling, and may transmit the actual communication in the case of telecom operators. Trust between users and CSs is thus required in order to ensure the safety of the conversation. Telecom operators rely on a Circle of Trust model to allow CS interoperability. Joining the Circle of Trust ensures an *implicit initial level of trust* [1] to its members. On the contrary, Web CSs are usually organized in a silo model: to communicate together, two users must be using the same CS. Large identity federation are difficult to build and present complex issues [2]. As a result users calling outside of their enterprise's domain are often stranded in a silo situation, with the fallback option of using self-asserted identities.

WebRTC [3] is a Web API, specified by the W3C and the IETF, which supports Peer-to-Peer (P2P) audio-video calling and data sharing. It allows real-time communications between browsers, without additional plugins. The IETF-draft by Rescorla et al. [4], now part of the specification, offers a mechanism for explicit authentication, decoupling the identity and signaling functions.

© Springer International Publishing AG 2017
J. Cabot et al. (Eds.): ICWE 2017, LNCS 10360, pp. 412–419, 2017.
DOI: 10.1007/978-3-319-60131-1_27

It is expected to see the emergence of numerous WebRTC-enabled Web sites, resulting in a large number of security configurations. Even more if services and applications start being interoperable with each others. Designing new P2P service architecture enabling dynamic trusted relationships among distributed applications is currently explored within the Matrix[1] specification or within the reThink project[2]. However, in such kinds of distributed, dynamic and heterogeneous architecture, it may become difficult for users to understand if their communications are secured enough.

One of the challenge is then to build a trust and security model for communication services, capable of returning a single metric encompassing different aspects of the communication's security configuration. It would help users in making trust decision, and in return allow negotiation to raise trust in the communication's security.

In this paper we report and discuss on our implementation of the WebRTC identity specification. We first present the WebRTC identity specification in Sect. 2. In Sect. 3 we detail our implementation, and in particular the change we did to two OpenID Connect (OIDC) [5] server implementations to support WebRTC user to user authentication. A report of a similar work for the SAML protocol was already published in 2015[3], but our work is to our knowledge the first implementation of a WebRTC IdP Proxy for OIDC. The WebRTC identity architecture gives an initial level of trust in the other party's identity. However, we show that users need more information and control in order to trust Identity Providers (IdP) used in the session establishment. In order to solve this issue, in Sect. 4, we propose a WebRTC extension to negotiate over the other-party authentication during a call setup. We implemented the negotiation, demonstrates it to just decorate the current protocol and remain compatible with existing user-agent and report on the limitations imposed by the current WebRTC specification. After discussing possible future work in Sect. 5, we conclude in Sect. 6.

2   WebRTC Overview

WebRTC communication setup can be decomposed into three different paths as shown in Fig. 1a. The Signaling path initializes the communication between Alice and Bob User-Agents (UA) through one or more Communication Service (CS) servers. On this path, Alice and Bob negotiate communication parameters by exchanging Session Description Protocol (SDP) [6] call offer and answer messages in order to setup the Media path. The Media path is a peer-to-peer encrypted connection between Alice and Bob used to exchange audio, video, or data. Optionally, one or two Identity paths can be used to generate and verify Identity Assertions. Identity Assertions bind the identity of the authenticated

---

[1] https://matrix.org.
[2] https://rethink-project.eu/.
[3] https://www.terena.org/activities/tf-webrtc/meeting2/slides/
20150519-webrtc-identity.pdf.

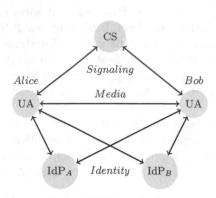

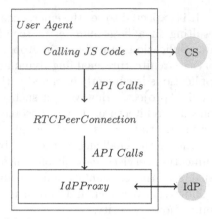

(a) A call with IdP-based identity and a single communication service provider. The Signaling path join Alice and Bob through a CS. The Media path is the P2P session. Each Identity path allows a user to authenticate to the other user through an IdP.

(b) API calls are the RTCPeerConnection's setIdentityProvider function, and the generateAssertion and validateAssertion functions, offered by the IdP Proxy to the RTCPeerConnection. The protocol used between the IdP Proxy and the IdP is unspecified.

Fig. 1. WebRTC call setup

user to the media session, and are included in SDP signaling messages. Their role is to assert that media path encryption keys are used by users that are authenticated by IdPs.

The WebRTC security architecture [4] specifies the IdP Proxy component. This component serves as an interface between the WebRTC RTCPeerConnection object and an IdP. The IdP Proxy is supposed to be available at a standardized location on each compatible IdP domain. Before making a SDP call offer or answer, the RTCPeerConnection calls the IdP Proxy to generate an assertion covering a set of keys. Once the IdP authenticated the user, an identity assertion is returned. This assertion is then included in the SDP message with the IdP Proxy origin, i. e., the IdP domain's URL. This allows users to discover an IdP Proxy location without prior knowledge or relationship with the other party's IdP. On receiving a SDP call offer or answer containing an Identity Assertion and the associated IdP domain-name, the User-Agent downloads the IdP Proxy from the specified location. It then calls its verifyAssertion function. If successful, the IdP Proxy returns the key fingerprint and the user identifier, for instance an email address. This bounds the key used on the media channel to the identity authenticated by the verifying IdP and ensures that no MitM attack is being set. The WebRTC Identity Architecture is shown in Fig. 1b.

Richard L. Barnes and Martin Thomson gives a clear overview of the WebRTC security architecture in their 2014 IEEE article [7]. They describe the trust model on which the specification is based and explain why the CS may be considered as an adversary in this trust model. Their paper gives a good understanding, on how authenticated WebRTC communication are protected against network and signaling attack. Considering the CS to be untrusted, and at the same time relying on the same CS to offer protection through the IdP Proxy mechanism, may be considered quite restrictive. Besides, the question of trust in the other party IdP is only addressed from the perspective of verifying the IdP's origin. The strength of the authentication process is not considered, and it is left to the user to decide whether to trust the received identity by using an address book.

3 Implementing the WebRTC Identity Specification

To implement the WebRTC identity specification, we develop a simple WebRTC service[4] offering communication for two users in a single room. The session signaling is done over WebSockets and through the communication server. A client code in Javascript manages the call session. Tests are conducted on Firefox version 50.1.0.

We implement IdP Proxies in conformance with the WebRTC specification. These proxies are added to two OIDC servers: a reference implementation by Nat Sakimura[5] and an implementation in NodeJS[6].

3.1 IdP Proxy Implementation

In its simplest form, our IdP Proxy uses a simple REST API on the IdP server, matching the *generateAssertion* and *validateAssertion* functions. The GET request to the *generateAssertion* endpoint use the session cookie to prove that the user has been authenticated. The content parameter is passed in the request. If the user does not have an active session, the IdP login URL is returned by the IdP Proxy. A successful login following this URL is signaled to the CS client page by a "LOGINDONE" message from the login page. The *generateAssertion* returns an opaque token as a string. This token can then be used on the *validateAssertion* endpoint to retrieve the user identity and the associated content parameter, i. e. the media key fingerprint declared by the user.

The resulting IdP Proxy is a small piece of Javascript that can easily be deployed. And could even be provided by CS if an interoperability need arises.

3.2 IdP Proxy with OpenID Connect

To map the WebRTC Identity architecture to our OIDC server implementations, we consider the IdP Proxy to be a OIDC client using the *implicit flow*. Our implementation however requires additional modifications to the specification. As the *implicit flow* is used, and the URL fragment response is inaccessible from a Fetch API request[7], we request the ID Token to be returned in the

---

[4] https://github.com/Sparika/ACOR_SDP.
[5] https://github.com/reTHINK-project/dev-IdPServer-phpOIDC.
[6] https://github.com/reTHINK-project/dev-IdPServer.
[7] https://developer.mozilla.org/en/docs/Web/API/Fetch_API.

response body, using the `response_mode` parameter. Additionally, we introduce a new `rtcsdp` claim, to have the media key fingerprint included in the ID Token. Finally, the `redirect_uri` of the request is set to a particular page, sending the "LOGINDONE" message to the CS client page.

OpenID Connect is based on OAuth2, an authorization protocol. As such the concept of user consent is central to the protocol. However, in the user-to-user authentication use case, it is not clear who is the intended *audience* for the ID Token. As our IdP Proxy acts as the OIDC client, it is the effective *audience*. In practice, the identity assertion, i.e. the ID Token, is visible to each CS on the signaling path, and is ultimately verified by the peer's user-agent. The user should thus be asked for consent to share the ID Token to any of these actors. However, these are not all known in advance, and by design, any actor could instantiate an IdP Proxy to validate and decode the identity assertion.

Actually, without appropriate protection, a web page could look for user's identity assertion without triggering any warning. This could reveal user identities, or at least existing user accounts on IdPs. For instance, given a list of WebRTC compatible IdPs, running the following script would reveal identity assertion from unprotected IdPs with active sessions.

```
IdPArray.forEach(function(idp){
    var pc = new RTCPeerConnection()
    pc.setIdentityProvider(idp.domain, idp.proxy)
    pc.getIdentityAssertion()
    .then(res => {alert('Got your ID token: '+res)})
})
```

This is a major issue, and implementors should take appropriate measure to protect their interfaces against such vulnerabilities. Protections could consist in requesting user consent for every request, limiting the access to known web page origin, or managing user consent and authorization based on web page origin.

In conclusion, although OIDC offers an existing framework to generate and validate identity assertion, several modifications to existing implementations are still required in order to support WebRTC IdP Proxy peer-to-peer authentication. The resulting solution is more complex than a simple cookie based IdP Proxy, but would also easily handle authorization flow to check user consent.

4   Going Further: Trust Issue

Although the WebRTC identity specification allows to bind the media session to a validated peer identity, it does not offer a clear measure of trust to the user. In particular, a user may ask himself the following questions: should I trust my peer's IdP and what is the strength of my peer's authentication? In order to require a particular trust level, it would be necessary to act on the communication setup, i.e. conduct negotiation on the peer authentication process.

4.1 SDP Negotiation

Two parameters could be negotiated: the other-party IdP origin and the authentication level [8]. To convey these requests we define two parameters. The Authentication Class Request (ACR): List<ACRValue>, a list, ordered by preference, of accepted authentication class value. And the Origin Request (OR): List<Origin>, a list, ordered by preference, of accepted IdP's origins.

WebRTC conveys identity assertions as SDP session level attributes [4]. Extension to the attribute are possible, but none are defined by the specification. Due to the identity attribute grammar, identity assertion would be mandatory to negotiate identity parameters.

We instead propose to define a new type of SDP session-level attribute to negotiate these identity parameters. The Authentication Class and Origin Request (ACOR) SDP attribute defines a list of accepted authentication class and IdP domain for the other peer identity.

- a=acor:LIST<ACRValue> ; List<Origin>

In SDP, a negotiation is a sequence of offer and answer exchanges, with an offer always followed by an answer. To accept the requested ACOR attribute, a peer must thus reply a SDP message with a compatible identity assertion.

4.2 Implementation

We implemented our solution[8] to negotiate identity parameters over SDP exchange. SDP messages are returned by the createOffer and createAnswer functions offered by the PeerConnection object. Once generated, the client code appends an ACOR attribute to the generated SDP offer or answer. The SDP is then sent to the other peer's client. On receiving a message, the client code reads the requested ACOR attribute. It also verifies that the received peer-identity follows the ACOR it previously requested.

The resulting negotiation solution is implemented in under 100 Javascript code lines, for a very simple client. Renegotiation allows both clients to make new offer once the session has been established. For instance, this allows to ask to an anonymous user to authenticate itself. We however identify some important limitations, mostly due to the specifications.

The *generateAssertion* function from the IdP Proxy has for parameters *contents, origin,* and *usernameHint*. Request for a particular authentication class to the IdP is not defined. We use the *usernameHint* parameters to pass ACR parameters to the IdP. The IdP is modified accordingly to understand this parameter. However, the browser generated WebRTC *IdentityValidationResult* do not represent the ACR. It is thus impossible for the validating IdP Proxy to return a certified ACR value to the client. The client could directly read the identity assertion contained by the SDP message, but this solution would loose the benefits of the WebRTC identity abstraction.

---

[8] Available with sequence diagrams at https://github.com/Sparika/ACOR_SDP.

It also appears impossible to change identity at call runtime or use multiples identities simultaneously. The WebRTC specification states that if the PeerConnection object has *"previously authenticated the identity of the peer [...], then this also establishes a target peer identity. The target peer identity cannot be changed once set"* [3]. Our tests demonstrate that modifying the remote peer identity effectively close the connection. Once a first identity has been set, it cannot be changed. If this is an issue and if several IdPs are available, the client should wait to receive an ACOR request from the peer before setting an IdP.

In the end, we are able to establish two anonymous sessions and then request the other peer to authenticate with a particular identity domain. We are however unable to control the strength of the authentication. In addition, our modified SDP messages are effectively ignored by other services and by the user-agent. As a result, interoperability with other services should not be compromised by this new attribute.

5   Discussion and Future Work

Identity Parameters Negotiation. We evaluate the possibility to deploy negotiation over ACR and IdP's origin with current WebRTC specifications. Our conclusion shows that it is not possible to request ACR to IdP Proxy when calling the *generateAssertion* function. As a result, the specification would need to be updated to support ACR negotiation. In particular: the *generateAssertion* function could be extended to accept additional parameters, and the *IdentityValidationResult* could be left open to extensions.

Recommendation Source. We implement our identity parameters negotiation solution on the CS side as it was the simplest solution. As identity negotiation is most useful in scenarios of inter-operable communication services, such services could be acting as the identity recommendation source. This may seems to contradict the WebRTC trust model with untrusted CS. However, in the inter-operable scenario, we may want to relax the trust model and consider that a CS may be trusted by its own user. In this situation a CS could be well-suited to provide recommendation and evaluation of the other-peer's authentication.

Formalizing the Trust Model. We are working on a model of the full WebRTC communication setup, from a security configuration point of view. We use security configuration inputs from the three communication setup paths described in Fig. 1a. Knowing and negotiating the other peer's authentication strength and IdP, as presented in Sect. 4, is a first step towards building an explicit trust model for the identity path. User-agents could use this model to display a security indicator for the overall WebRTC communication.

6   Conclusion

Decoupling authentication from the signaling process by providing an explicit identity assertion to the user is an interesting new paradigm. Especially true for the design of interoperable communication services on the web.

In this paper, we described our implementation of the WebRTC identity architecture and the issues we encountered in our OpenID Connect implementation. Although WebRTC identity specification provides users with an initial trust level in their peer's authentication, it does not offer a clear measure of trust. To solve this issue we propose an extension to WebRTC standard based on a new SDP attribute for users to negotiate over the authentication of the other party during the signaling process. Our implementation highlights that this solution could be readily used by communication services, but it also demonstrates actual limit of the current specifications in some scenarios, such as a multiple identities for a user, e. g. personal and professional identities.

Users are not best fitted to analyze at runtime a complex security configuration, such as a WebRTC setup configuration. They need help from a recommendation source that would conduct the negotiation and notify them of the current trust and security level. This recommendation source would use actual session parameters at runtime to evaluate trust in the communication. In our future work we will build such trust model.

Acknowledgment. This work has received funding from the European Union's Horizon 2020 research and innovation program under grant agreement No. 645342, project reTHINK.

References

1. Boursas, L., Danciu, V.A.: Dynamic inter-organizational cooperation setup in circle-of-trust environments. In: NOMS 2008–2008 IEEE Network Operations and Management Symposium, pp. 113–120. IEEE (2008)
2. Jøsang, A., Fabre, J., Hay, B., Dalziel, J., Pope, S.: Trust requirements in identity management. In: Proceedings of the 2005 Australasian Workshop on Grid Computing and E-research, vol. 44, pp. 99–108. Australian Computer Society Inc. (2005)
3. Jennings, C., Narayanan, A., Aboba, B., Bergkvist, A., Burnett, D.: WebRTC 1.0: Real-time communication between browsers, W3C, Working Draft, March 2017
4. Rescorla, E.: WebRTC security architecture. IETF Secretariat, Internet-Draft draft-ietf-rtcweb-security-arch-12, June 2016
5. Sakimura, N., Bradley, J., Jones, M., de Medeiros, B., Mortimore, C.: OpenID connect core 1.0. The OpenID Foundation, OpenID Specification, 2014. http://openid.net/specs/openid-connect-core-1_0.html
6. Handley, M., Jacobson, V., Perkins, C.: SDP: Session Description Protocol. Network Working Group, RFC 4566, July 2006
7. Barnes, R.L., Thomson, M.: Browser-to-browser security assurances for WebRTC. IEEE Int. Comput. **18**(6), 11–17 (2014)
8. ISO/IEC 29115:2013 - Information technology - Security techniques - Entity authentication assurance framework

A UML Profile for OData Web APIs

Hamza Ed-douibi[1]([✉]), Javier Luis Cánovas Izquierdo[1], and Jordi Cabot[1,2]

[1] UOC, Barcelona, Spain
{hed-douibi,jcanovasi}@uoc.edu
[2] ICREA, Barcelona, Spain
jordi.cabot@icrea.cat

Abstract. More and more individuals and organizations are making their data available online publicly, resulting in a growing market of technologies and services to help consume data and extract its real value. One of the several ways to publish data on the Web is via Web APIs. Unlike other approaches like RDF, Web APIs provide a simple way to query structured data by relying only on the HTTP protocol. Standards and frameworks such as Open API or API Blueprint offer a way to create Web APIs but OData stands out from the rest as it is specifically tailored to deal with data sources. However, creating an OData Web API is a hard and time-consuming task for data providers as they have to choose between relying on commercial solutions, which are heavy and require a deep knowledge of their corresponding platforms, or create a customized solution to share their data. We propose an approach that leverages on model-driven techniques to facilitate the development of OData Web APIs. The approach relies on a UML profile for OData allowing to annotate a UML class diagram with OData stereotypes. In this paper we describe the profile and show how class diagrams can be automatically annotated with such profile.

Keywords: UML · OData · Web API

1 Introduction

Recent years have seen an explosion of data available online via Web APIs, coming from both the public sector and private sources. Unlike other approaches like RDF, Web APIs provide a simple way to query structured data by relying only on the HTTP protocol. The increasing number of Web APIs has actually led to an explosion of specialized applications that combine data from different sources to provide insights on specific topics not visible at first glance, thus contributing to the growth of data economy.

While standards and frameworks such as Open API[1] or API Blueprint[2] offer a way to create Web APIs, the Open Data Protocol (OData)[3] is specifically

---

[1] https://www.openapis.org/.
[2] https://apiblueprint.org/.
[3] http://www.odata.org/.

© Springer International Publishing AG 2017
J. Cabot et al. (Eds.): ICWE 2017, LNCS 10360, pp. 420–428, 2017.
DOI: 10.1007/978-3-319-60131-1_28

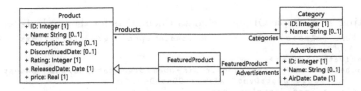

Fig. 1. UML class diagram of the running example.

tailored to deal with data sources. Thus, in the last years, OData protocol has been accepted as the favored standard to publish datasets as Web APIs. As a result, many commercial infrastructures have integrated OData to their products (e.g., SAP, IBM WebSphere, JBoss Data Virtualization).

OData enables the creation of data-centric Web APIs, which allow resources, identified using Uniform Resource Locators (URIs) and defined in a data model, to be published and edited by Web clients using simple HTTP messages. It defines also a small URL-based query language to identify and query the data described in the data model. The current version of OData (version 4.0) has been approved as OASIS standard [4]. However, creating an OData Web API is still a hard and time-consuming task for data providers as they have to choose between relying on commercial solutions, which are heavy and require a deep knowledge of their corresponding platforms, or create a customized solution to share their data.

Model-Driven Engineering (MDE) is a paradigm which emphasizes the use of models to raise the level of abstraction and automation in software development [10]. MDE aims to address platform complexity by using models and model transformations for the specification/generation of software artifacts. Thus, MDE techniques have been increasingly used to automate the generation of Web applications [2,3,6–9,11,12]. While these existing MDE approaches cover a variety of technologies (e.g. web services and ubiquitous applications), they lack of specific support for OData (and REST APIs in general, with very few exceptions [2,6,9]).

In this sense, our goal is to advance towards the definition of an MDE infrastructure for the generation (and reverse engineering) of OData applications. As a first step towards this vision, this paper presents a UML profile for OData that enables an easy definition of OData sources at the model level.

The remainder of this paper is structured as follows. Section 2 shows the running example used along the paper. Section 3 presents the OData profile and Sect. 4 presents the rules to generate default profile definitions. Finally, Sect. 5 concludes the paper and presents some future work.

2   Running Example

We define a simple OData Web API of an online store as running example. This example is inspired in the official reference example of the OData community[4]. Figure 1 shows an excerpt of the UML class diagram for the Web API data model,

---

[4] http://services.odata.org/V4/OData/OData.svc/$metadata.

Listing 1. A simple OData Metadata Documents for the products service

```
1  <edmx:Edmx xmlns:edmx="http://docs.oasis-open.org/odata/ns/edmx" Version
      ="4.0">
2    <edmx:DataServices>
3      <Schema xmlns="http://docs.oasis-open.org/odata/ns/edm" Namespace="com
         .example.ODataDemo" Alias="ODataDemo">
4        <EntityType Name="Product">
5          <Key><PropertyRef Name="ID"/></Key>
6          <Property Name="ID" Type="Edm.Int32" Nullable="false"/>...
7          <NavigationProperty Name="Categories" Type="Collection(ODataDemo.
            Cotegory)" Partner="Products"/>
8        </EntityType>
9        <EntityType Name="Category">
10         <Key><PropertyRef Name="ID"/></Key>
11         <Property Name="ID" Type="Edm.Int32" Nullable="false"/>
12         <Property Name="Name" Type="Edm.String"/>
13         <NavigationProperty Name="Products" Type="Collection(ODataDemo.
            Product)" Partner="Categories"/>
14       </EntityType>
15       <EntityType Name="FeaturedProduct" baseType="Product">...</
            EntityType>
16       <EntityType Name="Advertisement">...</EntityType>
17       <EntityContainer Name="ODataDemoService">
18         <EntitySet Name="Products" EntityType="ODataDemo.Product">
19           <NavigationPropertyBinding Path="Categories" Target="Categories
              "/>
20         </EntitySet>
21         ...
22       </EntityContainer>
23     </Schema>
24   </edmx:DataServices>
25  </edmx:Edmx>
```

which includes the classes: `Product` to represent products, `Category` to classify products, `FeaturedProduct` for premium products to be featured in commercials, and `Advertisement` which records the data about those commercials.

In OData data models are not expressed in UML but as XML metadata documents describing an Entity Data Model (EDM) using the Conceptual Schema Definition Language (CSDL) [5]. Web clients use this document to understand how to query and interact with the API using standard HTTP methods. Listing 1 shows an excerpt of the metadata document for the data model shown in Fig. 1. The `Schema` element describes the entity model exposed by the OData Web APIs and includes the entity types `Product` and `Category`, `FeaturedProduct`, and `Advertisement`, which also includes properties and navigation properties to describe primitive attributes and associations, respectively. The `Schema` element includes also an `EntityContainer` element defining the entity sets exposed by the service and therefore the entities that can be queried. In the following section, we will describe the OData profile which enables the generation of such file.

3   A UML Profile for OData

To formalize domain-specific knowledge, we can either create a new metamodel or extend an existing modeling language. Given the similarities between OData and available concept in UML (specifically, UML class diagrams), we opted to use the UML extension mechanisms (providing stereotypes, tagged values, and constraints to adapt the UML metamodel to different platforms or domains) as the basis for our OData modeling language. Therefore, this section presents our OData profile for UML.

We organize the OData profile into two parts, namely: (i) the Entity Data Model (EDM) which describes the data exposed by an OData Web API, and (ii) the advanced configuration model, which defines additional characteristics or capabilities of OData Web APIs (i.e. what parts of the EDM can be modified, what permissions are needed,...).

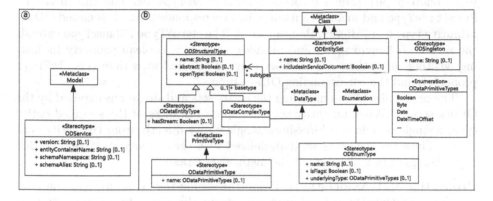

Fig. 2. OData profile: (a) the service wrapper and (b) data types elements.

3.1 The Entity Data Model

OData Service Wrapper. An OData Web API exposes a single entity model which may be distributed over several schemas, and should include an entity container that defines the resources exposed by the Web API. Figure 2a shows the extension of UML to define these elements. We consider that the entity model is defined in one schema, represented by the element *Model* of a UML class diagram. Thus, `ODService` stereotype extends the metaclass `Model` and includes: the version the OData specification, the namespace of the schema (e.g., `com.example.ODataDemo`), an alias for the schema namespace (e.g., `ODataDemo`), and the name of the entity container (e.g., `ODataService`).

Data Types. An OData entity model defines data types in terms of structural types, enumerations, and primitive types. There are two kinds of structural types: *entity types* and *complex types*. An *entity type* is a named structured type which defines the properties and relationships of an entity. *Entity types* are mapped to the concept *Class* in a UML model. A *complex type* is a named structural type consisting of a set of properties. *Complex types* are mapped to the concept *Data type* in a UML model.

Figure 2b shows the stereotypes related to data types and their mapping with UML concepts. The abstract stereotype ODStructuralType defines the common features of all the structural types and includes a name, a property indicating whether the structural type cannot be instantiated (i.e., abstract property), and a property indicating whether undeclared properties are allowed (i.e., openType property[5]).

ODStructuralType supports also the concept of inheritance by allowing the declaration of a base structural type (i.e., basetype association). The stereotypes ODEntityType and ODComplexType inherit from ODStructureType and extend the metaclasses Class and DataType, respectively. Additionally ODEntityType includes the hasStream property, indicating if the entity is a media entity (e.g., photograph). The stereotype ODPrimitiveType extends the metaclass PrimitiveType and includes a name which corresponds to the associated OData primitive type (e.g., Binary, Boolean, etc.). The stereoType ODEnumType extends the metaclass Enumeration and includes a name, a boolean property indicating whether more than one member may be selected at a time (i.e., IsFlags property), and the corresponding OData type.

The profile also allows modeling the entity sets and singletons exposed by the OData service. While an entity set can expose instances of the specified entity type, a singleton allows addressing a single entity directly from the entity container. These two concepts are materialized with the stereotypes ODEnitity Set and ODSingleton which extend the metacalass Class.

Properties and Associations. Properties define the structure and the relationships in OData. Structural properties define the attributes of an entity type or a complex type where as navigation properties define associations between entity types. In UML the element *Property* is a *StructuralFeature* which, when related by *ownedAttribute* to a *Classifier* (other than *Association*), represents an attribute, and when related by *memberEnd* of an *Association*, represents an association end. Both structural properties and navigation properties are mapped to the concept *Property* in a UML model.

Figure 3 shows the stereotypes defining properties and associations. The stereotypes ODProperty and ODnavigationProperty represent a structural property and a navigation property, respectively. They both extend the metaclass Property. The stereotype ODProperty includes a name and several constraints to provide additional constraints about the value of the structural property (e.g., nullable, maxLength properties). Additionally, the stereotype ODKey inherits from ODProperty and defines a property as the key of the entity type (required for a an OData entity type). The stereotype ODNavigationProperty includes a name, a containment property, and a nullable property. The stereotype ODNavigationPropertyBinding extends also the metaclass Property and defines a navigation binding for the corresponding entity set.

---

[5] Open types entities allows clients to persist additional undeclared properties.

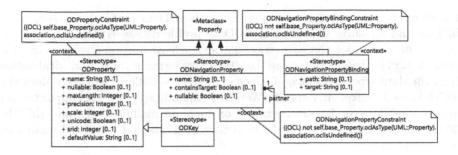

Fig. 3. Properties and associations stereotypes.

To ensure the validity of the applied stereotypes, we have enriched the profile with a set of constraints written using the Object Language (OCL) [1]. For instance, since the stereotypes related to properties and navigations properties extend all the metaclass `Property`, `ODPropertyConstraint` ensures that the stereotype `ODProperty` is applied to a UML property element representing an attribute.

3.2 Advanced Configuration of OData Web APIs

OData defines annotations to specify additional characteristics or capabilities of a metadata element (e.g., entity type, property) or information associated with a particular result (e.g., entity or property). For example, an annotation could be used to define whether a property is read-only. Annotations consist of a term (i.e., the namespace-qualified name of the annotation), a target (the element to which the term is applied), and a value. A set of related terms in a common namespace comprises a vocabulary. Our profile supports the three standardized vocabularies defined by OData, namely: the core vocabulary, capacity, and measures.

Figure 4 shows an excerpt of the profile defined for representing annotations. The stereotype `ODAnnotations` extends the metaclasses `Model`, `Class`, and `Property`, and has an association with `ODVocabulary`, thus allowing adding annotations according to the vocabularies. `ODVocabulary` is the root class of

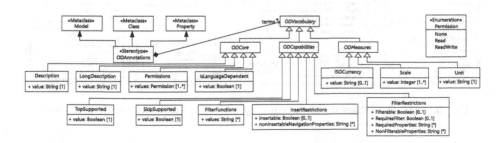

Fig. 4. Annotation and vocabulary stereotypes.

the hierarchy of vocabularies supported by the OData profile (i.e., core, capabilities, and measures vocabularies). OData profile defines (i) the `ODCore` hierarchy which includes the core vocabularies such as documentation (e.g., the class `Description`), permissions (i.e., the class `Permissions`, and localization (i.e., the data type `IsLanguageDependent`); (ii) the `ODCapabilities` hierarchy which is used to explicitly specify Web API capabilities (e.g., `TopSupported` for query capabilities or `InsertRestriction` for data modification); and (iii) the `ODMeasures` hierarchy to describe monetary amounts and measured quantities (e.g., `ISOCurency`).

4   Default Profile Generation

Our OData profile can be used to annotate any new or preexisting UML class diagram. Nevertheless, to simplify the application of our profile, we have also developed a model-to-model transformation that given an standard UML model, returns an annotated one by relying on a set of default heuristics that embed our knowledge on typical uml-to-odata design decisions. This annotated model can be regarded as just an initial option to bootstrap the process that the designer can then modify at will.

Table 1 summarizes our mapping strategies. From left to right, the columns of the table show (1) the involved UML element; (2) the conditions to apply an stereotype (if any), (3) the stereotype to be applied; and (4) the values of the stereotype properties. In a nutshell, each class is mapped to an entity type and is exposed as entity set, each attribute is mapped to a property, and each navigable association is mapped to a navigation property. Figure 5 shows the running example including some of the generated OData profile annotations. This first version of the class diagram can later be customized and used in other model-driven processes to fast prototyping OData Web APIs.

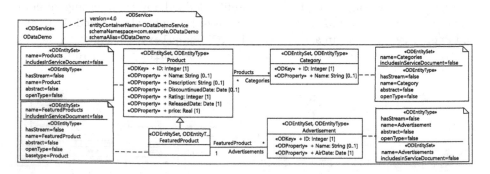

Fig. 5. UML class diagram of the running example annotated by the generator.

Table 1. Rules of the OData profile annotation generator.

UML ELEMENT	CONDITION	STEREOTYPE	VALUE
m: Model	-	ODService s	- s.version = "4.0"
			- s.entityContainerName= m.name+"Service"
			- s.schemaNamespace = "com.example."+m.name
			- s.schemaAlias = m.name
c: Class	-	ODEntityType et	-et.name = c.name
			-if c.abstract == true then et.abstract = true
			-if c.generalization contains t then et.basetype=ot (ot is the entity type of t)
		ODEntitySet es	- es.name = the plural form of c.name
p: Property	p is an class attribute OR a data type attribute	ODProperty op	- op.name = p.name
			- if p.lower == 1 then op.nullable = false
			-op.defaultValue = p.default
	p is an class attribute marked as key	ODKey ok	- ok.name = p.name
	p is an navigable association end	ODNavigationProperty np	- np.name = p.name
			- if p.lower == 1 then np.nullable = false
			- if p.aggregation ==composite then np.containsTarget = true
		ODNavigationPropertyBinding npb	- npb.path = p.name
			- npb.target = the name of the corresponding entity set of the association end
dt: DataType	-	ODComplexType ct	- ct.name = dt.name
			- if dt.abstract == true then ct.abstract = true
			- if dt.generalization contains t then ct.base=ot (ot is the complex type of t)
pt: PrimitiveType	-	ODPrimtiveType opt	- opt.value = the corresponding primitive type of pt.name
e: Enumeration	-	ODEnumType oe	- oe.name = e.name

5   Conclusion

In this paper we have presented a UML profile to model OData Web APIs and the corresponding annotation generator for any UML class diagram. We believe our approach is the first step to boost the model-based development of OData Web APIs, offering developers the opportunity to leverage on the plethora of modeling tools to define forward and reverse engineering to generate, visualize and manipulate OData sources. The OData profile along with the default profile generator are available as an Open Source Eclipse plugin[6]. The plugin repository includes also the steps to reproduce the running example.

As future work we aim at extending our profile in order to capture additional OData behavioral concepts such as functions and actions. We would also

---

[6] https://github.com/SOM-Research/OData.

like to integrate this profile with other web-based modeling languages like IFML (e.g. by enabling the use of OData-like data sources as part of the interface modeling components) in order to create a rich modeling environment combining front-end and back-end development. Finally, we plan to complement the profile support with model-to-text and model-to-model transformations to offer, for instance, the (semi)automatic code-generation of OData services from the annotated models.

Acknowledgment. This work has been supported by the Spanish government (TIN2016-75944-R project).

References

1. Cabot, J., Gogolla, M.: Object constraint language (OCL): a definitive guide. In: Formal Methods for Model-Driven Engineering, pp. 58–90 (2012)
2. Ed-Douibi, H., Izquierdo, J.L.C., Gómez, A., Tisi, M., Cabot, J.: EMF-REST: generation of restful APIs from models. In: SAC Symposium, pp. 1446–1453 (2016)
3. Fraternali, P.: Tools and approaches for developing data-intensive web applications: a survey. CSUR **31**(3), 227–263 (1999)
4. Pizzo, M., Handl, R., Zurmuehl, M.: Odata version 4.0 part 1: protocol. Technical report, OASIS (2014)
5. Pizzo, M., Handl, R., Zurmuehl, M.: Odata version 4.0 part 3: Common Schema Definition Language (CSDL). Technical report, OASIS (2014)
6. Rivero, J.M., Heil, S., Grigera, J., Gaedke, M., Rossi, G.: MockAPI: an agile approach supporting API-first web application development. In: ICWE Conference, pp. 7–21 (2013)
7. Rossi, G., Pastor, O., Schwabe, D., Olsina, L. (eds.): Web Engineering: Modelling and Implementing Web Applications. Human-Computer Interaction Series. Springer, London (2008)
8. Schwinger, W., Retschitzegger, W., Schauerhuber, A., Kappel, G., Wimmer, M., Pröll, B., Cachero Castro, C., Casteleyn, S., De Troyer, O., Fraternali, P., et al.: A survey on web modeling approaches for ubiquitous web applications. IJWIS **4**(3), 234–305 (2008)
9. Segura, Á.M., Cuadrado, J.S., de Lara, J.: ODaaS: towards the model-driven engineering of open data applications as data services. In: EDOCW Workshop, pp. 335–339 (2014)
10. Selic, B.: The pragmatics of model-driven development. IEEE Softw. **20**(5), 19–25 (2003)
11. Valderas, P., Pelechano, V.: A survey of requirements specification in model-driven development of web applications. TWEB **5**(2), 10 (2011)
12. Vallecillo, A., Koch, N., Cachero, C., Comai, S., Fraternali, P., Garrigós, I., Gómez, J., Kappel, G., Knapp, A., Matera, M., Meliá, S., Moreno, N., Pröll, B., Reiter, T., Retschitzegger, W., Rivera, J.E., Schauerhuber, A., Schwinger, W., Wimmer, M., Zhang, G.: MDWEnet: a practical approach to achieving interoperability of model-driven web engineering methods. In: MDWE Workshop, @ICWE (2007)

A Query Log Analysis of Dataset Search

Emilia Kacprzak[1,2]([✉]), Laura M. Koesten[1,2], Luis-Daniel Ibáñez[1],
Elena Simperl[1], and Jeni Tennison[2]

[1] University of Southampton, Southampton SO17 1BJ, UK
{l.d.ibanez,e.simperl}@soton.ac.uk
[2] The Open Data Institute, London EC2A 4JE, UK
{e.kacprzak,laura.koesten,jeni.tennison}@theodi.org

Abstract. Data is one of the most important digital assets in the world
and its availability on the web is increasing. To use it effectively, we need
tools that can retrieve the most relevant datasets to match our information needs. Web search engines are not well suited for this task, as
they are designed primarily for documents, not data. In this paper, we
present the first query log analysis for dataset search, based on logs of
four national open data portals. Our aim is to gain a better understanding of the typical users of these portals and the types of queries they issue,
and frame the findings in the broader context of dataset search. The logs
suggest that queries issued on data portals differ from those issued to
web search engines in their length and structure. From the analysis we
could also infer that the portals are used exploratively, rather than to
answer focused questions. These insights can inform the design of more
effective dataset retrieval technology, and improve the user experience of
data portals.

Keywords: Query log analysis · Dataset search · User behaviour

1 Introduction

Data has become one of the most important digital assets in the world with
its potential for generating business value and social impact. As we advance in
the digital age, more and more of the data we generate can be accessed (or
purchased) online. Cafarella estimates more than one billion sources of data on
the web as of February 2011, counting structured data extracted from Web pages
[3]; and The Web Data Commons project recently extracted 233 million data
tables from the Common Crawl [12].

A growing number of organisations, mostly in the public sector, have set
up their own data portals to publish datasets related to their activities. The
European Data Portal[1] indexes to date 638, 817 datasets published by regional
and national authorities in 28 EU countries. Similar trends can be observed in
the commercial sector and in science, as open access and reproducibility become

---

[1] https://www.europeandataportal.eu/data/en/dataset.

© Springer International Publishing AG 2017
J. Cabot et al. (Eds.): ICWE 2017, LNCS 10360, pp. 429–436, 2017.
DOI: 10.1007/978-3-319-60131-1_29

mainstream across subjects and research communities. Data is used in a variety of professional roles - whether it is a business analyst searching for evidence to substantiate their report, or a scientist replicating an experiment, the first and foremost step is to find, or *retrieve* the most relevant datasets for their needs. In our previous work we asked people to share their experiences when carrying out this task [10]. Using conventional search engines is not ideal, as these have been designed primarily for documents, not data [3]. Kunze et al. have recently introduced the concept of *dataset retrieval*, as a branch of information retrieval applied to data instead of documents focused on determining the most relevant datasets according to a user query [11]. Additionally, there have been proposals for special-purpose search systems, tailored for datasets in a given domain, including hydrology [1]; earth sciences [4]; or datasets from scientific experiments [15]. However, they are mostly restricted to their specific domains and rely on metadata that is created manually.

We advocate for a user-driven approach, where the needs of the dataset searchers are analysed and understood, serving as a basis for the research and development of search functionalities for data. Through the analysis of query logs we explore how people use existing data portals to look for data. We analysed query logs from four open data portals: three belong to the national governments of the UK, Australia, and Canada, the fourth one from the UK's Office for National Statistics. Together, the logs include more than 2.2 million queries (of which 1.3 million are unique queries), issued between 2013 and 2016.

Our study considered the following questions: (ii) What are the main characteristics of queries in terms of length, distribution, and structure? (ii) Can we identify types of queries? Which ones are the most common? (iii) How these metrics differ between dataset and web search?

2   Background and Related Work

Analysing query logs serves as a proxy to analyse the search behaviour of users [9]. The first query log analysis on the web was made for the Altavista search engine [14], and the technique has been used since to study several aspects of web search (see [7] for a survey). As shown in [6], which reports transaction log analysis of nine search engines, results of different search log analyses are not directly comparable. Vertical search engines have also used query log analysis for example, in people search engines [18] and digital libraries [8]. To the best of our knowledge, this is the first query log analysis study for dataset search.

Various metrics for analysing query logs were developed in the area of general web search; a summary of the ones used in this study is shown in Table 1.

3   Data and Methods

Query logs. Four well-known data portals provided their query logs to us. All portals collect log data using Google Analytics, but are using different settings. As a consequence, the collected information and time frame per portal vary.

Table 1. Metrics from web search studies used in this study

Analysis	Metrics used
Query length & distribution	Average length, distribution, percentage of 1, 2 and 3 words queries. These are the most commonly presented statistics and are part of the analysis in our study. E.g., [17] shows trend in web search - an increase in query length through the years. [2,17]
Query structure	Types of queries: question, operator, composite and non-composite [3]. Question queries: starting with words that indicate questions, eg. what or how; operators:containing boolean operators like AND, OR and NOT or special web search operators eg. url, site or filetype. We found a negligible use of special operators (e.g. AND, OR) in our dataset, so we did not conduct further analysis on these structures The composite and non-composite characteristics studied in [3] are relevant for long queries. As dataset search queries are typically short, we did not find this analysis suitable for our case.
Query types Classification	Broder et al. created a taxonomy of web search queries based on user needs [4]. We believe this taxonomy is not directly applicable to dataset search as the information need is finding data and could so be seen as predominantly informational. We chose queries containing specific types of information that have been studied in other search contexts: acronyms, geographic location, temporal indication and numeric values. Understanding the proportion of queries related to these dimensions can help shape indexing strategies

In Table 2 we present the number of queries for each portal and the time frame in which those were collected. We had available data on three types of information objects for analysis: *queries, sessions* and *users*. A query object is comprised of the following fields: *search terms* and *total unique searches*. The search terms of a query are made out of the string typed into the search box of the portal. Neither site had any additional event tracking, e.g., click-through data, configured.

Pre-processing. We filtered out potentially identical queries, for example, those that differ only in lettercase (*LIDAR* vs. *lidar*) or spelling mistakes (*London* vs. *Londno*). We applied two clustering methods on the raw data, Fingerprint and N-Gram Fingerprint, using the OpenRefine framework[2]. Next, we discarded

Table 2. Summary of search log data after pre-processing. Column *all queries* refers to the total number of queries per portal, while column *unique queries* refers to the number of unique queries.

Portal	All	Unique	Time ranges
(DGU) UK Gov Open Data portal	1,058,197	332,823	30/01/2013 – 31/08/2016
(ONS) UK Office for National Statistic[a]	950,593	342,054	28/02/2016 – 31/08/2016
(CAN) Canadian Gov Open Data portal[b]	231,473	46,661	23/08/2015 – 23/08/2016
(AUS) Australian Gov Open Data portal[c]	5,311	2,557	01/08/2016 – 31/08/2016

[a] https://www.ons.gov.uk
[b] https://www.data.gov.au
[c] https://www.open.canada.ca

---

[2] https://www.openrefine.org.

outliers in terms of length. 99.9% of all queries had less than 19 words. We considered longer queries to be likely the result of accidental pasting of text into the search box and discarded them from our analysis.

Analysis. Statistic performed in this analysis include: *Query length* including average query length and distribution for all and unique queries, *Query characteristics* queries containing keywords describing: location; time frame; file and dataset type; numbers; abbreviations - described in detail in Table 4 and *Question queries*. To recognise question queries we counted queries containing the words: *what, who, where, when, why, how, which, whom, whose, whether, did, do, does, am, are, is, will, have, has* as done in [2].

4   Results

This section describes the results of our analysis, reporting on query length and query types. Where applicable we provide related results from web search log analyses. However, as detailed in Sect. 6 on limitations, our results cannot be directly compared with web search - due to the different nature of data - but only serve as an indication of differences for the reader.

Query Length. Table 3 shows the average query length. Almost 90% of all queries have between one and three words, with an average of 2.03 words for all queries and 2.67 for the unique ones. Figure 1 shows the average percentage of queries according to their length, for all portals, for both all and unique internal queries. When considering all queries, single word queries represent almost half of the entire corpus. When focusing just on unique queries, this metric falls down to 25%. The distribution for unique queries is very similar to the results reported for web search engines in 2001 by Spink [16]. It could be that advances in dataset search will lead to similar behaviour patterns as observed in web search today (as reported by [17] they were steadily growing in length over time), which would mean longer queries, closer to natural language as technologies and tools improve.

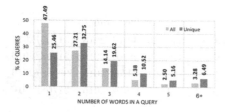

Fig. 1. Percentage of internal queries according to average number of words (all and unique queries, all portals)

Table 3. Average number of words per query

Portal	All	Unique
DGU	2.04	2.78
ONS	2.52	3.42
DGA	1.63	2.31
OCC	1.93	2.17

Query Types. Having learned that people seem to be interested in the provenance of the datasets they are looking for, we explored whether location, time, or specific publishing formats are in their focus.

Table 4. Definition of query characteristic metrics. Percentage of queries computed for four portals

Metric	Definition	% queries
Location	The name of a city or geographical area (either town, city, county, region or countries)	5.44%
Time frame	Years (1000 to 2017), names of months, days of a week and the words *week(ly)*, *year(ly)*, *month(ly)*, *day(ly)*, *date*, *time* and *decade*	7.29%
File and dataset type	File types: *csv, pdf, xls, json, wfs, zip, html, api* and keywords denoting a type of dataset: *data, dataset, average, index, graph, table, database, indice, rate, stat*	6.25%
Numbers	The number of queries including numbers excluding those indicating time frames	5.23%
Only numbers	Queries that contain only numbers	0.38%
Abbreviations	72 most popular, manually identified acronyms	5.11%
Question queries	Queries containing the words: *what, who, where, when, why, how, which, whom, whose, whether, did, do, does, am, are, is, will, have, has* as done in [2]	<1%

Location, temporal, filetype, numerical, and acronym queries. Table 4 summarises the percentage of queries for the individual metrics. 5.44% of queries contain location keywords. A previous study on general web search [5] reports this figure at 12.01%. This difference could be caused by the fact that data portals are already area-bound and users do not need to specify the location as frequently as in web search.

7.29% of the queries contain temporal information. This number is much higher than the 1.5% reported for general web search [13]. This may mean that, for datasets, users have more interest in the time frame in which the data was created or the frequency of data releases and updates.

6.25% of queries included common filetypes. We note that the governmental portals represented in this study offer filtering options for file types that are not reflected in our data - the actual number of queries filtering by a file format could be higher. From an interface point of view, this suggests the filters might not be prominent enough. From a data point of view, this figure is an indication that users search for particular filetypes and formats and that publishers need to be able to support different and popular formats for their data.

5.23% of queries contain numerical values other than temporal information, notably, we discovered searches age and distance intervals associated to datasets (e.g., *wage age 18–24* or *LIDAR 25 meters*) that might indicate a need for retrieving subsets or slices of datasets.

In our analysis, we also identified that users frequently use abbreviations in their queries, as many datasets use acronyms like *rpi* for *Retail Price Index.*

5.11% of queries contained at least one acronym. However, the full expansion of those acronyms is also used in queries.

Question queries. These are increasingly common in web search queries, thanks to advances in speech recognition and conversational search interfaces. This is not yet the case for dataset search - less than 1% of queries in our logs are question queries, significantly below the 7.49% reported by [2] for web search. We believe this is mostly due to the way portals are used: as a source of data to be downloaded for further use, and not as a question answering engine.

5   Discussion and Implications

Our analysis shows differences in the ratio of question queries in dataset search and general web search and that they are generally short, on average one word shorter than web search queries, as per the 2011 report by [17]. We believe short queries potentially indicate a lack of trust that the search functionality will be able to provide relevant data for longer queries. It appears that users currently tend to treat the search field of a data portal as a starting point for further exploration. The categories and metadata attributes used in data portals, and the ability to link from one dataset to another becomes key in enabling users to find what they are looking for. Our analysis of query characteristics offers a starting point for data portal designers to refine their schemas and vocabularies.

We believe both temporal search, which is more prevalent than in document search, and geospatial search require better support. In both cases, relevant keywords are at different levels of granularity (e.g., months vs years, cities vs regions or countries), which is not always matched by the publishing practices of the data owners. While the data portals we analysed are location-bound to a country and most datasets hold national data, supporting question-answering and dataset search scenarios will require more advanced geospatial indexing and reasoning features. Queries including some indication of time were almost five times more frequent than in web search, suggesting that datasets have a stronger relationship to time than documents. This could be due, for instance, to periodic updates or temporal coverage of the data. Currently time references are recorded in the title of a dataset or as metadata attributes. This motivates the indexing of data along time dimensions, and the generation of temporal information from datasets that do not have it. The same is true about queries using numbers, abbreviations and named entities such as economic indicators. These kinds of information are likely to be found either in well-maintained, rich metadata descriptors or if data portals would expand their search capabilities, in the content of datasets. To get more accurate results for users' information needs, portals could encourage users to ask more specific or longer queries. One solution might be implementing query recommendations based on the strongest co-occurrences of words in the datasets or past search queries. However, to achieve better results in dataset search, we need to improve the way datasets are indexed. We believe that automatically generated descriptions of the content of datasets, together with encouraging users to issue longer queries, could improve this process.

6 Limitations of Our Study

Comparisons of search log analysis present difficulties as concluded by [6] in their study comparing nine search engines by their transaction logs (over 1 billion queries in total). Even within web search, it is stated that findings resulting from the analysis of one search engine cannot be applied to all web search engines. Even more so, the comparison of our results with web search needs to be seen with caution, due to the different nature of the collected data. However, we believe that including data from several countries and different audiences increases the generalisation of our analysis.

Our study is based on dataset search engines that are part of governmental open data portals. Further studies with other kinds of dataset search engines are required before drawing general conclusions.

As we did not have control on the analytics being collected by each data portal, the time frames was different for each. In cases where all queries were considered, there is a bias towards DGU, as the one with more available data.

7 Conclusions and Future Work

We have presented the first analysis of query log data for the search vertical of dataset retrieval, based on query logs of four national data portals. Our findings can be summarised as: (i) Dataset queries are generally short. (ii) The portals are used exploratively, rather than to answer focused questions. (iii) There is a difference in topics between dataset queries issued directly to data portals and general web search queries.

As future work, we would like to (i) Analyse query log data from commercial dataset search engines, to identify differences and similarities with this study. (ii) Extend our study to click-through data: knowing which dataset pages users visited after performing a search and if a user downloaded them can prove invaluable to evaluate the effectiveness of the dataset search. (iii) Create a dataset search corpus in order to evaluate dataset search engines. (iv) Develop metrics specifically tailored for the analysis of dataset search logs, due to the unique characteristics of this vertical.

Acknowledgement. This project is supported by the European Union Horizon 2020 program under the Marie Sklodowska-Curie grant agreement No. 642795.

References

1. Ames, D.P., Horsburgh, J.S., Cao, Y., Kadlec, J., Whiteaker, T.L., Valentine, D.: HydroDesktop: web services-based software for hydrologic data discovery, download, visualization, and analysis. Environ. Model. Softw. **37**, 146–156 (2012)
2. Bendersky, M., Croft, W.B.: Analysis of long queries in a large scale search log. In: Proceedings of the 2009 Workshop on Web Search Click Data, pp. 8–14. ACM (2009)

3. Cafarella, M.J., Halevy, A., Madhavan, J.: Structured data on the web. Commun. ACM **54**(2), 72–79 (2011)
4. Devarakonda, R., Palanisamy, G., Wilson, B.E., Green, J.M.: Mercury: reusable metadata management, data discovery and access system. Earth Sci. Inform. **3**(1), 87–94 (2010)
5. Gan, Q., Attenberg, J., Markowetz, A., Suel, T.: Analysis of geographic queries in a search engine log. In: Proceedings of the First International Workshop on Location and the Web, pp. 49–56. ACM (2008)
6. Jansen, B.J., Spink, A.: How are we searching the world wide web? a comparison of nine search engine transaction logs. Inf. Process. Manag. **42**(1), 248–263 (2006)
7. Jiang, D., Pei, J., Li, H.: Mining search and browse logs for web search: a survey. ACM Trans. Intell. Syst. Technol. **4**(4), 57:1–57:37 (2013)
8. Jones, S., Cunningham, S.J., McNab, R., Boddie, S.: A transaction log analysis of a digital library. Int. J. Digit. Libr. **3**(2), 152–169 (2000)
9. Kelly, D.: Methods for evaluating interactive information retrieval systems with users. Found. Trends Inf. Retrieval **3**(1–2), 1–224 (2009)
10. Koesten, L.M., Kacprzak, E., Tennison, J., Simperl, E.: The trials and tribulations of working with structured data - a study on information seeking behaviour. In: Proceedings of the 2017 CHI Conference on Human Factors in Computing Systems, CHI 2017. ACM (2017, to appear)
11. Kunze, S.R., Auer, S.: Dataset retrieval. In: 2013 IEEE Seventh International Conference on Semantic Computing, September 2013
12. Lehmberg, O., Ritze, D., Meusel, R., Bizer, C.: A large public corpus of web tables containing time and context metadata. In: Proceedings of the 25th International Conference Companion on World Wide Web, pp. 75–76 (2016)
13. Nunes, S., Ribeiro, C., David, G.: Use of temporal expressions in web search. In: Macdonald, C., Ounis, I., Plachouras, V., Ruthven, I., White, R.W. (eds.) ECIR 2008. LNCS, vol. 4956, pp. 580–584. Springer, Heidelberg (2008). doi:10.1007/978-3-540-78646-7_59
14. Silverstein, C., Marais, H., Henzinger, M., Moricz, M.: Analysis of a very large web search engine query log. ACM SIGIR Forum **33**(1), 6–12 (1999)
15. Singhal, A., Kasturi, R., Sivakumar, V., Srivastava, J.: Leveraging web intelligence for finding interesting research datasets. In: IEEE/WIC/ACM International Joint Conferences on Web Intelligence and Intelligent Agent Technologies (2013)
16. Spink, A., Wolfram, D., Jansen, M.B., Saracevic, T.: Searching the web: the public and their queries. J. Am. Soc. Inf. Sci. Technol. **52**(3), 226–234 (2001)
17. Taghavi, M., Patel, A., Schmidt, N., Wills, C., Tew, Y.: An analysis of web proxy logs with query distribution pattern approach for search engines. Comput. Stand. Interfaces **34**(1), 162–170 (2012)
18. Weerkamp, W., Berendsen, R., Kovachev, B., Meij, E., Balog, K., de Rijke, M.: People searching for people: analysis of a people search engine log. In: Proceedings of the 34th international ACM SIGIR Conference on Research and Development in Information Retrieval (2011)

Recruiting from the Network: Discovering Twitter Users Who Can Help Combat Zika Epidemics

Paolo Missier[1], Callum McClean[1], Jonathan Carlton[1(✉)],
Diego Cedrim[2], Leonardo Silva[2], Alessandro Garcia[2], Alexandre Plastino[3],
and Alexander Romanovsky[1]

[1] School of Computing Science, Newcastle University, Newcastle upon Tyne, UK
j.carlton@ncl.ac.uk
[2] PUC-Rio, Rio de Janeiro, Brazil
[3] Universidad Federal Fluminense, Niterói, Brazil

Abstract. Tropical diseases like *Chikungunya* and *Zika* have come to prominence in recent years as the cause of serious health problems. We explore the hypothesis that monitoring and analysis of social media content streams may effectively complement institutional disease prevention efforts. Specifically, we aim to identify selected members of the public who are likely to be sensitive to virus combat initiatives. Focusing on Twitter and on the topic of Zika, our approach involves (i) training a classifier to select topic-relevant tweets from the Twitter feed, and (ii) discovering the top users who are actively posting relevant content about the topic. In this short paper we describe our analytical approach and prototype architecture, discuss the challenges of dealing with noisy and sparse signal, and present encouraging preliminary results.

1 Introduction

Mosquito-borne disease epidemics such as *Chikungunya* and *Zika* viruses are becoming more frequent in subtropical areas around the world [8], and are responsible for thousands of deaths every year [3]. In Brazil, the regional focus of our research, disease prevention programs led by health government authorities have not been particularly effective. It is therefore natural that Brazilians have become heavy users of social channels to share mosquito-related information. This includes complaints about personal health, dissemination of public news, but also, importantly, details about the discovery of mosquito breeding sites in public locations. This presents an opportunity to complement existing disease prevention programs, as real-time social media is potentially a much faster vehicle for information than traditional channels, and we also hope to discover a few users who stand out for the quality and relevance of their contribution to the social media. These users are referred to as *social sensors* [12], as they spontaneously contribute with information on social media channels, which is relevant to a particular topic.

© Springer International Publishing AG 2017
J. Cabot et al. (Eds.): ICWE 2017, LNCS 10360, pp. 437–445, 2017.
DOI: 10.1007/978-3-319-60131-1_30

In this short paper we present our initial investigation into techniques to identify target users who show to be good social sensors, with the aim to engage them in disease prevention programs within their community. Our approach, summarised in the dataflow diagram of Fig. 1, combines content-based automated classification of tweets, aimed at isolating the relevant signal out of generally noisy chatter about Zika (training phase indicated in the left of the figure), followed by a ranking of the users who author such relevant content (online phase, in the right of the figure). Note that *Zika* is a common slang word in Brasilian Portuguese, often used out of context, resulting in a high-recall but particularly noisy harvest from the Tweeter feed. In this "needle in the haystack" problem, the main challenge is to filter out the large proportion of noise and irrelevant news items about Zika (relevant tweets are less than 10% of a typical harvest), as well as the identify the very few target users who consistently tweet relevant content. We present preliminary results on comparing three user ranking metrics, including our own single-topic adaptation of TwitterRank [14], computed from a set of about 200,000 tweets and 180 active users, harvested during 4 months in 2016. Many details are omitted for space reasons. Please refer to our Technical Report [9] for a more complete account.

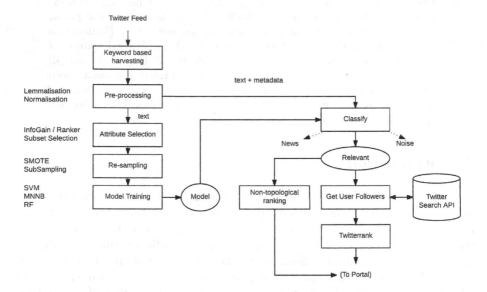

Fig. 1. Dataflow diagram for content classification and user ranking

This work follows on from [10], by re-focusing it on Zika content, extending it to user ranking, and providing a prototype implementation of our streaming architecture. Also, the prototype has been integrated with our VazaZika portal.[1] VazaZika works as an entomological surveillance system in order to combat the

---

[1] Available at http://vazadengue.inf.puc-rio.br/.

mosquito that transmits Zika, Chikungunya, and Dengue. The portal and a mobile app allow users to report and visualize occurrences of the mosquito or cases of sick people. VazaZika is integrated to social medias in order to reveal social sensors in such medias. Our solution plays an important role to popularize the surveillance system and the engagement programs provided by the VazaZika portal.

Related work. TURank [15] uses link structure analysis on the user-tweet graph to rank Twitter users, including *follow* and *retweet* relationships. While our approach, based on TwitterRank, does not include retweets, unlike TURank we do analyse tweet content.

Another related approach [5] aims to find influential users such as disseminator, expert, etc. However there is an assumption that the follower of someone who is an expert on a topic is also interested in that topic. TwitterRank, in contrast, only considers followers of Relevant tweets, who also authored other Relevant tweets.

Wei *et al.* [13] use a combination of Twitter lists (a grouping of followers per a criterion), the follower graph and the users profile information to produce a global authority score for each user in their data set. We may experiment with incorporating user profile metadata into our analysis in our future work.

Finally, [6] aims to discover expert uses on Twitter, assuming that experts will exhibit different Twitter usage patterns than non-experts. Our work differs as we aim at seeking out users who stand out not because of their expertise but because of their demonstrated interest in engaging with a specific topic.

2   Tweets Classification and User Ranking

User ranking requires firstly the capability to identify with high precision the few tweets that are relevant to the Zika topic, amongst a large amount of Twitter noise. For this, we tuned a harvester on a set of relevant keywords and then trained a supervised classifier on an initial set of about 10,000 tweets manually annotated by an expert. This set was collected over three months in 2016. In order to check the consistency of the manual annotations, a representative random sample of the set (margin of error $\pm 5\%$ at the 95% confidence level) was manually re-annotated by a second expert. The agreement between the experts was then assessed by calculating Cohen's Kappa coefficient [7]. We found a substantial agreement (0.70, p-value $= 0.000$), reaching an almost perfect agreement for the relevant class (0.82, p-value $= 0.000$).

The need to use the keyword *Zika*, which is a common slang word in Brasilian Portuguese, often used out of context, makes the harvesting and initial filtering difficult and results in a particularly noisy dataset. The process of fine-tuning the data pre-processing and model training pipeline is described in detail in our Report [9]. Here we only report on the final configuration and its performance.

Harvesting content from Twitter (top of Fig. 1) provides content both for manual annotation and model training, as well as for classification and then

Table 1. Classifier accuracy for various choices of N-grams and over- and sub-sampling

	RF			MNNB		
	1-grams	1+2-grams	1+2+3-grams	1-grams	1+2-grams	1+2+3-grams
SMOTE over-sampling	83.5	83.1	**84.1**	81.2	80.9	81.2
Sub-sampling (Spread)	75.8	76.3	76.1	77.5	78.9	79.95
Over- and sub-sampling	82.5	82.7	83.6	80.6	80.0	80.95
+600 Relevant samples	80.8	80.5	80.4	80.5	81.0	81.2

user ranking. High recall is important in the initial filtering, as the relevant tweets we seek to isolate are no more than about 10% of the feed. Filtering keywords were selected in two steps, following an approach similar to that suggested in [11] and using a short list of expert-chosen *seed* keywords, for bootstrapping the process: dengue, combateadengue, focodengue, todoscontradengue, aedeseagypti, zika, chikungunya, virus.

Those keywords were then used to harvest an initial corpus of tweets, whose terms were then ranked according to their TF-IDF score. The top 10 of those were added to the initial seed set: microcefalia, transmitido, epidemia, transmissao, doenca, eagypti, doencas, gestantes, infeccao, mosquitos.

2.1   Learning a Relevance Model

Having broadly harvested Zika-related content, the purpose of the tweet classifier is to accurately separate the few interesting tweets, i.e., the Relevant class, from a majority of content carrying News about Zika, and the general Noise class. Based on our prior experience comparing traditional supervised learning (Naive Bayes) with unsupervised topic modelling (LDA [1]), here we focus solely on classification.

The data preparation steps are listed in Fig. 1-left. Firstly, tweet content is reduced to a bag-of-words with N-grams (N = 1,2,3) representation, using POS tagging and lemmatisation and removing common abbreviations, as well as all emoticons and non-verbal forms of expressions. Secondly, we tested two attribute selection approaches, namely Ranking with Information Gain vs Subset Selection, but concluded that neither resulted in improved overall performance. Finally, noting imbalance in the distribution of examples over the classes in the training set: 50.6% News, 37.3% Noise, 12.1% Relevant, we added an extra 600 annotated examples to the Relevant class, and applied the SMOTE algorithm [4] to over-sample the Relevant class, boosting the examples from 1,214 to 2,428 (12.1% to 24.3%).

We experimented with three popular classification models that have proved effective for short text classification [2]: Support Vector Machines (SVM), Multinomial Naive Bayes (MNNB), and Random Forest (RF). After ruling out SVM due to poor performance (details in [9]), we mapped a space of pre-processing configurations as shown in Table 1. Interestingly, using SMOTE provides equivalent performance to that obtained by investing extra human annotation effort. Note that the results also show that down-sampling the majority class (News) is not as beneficial.

The best overall accuracy figure, 84.1%, is obtained using a Random Forest learner (using an ensemble of 100 trees), with 1,2,3-grams, no attribute selection, and SMOTE-based boosting (weighted average F-measure = 0.84, RMSE = 0.28). This is the classifier we used for the online content relevance detection phase in combination with user ranking, described next.

2.2 User Ranking Metrics

This sparsity of users suggests that TwitterRank may not be effective on this datasets, as it assumes knowledge of the social graph neighbourhood for each candidate user, and requires that meaningful social connections exist within those neighbourhoods. At the same time, note that no ground truth, i.e., explicit knowledge of the top users, is available for evaluation, as our content harvesting was performed purely "in the wild" (the same occurred in the original Twitter-Rank research [14]).

Our approach is therefore to compare user ranking from TwitterRank with two simple additional, non-topological ranking criteria. For a user u and a set K of keywords, let T_K denote the entire harvest, $T_K(u)$ the number of tweets in T_K that are attributed to u, $R_K(u)$ the number of Relevant tweets in $T_K(u)$, and $T(u)$ the total number of tweets posted by u during the harvest period. We experiment with **Topic Focus** per user, defined as $TF(u) = \frac{R_K(u)}{T_K(u)}$, and with **Overall Focus** per user as $TF(u) = \frac{R_K(u)}{T(u)}$. These count the Relevant fraction of u's tweets in the harvest, as an indication of how often user u used the keywords K to express relevant content; and the Relevant fraction of u's total tweets in the harvest period, i.e., relative to the user's global interests when posting on Twitter. Note that TwitterRank is agnostic to the set of topics users are interested in. Thus, we modify it to operate on a single topic, and we adapt the original definition of topical differences between two users to work in our context. Please refer to [9] for details.

3 Results

Our experimental dataset consists of a harvest of 278,351 tweets, collected and classified through our online pipeline (Fig. 1) using the keywords presented earlier during a period of 4 months (9–12) in 2016. Using our classifier, we found 15,124 Relevant tweets in this set. The distribution of tweets per user is very skewed, with the vast majority of users producing very few Relevant tweets during the harvest period: 2 users authored ≥20 Relevant tweets, 42 between 5 and 20 tweets; 57 : 4 tweets; 209 : 3 tweets, and 12,918 users only one tweet. In practice, the 13,228 candidate users produce a very weak signal both in terms of generated content and in terms of their social connections to other candidate users.

To deal with this long tail and to strike a balance between strength of content signal and numerosity of candidate users, we only considered users who posted at least 3 `Relevant` tweets. Out of these 310 users, however, we had to exclude a further 139 whose followers could not be obtained due to privacy settings, leaving 171 *candidate* users for ranking. The results presented below concern these users. Tables 2, 3, and 4 show the top 10 users ranked according to each of our three criteria (TwitterRank, Topic Focus, and Overall Focus), respectively. For each of these users, each table also shows the values for the other two metrics, and the position of that user when ranked according to those metrics.

Regarding TwitterRank, we note firstly that the small absolute figures are not indicative, as the original paper [14] does not provide any reference figures

Table 2. Top 10 TwitterRank candidate users

Screenname	Twitterrank (x100)	Relevant count	Overall focus (x100)	OF Rank	Topic focus	TF Rank
FlorzinhaSimoes	0.84	20	14.28	3	71.428	15
Lorrayn54837060	0.64	3	0.1708	142	75	14
pelotelefone	0.41	7	6.1947	7	87.5	7
SEIZETHEHEAVEN	0.39	7	0.3693	65	100	1
macabia	0.39	3	0.44	55	100	5
gushfsc	0.37	6	0.30	85	60	18
tiiancris	0.37	3	0.19	128	50	24
scomacinha	0.35	3	0.13	164	33.33	28
sophiaboggiano	0.35	3	0.14	160	75	14
mariabarrozoo	0.34	3	0.11	169	60	19

Table 3. Top 10 topic focus candidate users

Screenname	Topic focus	Relevant count	All tweets count	Overall focus (x100)	OF rank	TR (x100)
SEIZETHEHEAVEN	100	7	1895	0.3693	65	0.39
LairaMaia	100	6	799	0.7509	35	0.07
llGueto	100	6	1427	0.4204	58	0.07
Giovannacoosta	100	5	960	0.5208	45	0.06
pakito_lucas	100	5	2149	0.2326	111	0.06
Lorranna_Castro	100	5	1573	0.3178	84	0.06
laricrvlh	100	5	951	0.5257	43	0.06
mauriciooasn	100	4	495	0.8080	33	0.04
masoqmath_	100	4	2412	0.1658	145	0.04
isaah13_ferreir	100	4	272	1.4705	19	0.04

Table 4. Top 10 overall focus candidate users

Screenname	Relevant count	Keyword count	All tweets count	Overall focus (Rel/All)	Topic focus	TF rank	TR	TR position
Leilaquintsepe	4	4	19	21	100	=4	0.04	70
DCGRodrigues	3	3	18	16.6	100	=5	0.03	169
FlorzinhaSimoes	20	28	140	14.2	71.4	15	0.8	1
RobelioValle	3	4	31	9.6	75	=14	0.03	156
iaedayana	3	3	37	8.1	100	=5	0.03	125
iPedersoly	4	5	51	7.8	80	=10	0.04	81
pelotelefone	7	8	113	6.1	87.5	=7	0.4	3
tacianebielinki	6	10	136	4.4	60	=18	0.07	32
isaldcunha	3	4	98	3	75	=15	0.03	147
onelastovada	7	9	285	2.4	77.7	11	0.1	24

at all. However we note a significant spread (150%) between the top and bottom ranks in the top-10 list. As noted earlier, however, the significance of this ranking is questionable, because our candidate users have very few connections amongst each other. Despite this, looking at the social connections amongst some of our candidate users, as in Fig. 2 (Appendix A) reveals few but interesting connections, and indeed even a few friends (shown with the double arrow). Note also that all of our top-10 TwitterRank users appear in some connected component of the graph, which is natural as it is their connectivity that contributes to their TwitterRank. On the other hand, the number of followers of any user who actually influence the user's rank is very small.

Comparing this ranking with the other two metrics, we note (Table 3) that for each of the top-10 Topic Focus users, *all* of their tweets in the harvest ($T_K(u)$), however few (<10), are `Relevant`. Furthermore, the `TF Rank` column in Table 2 shows that all top 10 TwitterRank users are top-30 Topic Focus users, suggesting that high TwitterRankx may correlate well with high Topic Focus. Interestingly, in Table 4 we see that the top-10 Overall Focus users also have a high Topic Focus, and rank within the top-20. Again in this list we find users that rank high in other lists: `FlorzinhaSimoes` and `pelotelefone`.

4   Conclusions

We have begun exploring the hypothesis that social media analytics can be used to identify individuals who are actively contributing to social discourse on the specific topic of the Zika virus and its consequences, and are thus likely to be sensitive to health promotion campaigns. We tested this hypothesis by focusing on Twitter content related to the Zika virus and its effect on people. We trained a classifier to separate the very sparse interesting signal from large amounts of noise in the feed, and then applied multiple ranking criteria to the set of candidate users who authored such interesting content.

Given the sparsity of the contributors and their limited connections within the social graph, we found that the very popular TwitterRank algorithm [14] is not very effective. Despite facing a "needle in the haystack" problem, however, we report promising results which indicate that non topology-based metrics that count relevant tweets by user appear to be equally effective, and that a few interesting connections indeed exist in the graph amongst the top ranked users. We are currently experimenting with larger datasets which we continually harvest from the live twitter feed. We have developed a public-facing portal where Relevant tweets that are also geo-located are placed on a map of Brasil, and soon the top-k users computed using our metrics will be shown and continually updated.

A   Social Graph Fragment

See Fig. 2.

Fig. 2. Fragment of followers and friends graph for candidate users in our experimental dataset. Green nodes are in the top 10 TwitterRank. Blue nodes are in top 10 TwitterRank *and* top 10 Overall Focus. Red nodes are in top 10 TwitterRank *and* top 10 Topic Focus.

References

1. Blei, D.M., Ng, A.Y., Jordan, M.I.: Latent dirichlet allocation. J. Mach. Learn. Res. **3**, 993–1022 (2003)
2. Carvalho, J., Plastino, A.: An assessment study of features and meta-level features in twitter sentiment analysis. In: ECAI 2016–22nd European Conference on Artificial Intelligence, 29 August-2 September 2016, The Hague, The Netherlands, pp. 769–777 (2016)
3. CDC: Centers for Disease Control and Prevention (2015). http://www.cdc.gov/dengue/. Accessed 15 Dec 2015

4. Chawla, N.V., Bowyer, K.W., Hall, L.O., Kegelmeyer, W.P.: SMOTE: synthetic minority over-sampling technique. J. Artif. Intell. Res. **16**, 321–357 (2002)
5. Chen, C., Gao, D., Li, W., Hou, Y.: Inferring topic-dependent influence roles of twitter users. In: Proceedings of the 37th International ACM SIGIR Conference on Research and Development in Information Retrieval, SIGIR 2014, NY, USA, pp. 1203–1206. ACM, New York (2014)
6. Horne, B.D., Nevo, D., Freitas, J., Ji, H., Adali, S.: Expertise in social networks: how do experts differ from other users? In: Proceedings of the Tenth International AAAI Conference on Web and Social Media, vol. 10, pp. 583–586 (2016)
7. Landis, J.R., Koch, G.G.: The measurement of observer agreement for categorical data. Biometrics **33**, 159–174 (1977)
8. Miles, T., Hirschler, B.: Zika virus set to spread across americas, spurring vaccine hunt, January 2016
9. Missier, P., Mcclean, C., Carlton, J., Cedrim, D., Silva, L., Garcia, A., Plastino, A., Romanovsky, A.: Recruiting from the network: discovering Twitter users who can help combat Zika epidemics. Research report, School of Computing Science, Newcastle University (2017). https://arxiv.org/abs/1703.03928
10. Missier, P., Romanovsky, A., Miu, T., Pal, A., Daniilakis, M., Garcia, A., Cedrim, D., Silva Sousa, L.: Tracking dengue epidemics using twitter content classification and topic modelling. In: Casteleyn, S., Dolog, P., Pautasso, C. (eds.) ICWE 2016. LNCS, vol. 9881, pp. 80–92. Springer, Cham (2016). doi:10.1007/978-3-319-46963-8_7
11. Nagarajan, M., Gomadam, K., Sheth, A.P., Ranabahu, A., Mutharaju, R., Jadhav, A.: Spatio-temporal-thematic analysis of citizen sensor data: challenges and experiences. In: Vossen, G., Long, D.D.E., Yu, J.X. (eds.) WISE 2009. LNCS, vol. 5802, pp. 539–553. Springer, Heidelberg (2009). doi:10.1007/978-3-642-04409-0_52
12. Sakaki, T., Okazaki, M., Matsuo, Y.: Earthquake shakes Twitter users: real-time event detection by social sensors. In: Proceedings of WWW 2010, p. 851 (2010)
13. Wei, W., Cong, G., Miao, C., Zhu, F., Li, G.: Learning to find topic experts in twitter via different relations. IEEE Trans. Knowl. Data Eng. **28**(7), 1764–1778 (2016)
14. Weng, J., Lim, E.P., Jiang, J., He, Q.: TwitterRank: finding topic-sensitive influential twitterers. In: Proceedings of the Third ACM International Conference on Web Search and Data Mining, pp. 261–270. ACM (2010)
15. Yamaguchi, Y., Takahashi, T., Amagasa, T., Kitagawa, H.: TURank: twitter user ranking based on user-tweet graph analysis. In: Chen, L., Triantafillou, P., Suel, T. (eds.) WISE 2010. LNCS, vol. 6488, pp. 240–253. Springer, Heidelberg (2010). doi:10.1007/978-3-642-17616-6_22

Towards Automatic Generation of Web-Based Modeling Editors

Manuel Wimmer[1]($\boxtimes$), Irene Garrigós[2], and Sergio Firmenich[3,4] (iD)

[1] BIG, TU Wien, Vienna, Austria
wimmer@big.tuwien.ac.at
[2] WaKe Research, University of Alicante, Alicante, Spain
igarrigos@dlsi.ua.es
[3] LINVI, Universidad Nacional de la Patagonia San Juan Bosco,
Puerto Madryn, Argentina
sergio.firmenich@lifia.info.unlp.edu.ar
[4] LIFIA, Universidad Nacional de La Plata and CONICET Argentina,
La Plata, Argentina

Abstract. With the current trend of digitalization within a multitude of different domains, the need raises for effective approaches to capture domain knowledge. Modeling languages, especially, domain-specific modeling languages (DSMLs), are considered as an important method to involve domain experts in the system development. However, current approaches for developing DSMLs and generating modeling editors are mostly focusing on reusing the infrastructures provided by programming IDEs. On the other hand, several approaches exist for developing Web-based modeling editors using dedicated JavaScript frameworks. However, these frameworks do not exploit the high automation potential from DSML approaches to generate modeling editors from language specifications. Thus, the development of Web-based modeling editors requires still major programming efforts and dealing with recurring tasks.

In this paper, we combine the best of both worlds by reusing the language specification techniques of DSML engineering approaches for generating Web-based modeling editors. In particular, we show how to combine two concrete approaches, namely Eugenia from DSML engineering and JointJS as a protagonist from JavaScript frameworks, and demonstrate the automation potential of establishing Web-based modeling editors. We present first results concerning two reference DSML examples which have been realized by our approach as Web-based modeling editors.

1 Introduction

With the current trend of digitalization in a multitude of domains, effective approaches to capture domain knowledge are a must. Modeling languages, especially domain-specific modeling languages (DSMLs) [1], are considered an important foundation to involve domain experts in the system development. A DSML consists of *(i)* an abstract syntax that defines the concepts of a language and the relationships between them, as well as the rules that establish when a model is well formed, *(ii)* a concrete syntax that establishes the language notation which is used by the users of the

© Springer International Publishing AG 2017
J. Cabot et al. (Eds.): ICWE 2017, LNCS 10360, pp. 446–454, 2017.
DOI: 10.1007/978-3-319-60131-1_31

language, and *(iii)* the semantics, i.e., how the modeling concepts are interpreted. In model-driven engineering (MDE) [2], the abstract syntax of a DSML is defined in terms of a metamodel. The concrete syntax can be both; textual and graphical; or even a mixture of both. To support the development of DSMLs as well as supporting tools, various metamodeling tools have emerged that allow to create textual DSMLs (e.g., consider EMFText [3] and Xtext [4]) and graphical DSMLs (e.g., GMF, MetaEdit+ [5, 6], Eugenia [8], and DSL Tools [7]).

Regarding graphical DSMLs, current approaches for developing them and generating modeling editors [8, 9] are mostly focusing on reusing the infrastructures provided by programming IDEs. On the other hand, several approaches exist [11, 12] for developing Web-based modeling editors using dedicated JavaScript (JS) frameworks. Compared to DSML-aware editors such as developed with EMF/xText, a Web-based editor may allow a much richer graphical representation of the DSML. Another advantage is that a Web-based modeling editor is very lightweight and simple to access. In cases where is not possible or not desired to use a modeling framework such Eclipse or MPS [10], you might still be able to integrate a Web-based modeling editor. However, these frameworks do not exploit the high automation potential from DSML approaches to generate modeling editors from language specifications. Thus, the development of Web-based modeling editors requires still major programming efforts.

In this paper, we combine the best of both worlds by reusing the language specifications for generating Web-based editors. In particular, we show how to combine two concrete approaches, namely EuGENia [8] from DSML engineering and JointJS [11] as a protagonist from JS frameworks and demonstrate the automation potential of establishing Web-based modeling editors. Finally, we discuss the results of using our approach for two existing DSMLs.

The outline of the paper is as follows. Section 2 introduces and compares the two approaches we are connecting in our proposal (i.e., Eugenia and JointJS). Section 3 presents our approach based on code generation and two concrete cases realized by our approach. Finally Sect. 4 presents the related work, before Sect. 5 concludes and outlines future work.

2 Background

In this section, we explain the basics of graphical modeling languages as well as the two worlds we connect with our approach. In particular, we present EuGENia as a concrete approach to specify DSMLs and automatically generate graphical modeling editors and JointJS for implementing graphical Web-based modeling editors.

2.1 Anatomy of Graphical Modeling Languages

A graphical concrete syntax (GCS) [2] has to define the following elements: *(i) graphical symbols*, e.g., lines, areas, complete figures such as SVG graphics, *(ii) labels* for representing textual information, e.g., for visualizing the names of modeling elements; *(iii) compositional rules*, which define how these graphical symbols are nested and combined, e.g., a label visualizing the name of a model element is centered within a rectangle representing the model element; and *(iv)* *mapping* of the *graphical symbols* to the elements of the *abstract syntax* for stating which graphical symbol should be used for which modeling concept, e.g., a specific model element type is visualized by a rectangle.

Current graphical modeling editors use modeling canvases which allow the positioning of model elements in a two-dimensional raster. Each element has an assigned *x,* *y* coordinate which normally stands for the upper-left corner of the graphical symbol. The model elements are mostly arranged as a graph which is contained in the modeling canvas. This graph is called diagram and represents a graphical view on the model. Please note that not all model information has to be actually shown in the modeling canvas. Several property values may be shown and may be editable in an additional property view. This, on the one hand, allows accessing and editing every property of a model element, while, on the other hand, avoids overloading the diagram with too much information.

2.2   Eugenia

We selected EuGENia for demonstrating the bride between DSML engineering approaches and JS-based modeling editors, because it allows to introduce a GCS on an appropriate level of abstraction and complements the Eclipse Modeling Framework (EMF) in this respect. In particular, EuGENia provides several annotations for specifying the GCS for a given Ecore-based metamodel which describes the abstract syntax of a modeling language, i.e., the concepts and their properties without describing the concrete notation for the users of the language. In the following, the main annotations[1] are first enumerated and subsequently applied for an application example.

Diagram: The root element of the abstract syntax representing the model, i.e., the element containing (directly or indirectly) all other elements, is a perfect match for representing the modeling canvas.

Node: Instances of metamodel classes are often visualized as nodes within the diagrams. Thus, EuGENia allows annotating classes with the *Node* annotation. This annotation has several features, such as selecting the attribute of the annotated class which should be used as the label for the node, layout information such as border styles, colors, and either an external figure (e.g., provided as a SVG graphic) or a predefined figure by EuGENia (e.g., rectangle or ellipse) may be used to render the node.

Link: This annotation is applicable to classes as well as to non-containment references that should appear in the diagram as edges. This annotation provides attributes for setting the style of the link, e.g., if it is dashed, and the decoration of the link end, e.g., if the link end should be visualized as an arrow.

---

[1] More information on EuGENia annotations is provided at: http://www.eclipse.org/epsilon/doc/articles/eugenia-gmf-tutorial.

Compartment: Containment references may be marked with this annotation. It defines that the containment reference will create a compartment where model elements that conform to the type of the reference can be placed within.

Label: Attributes may be annotated with this annotation which implies that these attributes are shown in the diagram for nodes or links.

Figure 1 exemplifies the usage of EuGENia for a simple *hypertext modeling language* (HML). In the upper part there is the definition of HML by stating the three modeling concepts, i.e., the *hypertext* model is composed of *pages* and *links*. Furthermore, with annotations shown in comments notation, the concrete syntax of the modeling concepts is described. The hypertext models are represented by the *diagram* which is used to contain pages and links of the hypertext system. Furthermore, pages are shown as *rectangles* and links are shown as *arrows* pointing from the source page to the target page. The bottom part of the figure shows an example model using the concrete syntax of HML.

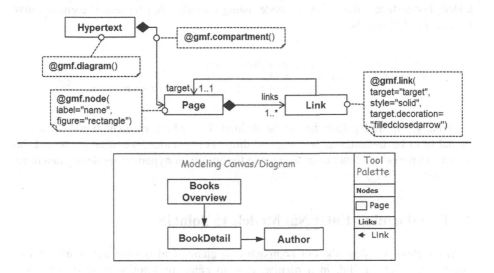

Fig. 1. GCS definition with EuGENia by-example.

2.3 JointJS - JavaScript Diagramming Library

JointJS is an open source library for building interactive diagram-based modeling editors that run in Web browsers. JointJS comes with a commercial extension called Rappid which provides out-of-the-box UI components. Both are based on standard Web technologies such as SVG, HTML5, CSS3, and JavaScript, and follow a MVC architecture. This means, the model content is also separated from the model visualization as it is done by most DSML engineering approaches such as by EuGENia.

A modeling language which is supported by a JointJS-based editor is mostly defined with stencils. Within stencils, first, the modeling concepts have to be defined including the concrete syntax notation. For instance, consider the following code listing excerpt for defining the Hypertext modeling concept.

```
joint.shapes.Hypertext = joint.shapes.Hypertext.extend({
        markup: '<g class="rotatable">
                <g class="scalable"><rect/></g>
                <image/><text/><line/>
                </g>',
        defaults: joint.util.deepSupplement({
                type: 'Hypertext',
                paperWidth: pWFolder,
                paperHeight: pHFolder,
                position:{x: 0, y:0},
                ...
        }... });
```

After having defined the concepts and their notational appearance, the composition rules such as before done with containment structures within the metamodel, have to be added. For instance, the following code listing specifies that *Hypertext* elements may contain *Page* elements.

```
var folder = new joint.shapes.Hypertext();
folder.prop({ inherit: { container: true,
                        canContain: [ joint.shapes.Pages ] }});
```

Finally, the tool palette has to be defined, i.e., which element types should be available to be instantiated by drag-and-drop. For instance, we want to be able to instantiate pages and links from the palette for our given hypertext modeling language.

```
modeler.stencil = [page, link];
```

3   Transforming EuGENia Models to JointJS

In this section, we describe our approach at a glance and outline the results of two experiments that we did, in particular, how to generate JointJS-based editors from existing EuGENia models of structural and behavioral modeling languages.

3.1   Overview

Our approach how to bridge current DSML engineering approaches and Web-based modeling editor programming approaches is outlined in Fig. 2. As it is currently possible to generate from EuGENia models, Java code which runs on top of the Graphical Modeling Framework (GMF) in Eclipse, we developed a Model-to-Text (M2T) transformation to generate stencils for the JointJS platform which provide the definitions which we before discussed in Sect. 2. By this, we can follow a systematic language engineering approach based on metamodeling and at the same time, exploit rich Web-based modeling platforms without having to re-invent the wheel. The M2T transformation is implemented with Acceleo[2] which reads in the annotated Ecore models representing the abstract syntax as well as the graphical concrete syntax for the DSML contained in the annotations.

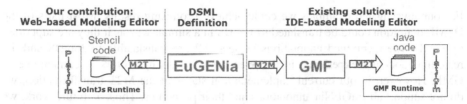

Fig. 2. Extending EuGENia with an additional M2T transformation to generate Stencils for JointJS.

As Fig. 2 outlines, our contribution is orthogonal to the existing support that comes with EuGENia. This means, we can now have a modeling editor inside the programming IDEs based on the GMF runtime, but at the same time, we can generate a Web-based modeling editor which runs on top of the JointJS runtime from the same DSML definition. However, an important requirement was to have a similar modeling experience in the Web-based modeling editor as in the IDE-based one. The next subsection describes how we approached this issue.

3.2   Development Methodology for the M2T Transformation

As development methodology for the M2T transformation we followed a reference system based approach. We investigated several existing EuGENia-based DSML definitions and how they are realized on the GMF platform. Based on this study, we re-implemented these projects directly with JointJS and aimed for having a similar modeling canvas, tool palette and graphical appearance of the modeling concepts. In particular, we used for this structural modeling languages such as the filesystem DSML[3] as well as behavioral modeling languages such as Petri nets[4].

Based on these projects, we developed the M2T transformation as an Acceleo template. The goal was to produce the before manually written JS code. Of course, this was an iterative process which required code adaptations in the different manually created projects to have one common template to generate the given code structures. To ensure the correctness and compatibility of the output files of the code generation process, a dedicated testing phase was required. Testing was an important part of every phase of development. Not only was the output JS file compared to the one manually

---

[2] https://www.eclipse.org/acceleo.

[3] http://www.eclipse.org/epsilon/doc/articles/eugenia-gmf-tutorial.

[4] https://profesores.virtual.uniandes.edu.co/ ~ isis4712/dokuwiki/doku.php?id=tut_eugenia.

written and included in the HTML5 project to see how it behaves, there was also an extensive testing phase at the end of the development cycle of the M2T transformation.

3.3 The Resulting Web-Based Modeling Editors

For our investigated examples we could achieve promising results. Most parts of the DSML definition could be translated to JointJS in a similar way as it they are supported by GMF. The mentioned annotations in Sect. 2.2 are translatable to JointJS and the resulting modeling experience in the Web-based modeling editors is comparable to GMF. However, in our current implementation status of the M2T transformation, we do not support all EuGENia annotations and their properties. Thus, as future work, we have to further investigate if there are definitions which cannot be supported by JointJS, e.g., more complex label computations for elements which are supported by EuGENia. Thus, we see our current work as a baseline to further compare modeling editor support in IDEs and Web browsers and to learn if there are fundamental differences or not between the current approaches which emerged in different development branches in different communities.

Concerning the development efforts, we compared the Lines of Code (LoC) of the EuGENia solutions as there is also a textual concrete syntax to define such models based on Emfatic[5] and the JointJS stencil solutions. For the given examples, the LoC for the EuGENia solutions are between 30 and 40. However, for the JointJS stencil solutions 250 to 280 LoC are needed to realize the languages. This shows that there is a potential effort reduction in defining the modeling languages directly on the EuGENia level.

One important difference is of course how the models are stored in EuGENia/EMF and JointJS. In EMF the standard storage format is based on the XML Metadata Interchange (XMI) format. In JointJS models are stored as JSON files. In order to have model exchange capabilities between the Web-based modeling editors and the IDE-based modeling editors, a dedicated transformation has to be developed to convert models represented in XMI into models represented in JSON, and vice versa.

4 Related Work

The Web may be considered a natural modeling platform given its straightforward support for collaboration, portability, and its very powerful, interactive and advanced UIs. As pointed out in existing literature [14], the Web became a platform where model engineering may be fully exploited in practice. The facts have shown this. A proliferation of model editor libraries based on Web technologies has been happening in the last years.

These libraries facilitate the development of Web-based modeling editors using dedicated JS frameworks. A Web-based editor is very lightweight and simple, and allows a much richer graphical representation of the DSL, as some products show [11, 13]. However, these frameworks do not exploit the high automation potential from DSML approaches to generate modeling editors from language specifications. Instead

---

[5] https://www.eclipse.org/emfatic.

of this, they usually offer a low-level API that allows developers to code the editor behavior, model constraints, define model elements, etc. Thus, the development of Web-based modeling editors requires still significant programming efforts.

Some research works have early emerged for conducting the collaboration and groupware concern in modeling editors based on Web technologies [12, 15]. Nevertheless, this issue is not related with the goal of this paper. Note that the collaboration concern in this kind of applications depends on the features of the underlying JS library being used rather than in how JS code for that library is generated automatically given a specific metamodel. As we mentioned before, the main goal of this paper is the generation of Web-based model editors starting from the design of a metamodel, making it compatible with our HTML5 modeling tool to create proper models. This whole concept as a unit seems to be unaddressed by the current state of technology, in spite that there exist some works proposing the generation of graphical editors [8, 9].

However, some existing work is more specifically related to defining the modeling language in itself, such as Clooca [16]. Clooca allows developers to define DSMLs, and the corresponding software that generates the code for a particular model instance. Although Clooca proposes a tool developed with Web technologies, this is not oriented to create the Web-based model editors for the DSML specified.

5   Conclusions

The Web is currently a platform where users perform daily tasks, and with this in mind, modeling in the context of the Web browser may be useful in several ways. Some well-known Web modeling editors have been arising for letting users define different software artifacts, such as the case of Node-RED [17] for modeling IoT application flows, which support this claim.

Although it is true that there was a proliferation of libraries for developing Web-based editors, they require advanced programming skills, which could be error prone at the moment of defining modeling constraints. The importance of having reliable Web-based modeling editors depends strongly on the possibility of specifying particular behavior for these editors regarding how model elements will be managed, their relationships, constraints, properties, etc. Our approach makes the creation of a Web-based modeling editor simpler and more guided, even without requiring programming skills on Web technologies. It reduces the possibility of introducing errors when programming the editor but also improves the editor maintenance when the underlying metamodel evolves. The next step in our research is to perform an evaluation of the Web-based editor generation process for larger languages such as UML, SysML, or BPMN and to evaluate how to generate animation code from the operational semantic definitions of the modeling languages.

Acknowledgements. We thank Richard Sevela for his work on the M2T transformation implementation. This work has been funded by the Austrian Federal Ministry of Science, Research and Economy and by the National Foundation for Research, Technology and Development and the project TIN2016-78103-C2-2-R of the Spanish Ministry of Economy, Industry and Competitiveness.

References

1. Kelly, S., Tolvanen, J.-P.: Domain-Specific Modeling - Enabling Full Code Generation. Wiley, New York (2008)
2. Brambilla, M., Cabot, J., Wimmer, M.: Model-Driven Software Engineering in Practice. Morgan & Claypool, San Rafael (2012)
3. http://www.emftext.org
4. http://www.eclipsc.org/Xtext
5. http://www.metacase.com/mep
6. Baetens, N.: Comparing graphical DSL editors: AToM3, GMF, MetaEdit+. Technical report, University of Antwerp (2011)
7. Cook, S., Jones, G., Kent, S., Wills, A.: Domain-Specific Development with Visual Studio DSL Tools. Addison-Wesley, Boston (2007)
8. Kolovos, D.S., García-Domínguez, A., Rose, L.M., Paige, R.F.: Eugenia: towards disciplined and automated development of GMF-based graphical model editors. Softw. Syst. Model. 16(1), 229–255 (2017)
9. Temate, S., Broto, L., Tchana, A., Hagimont, D.: A high level approach for generating model's graphical editors. In: Proceedings of ITNG (2011)
10. https://www.jetbrains.com/mps
11. https://www.jointjs.com
12. Rose, L.M., Kolovos, D.S., Paige, R.F.: EuGENia live: a flexible graphical modelling tool. In: Proceedings of the Extreme Modeling Workshop (XM) @ MODELS (2012)
13. http://concrete-editor.org
14. Wimmer, M., Schauerhuber, A., Strommer, M., Flandorfer, J., Kappel, G.: How web 2.0 can leverage model engineering in practice. In: Proceedings of DSML Workshop (2008)
15. Thum, C., Schwind, M., Schader, M.: SLIM—a lightweight environment for synchronous collaborative modeling. In: Schürr, A., Selic, B. (eds.) MODELS 2009. LNCS, vol. 5795, pp. 137–151. Springer, Heidelberg (2009). doi:10.1007/978-3-642-04425-0_11
16. Hiya, S., Hisazumi, K., Fukuda, A., Nakanishi, T.: Clooca: web based tool for domain specific modeling. In: Proceedings of Demos/Posters/StudentResearch@ MoDELS (2013)
17. Node-Red. http://nodered.org

Application Papers

Harvesting Forum Pages from Seed Sites

Luciano Barbosa[✉]

Universidade Federal de Pernambuco,
Av. Jornalista Anibal Fernandes, s/n, Recife, Brazil
luciano@cin.ufpe.br

Abstract. Web forums are rich sources of conversational content. Many applications, such as opinion mining and question answering, can greatly benefit from mining and exploring such useful content. A key step towards making this content more easily available is to collect conversational pages on forum sites – so-called thread pages. In this paper, we propose a two-step crawling solution for the problem of collecting thread pages in large scale. First, since thread pages are located within forum sites, we propose an inter-site crawler that locates forum sites on the Web. To do that, the inter-site crawler focuses on the Web graph neighbourhood of forum sites, and explores the content patterns of the links in this region to guide its visitation policy. Next, to collect thread pages within the discovered forum sites, we propose an intra-site crawler that finds thread pages by learning the context of links that lead to those pages and, to detect them, relies on their content and structural features. Experimental results demonstrate that both the inter-site and the intra-site crawlers are effective and obtain superior performance in comparison to their baselines.

1 Introduction

There is a great variety of social data available on the Web. Internet forums are social data where users hold conversations about particular subjects or topics. There are forums in diverse topics such as movies[1], agriculture[2] and health[3]. Forums are also very popular: to give some numbers, as of September 2015, a big forum website – ConceptArt.org[4] – had more than 376 thousand users and more than 8 billion posts, another forum – Gaia Online[5] – had about 26 million users and 2 billion messages (source: The Biggest Boards[6]). This huge amount of diverse human-generated content is very helpful for a variety of applications such as opinion mining [11], question answering [17] and forum search [13].

To take advantage of such rich content, methods to collect and process forum data have been previously proposed [2,5,9,15]. In this paper, we focus on the

---

[1] http://www.empireonline.com/forum/.
[2] http://community.agriculture.com/.
[3] http://ehealthforum.com/.
[4] http://www.conceptart.org/.
[5] http://www.gaiaonline.com/.
[6] http://www.thebiggestboards.com/.

© Springer International Publishing AG 2017
J. Cabot et al. (Eds.): ICWE 2017, LNCS 10360, pp. 457–468, 2017.
DOI: 10.1007/978-3-319-60131-1_32

Fig. 1. Overview of our two-step strategy to collect thread pages from forums.

particular problem of collecting conversational pages of forums, also known as thread pages. Previous approaches in the area of forum crawling have mainly focused on collecting thread pages within forum sites [4,7,9,16]. To avoid visiting unproductive regions of those sites, they learn regular expression patterns of URLs in the navigational paths that lead to thread pages, and use these patterns to guide the crawler's visitation policy.

In this work, we aim to perform a broader harvest of thread pages on the Web. More specifically, we are interested not only in collecting thread pages in a particular forum site, as previous approaches [4,7,9,16], but also to gather these pages from as many sites as possible. For that, we propose a two-step approach that first finds forum sites on the Web and, subsequently, collects thread pages within those sites. Figure 1 gives an overview of our solution. For the first step, we propose the Inter-Site Crawler that focuses on the Web neighbourhood of already known forum sites to discover new forum sites on the Web. Since not all links in the neighbourhood of forums lead to relevant information, we apply machine learning techniques to learn the patterns of relevant links in it. Once forum sites are found, the next step is to collect thread pages within them. This is the role of the Intra-Site Crawler. From the homepage of a given forum site, the Intra-Site Crawler navigates through the link structure of the site to locate thread pages. To focus its visitation policy on promising regions of the site, the crawler explores the context of the link neighbourhood of thread pages by using classifiers, instead of using regular expressions as previous approaches [9,14].

The remainder of the paper is organized as follows. Section 2 describes in details the Inter-Site Crawler, and Sect. 3 the Intra-Site Crawler. In these sections, we also provide experimental evaluation of these solutions. Finally in Sect. 4, we conclude the paper.

2   Inter-Site Crawler

The goal of the Inter-Site Crawler is to discover forum sites on the Web. To avoid visiting unproductive regions of the Web, the Inter-Site Crawler must focus on the region of the Web where forum sites are located (site discovery) and, to collect a high-quality set of forum sites with low cost, it needs to effectively and efficiently detect forum sites (site detection). In the remaining of this section, we explain in details these two tasks.

2.1   Site Detection

The Site Detector is the component of the crawler responsible for identifying forum sites. Given a website, the Site Detector needs to perform the detection

Fig. 2. Word cloud created from a set of entry-pages of forums.

with high-quality and low-cost, i.e., visiting few pages as possible of the website. Our strategy of detecting forum sites is based on two observations: (1) sites containing forums usually have an entry page to the forum content, which gives an overview of the current discussions in the forum; and (2) the forum entry pages are either the initial page of the site, or located at a shallow depth. Based on those, the crawler performs the detection of forum sites by doing a shallow crawling in the websites looking for the forum entry page if it exists.

A previous approach [9] proposed a heuristic method to identify the URL of the forum entry page given a forum site. We can not apply this strategy directly to our problem of Site Detection because we do not assume the input of the algorithm is a forum site. In fact, given any site on the Web, we want to verify whether the site is a forum site or not. Thus, to perform the detection of entry pages, we build a classifier based on the content of entry pages. Usually, entry pages of forums have a common vocabulary. Figure 2 shows a word cloud from a set of entry pages of forums to illustrate that. Words as "forum", "post" and "topics", which are not associated to a particular domain, have high frequency in this set. Based on this observation, we built a generic classifier, the Entry-Page Classifier, that uses as features the content (words) in entry pages (positive examples) as well as in non entry pages (negative examples) to detect entry pages of forums.

The Entry-Page Classifier is used in the site detection as follows. Given the homepage of a website, it verifies whether this page is an entry page to a forum using the Entry-Page Classifier. If so, the site is classified as a forum site. Otherwise, the outlink pages of the homepage are checked by the classifier. To avoid downloading too many pages, only outlink pages whose links contain indicative words of forums such as "forum", and "community" are visited. At the end of this process, a site is classified as a forum site if the entry-page classifier considers relevant one the visited pages of the site in the shallow crawling.

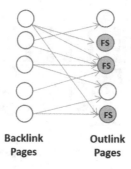

**Backlink   Outlink
Pages   Pages**

Fig. 3. Example of bipartite graph used by the Inter-Site crawler to locate forum sites (FS stands for forum sites).

2.2 Site Discovery

To the best of our knowledge, no previous work has been proposed to locate forum sites on the Web. The main challenge in performing this task is that forum sites are sparsely distributed on the Web. As a result, a simple crawling strategy that randomly follows outlinks obtains a poor performance in finding forum sites, as the results in Sect. 2.3 suggest. To find forum sites more efficiently, we propose a crawling strategy that focuses its visitation policy on Web neighbourhood of forum sites. More specifically, the crawler explores the neighborhood graph defined by the bipartite graph composed by the backlink pages (BPs) of URLs of forum sites, and the pages pointed by BPs (outlink pages), as shown in Fig. 3. The intuition behind this strategy is that a single backlink page might refer to many related pages (forum sites in our context). We call this link-rich backlink as hub page. Thus, once the crawler finds a hub page, it is only one step away from multiple forum sites. A previous work [1] used a similar strategy to locate multilingual sites on the Web.

The strategy works as presented in Algorithm 1. Initially, the user provides a set of seed URLs of forum sites. The crawler then retrieves the backlinks of these sites (line 9) using a backlink API available online[7], adding them to the backlink frontier (line 10). One of the backlinks is selected from the backlink frontier (line 12) and the page that the backlink points to (backlink page) is downloaded (line 13). The outlinks of this backlink page are extracted from the page and inserted into the outlink frontier (15). Next, a link from the outlink frontier is picked (line 6) and the Site Detector verifies whether it is a forum site or not (line 8). If so, the backlinks of this link are retrieved and added to the backlink frontier, and the process continues as described before. Notice that the crawler does not explore the outlinks of outlink pages, and only explores backlinks of forum sites, detected by the Site Detector.

Since backlinks in the backlink frontier not necessarily lead to hub pages, and outlinks in the outlink frontier not necessarily lead to forum sites, we apply

---

[7] In our experiments, we used the Mozcape API: https://moz.com/products/api.

Algorithm 1. Site Discovery

1: **Input:** *seeds*, *detector*
   {*seeds* : URLs of forum sites, *detector*: Forum Site Detector.}
2: *backlinkFrontier* = ∅
3: *outlinkFrontier* = ∅
   {Create the empty frontiers.}
4: *outlinkFrontier.addLinks(seeds)*
   {Add the seeds to the outlink frontier.}
5: **repeat**
6:   *outlink* = *outlinkFrontier.next()*
     {Retrieve from the outlink frontier the next link to be visited.}
7:   *page* = *download(outlink)*
     {Download the content of the page.}
8:   **if** *detector.isRelevant(page)* **then**
9:     *backlinks* = *collectBacklinks(page)*
       {Collect the backlinks to the given page provided by a search engine API.}
10:    *backlinkFrontier.addLinks(backlinks)*
       {Add backlinks to the backlink frontier.}
11: **end if**
12: *inlink* = *backlinkFrontier.next()*
     {Retrieve from the backlink frontier the next link to be visited.}
13: *inpage* = *download(inlink)*
     {Download the content of the page.}
14: *outlinks* = *extractOutlinks(inpage)*
     {Extract the outlinks of a backlink page.}
15: *outlinkFrontier.addLinks(outlinks)*
     {Add outlinks to the outlink frontier.}
16: **until** *frontier.isEmpty()*

---

machine learning techniques to rank links in those frontiers. More specifically, the crawler uses the Backlink Classifier that predicts the likelihood of a backlink *b* being a hub page, given features of *b* such as the tokens in the URL and in the title of the page[8]. Likewise, the crawler uses the Outlink Classifier to predict the likelihood of a given outlink to point to a forum site based on tokens in the URL, in the anchor and around the link. The two classifiers are automatically created during the crawling process. Initially, the crawler starts with no link prioritization. After a specified number of crawled pages, a learning iteration is performed by collecting the link neighbourhood of the links that point to relevant and non-relevant pages in each set. The result of this process is used as training data for the Backlink and Outlink classifiers.

2.3   Experimental Evaluation

In this section, we evaluated the two tasks performed by the Inter-Site Crawler – site detection and discovery – presented previously.

---

[8] The title of the page is usually provided by the backlink apis.

Table 1. Results from different machine learning algorithms.

Technique	Prec	Rec	F-1	Acc
Naive Bayes	0.705	0.925	0.8	76.8
Logistic Regression	0.781	0.851	0.814	80.5
SVM	0.855	0.791	0.822	82.8

Table 2. Results of varying the minimum likelihood of a page being considered relevant by the entry page classifier.

Min. Likelihood	Prec	Rec	F-1
0.5	0.855	0.791	0.822
0.6	0.91	0.77	0.83
0.7	0.94	0.76	0.84
0.8	0.96	0.74	0.83
0.9	0.96	0.73	0.83

Site Detection. To measure the quality of the Site Detector, we manually labeled 380 Web pages (182 positive and 200 negative) from a variety of sites on the Web. Positive examples are entry pages to forums in the sites. We used two thirds of the data for training and one third for test.

Table 1 presents the precision, recall, F-1 and accuracy values for 3 different machine learning algorithms. The classifiers were created using the Weka package [8] with default values. The numbers show that Support Vector Machine (SVM) obtained the best results (precision $= 0.855$, recall $= 0.791$, F-1 $= 0.822$, accuracy $= 82.8$). The SVM version used was the probabilistic SVM [12], since we are interested in the class likelihood of the instances. A threshold over the class likelihood can be used, for instance, as a filter to improve the precision of the Site Detector. This filter is very useful in this environment whereby the proportion of negative examples is much higher than the positive ones. For this purpose, we varied the minimum likelihood for a page be considered relevant from 0.5 to 0.9, as presented in Table 2. For each value, we measured its quality (precision, recall and F-1). As expected, the minimum likelihood is directly proportional to the precision and inversely proportional to the recall. For instance, when the minimum likelihood is 0.9, i.e., only pages with likelihood higher 0.9 are considered relevant by the detector, the precision is 0.96.

Site Discovery. To evaluate our strategy to locate forum sites, we executed the following crawling configurations:

- Forward Crawler (Forward): The forward crawler randomly follows the forward links. Only out-of-site links are considered, i.e., it excludes from the crawling links to internal pages of the sites;

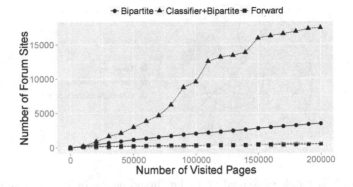

Fig. 4. Performance of the 3 strategies of inter-site crawling.

- Bipartite-Graph Crawler (Bipartite): our strategy focusing on the bipartite graph composed by backlink and outlink pages without any prioritization of links in the graph;
- Classifier-Based Bipartite-Graph Crawler (Bipartite+ Classifiers): our strategy using classifiers to prioritize links in the bipartite graph.

Each configuration visited 200,000 pages. Only 2 forum sites were provided as seeds to start the crawl[9]. The performance of the crawling strategies was measured by the total number of forum sites collected. The minimum likelihood used by the Site Detector to consider a site as relevant was 0.9, since we are interested in obtaining a high-quality collection of forum sites.

Figure 4 presents the number of collected forums versus the number of visited pages for each approach. At the end of the crawling processes, the Bipartite + Classifiers approach collected the highest number of forum sites (17,429 sites). That is, only from 2 seed URLs, our best strategy discovered more than 17 K forum sites. The Bipartite crawler located fewer sites (34,515) followed by the Forward crawler (427 sites). These results show that (1) the bipartite graph strategy is in fact effective: the Bipartite crawler found 8 times more sites than the baseline (the Forward crawler); and (2) the classifiers (Backlink and Forward classifiers) used to prioritize the links in the bipartite graph hugely improved the performance of the bipartite crawler: Bipartite + Classifiers crawler discovered 5 times more for forum sites than the Bipartite crawler.

3 Intra-Site Crawler

The pages in forum sites that contain users' conversations are called thread pages. The goal of the Intra-Site Crawler is to locate thread pages in a given forum site. To achieve that, it needs to locate and detect thread pages in forum

---

[9] Seeds: http://www.woodworkforums.com/ and http://ubuntuforums.org/index. php.

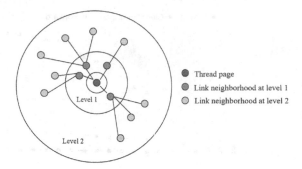

Fig. 5. Example of context graph for thread pages (adapted from [6]).

sites. For the first task, it is important that the crawler avoids unproductive regions of the sites that might not lead to thread pages. For the second one, the crawler needs to perform a high-quality detection, otherwise the repository of thread pages collected by the crawler might have poor quality. We give further details about these tasks in the remaining of this section.

3.1   Locating Thread Pages

Thread pages are only a subset of the pages in forum sites. In addition to them, forum sites might contain pages related to documentation, news etc. The Intra-Site crawler then needs to focus on the region of the website where thread pages are located to collect the maximum number of thread pages visiting the lowest number of pages as possible. For that, the crawler explores the patterns of links inside forum websites using context graphs [6]. Figure 5 presents an example of a context graph of two levels. The thread page is located in the center of the context graph. The main assumption of context graphs is that links at the same distance to the thread page in this graph have similar context. For instance, in the site ubuntuforums.org, URLs that point to thread pages (one step away) have the string "showthread" in common, whereas URLs two steps away have the string "forumdisplay" in common. Context graphs have been used to locate pages in other contexts such as pages in a given topic [6] or pages containing Web forms [3]. The Link Classifier is the component in the intra-site crawler that leverages context graphs to locate thread pages. More specifically, the Link Classifier estimates the distance (number of edges) of a given link to a thread page based on its context. The context is composed by the tokens in the URL, anchor and words around the anchor, which are the features used by the classifier.

The Link Classifier is automatically built as the crawl progresses. Initially, the crawler starts with no link prioritization. After a specified number of crawled pages, the links visited that led to thread pages collected so far, which compose the context graph, are used as training data to build the Link Classifier.

(a) Thread page

(b) Index page

Fig. 6. Examples of an index and a thread page.

3.2   Detecting Thread Pages

In order to collect a high-quality set of thread pages from a given site, the intra-site crawler needs to perform an effective thread page detection. Thread pages are composed of user posts (see Fig. 6(a)). Posts are usually within records, which contain, in addition to posts, meta-information about the posts such as the user who posted the information, the date of posting etc. In a previous work [2], we implemented a method to extract records from forum pages. Records also appear in other types of pages in forum sites such as the ones that point to thread pages (a.k.a. index pages). Figure 6(b) presents an example of index page.

Based on these observations, our thread page detection relies on the record extraction to obtain features for the classification. Only pages with extracted records are considered for classification. Thus, given a forum page, the first step of the detection is to extract records from the page. If records are extracted,

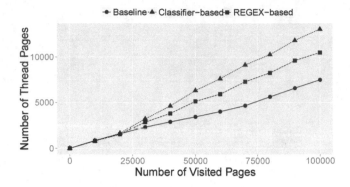

Fig. 7. Performance of the 3 strategies of intra-site crawling.

thread classification is applied over the records to verify whether the page is a thread page or not. Records of index pages have different layouts from records of thread pages as one can see in Fig. 6(a). For instance, records in thread pages usually have longer texts than in index pages; and records in index pages contain internal links to thread pages which is not always the case for records from thread pages. Based on that and similar to [9], we employ the following set of features based on the layout of the records:

- Average number of date and user information for all records in the page. To detect information about date and user in a record, we used detectors based on regular expressions [2];
- Average size, variance and noise to signal (standard deviation/average) of the texts in the records;
- Average number of images, internal and external links.

3.3   Experimental Evaluation

In this section, we evaluate the thread-page detection and discovery presented previously.

Thread-Page Detection. To build the training data, we labeled 1280 instances (50% positive and 50% negative) and split in 722 for training and 558 for test. We tried different machine learning techniques available in Weka package [8] with their default values. The multi-layer perceptron obtained the best results: Accuracy = 92.2%, Precision = 0.91, Recall = 0.87 and F-measure = 0.89.

Thread-Page Discovery. For the problem of thread-page discovery, we implemented 3 different strategies of prioritization of links:

- Baseline: the baseline randomly follows the internal outlinks of the pages;
- REGEX-based: the regex-based strategy follows links using regular expressions learned from URLs of thread pages, as proposed by [10] and used by Jiang et al. [9] in their forum crawler;

- Classifier-based: the classifier-based approach is the one proposed in this paper for the Intra-Site crawler, which builds a Link Classifier learned from the context graph as we described previously. Probabilistic SVM [12] was the learning algorithm used by the Link Classifier, and it was created with default values for SVM provided in Weka [8].

The 3 approaches collected thread pages within 586 forum sites, and each one visited a total of 100,000 pages of these sites. We evaluated the approaches based on the number of thread pages collected by each approach, identified by the thread page detector. The minimum likelihood used by the thread page detector to consider a page as a thread page was 0.8. Both the REGEX-based and the Classifier-based crawlers start with the same link prioritization than the baseline. The learning process to create the regular expressions, for the REGEX-based crawler, and the link classifier, for the Classifier-based crawler, was performed every 20,000 visited pages.

Figure 7 presents the results for the 3 strategies. As expected, until 20,000 visited pages all approaches had similar behaviour. After that, the approaches presented different performance since the Classifier-based and REGEX-based approaches started running the learning process. At the end of the crawling process, our approach outperformed the other two: the Classifier-based crawler collected 13,063 thread pages, whereas the REGEX-based collected 10,483 and the Baseline 7,486. A possible reason why our strategy obtained better results than the REGEX-Based one is that the REGEX-Based crawler learns patterns only from the URLs of relevant pages, whereas our strategy leverages patterns using machine learning not only from URLs but also from the anchor of these links and the context around them.

4 Conclusion

We presented in this paper a two-step crawling approach to collect conversational pages on the Web. First, the Inter-Site Crawler discovers forum sites from seed sites. For that, it restricts its crawl to the link neighbourhood composed of the backlink pages of forum sites, and the outlink pages of backlink pages. To prioritize links in this graph, we apply machine learning techniques. The URLs of the discovered forum sites are provided to the Inter-Site Crawler that collects the conversational pages in these sites. To more efficiently locate those pages, it uses a link classifier that explores the context of the links around conversational pages. Our experimental evaluation shows that our crawling approaches harvested high-quality collections and are more efficient than the baselines. As future work, we plan to apply this strategy to specific domains, instead of using it in a generic manner as we presented in this paper.

References

1. Barbosa, L., Bangalore, S., Sridhar, V.K.R.: Crawling back and forth: using back and out links to locate bilingual sites. In: Proceedings of the 5th International Joint Conference on Natural Language Processing, pp. 429–437 (2011)

2. Barbosa, L., Ferreira, G.: Extracting records and posts from forum pages with limited supervision. In: Wang, J., Cellary, W., Wang, D., Wang, H., Chen, S.-C., Li, T., Zhang, Y. (eds.) WISE 2015. LNCS, vol. 9419, pp. 233–240. Springer, Cham (2015). doi:10.1007/978-3-319-26187-4_19

3. Barbosa, L., Freire, J.: Searching for hidden-web databases. In: WebDB, pp. 1–6 (2005)

4. Cai, R., Yang, J.-M., Lai, W., Wang, Y., Zhang, L.: iRobot: an intelligent crawler for web forums. In: Proceedings of the 17th International Conference on World Wide Web, pp. 447–456. ACM (2008)

5. Cong, G., Wang, L., Lin, C.-Y., Song, Y.-I., Sun, Y.: Finding question-answer pairs from online forums. In: Proceedings of the 31st Annual International ACM SIGIR Conference on Research and Development in Information Retrieval, pp. 467–474. ACM (2008)

6. Diligenti, M., Coetzee, F., Lawrence, S., Giles, C.L., Gori, M., et al.: Focused crawling using context graphs. In: VLDB, pp. 527–534 (2000)

7. Guo, Y., Li, K., Zhang, K., Zhang, G.: Board forum crawling: a web crawling method for web forum. In: Proceedings of the 2006 IEEE/WIC/ACM International Conference on Web Intelligence, pp. 745–748. IEEE Computer Society (2006)

8. Hall, M., Frank, E., Holmes, G., Pfahringer, B., Reutemann, P., Witten, I.H.: The weka data mining software: an update. ACM SIGKDD Explor. Newsl. 11(1), 10–18 (2009)

9. Jiang, J., Song, X., Yu, N., Lin, C.-Y.: Focus: learning to crawl web forums. IEEE Trans. Knowl. Data Eng. 25(6), 1293–1306 (2013)

10. Koppula, H.S., Leela, K.P., Agarwal, A., Chitrapura, K.P., Garg, S., Sasturkar, A.: Learning URL patterns for webpage de-duplication. In: Proceedings of the Third ACM International Conference on Web Search and Data Mining, pp. 381–390. ACM (2010)

11. Pang, B., Lee, L.: Opinion mining and sentiment analysis. Found. Trends Inf. Retrieval 2(1–2), 1–135 (2008)

12. Platt, J., et al.: Probabilistic outputs for support vector machines and comparisons to regularized likelihood methods. Adv. Large Margin Classifiers 10(3), 61–74 (1999)

13. Seo, J., Croft, W.B., Smith, D.A.: Online community search using thread structure. In: Proceedings of the 18th ACM Conference on Information and Knowledge Management, pp. 1907–1910. ACM (2009)

14. Vidal, M.L., da Silva, A.S., de Moura, E.S., Cavalcanti, J.: Structure-driven crawler generation by example. In: Proceedings of the 29th Annual International ACM SIGIR Conference on Research and Development in Information Retrieval, pp. 292–299. ACM (2006)

15. Wang, H., Wang, C., Zhai, C., Han, J.: Learning online discussion structures by conditional random fields. In: Proceedings of the 34th International ACM SIGIR Conference on Research and Development in Information Retrieval, pp. 435–444. ACM (2011)

16. Wang, Y., Yang, J.-M., Lai, W., Cai, R., Zhang, L., Ma, W.-Y.: Exploring traversal strategy for web forum crawling. In: Proceedings of the 31st Annual International ACM SIGIR Conference on Research and Development in Information Retrieval, pp. 459–466. ACM (2008)

17. Webber, B., Webb, N.: Question answering. In: The Handbook of Computational Linguistics and Natural Language Processing, pp. 630–654 (2010)

Decentralised Authoring, Annotations and Notifications for a Read-Write Web with dokieli

Sarven Capadisli[1]([✉]), Amy Guy[2], Ruben Verborgh[3],
Christoph Lange[1,4], Sören Auer[1,4], and Tim Berners-Lee[5]

[1] University of Bonn, Bonn, Germany
info@csarven.ca, {langec,auer}@cs.uni-bonn.de
[2] School of Informatics, University of Edinburgh, Edinburgh, UK
amy@rhiaro.co.uk
[3] Ghent University – Imec, Ghent, Belgium
ruben.verborgh@ugent.be
[4] Fraunhofer IAIS, Sankt Augustin, Germany
[5] Decentralized Information Group, CSAIL, MIT, Cambridge, USA
timbl@w3.org

Abstract. While the Web was designed as a decentralised environment, individual authors still lack the ability to conveniently author and publish documents, and to engage in social interactions with documents of others in a truly decentralised fashion. We present dokieli, a fully decentralised, browser-based authoring and annotation platform with built-in support for social interactions, through which people retain ownership of and sovereignty over their data. The resulting "living" documents are interoperable and independent of dokieli since they follow standards and best practices, such as HTML+RDFa for a fine-grained semantic structure, Linked Data Platform for personal data storage, and Linked Data Notifications for updates. This article describes dokieli's architecture and implementation, demonstrating advanced document authoring and interaction without a single point of control. Such an environment provides the right technological conditions for independent publication of scientific articles, news, and other works that benefit from diverse voices and open interactions. To experience the described features please open this document in your Web browser under its canonical URI: http://csarven.ca/dokieli-rww.

Keywords: Decentralisation · Human-computer interaction · Linked Data · Semantic publishing · Social machine · Social web

1   Introduction

While the Web was originally conceived as a decentralised platform where every organisation and individual can participate, it became increasingly centralised with less than 1% of the servers serving more than 99% of the content. The main reason for this is rooted in technology: it is currently much easier and

© The Author(s) 2017
J. Cabot et al. (Eds.): ICWE 2017, LNCS 10360, pp. 469–481, 2017.
DOI: 10.1007/978-3-319-60131-1_33

more efficient to author, manage, publish, and search large amounts of similarly structured content using a centralised platform. Blogger, YouTube and Facebook, for example, are centralised authoring, publishing, and search platforms for blog posts, videos or social network content respectively.

However, independence of centralised platforms is a necessity for ownership of published ideas, and to establish a relation of trust. For example, Facebook has been accused of bias, false information, and censorship–but rather than blaming this on any particular platform, we identify it as an unavoidable result of centralisation. After all, there is a continued tension between unrestricted publication rights on the one hand, and a guarantee of balanced, verified information on the other. In a fully decentralised setting, each source is filterless and responsible for its own quality and reputation, while others are free to selectively (dis-)trust certain sources using any mechanism they desire.

Decentralised authoring, publication, and annotation furthermore have the potential to impact areas in which centralisation currently determines the pace of evolution. Scientific publishing, for instance, is often bound to centralised review and dissemination processes. Instead, rigorous scientific discourse could be realised with an open, decentralised environment for the annotation of manuscripts, which has the potential to engage more people sooner. Trust then no longer stems from a finite process with limited transparency, but is rather continuously assessed by repeated independent validation. Publication thereby becomes the starting point rather than the end point.

If we want to strengthen the decentralised nature of the Web again, we need to develop technologies to simplify the decentralised authoring, management, exploration, and search of Web content.

In this article we present the principles and architecture for a fully distributed authoring and publishing system in Sects. 2 and 4 respectively. We describe the dokieli implementation of this architecture as well as an overview on its current adoption in Sect. 5 before we conclude with an outlook on challenges and future work in Sect. 6.

2 Principles

This describes the principles against which decentralised approaches for authoring, annotation and notifications should be designed. These principles are derived from current literature on decentralisation, and Web development best practices.

Data storage independent of service providers: Users should have a choice in where they store their data and full control over it e.g. with regard to who is allowed to access it. The *Industrial Data Space* [1] initiative calls this "data sovereignty".

Interoperability: By allowing the application logic to be decoupled from the data, users can switch between applications and personal data storage servers, thereby avoiding a *vendor lock-in*. To achieve maximum interoperability, applications should conform to well-defined Web standards and protocols (rather than properietary software implementations). Dangers of data silos and some example standards to use to decentralise are given in [2].

Separation of concerns: A *progressive enhancement* strategy to connect the structural, presentational, and behavioural layers allows content and base functionality to be accessible through different media and devices (as described in [3]).

Accessibility: To lower the entry barrier for all forms of participation, enhanced functionality should be accessible to users based on the capabilities of their user-agents, storage availability, network access or personal preferences (we consider this to be self-evident, and there are a plethora of Web best practices in this area).

Freedom of expression: Because there are no central authorities, we must assume applications follow the open-world principle, where "any author can say anything about anything". Identifying everything using [de]referenceable IRIs allows any distributed authoring or annotation application to reference and link to previously published content (this overlaps with Principles 1, 3 and 4 in the *W3C Semantic Web Activity* [4] charter).

Web of Trust: The Web as a collaborative medium makes it possible for people to take responsibility (or be accountable) for their contributions. It should be possible for people to publish, share, and annotate information while ensuring their provenance, authenticity and integrity [5–7].

3   Related Work

An overview on relevant related work is given in [8]. A range of quality attributes such as collaboration, interoperability, and scalability, while relevant to our work, we also consider systems and tools on dimensions based on the principles that we have outlined.

Centralised authoring and annotation platforms: *Google Docs*, *Medium*, and *Authorea* are examples of Web applications for collaborative creation and publication of content which require account creation and data storage with respective centralised services. They allow multiple participants to annotate and hold discussions around the primary content; users must access their accounts to be notified of updates to conversations, and data from both the main content and related discussion is confined to the service which was used to create it.

WordPress is a free and open-source platform for article publication which can be self-hosted on a server controlled by the user. Visitors may sign-in with their WordPress accounts to leave comments on others' articles, however they are typically under the hosting site' database.

Hypothesis makes it possible for users to leave annotations on different types of documents on the Web using a browser plugin or via a proxy. Annotations may be private or public, and can be threaded to form conversations around a piece of content. Despite allowing the attachment of annotations to resources hosted anywhere, they depend on centralised account creation and storage for the annotations themselves. Hypothesis is open source, with an API that uses Web Annotations data model, and may be self-hosted, but currently it is not possible to federate between different instances.

Pundit is a set of tools that allow web annotation with highlights, comments and semantic annotations. It is similar to Hypothesis in its architecture and deployment, i.e., annotations made through the pundit client require it to be saved on its corresponding annotations server.

Decentralised authoring and annotation systems: Some authoring and publishing systems already go into a decentralised direction. However, they only realize a relatively small subset of the principles outlined in the last section. *LibreOffice Online*, for example, allows collaborative editing of office documents (e.g., Writer) from the Web browser. Content can be stored under different CMSs in the cloud. The document's interface consists of image tiles which are sent from the server and rendered in the browser. However, it hardly provides accessibility, rich interlinking and annotations or separation of concerns.

The *Smallest Federated Wiki* allows pages to be forked; users can maintain personal copies.

Amaya is a desktop Web editor application (to create and update documents) as well as a lightweight browser developed by W3C to test its technologies.

The tools that provide good collaborative editing UIs appear to do so at the expense of data ownership and interoperability; those which promote data creation and publication in open reusable formats are lacking facilities for linking discourse and conversation to concepts published. Decentralised creations also mean that each author can choose their own semantics (e.g. their own vocabulary to annotate RDF), and then such decentralised documents can link to each other and their schemas can also be mapped to each other, whereas in centralised platforms this is (if they support semantics at all) often prescribed, either technically enforced, or encouraged by social convention.

4   Architecture and Technologies

In this section we discuss an architecture to bridge the gaps in existing work for a decentralised authoring and semantic annotation client-side application, which decouples itself from data and specific server requirements.

4.1   Architectural Overview

Decentralised read-write environments make it possible for different actors (e.g., authors, reviewers) to have their own personal online storages where they can: manage their data; have socially-aware access controls on the data (e.g., who gets to see and update what); send notifications based on their interactions; and permit different applications to operate on the data, including moving the data from one server to another seamlessly. Figure 1 depicts the contrast between typical centralised and decentralised architectures.

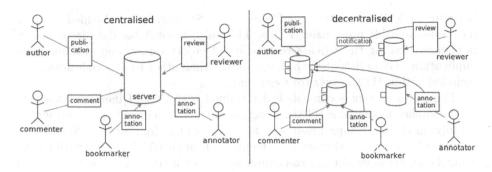

Fig. 1. Entities and relations within dokieli's architecture

dokieli as a client-side application can be deployed on a single-page or through a browser extension, which can consume and interact with Linked Data anywhere on the Web. We consider an HTML document with embedded JavaScript as the default UI of a document. It is independent from specific server-side software, proprietary APIs or the requirement to have an account.

On the other hand, if desired and available, users can participate using their own profiles (WebIDs) located anywhere on the Web, and get to store and make their annotations in their own personal storage, as well as assign access controls to documents. Similarly, a decentralised communications protocol, *Linked Data Notifications* [9] (*W3C Proposed Recommendation* [10]), is used get past the limits of centralisation by enabling communication to happen across independent servers. Figure 2 depicts relations between the kinds of entities which underlie dokieli's architecture, where nodes are under different domains and authority.

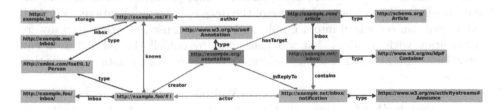

Fig. 2. Typical centralised and decentralised architectures

dokieli is *self-replicating*, in that the reader of a dokieli document can spawn an instance–either a copy or a brand new empty document–into their own storage space at the click of a button.

4.2   Creating Documents

Documents use HTML5 *Polyglot Markup* [11] to ensure that when served as (X)HTML respectively, they can be processed as either HTML or XML, which is useful in XML ecosystems and toolchains. Semantics is embedded directly into human-visible prose using RDFa. The machine-readable data is thus kept in context, reusing the article's text as literal object values, and avoiding data duplication or data 'islands', which can occur when other RDF serialisations are included within HTML <script> elements.

The appearance of documents is determined with CSS3. Different stylesheets can be applied to the same HTML structure so that a document can be presented flexibly, in the most appropriate way for a particular circumstance. Stylesheets can be switched from either dokieli's menu or through Web browsers with native controls, for example from a two-column layout required by an academic journal to a design in keeping with the author's blog.

When JavaScript is enabled, dokieli provides a rich editing interface which includes visual and structural formatting of text as well as embedding machine-readable semantics, media, dynamic citations, and inclusion of statistical charts from live endpoints. An external personal data store, or even internet connection, are not needed at this stage as modifications to a document made in the browser this way can be persisted to a local filesystem using the dokieli menu *export* function (or the browser's *save as*).

4.3   Consuming Documents

Documents can be retrieved from Web servers with a single HTTP GET request, by either a browser (for human-readable HTML) or script (machine-readable RDF). Through the use of progressive enhancement, document contents are available in text-only browsers, and further functionality of CSS and JavaScript are layered on according to the user agent's abilities.

dokieli's approach to marking human-visible content in RDFa makes it possible to further decouple itself, the application that produced the data, from the data itself, facilitating potential reuse of the data by other applications. A dokieli document can be parsed into a graph, and users can use any other RDF-aware application with the data that was generated by dokieli. Thereby, dokieli can effectively remove itself as a dependency when it comes to data consumption and reuse.

dokieli is able to authenticate users via WebID-TLS if they provide a WebID. This enables further functionality: the user can use dokieli to access protected resources and write to non-public data storage containers if their WebID is authorised to do so. For authenticated users leaving annotations on other documents dokieli fetches their name and display picture from their online profile if available, to display alongside their comment.

dokieli uses the following vocabularies as standard: *schema.org* to describe the general-purpose relations about the document as well as profiles, the *SPAR Ontologies* for scholarly articles and referencing, *Web Annotations* for annotations (with motivations e.g., replying, bookmarking, commenting, assessments), *LDP* for personal storage management, *WebAccessControl/ACL* for access control, *LDN Inbox* and *ActivityStreams* for social notifications, *Creative Commons* for rights and licensing, *PROV Ontology* for provenance, and the *RDF Data*

Cube vocabulary to consume multi-dimensional data from SPARQL endpoints. Authors can optionally include other vocabularies to mark up specific concepts through dokieli's UI.

4.4   Publishing Documents

Documents can of course be published on ordinary Web servers, as ordinary Web pages. The next layer of enhancement is for authors who wish to edit documents on a Web server directly rather than locally; they can make use of dokieli's write operations. This uses JavaScript, and moves the burden of processing user input from servers (i.e., offering HTML forms and processing of the form submission) to the client. In essence, the expectation is that dokieli should be a "smart client".

dokieli implements the *Linked Data Platform* (LDP) protocol for creating, updating and deleting documents. As such, personal data stores or servers which implement the server portion of the protocol can be used to store and edit dokieli documents directly. An HTTP PUT request to a URL is used to create a *new* document, to clone an existing one with *save as*, to *save* changes, and so that readers of a document can create their own document in *reply*. All of these operations are available through the dokieli menu.

4.5   Social Interactions and Annotations

Interactions with a document take the form of: a comment or (dis)like about the document as a whole, or an comment or (dis)like of a selection of text, i.e. an annotation. In both cases, the dokieli menu presents an input to the commenter, and then, adding additional semantic markup where necessary, sends the data off for appropriate storage so it can be retrieved and re-displayed on future document loads.

Document authors can point to a storage service (using the Web Annotations annotationService property) which lets readers without their own personal storage comment nonetheless. Readers who have a preferred storage location against which they can authenticate are able to direct dokieli to store their input there instead (or in addition). Conforming to LDP, dokieli allows users to remove their annotations with an HTTP DELETE operation. In this way, dokieli does not impose a centralised mechanism for social interactions, and allows users to effectively 'own' their comments, annotations, and reviews, in their own space.

4.6   Notifications

When readers interact with a document, the author is notified by means of the *Linked Data Notifications* (LDN) protocol. A notification, composed of the data from the interaction or annotation, is sent to the *inbox* advertised by the document or arbitrary parts, thereof. This inbox may be on the same server as the document itself, or may be elsewhere. dokieli subsequently reads this inbox to display interactions and annotations on the document.

As the *author* of the document has control of the inbox, they can remove notifications for interactions they find inappropriate, *without* needing to worry about their inability to access the original source of the interaction. Conversely, annotators do not lose control or authority over their contributions, even if the object of their interaction wishes to disassociate itself. Each contributor retains their own respective rights over the entities they create on the Web.

5   Implementation

dokieli is open source (https://github.com/linkeddata/dokieli) and available to try at https://dokie.li/ (or at any instance on the Web, see Adoption).

5.1   Components

dokieli's components include data (for structure and semantics), stylesheets (for presentation) and scripts (for interaction). All data (articles, annotations, notifications) are represented in HTML and RDF with vocabularies expressing the underlying content, resources are *self-descriptive* to increase their reuse, and contain relations to related external resources to foster *follow-your-nose* type of exploration.

Several stylesheets provide alternative views for consumption (e.g., stylesheets for different media: screens, print, slideshow). dokieli's JavaScript includes: a library for editing (*MediumEditor*) in its *authoring environment*; features to *fetch and display statistical data* from SPARQL endpoints (*Sparqlines* [12]); retrieval of profile information, as well as means to *sign in* with *WebID Authentication over TLS* [13]; functionality for write-operations, which includes checking *authorisation* against access-control level settings on the server with the authenticated user's WebID and personal certificate; creation and consumption of *Web Annotations* and *Linked Data Notifications*; and fetching information from remote articles when adding a citation.

The scope of dokieli includes documents and the interactions around them. The creation and maintenance of user profiles, personal storage spaces, and access control rules are not managed by dokieli; since they are all standard mechanisms, users are expected to be able to accomplish this using other specialised applications.

5.2   Deployment

dokieli employs two complementary deployment approaches: *single-page application* and *Web browser extension*.

dokieli's presentational and behavioural code layers can be included in Web pages in order to trigger them as active single-page applications. It is a smart client that allows different kinds of articles e.g., academic, blog posts, news, to be authored and annotated from within Web browsers, without necessarily having them deployed from a server, i.e., it can be used offline or on localhost.

dokieli internally handles its content and well-formed structural and semantic representation based on user's interactivity. Articles, profiles and their contact information, notifications, annotations with different motivations, for instance, can be read and written ubiquitously to any Web space with standard LDP and access control mechanisms.

The Web browser extension is a thin wrapper around dokieli's core code in order to embed itself in any HTML-based Web page on the Web. It inherits all of the features of a single-page application. While HTML based documents on the Web vary in their quality, dokieli's write operations generate well-formed HTML+RDFa. One of the primary utilities for the extension is to have a consistent interface for annotating (comment, bookmark, like) any text selection on a Web page, as well as sharing parts of pages with ones contacts via notifications, without having a service dependency or being limited by the Web page's UI.

5.3   Interactions

Screencasts for the following use-cases showcase dokieli's social features where users interact by creating and sharing information are at https://dokie.li/.

Annotations: A core feature to facilitate collaboration is the possibility to annotate arbitrary parts of a Web document. Users can select an entity or a span of text of interest, a context menu is presented to input their annotations along with the choice to select a license for their contribution. If the user is signed-in with their WebID, and provided they have a personal storage space, dokieli discovers this through their online profile and saves the annotations to that location. Once the annotation is submitted by the user, dokieli proceeds with three operations: (1) the annotation is requested to be saved at the user's personal storage, and if it is access controlled, the user will be prompted to authenticate themselves against that server, before the annotation is saved and is assigned its own URL, (2) if the article or any identifiable statement or segment has its own inbox, dokieli sends a notification to the inbox indicating that an annotation was made and with its retrievable location, and accompanying metadata like creator, date, license, etc., (3) the annotation is fetched from its canonical location, and integrated into the article e.g., in marginalia. If the article has a reference to a public annotation service (a writeable space adhering to the *Web Annotation Protocol* [15]), the user has the option to send a copy of the annotation there as well. In cases where the user does not want to have the canonical copy of the annotation on their server, or if a user does not have write access to a storage, they can use this option to engage with the article (Figs. 3 and 4).

Social Sharing: A key aspect of the Social Web is sharing with others. After (optionally) authenticating with a WebID, dokieli documents can be shared with contacts, which are discovered from the user's WebID profile. Contacts whose profiles advertise an LDN Inbox will receive a notification of the share. The notification contains Activity Streams 2.0 vocabulary terms, and recipients can

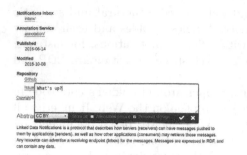

Fig. 3. dokieli Web Annotation

Fig. 4. dokieli Share

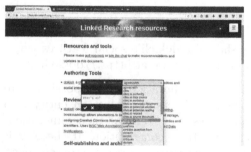

Fig. 5. Semantic inline citations in dokieli

Fig. 6. Sparqlines interaction in dokieli

use any LDN-compatible application to view the notification, without needing to have ever used dokieli before.

Inline Citations: Rich semantic links can also be established between dokieli documents themselves. The author selects a text fragment and inserts the URL of the document to be linked to as well as a semantic link type (e.g. "agrees with", "confirms", "cites as evidence") from the CiTO and schema.org ontologies. dokieli automatically retrieves metadata (e.g. title, authors) from the linked document and adds a proper scientific endnote reference. If the linked document advertises a standard LDN Inbox, a notification of the citation is sent as well, thus allowing bi-directional linking (Figs. 5 and 6).

Statistical Data and Diagrams: Embedding dynamically generated diagrams and charts is possible: after selecting text, the dokieli context menu offers results of a search across registered SPARQL endpoints for the keywords in the selected text, and presents a list of available data series to visualise. The result is an inline sparkline diagram.

5.4   Adoption

The W3C Working Group Note *Embedding Web Annotations* [16] in HTML includes examples from dokieli's use of the Web Annotation data *Model* [14] and *Vocabulary* [17] with motivations for example for "Lightweight, decentralised Annotation Tools". The *Linked Data Notifications* specifications use dokieli's HTML+RDFa template, and the Editor's Draft showcase dokieli as a consumer of LDN and Web Annotations. The LDN tests suite also uses dokieli's templates and stylesheets.

The academic workshop SemStats series use dokieli in its Website templates, including the call for contributions. CEUR-WS.org, an "Online Proceedings for Scientific Conferences and Workshops" offers the ceur-make tool to help organisers generate proceedings using dokieli's HTML+RDFa template. We list a community (of academics) who self-publish their articles and thesis using dokieli with different stylesheets and derived scripts under its examples in the wild. The conference series: WWW e.g., LDOW, WOW, ISWC, and ESWC propose dokieli as one tooling in which authors can use to for their contributions.

https://linkedresearch.org/ uses dokieli in its templates on the site as well as workshop proposals and call for contributions. http://csarven.ca/ uses dokieli in full, where some articles (like this article) offer pointers to a public annotation service which users may wish to use for their annotations. Articles also dynamically embed annotations from personal storage spaces.

6   Conclusions

The Web's design stands out because of its absence of centralised control, both for technical reasons of scalability and resilience as well as a societal need for freedom of expression. A challenge in such large-scale decentralised networks is how related publications can be semantically interlinked, even if they are authored and published by different parties. Centralising their publications is practiced by the majority of authoring networks today, demanding authors to give up some or all of their control in exchange for technical simplicity.

dokieli shows it is possible to build a social machine wherein people interact with each other without the need of centralised coordination. Users can choose storage space for their content independently of the applications with which they edit and view that content. Documents are connected statically through links and dynamically through Linked Data Notifications. This is a proof for the viability of a decentralised authoring and annotation environment built with Web standards.

On the other hand, dokieli's use of standards shows that dokieli itself is only one means to an end: once the document has been created, it lives on as an independent Web citizen. The social machine consists of people and documents, connected by Web standards, with dokieli acting as just one possible catalyst. Different Web applications can incorporate any of dokieli's functions and implement the principles to varying extents. Since the data is loosely coupled to the application, we avoid the *walled garden* problem of many current social platforms today.

A couple of important socio-technical challenges remain. Resources might want to indicate in a granular way which actions they support or encourage, such as liking, bookmarking, or sharing, and perhaps conditions about which notifications should be sent when any of these events take place. In order to encourage positive behaviour, we might want ways to provide moderation, and solutions to prevent harassment and abuse. Closely related is the issue of identity, pseudonymity and anonymity, and its relation with trust and verification. While there is likely no final solution to these issues in an open ecosystem, it is worthwhile exploring within dokieli or other tools.

Future work can examine how additional features can be realised on top of existing Web standards, or where more development is required. Real-time collaborative editing is often realised with centralised communication (even though some p2p alternatives exist). Services like top-down annotations or automated entity marking can improve the discoverability of a publication, yet the question of how to offer these without being tied to certain servers needs to be still solved. We invite you to try dokieli yourself. Annotate this article or spawn a new or a copy that you can edit yourself at http://csarven.ca/dokieli-rww.

Acknowledgements. Special thanks to our colleagues at MIT/W3C; Nicola Greco, Dmitri Zagidulin, Andrei Sambra, Sandro Hawke, as well as Henry Story and Melvin Carvalho for their contributions. We are also thankful to collaborate with colleagues at QCRI. This research was supported in part by Qatar Computing Research Institute, HBKU through the Crosscloud project from 2015-10 to 2016-09. Kingsley Idehen and OpenLink Software for their support and contributions to the browser extension. The OSCOSS project (DFG grant AU 340/9-1) for supporting the publication. Last but not least, the contributors to the dokieli code, issues, and discussion.

References

1. Otto, B., Auer, S., et al.: Industrial Data Space, TR (2016). https://www.fraunhofer.de/content/dam/zv/en/fields-of-research/industrial-data-space/whitepaper-industrial-data-space-eng.pdf
2. Berners-Lee, T.: Socially-aware Cloud Storage (2009). https://www.w3.org/DesignIssues/CloudStorage.html
3. Champeon, S.: Progressive Enhancement and the Future of Web Design (2003). http://hesketh.com/publications/progressive_enhancement_and_the_future_of_web_design.html
4. Koivunen, M.-R., Eric Miller, E.: W3C Semantic Web Activity (2001). https://www.w3.org/2001/12/semweb-fin/w3csw
5. Golbeck, J.: Weaving a Web of Trust, Science Magazine, vol. 321 (2008). http://hcil.cs.umd.edu/trs/2008-41/2008-41.pdf
6. Richardson, M., Agrawal, R., Domingos, P.: Trust management for the semantic web. In: Fensel, D., Sycara, K., Mylopoulos, J. (eds.) ISWC 2003. LNCS, vol. 2870, pp. 351–368. Springer, Heidelberg (2003). doi:10.1007/978-3-540-39718-2_23. https://link.springer.com/content/pdf/10.1007%2F978-3-540-39718-2_23.pdf

7. Golbeck, J., Parsia, B., Hendler, J.: Trust networks on the semantic web. In: Klusch, M., Omicini, A., Ossowski, S., Laamanen, H. (eds.) CIA 2003. LNCS (LNAI), vol. 2782, pp. 238–249. Springer, Heidelberg (2003). doi:10.1007/978-3-540-45217-1_18. http://sir-lab.usc.edu/cs586/20151readings/w13-1.pdf

8. Khalili, A., Auer, S.: User interfaces for semantic authoring of textual content: a systematic literature review. Web Semant. **22**, 1–18 (2013). http://svn.aksw.org/papers/2011/JWS_SemanticContentAuthoring/public.pdf

9. Capadisli, S., Guy, A., Lange, C., Auer, S., Sambra, A., Berners-Lee, T.: Linked Data Notifications: a resource-centric communication protocol. In: ESWC (2017). http://csarven.ca/linked-data-notifications

10. Capadisli, S., Guy, A.: Linked Data Notifications, W3C Proposed Recommendation (2017). https://www.w3.org/TR/ldn/

11. Graff, E., Silli, L.: Polyglot Markup: a robust profile of the HTML5 vocabulary, W3C Working Group Note (2015). https://www.w3.org/TR/html-polyglot/

12. Capadisli, S.: Sparqlines: SPARQL to Sparkline, ISWC SemStats (2016). http://csarven.ca/sparqlines-sparql-to-sparkline

13. Story, H., Corlosquet, S., Sambra, A.: WebID Authentication over TLS, W3C Editor's Draft (2014). https://www.w3.org/2005/Incubator/webid/spec/tls/

14. Sanderson, R., Ciccarese, P., Young, B.: Web Annotation Data Model, W3C Recommendation (2017). https://www.w3.org/TR/annotation-model/

15. Sanderson, R.: Web Annotation Protocol, W3C Recommendation (2017). https://www.w3.org/TR/annotation-protocol/

16. Cole, T., Capadisli, S., Young, B., Herman, I.: Embedding Web Annotations in HTML, W3C Working Group Note (2017). https://www.w3.org/TR/annotation-html/

17. Sanderson, R., Ciccarese, P., Young, B.: Web Annotation Vocabulary, W3C Recommendation (2017). https://www.w3.org/TR/annotation-vocab/

Evaluating Genomic Big Data Operations on SciDB and Spark

Simone Cattani, Stefano Ceri[(✉)], Abdulrahman Kaitoua, and Pietro Pinoli

Dip. Elettronica, Informazione e Bioingegneria, Politecnico di Milano, Milano, Italy
{Simone.Cattani,Stefano.Ceri,Abdulrahman.Kaitoua,Pietro.Pinoli}@polimi.it

Abstract. We are developing a new, holistic data management system for genomics, which provides high-level abstractions for querying large genomic datasets. We designed our system so that it leverages on data management engines for low-level data access. Such design can be adapted to two different kinds of data engines: the family of scientific databases (among them, SciDB) and the broader family of generic platforms (among them, Spark). Trade-offs are not obvious; scientific databases are expected to outperform generic platforms when they use features which are embedded within their specialized design, but generic platforms are expected to outperform scientific databases on general-purpose operations.

In this paper, we compare our SciDB and Spark implementations at work on genomic abstractions. We use four typical genomic operations as benchmark, stemming from the concrete requirements of our project, and encoded using SciDB and Spark; we discuss their common aspects and differences, specifically discussing how genomic regions and operations can be expressed using SciDB arrays. We comparatively evaluate the performance and scalability of the two implementations over datasets consisting of billions of genomic regions.

1 Introduction

Next Generation Sequencing (NGS) is a technology for reading the DNA that is changing biological research and will change medical practice; thanks to the availability of millions of whole genome sequences, genomic data management may soon become the biggest and most important "big data" problem of mankind, and bringing genomics to the cloud is becoming more and more essential. In this context, we are currently developing a new, holistic approach to genomic data modelling and querying that uses cloud-based computing to manage heterogeneous data produced by NGS technology [11][1]. Our approach is based on a new, high-level query language, called GenoMetric Query Language (GMQL) [12], which enables building new datasets from a repository of existing datasets, using algebraic operations.

---

[1] Advanced ERC Grant, http://www.bioinformatics.deib.polimi.it/geco/.

© Springer International Publishing AG 2017
J. Cabot et al. (Eds.): ICWE 2017, LNCS 10360, pp. 482–493, 2017.
DOI: 10.1007/978-3-319-60131-1_34

The current implementation of GMQL, described in [11], uses Flink [3] and Spark [4]. We recently considered another target for GMQL, opting for a scientific data management system. Scientific databases are known to for their efficient support of data aggregation over several dimensions, which is crucial for genomics. Among the various alternative systems, we selected SciDB, because it supports an add-on specifically dedicated to tertiary data analysis for genomics [1]; thus, it is an ideal alternative implementation framework for GMQL.

In this paper, we closely compare Spark and SciDB at work on genomic queries. We describe four widely used genomic abstractions: region selection, aggregation, histogram and mapping; by composing them, we obtain a significant subset of domain-specific operations of GMQL and of the other tools for genomic region management. We built a big data benchmark with a large dataset of regions and samples: the largest test compares half million regions to hundred millions regions, scattered over two thousand samples, corresponding to 50 trillion potential region intersections.

This paper demonstrates that both Spark and SciDB can manage such workload and qualify as relevant candidates for hosting tertiary genomic data analysis. Our benchmark demonstrates the superiority of SciDB in computations which perform selections and aggregations, but also shows that Spark outperforms SciDB in computations that perform genome-wise region comparisons; in such cases, both the SciDB and Spark computations use **binning**, a method for partitioning the genome into disjoint portions so as to enable parallelism.

The organization of this paper is as follows. Section 2 briefly introduces SciDB and Spark; Sect. 3 explains the Genomic Data Model and its mapping to an array database such as SciDB. Then, Sect. 4 provides a high-level description of the encoding of genomic abstractions using Spark and SciDB, and Sect. 5 provides their benchmark. Conclusions summarize our findings.

2   Platform Features

Apache Spark [17–19] is a general-purpose data processing engine providing high-level data operators and making a more efficient use of memory as compared with low-level map-reduce programming. SciDB [6,15,16] is a computational multi-dimensional database engine optimized for fast data selection and aggregations, required by most scientific applications.

Spark. The programming model of Spark is based on an abstraction called *resilient distributed datasets* (RDDs); a RDD is a distributed, fault tolerant data collection which can be processed on large servers or clusters. RDDs empower Spark with the support for in-memory data processing by allowing the state of the memory to be shared across different jobs. On contrary, for conventional MapReduce systems, sharing of intermediate data is only possible through write and read on persistent storage (e.g., the distributed file system), incurring significant cost for loading the data and writing it back at each stage.

Spark includes a rich set of operators, including *map, flatMap, mapPartition, reduce, repartition, filter, union, cartesian, coGroup, sortByKey, countByKey,* the above operations are also denoted as *transformations*, as they produce RDDs from either RDDs or input files, whereas other operations are denoted as *actions*, as they do not produce RDDs, but instead they either pass a result set to the embedding program or write data to the disk We opted for RDDs over Spark's DataFrame so as to keep the Spark and Flink deploys similar, as Flink operators are similar to RDD operators.

SciDB. The database engine of SciDB is based on a native array data model. Each array is described by a list of dimensions and a list of attributes. Dimensions have the integer type and each combination of them defines a cell in the array; attributes have arbitrary types, and consequently each cell is an vector of correspondingly typed values. Each array is implemented as a specific data structure, managed by the SciDB engine. Arrays are divided into chunks, where the chuck size is an important parameter under the control of the database designer. An hash function uses the dimension values associated to each chunk in order to assign it to a specific node of the cluster; by using this method, called *Multidimensional Array Clustering*, every query processing operation is mapped to specific chunks and executed in parallel at the nodes where such chunks are allocated.

SciDB queries are programmed using the Array Functional Language (AFL), a query language where each operation is defined as a function that receives as input either one or two arrays and produces as output one array; operations can be nested. The operations of AFL include: `filter` for selecting the array elements that satisfy a Boolean condition, `between` for extracting a rectangular region of the array, `cross_join` for pairing arrays (using equi-join on dimensions to speed up the computation), `redimension` to promote attributes to dimensions or to deprecate dimensions to attributes.

3   SciDB Representation of the Genomic Data Model

The Genomic Data Model (GDM), biologically motivated in [13], is based on the notions of datasets and samples. Datasets are collections of samples carrying the same region schema. Each sample consists of two parts, the *region data*, which describe portions of the DNA and their features, and the *metadata*, which describe general properties of the sample.

GDM Model. A genomic region r is a portion of the genome defined by the quadruple of values $\langle chr, left, right, strand \rangle$, called *region coordinates*, where chr is the chromosome, $left$ and $right$ are the two ends of the region along the DNA coordinates; strand is encoded as either $+$ or $-$, and can be missing. Formally, a *sample* s is a triple $\langle id, R, M \rangle$ where:

- id is the sample identifier of type *long*.
- R is the set of regions of the sample, built as pairs $\langle c, f \rangle$ of coordinates c and features f. Coordinates are arrays of four fixed attributes chr, $left$, $right$, $strand$ which are respectively typed *string, long, long, string*. Features are arrays of typed attributes; we assume attribute names of features to be

different, and their types to be arbitrary. The *region schema* of *s* is the list of attribute names and types used for the identifier, the coordinates and the features.
- *M* is the set of metadata of the sample, built as attribute-value pairs $\langle a, v \rangle$, where we assume the type of each value v to be *string*.

A *dataset* is a collection of samples with the same region schema and with features having the same types; sample identifiers are unique within each dataset.

SciDB Representation of GDM. We store metadata into a cube where the three dimensions are: *attribute name*, *value* and *sample id*. Given that SciDB does not permit *string* as dimension type, we introduced hashing for attribute name and value as dimensions. As the hashing of strings into 64-bit integers (standard dimension type in SciDB) introduces a high risk of collision errors, we opted for a double hashing, i.e. we used both for attribute name and value two different hashing functions, specifically selecting two orthogonal hash functions [9]. In conclusion, metadata are stored into a single 5-dimensional array.

```
DS_MD = <name:String,value:String>[sid, nid_1, nid_2, vid_1,vid_2]
```

where `sid` is the sample id and (`nid_1`, `nid_2`) and (`vid_1`, `vid_2`) are produced by the double hashing for attribute name and value. The schema of metadata arrays is identical for all the GDM datasets imported into SciDB.

Regions of a dataset are also stored into a single array; they are organized according to the relative sample id and genomic coordinate. In order to use region coordinates as dimensions for the SciDB data model, it is required to cast to integers the chromosome and strand values. For chromosomes, we defined a global codification map table that provides chromosome ids shared among all the datasets. This indexing operation is natively supported by SciDB through `uniq` and `index_lookup` operators. For strand, we applied a static conversion. Using these transformations, regions data are mapped to a 6-dimensional array, where attribute fields are based on the specific dataset feature schema provided by the user. The x dimension is an enumeration value, required because each GDM sample could have more than one region with the same coordinates.

```
DS_RD = < feature_schema > [sid, chr, left, right, strand, x]
```

Figure 1 shows the representation of the two arrays DS_MD and DS_RD; hash values are truncated for a better visualization. The physical arrays were designed using columnar storage with respect to each attribute, algebraic indexing (built by using a combination of hashing as well as lookup structures that are automatically maintained as the data sizes grow), and clustering of logically contiguous regions. This makes *slice* or *between* queries very efficient. Arrays are stored within fixed-size rectilinear **chunks** that partition the multidimensional space. Each chunk is then assigned to a computational node, using a hash function over the chunk's coordinates; the usage of region ends as coordinates allows their storage based on real region proximity, a fundamental property in order to speed up domain specific operations that use range intersection or range selection. According to [2], the optimal size for a chunk should be between 5 MB and

```
DS_MD                                                    DS_RD

{sid,nid_1,nid_2,vid_1,vid_2}  name,   value    {sid, chr, left, right, strand, x} score
{1,   1020, 6526, 8844, 3474}  'avg',  '0.0'    {1,    4, 1129,  3425,     1, 1} 0.5867
{1,   5139, 2589, 8864, 6221}  'type', 'chiapet'{1,    3, 5120,  9253,     1, 2} 0.7632
{2,   5139, 2589, 7534, 3123}  'type', 'chipseq'{1,    4, 3342,  3544,     0, 3} 0.3324
{1,   2984, 8763, 1123, 8232}  'cell', 'hela-s3'{2,    1, 4212,  7676,     1, 1} 0.9981
                                                {2,    2, 1112,  1745,     1, 2} 0.7783
                                                {2,    2, 1112,  1745,     1, 3} 0.7783
                                                {2,    3, 5142,  7435,     1, 4} 0.5741
```

Fig. 1. Metadata and regions arrays imported into SciDB.

50 MB. In our example, with a single attribute (and a size of about 8 Bytes), chunks with a million of regions have size of about 8–10 MB.

4   Genomic Operations

We next present four basic abstractions which are composed in several ways within GMQL operations; given that regions are several orders of magnitude greater than metadata, we focus on the operations which apply to regions.

4.1   Region Filtering

We consider three selection predicates: (a) by chromosome (coordinate), (b) by a region attribute and (c) by a conjunctive expression on both chromosome and score. Both implementations are straightforward.

A. Spark. In the Spark implementation, regardless of the condition type, we just need to filter the RDD of the regions:

```
regionsRDD.filter(predicate)
```

B. SciDB. The SciDB implementation is based on a simple AFL selection. In the SciDB model, chromosome is a dimension while the region attribute is placed within a cell. Therefore, case (a) is implemented simply as a dimension lookup and case (c) as a dimension lookup followed by a filter. The query of case (b) requires instead to scan all the chunks and then test the condition.

4.2   Region Aggregation

We consider an aggregate operation, such as COUNT, AVG, which is applied to all the regions of each sample of a dataset; the operation returns pairs <SampleId, Value>.

A. Spark. The *Spark* implementation is based on grouping samples based on SampleID and then calculate the aggregation for each sample separately. To calculate the count, we use the following code:

```
DS_RDD.groupBy(x=>x.sampleId).mapValues(x=>x.size)
```

B. SciDB. The AFL language of SciDB provides the `aggregate` operator which takes as input a SciDB array, a list of aggregate functions and (optionally) a list of dimensions along which to compute the aggregates. The mapping of Region Aggregation to `aggregate` is straightforward, where the dimension corresponds to the sample ID. As an example, the AFL code for computing the count aggregate looks like:

```
aggregate(dataset_array, count(*),SampleID);
```

where the first argument of the `aggregate` is the input dataset, the second is the aggregation function and the last one is the aggregation dimension.

4.3 Region Histogram

A classic operation in genomics is to compute the *accumulation index*, i.e. for each position in the genome the number of regions which overlap with that position; the operation applies to all the samples of a dataset. Note that the regions of the three input samples S1, S2, S3 have several overlaps, and the accumulation index ranges between 1 and 3, as shown in Fig. 2.

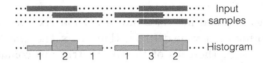

Fig. 2. Genomic histogram: computation of an accumulation index

A sequential algorithm for solving this problem consists of scanning the genome from left to right and maintain the accumulation count. Every time we meet the start of a region, we increment the count; conversely, every time we meet the stop of a region, we decrement the count. The result is made of all the consecutive couples of region ends (either starts or stops) between which the accumulation count is positive and does not change.

A. Spark. A parallel and distributed Spark implementation for computing the histogram is described in detail in [5]. The implementation of histogram (and of most operations on genomic regions) is based on **mono-dimensional binning**, consisting in partitioning the genome into equal size, disjoint and consecutive segments, called **bins**, originally introduced for fast genome viewing on a browser (https://genome.ucsc.edu/). For each chromosome, the *i-th* bin spans from i*BIN_SIZE to (i+1)*BIN_SIZE. Then the histogram is computed in parallel for each bin, through a functional style algorithm based on the `reduce` operation; results of all the bins are merged in order to produce the histogram of the whole genome.

B. SciDB. The computation of histograms by a SciDB program is complex and requires a good understanding of AFL primitives We provide a high-level description of the algorithm used for aggregating regions and of how such algorithm takes advantage of an array representation. Figure 3(a) shows that each INPUT is represented as 6-dimensional arrays (recall Sect. 3); for the purpose of histogram evaluation we do not consider strands, therefore each region within a sample is characterized by chromosome, right end, left end, and x value. Thanks to suitable redimension and apply operations, we build two 2D matrixes, respectively called LEFT and RIGHT, illustrated in Fig. 3(b), where one dimension is the chromosome and the other dimension is a projection of the region respectively on the left and right end; the tables contain +M at every region's left end and −N at every region's right end, where M and N respectively denote the number of regions starting and ending at each base.

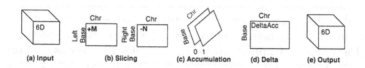

(a) Input   (b) Slicing   (c) Accumulation   (d) Delta   (e) Output

Fig. 3. Schematic representation of the steps for histogram computation in SciDB

Then, we merge the two matrixes into a 3D matrix, called ACCUMULATION, with a fictitious dimension valued $(0, 1)$, that pairs the two 2D matrixes (see Fig. 3(c)); at this point, the cells of the 3D matrix are collapsed by applying the sum aggregation on colliding cells, followed by a redimension; the output matrix, called DELTA, is shown in Fig. 3(d); its cell at a place (c, r) is an integer representing a count of regions starting or ending at position r of chromosome c (it can be a positive or negative value).

At this point, the actual histogram can be extracted for each chromosome and cell by applying the cumulate operation of SciDB to the DELTA matrix (it adds to each value within an array the sum of all its predecessors, e.g. $\mathtt{cumulate}([4, -2, 1, -2, 2] = [4, 2, 3, 1, 3])$; results are positive numbers reflecting the organization of regions in the genome; a final application of apply and redimension returns the result as 6-dimensional array (Fig. 3(e)), where regions are characterized by a new attribute ACCUMULATION storing their accumulation index.

4.4   Region Mapping

A region mapping operation applies to two datasets, called Reference and Experiment respectively. This operation performs the intersection of Experiment samples over each Reference and then computes an aggregate over such intersection. This behavior is explained in Fig. 4, where we show a simple case consisting of one sample of Reference and one sample of Experiment with overlapping regions, where we count the number of experiment regions which intersect with each reference region (e.g., the third region of the Reference

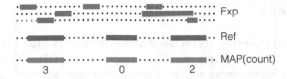

Fig. 4. Mapping experiments to references in genomics.

intersects with 2 regions of **Experiment** and therefore its count is 2). Region mapping requires the computation of a particular kind of join between regions, which is satisfied when two regions intersect. Joins with arbitrary distal conditions are discussed in [11].

A. Spark. The Spark implementation consists of two main steps: (a) binning and (b) checking for intersection. In the binning phase, the genome is divided into bins and every region of both the **Reference** and the **Experiment** datasets is assigned to all the bins it overlaps. Then, the datasets are left-joined on the key: (id,bin,chromosome). The cross product of the regions within the same bin is then computed and the intersection condition is checked, consisting of testing for overlap (by considering the start and the stop of both regions) and then testing if one of the regions starts on the current bin, thus creating just one result for each matching pair. This condition generalizes a binning method presented in [8]. Finally, adjacent regions on contiguous bins are aggregated using a reduce phase, producing the final result.

B. SciDB. In the SciDB implementation we adopt the above binning approach, but with an important difference. In SciDB it is not possible to dynamically split a region and distribute its replicas to an arbitrary number of adjacent bins, as we must apply identical operations to every cell in the array which stores the regions; thus, in order to apply a binning strategy, we must replicate all the regions an identical number of times. Such number is a function of the length M of the longest region in the **Reference** and **Experiment** datasets. In general, for given M and bin size S, each region will span to at most R bins, with: $R = \lceil M/S \rceil + 1$. This is a limitation w.r.t. Spark, which can manage variable region replication; region replication in Spark occurs only when the region spans across two or more contiguous bins.

Table 1. Features of the datasets used in the filtering operation.

Dataset	Size (MByte)	Regions (Million)	Samples
REF	2.3	0.506	1
DS_1	38	1.012	20
DS_2	375	10.120	200
DS_3	3832	101.2	2000
DS_4	38232	1012	20000

5   Benchmark

We performed our experiments on the Amazon Web Services (AWS) cloud, using a configuration with r3.4xlarge machine, 16 cores, 122 GB of RAM and 320 GB of SDD. For the experiments reported in this Section, we use synthetic data, so that we can trace performance scaling with controlled, growing data sizes (In Sect. 6 we also show experiments over real genomic datasets); synthetic datasets are similar to Encode peak datasets [10].

- The schema includes just a Score attribute. Chromosomes are 22, and each chromosome has 1 million bases.
- Regions in each chromosome are 2300, randomly distributed over the chromosome space; length is randomly distributed between 20 and 500 bases.

We then generate 4 datasets with an increasing number of samples (up to 20K) and regions (up to 1 billion); see Table 1.

5.1   Regions Filtering

We start comparing how Spark and SciDB execute the filtering operations discussed in Sect. 3. We consider three selection predicates: `Q1: chr='chr1'`, `Q2: score>0.9`, `Q3: (chr='chr1') and (score>0.9)`. Execution times of the operations in Spark and SciDB are reported in Fig. 5. We note that execution times for SciDB on `Q1` and `Q3` are much smaller than on `Q2`; in the former cases SciDB exploits the **between** operator and outperforms Spark. In `Q2`, instead, SciDB must read each single cell in order to apply the filtering operation, and in such case the execution time is similar to Spark, and it actually becomes worse with increasing data sizes.

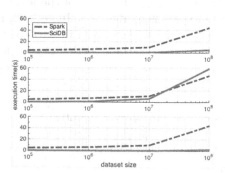

Test	DS_1	DS_2	DS_3	DS_4
Spark *Q1*	4.391	6.063	9.403	43.645
SciDB *Q1*	0.110	0.136	0.385	4.515
Spark *Q2*	4.640	6.447	10.299	46.049
SciDB *Q2*	0.161	0.581	5.673	58.137
Spark *Q3*	4.478	6.145	9.813	44.015
SciDB *Q3*	0.123	0.140	0.284	2.035

Fig. 5. Execution times (in seconds) for the filter operation.

5.2   Region Aggregation

Execution times of region aggregation Q4: `aggregate(count)` in Spark and SciDB are reported in Fig. 6. In this case we observe a huge difference between the two platforms performance: SciDB exploits the possibility to run in parallel the aggregation function in each chunk, and thus SciDB outperforms Spark.

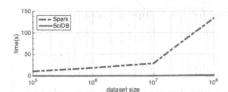

Test	DS_1	DS_2	DS_3	DS_4
Spark $Q4$	10.667	18.730	29.094	133.938
SciDB $Q4$	0.155	0.169	0.307	1.747

Fig. 6. Execution times (in seconds) for the aggregation operation.

5.3   Region Histogram

Execution times of region histogram Q5 in Spark and SciDB are reported in Fig. 7. In spite of rather different algorithms used in SciDB and Spark for computing the histogram, their overall performance is similar, especially considering the dataset DS_4 and the scale-up. However, for the small datasets DS_1 and DS_2, SciDB has better performance (respectively by factors 6 and 4).

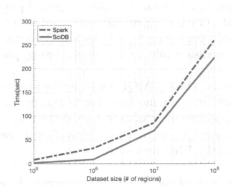

Test	DS_1	DS_2	DS_3	DS_4
Spark $Q5$	8.005	32.234	85.841	260.332
SciDB $Q5$	1.520	8.667	69.275	222.767

Fig. 7. Execution times (in seconds) for the histogram operation.

5.4   Region Mapping

Execution times of region mapping Q6 in Spark and SciDB are reported in Fig. 8. Region mapping is an operation of **quadratic complexity**, similar to the join; hence, execution times are much higher (expressed in minutes). In this case, we note that Spark outperforms SciDB, whose performance rises to about 1.5 h when comparing .5 million regions of the reference with 101 million regions of 2000 experiments (note that this is a *big data operation*, as it potentially requires 50 trillion comparisons).

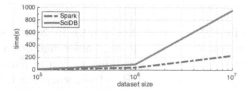

Test	DS_1	DS_2	DS_3
Spark $Q6$	0.12	0.57	3.82
SciDB $Q6$	0.28	3.29	95.33

Fig. 8. Execution times (in minutes) for the mapping operation.

6   Conclusions

Although a large number of benchmarks exist for comparing general purpose cloud-based engines such as Spark and Flink, including academic articles ([5,14]) and several posts[2], we are not aware of benchmarks comparing these engines with array-based scientific databases, such as SciDB. Our paper shows that this benchmark has no clear winner; as expected, SciDB performs better when it benefits from the array-based database organization (hence, on region filtering and aggregation), while Spark performs better on massive region mapping operations (similar to joins). The histogram operation, which does not fall in either categories, has very close performance in SciDB and Spark.

The best performances of Spark in massive operations (map) hints to preferring it over SciDB in the management of applications with billions of regions. However, consider that our design matches regions to arrays using a general purpose data design, that serves general data integration requirements; we expect that specific array-based data designs could perform very well in SciDB; among them, cases of variant analysis, gene expression mining and high-throughput screening are described in [1].

Our GMQL architecture, which includes three GMQL implementations to SciDB, Spark and Flink, appears even more strongly motivated after this benchmark; by supporting various implementation engines we will be able to match application requirements to the best target system and to closely follow the evolution of cloud-based platforms during the ERC project timeframe.

Acknowledgment. The authors would like to thank the SciDB support team for help during Simone Cattani's thesis [7] and for comments at his seminar, given at SciDB on July 19, 2016. This work is supported by the ERC Advanced Grant *GeCo (Data-Driven Genomic Computing)*.

References

1. Anonymous paper: Accelerating bioinformatics research with new software for big data to knowledge (BD2K). Paradigm4 (2015)
2. Anonymous paper: SciDB MAC Storage Explained, Paradigm4 (2015)

---

2 See: http://sparkbigdata.com/102-spark-blog-slim-baltagi/14-results-of-a-bench mark-between-apache-flink-and-apache-spark.

3. Apache Flink. http://flink.apache.org/
4. Apache Spark. http://spark.apache.org/
5. Bertoni, M., Ceri, S., Kaitoua, A., Pinoli, P.: Evaluating cloud frameworks on genomic applications. In: IEEE-Big Data Conference, pp. 193–202 (2015)
6. Brown, P.G.: Overview of SciDB: large scale array storage, processing and analysis. In: Proceedings of ACM-SIGMOD, pp. 963–968 (2010)
7. Cattani, S.: Genomic Computing with SciDB, a Data Management System for Scientific Computations. Master Thesis, Politecnico di Milano, July 2016
8. Chawda, B., et al.: Processing interval joins on map-reduce. In: Proceedings of EDBT, pp. 463–474, (2014)
9. Edelkamp, S., Sulewski, D., Yucel, C.: Perfect hashing for state space exploration on the GPU. In: Proceedings of ICAPS, pp. 57–64 (2010)
10. ENCODE Project Consortium: An integrated encyclopedia of DNA elements in the human genome. Nature **489**(7414), 57–74 (2012)
11. Kaitoua, A., Ceri, S., Bertoni, M., Pinoli, P.: Framework for supporting genomic operations. IEEE-TC (2016). doi:10.1109/TC.2016.2603980
12. Masseroli, M., et al.: GenoMetric Query Language: A novel approach to large-scale genomic data management. Bioinformatics (2015). doi:10.1093/bioinformatics/btv048
13. Masseroli, M., Kaitoua, A., Pinoli, P., Ceri, S.: Modeling and interoperability of heterogeneous genomic big data for integrative processing and querying. Methods (2016). doi:10.1016/j.ymeth.2016.09.002
14. Spangenberg, N., Roth, M., Franczyk, B.: Evaluating new approaches of big data analytics frameworks. In: Abramowicz, W. (ed.) BIS 2015. LNBIP, vol. 208, pp. 28–37. Springer, Cham (2015). doi:10.1007/978-3-319-19027-3_3
15. Stonebraker, M., Brown, P., Poliakov, A., Raman, S.: The architecture of SciDB. In: Bayard Cushing, J., French, J., Bowers, S. (eds.) SSDBM 2011. LNCS, vol. 6809, pp. 1–16. Springer, Heidelberg (2011). doi:10.1007/978-3-642-22351-8_1
16. Stonebraker, M., et al.: SciDB: a database management syatem for applications with complex analytics. Comput. Sci. Eng. **15**(3), 54–62 (2013)
17. Xin, R., et al.: Shark: SQL and rich analytics at scale. In: Proceedings of ACM-SIGMOD, June 2013
18. Zaharia, M., et al.: Resilient distributed datasets: a fault-tolerant abstraction for in-memory cluster computing. In: Proceedings of NSDI, pp. 15–28 (2012)
19. Zaharia, M., et al.: Discretized streams: fault-tolerant streaming computation at scale. In: Proceedings of SOSP, November 2013

Mining Worse and Better Opinions
Unsupervised and Agnostic Aggregation of Online Reviews

Michela Fazzolari[1(✉)] [iD], Marinella Petrocchi[1] [iD], Alessandro Tommasi[2] [iD],
and Cesare Zavattari[2] [iD]

[1] Institute of Informatics and Telematics (IIT CNR), Pisa, Italy
{m.fazzolari,m.petrocchi}@iit.cnr.it
[2] LUCENSE SCaRL, Lucca, Italy
{alessandro.tommasi,cesare.zavattari}@lucense.it

Abstract. In this paper, we propose a novel approach for aggregating online reviews, according to the opinions they express. Our methodology is unsupervised, due to the fact that it does not rely on pre-labeled reviews, and it is agnostic, since it does not make any assumption about the domain or the language of the review content. We measure the *adherence* of a review content to the domain terminology extracted from a review set. First, we demonstrate the informativeness of the adherence metric with respect to the score associated with a review. Then, we exploit the metric values to group reviews, according to the opinions they express. Our experimental campaign has been carried out on two large datasets collected from Booking and Amazon, respectively.

Keywords: Social web mining · Online reviews aggregation · Adherence metric · Domain terminology · Contrastive approach

1 Introduction

Online reviews represent an important resource for people to choose among multiple products and services. They also induce a powerful effect on customers' behaviour, therefore, they undertake an influential role on the performance of business companies. Since the information available on reviews sites is often overwhelming, both consumers and companies benefit from effective techniques to automatically analysing the good disposition of the reviewers towards the target product. To this aim, opinion mining [11,18] deals with the computational treatment of polarity, sentiment, and subjectivity in texts. However, opinion mining is usually context-sensitive [24], meaning that the accuracy of the sentiment classification can be influenced by the domain of the products to which it is applied [21]. Furthermore, sentiment analysis may rely on annotated textual corpora, to appropriately train the sentiment classifier, see, e.g., [8]. Also, most of the existing techniques are specialised for the English language: a cross-lingual adaptation is required in order to apply them to a different target language [10].

In this paper, we propose an original approach to aggregate reviews with similar opinions. The approach is unsupervised, since it does not rely on labelled

© Springer International Publishing AG 2017
J. Cabot et al. (Eds.): ICWE 2017, LNCS 10360, pp. 494–506, 2017.
DOI: 10.1007/978-3-319-60131-1_35

reviews and training phases. Moreover, it is agnostic, needing no previous knowledge on either the reviews domain or language. Grouping reviews is obtained by relying on a novel introduced metric, called *adherence*, which measures how much a review text inherits from a *reference terminology*, automatically extracted from an unannotated reviews corpus. Leveraging an extensive experimental campaign over two large reviews datasets, in different languages, from Booking and Amazon we first demonstrate that the value of the adherence metric is informative, since it is correlated with the review score. Then, we exploit adherence to aggregate reviews according to the reviews positiveness. A further analysis on such groups highlights the most characteristic terms therein. This leads to the additional result of learning the best and worst features of a product.

In Sect. 2, we define the adherence metric. Section 3 presents the datasets. Section 4 describes the experiments and their results. In Sect. 5, we report on related work in the area. Section 6 concludes the paper.

2 Review Adherence to Typical Terminology

We aim at proving that positive reviews, in contrast with negative ones, are generally more adherent to the emergent terminology of the whole review collection. This will provide us a form of alternative polarity detection: indeed, we might estimate the relative polarity of a review by measuring how adherent it is to the domain terminology. Because a meaningful comparison against terminology requires a sizeable chunk of text, the proposed approach best applies to a set of reviews. Here, we describe how the domain terminology is extracted and we define a measure of adherence of a piece of text against such terminology.

2.1 Extracting the Terminology

Every domain is characterised by key concepts, expressed by a *domain terminology*: a set of terms that are either specific to the domain (e.g., part of its *jargon*, such as the term "bluetooth" in the mobile domain) or that feature a specific meaning in the domain, uncommon outside of it (e.g., "monotonous" in the math domain). *Contrastive approaches* [2] to terminology extraction rely on sets of raw texts in the desired language: (1) a set belonging to the domain of interest and (2) a few others on generic topics (e.g., a collection of books, newspaper articles, tweets, etc.). The contrastive approach works by comparing the frequencies of the terms in the domain documents and in generic ones. The rationale is that generic, non-content words like "the", as well as non specific words, will be almost equally frequent in all the available sets, whereas words with a relevance to the domain will feature there much more prominently than they do in generic texts.

There are many sophisticated ways to deal with multi-words, but any statistics-based approach needs to consider that, for n-grams[1] to be dealt with

---

[1] Contiguous sequence of n words: "president of the USA" is a 4-gram.

appropriately, the data needed scales up by orders of magnitude. For our purposes, we stick to the simpler form of single-term (or 1-gram) terminology extraction.

Let $\mathcal{D}$ be a set of documents belonging to the domain of interest D, and let $\mathcal{G}_1 \ldots \mathcal{G}_M$ be M sets of other documents (the domain of each $\mathcal{G}_i$ is not necessarily known, but it is assumed not to be limited to D). All terms occurring in documents of $\mathcal{D}$ ($T_\mathcal{D}$) as candidate members of $\mathbb{T}_\mathcal{D}$, the terminology extracted from $\mathcal{D}$. For each term t, we define the *term frequency (tf)* of a term t in a generic set of documents $\mathcal{S}$ as:

$$\text{tf}_\mathcal{S}(t) = \frac{|\{d \in \mathcal{S} | t \text{ occurs in } d\}|}{|\mathcal{S}|} \tag{1}$$

(probability that, picking a document d at random from $\mathcal{S}$, it contains t). The tf alone is not adequate to represent the meaningfulness of a term in a set of documents, since the most frequent words are non-content words (such as "the", "be", "to", "of", "and", "a", "in", "that", "have", and "I" for the English language). Because of this, *inverse document frequency (idf)* [23] is often used to compare the frequency of a term in a document with respect to its frequency in the whole collection. In our setting, we can however simplify things, and just compare frequencies of a term inside and outside of the domain. We do this by computing the *term specificity (ts)* of a term t over domain set $\mathcal{D}$ against all $\mathcal{G}_i$'s, which we define as:

$$\text{ts}_\mathcal{G}^\mathcal{D}(t) = \frac{\text{tf}_\mathcal{D}(t)}{\min\limits_{i=1\ldots M} \text{tf}_{\mathcal{G}_i}(t)} \tag{2}$$

$\text{ts}_\mathcal{G}^\mathcal{D}(t)$ is effective at identifying very common words and words that are not specific to the domain (whose ts will be close to 1), as well as words particularly frequent in the domain, with a ts considerably higher than 1. Extremely rare words may cause issues: if $\mathcal{D}$ and $\mathcal{G}_i$'s are too small to effectively represent a term, such term will be discarded by default. We chose an empirical threshold $\theta_{\text{freq}} = 0.005$, skipping all terms for which $\text{tf}_\mathcal{D}(t) < \theta_{\text{freq}}$. This value is justified by the necessity to have enough documents per term, and 0.5% is a reasonable figure given the size of our datasets. We compute ts for all $t \in T_\mathcal{D}$. We define:

$$\mathbb{T}_\mathcal{D} = \{t | \text{ts}_\mathcal{G}^\mathcal{D}(t) \geq \theta_{\text{cutoff}}\} \tag{3}$$

To set the value of θ_{cutoff}, we might (i) choose the number of words to keep (e.g., set the threshold so as to pick the highest relevant portion of $T_\mathcal{D}$) or (ii) use an empirical value (higher than 1), indicating how much more frequent a term should be, being a reliably part of the terminology. Higher values include fewer terms in the terminology, improving precision vs. recall, whereas lower values include more terms, negatively affecting precision.

2.2 Adherence Definition

The adherence (adh) of a document d to a terminology $\mathbb{T}$ is defined as:

$$\text{adh}_\mathbb{T}(d) = \frac{|\{t | t \text{ occurs in } d\} \cap \{t \in \mathbb{T}\}|}{|\{t | t \text{ occurs in } d\}|} \tag{4}$$

It represents the fraction of terms in document d that belongs to terminology $\mathbb{T}$. This value will typically be much smaller than 1, since a document is likely to contain plenty of non-content words, not part of the domain terminology. The specific value of adherence is however of little interest to us: we show how *more adherent* reviews tend to be more positive than those with lower values of adherence, only using the value for comparison, and not on an absolute scale.

3   Datasets

The first dataset consists of a collection of reviews from the Booking website, during the period between June 2016 and August 2016. The second dataset includes reviews taken from the Amazon website and it is a subset of the dataset available at http://jmcauley.ucsd.edu/data/amazon, previously used in [13,14]. We also used a contrastive dataset to extract the domain terminology.

Booking Dataset. For the Booking dataset, we had 1,115,780 reviews, related to 1,025 hotels in 6 cities. We only considered hotels with more than 1,000 reviews, in any language. For each review, we focused on:

 - score: a real value given by the reviewer to the hotel, in the interval $[2.5, 10]$;
 - posContent: a text describing the hotel pros;
 - negContent: a text describing the hotel cons;
 - hotelName: the name of the hotel which the review refers to.

As review text, we took the concatenation of posContent and negContent.

Amazon Dataset. Reviews in the Amazon dataset are already divided according to the individual product categories. We chose two macro-categories, namely *Cell Phones & Accessories* and *Health & Personal Care* and we further selected reviews according to 6 product categories. For each review, we focused on:

 - score: an integer assigned by the reviewer to the product (range $[0, 5]$);
 - reviewText: the textual content of the review;
 - *asin*: the Amazon Standard Identification Number, that is a unique code of 10 letters and/or numbers that identifies a product.

Contrastive Terminology Dataset. In addition to the domain documents, originating from the above datasets, we used various datasets, collected for other purposes and projects, as generic examples of texts in the desired language, in order to extract the terminology as explained in Sect. 2.1. Table 1 resumes the data used to construct the *contrastive dataset*.

4   Experiments and Results

Each dataset $\mathcal{D}$ is organised in categories $\mathcal{C}_i$ and each category contains items that we represent by the set of their reviews $\mathcal{I}_j$. When performing experiments

Table 1. Outline of the contrastive dataset.

English	Italian	French
220k online newspaper articles	1.28M forum posts	198k tweets
15.98M tweets	7.37M tweets	

over $\mathcal{D}$, we extract the terminology of each category $\mathcal{C}_i$ ($\mathbb{T}_{\mathcal{C}_i}$). We then compute adh$_{\mathbb{T}_{\mathcal{C}_i}}(r)$ for each $r \in \mathcal{I}_j \in \mathcal{C}_i$ (r is the single review).

For the Amazon dataset, $\mathcal{C}_i$ are the product categories, whereas $\mathcal{I}_j$'s are the products (represented by their sets of reviews). For the Booking dataset, $\mathcal{C}_i$ are the hotel categories, whereas $\mathcal{I}_j$'s are the hotels (represented by their sets of reviews). We carried on experiments with and without *review balancing*. The latter has been considered to avoid bias: reviews with the highest scores are over-represented in the dataset, therefore the computation of the terminology can be biased towards positive terms. Thus, for each $\mathcal{C}_i$ and for each score, we randomly selected the same number of reviews. For page limits, we only report the results for the balanced dataset. The other results are available online [22]. We set $\theta_{\text{cutoff}} = 16$, which is the value used in the experiments presented in [7].

4.1 Adherence Informativeness

A first analysis investigates if there exists a relation between the adherence metric - introduced in Sect. 2 - and the score assigned to each review.

Amazon Dataset. For each product category, we extract the reference terminology, by considering all the reviews belonging to that category against the contrastive dataset, for the appropriate language. Then, we compute the adherence value for each review. To show the results in a meaningful way, we grouped reviews in 5 bins, according to their score, and compute the average of the adherence values on each bin.

For balancing the reviews in each bin, we set B as the number of reviews of the less populated bin and we randomly select the same number of reviews from the other bins. Then, we compute the average adherence values, obtaining the results in Fig. 1. The graph shows a line for each product category and it highlights that reviews with higher scores have higher adherence, in comparison to reviews with lower scores. Even if the Bluetooth Speakers and Oral Irrigators categories feature a slight decreasing trend in the adherence value, when passing from reviews with score 4 to reviews with score 5, the general trend shows that the adherence is informative of the review score.

Booking Dataset. We group the hotel reviews according to the city they refer to. For each city, we extract the reference terminology and we compute the adherence value for each review. To make the results comparable with the ones obtained for Amazon we re-arrange the Booking scoring system to generate a score evaluation over 5 bins. To this aim, we apply the score distribution suggested by Booking itself, since Booking scores are inflated to the top of the possible range of scores [15]. Thus, we consider the following bin distribution: very poor: reviews with a score ≤ 3; poor: score $\in (3, 5]$; okay: score $\in (5, 7]$; good: score $\in (7, 9]$; excellent: score > 9. Further, we consider a balanced number of reviews for each bin. The results are in Fig. 2. A line is drawn for each city, by connecting the points in correspondence to the adherence values. The graph suggests that the average adherence is higher for reviews with higher scores. Thus, the higher the score of the hotel reviews, the more adherent the review to the reference terminology.

4.2   Good Opinions, Higher Adherence

Interestingly, in the Booking dataset, the text of each review is conveniently divided into positive and negative content. Thus, we perform an additional experiment, by only considering positive and negative parts of reviews. For each city, we group positive and negative contents of reviews and we compute the adherence value for each positive and negative part, with respect to the reference terminology. Finally, we average the adherence values according to the score bins. The results are reported in Fig. 3, for the unbalanced dataset. In the graph, we report two lines for each city: the solid (dashed) lines are obtained by considering the positive (negative) contents of reviews. The same colour for solid and dashed line corresponds to the same city. We also perform the same calculation by considering a balanced dataset (Fig. 4). Both graphs highlight that there is a clear division between the solid and dashed lines. In particular, the average adherence obtained considering positive contents is, for most of the bins, above the average adherence computed considering negative contents. This separation is more evident when the review score increases (it does not hold for very poor scores). Overall, positive aspects of a hotel are described with a less varied language with respect to its negative aspects. Probably, this phenomenon occurs because unsatisfied reviewers tend to explain what happened in details.

   In addition to the average value, we also computed the standard deviation within each bin, that resulted to be quite high (detailed results are reported in

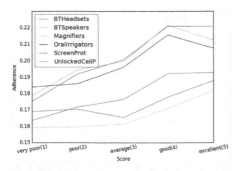

Fig. 1. Score vs adherence - Amazon dataset - balanced.

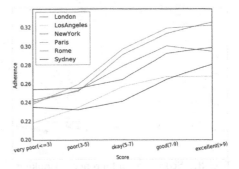

Fig. 2. Score vs adherence - Booking dataset - balanced.

the web page associated to the paper [22]). This suggests that, even correlated with the score, the adherence is not a good measure when considering a single review, but its informativeness should be rather exploited by considering an ensemble of reviews, as detailed in Sect. 4.4.

4.3 Extension to Different Languages

The experiments described so far were realised by considering a subset of reviews in English. To further evaluate the informativeness of adherence, we selected two additional review subsets, in Italian and in French. For each subset, we drawn two graphs, the first considering all the reviews content, the second the separation between positive and negative contents. We considered unbalanced bins, due to the limited number of reviews available in those languages. For page limits, the results are reported in the web page associated to the paper [22]. In both cases, it is confirmed that the higher the score, the higher the adherence when considering the overall text, and there is also a clear division between positive and negative adherence values, when the score increases.

4.4 Language and Domain-Agnostic Reviews Aggregation

We present an application of the outcome found in previous sections. Given a set of texts, we propose to aggregate texts with positive polarity and texts with negative polarity, without a priori knowing the text language and domain, and without using any technique of Natural Language Processing (NLP), while exploiting only the adherence metric. We apply the following methodology:

1. For each review $r \in \mathcal{I}_j \in \mathcal{C}_i$ we compute the adherence $\mathrm{adh}_{\mathcal{T}_{C_i}}(r)$.
2. Reviews $r \in \mathcal{I}_j$ are sorted in ascending order *w.r.t.* their adherence value.
3. Ordered reviews are split in bins with the same cardinality. We defined K_{bins} bins, each holding $|\{r \in \mathcal{I}_j\}|/K_{\mathrm{bins}}$ reviews in ascending order of adherence.

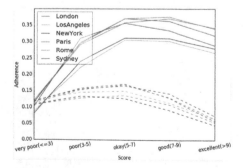

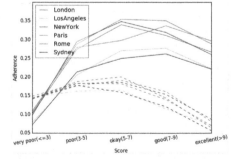

Fig. 3. Score vs adherence, Booking unbalanced dataset, considering positive and negative contents separately.

Fig. 4. Score vs adherence, Booking balanced dataset, considering positive and negative contents separately.

4. For each bin B_i, we compute the average of the adherence value of the reviews it contains: $\mathrm{Avg}_{\mathrm{adh},i} = \frac{1}{R} \sum \mathrm{adh}_{\mathcal{T}_{C_i}}(r)$, as well as, for the purposes of validation, the average score provided by those reviews, $\mathrm{Avg}_{\mathrm{score},i} = \frac{1}{R} \sum \mathrm{score}(r)$.
5. Finally, we aim at proving that, when the average adherence value of each bin increases, the average score value also increases. Thus, we compute the percentage of $\mathcal{I}_j \in \mathcal{C}_i$ for which we observe:

$$\mathrm{Avg}_{\mathrm{score},K_{\mathrm{bins}}} \geq \mathrm{Avg}_{\mathrm{score},1} \qquad (5)$$

$$\mathrm{Avg}_{\mathrm{score},i} \geq \mathrm{Avg}_{\mathrm{score},i-1} \qquad (6)$$

where $\mathrm{Avg}_{\mathrm{score},K_{\mathrm{bins}}}$ is the average score for the last bin, $\mathrm{Avg}_{\mathrm{score},1}$ is the average score for the first bin, and $i = 1, \ldots, K_{\mathrm{bins}}$.

Table 2 reports the results for the Amazon dataset. For each category $\mathcal{C}_i$, we apply the methodology three times, modifying the minimum number of reviews ($minRev$) for each item $\mathcal{I}_j$, in order to discard items with few reviews. We set $K_{\mathrm{bins}} = 3$ and we report the number of items (#$\mathcal{I}$) and the total number of reviews (#Rev) considered, plus the percentage of $\mathcal{I}_j \in \mathcal{C}_i$ for which (5) is true (%). This result shows that, considering 3 bins, the percentage of items for which the average score of the last bin is higher than the average score of the first bin is above 80% for each category (except for *Magnifiers* in case the minimum number of reviews is 20). Nevertheless, the percentage grows in almost all cases, when the minimum number of reviews increases. It exceeds 90% for every category, when the minimum number of reviews is, at least, 100. Therefore, in the majority of cases, it is true that, when the average adherence of reviews belonging to the last bin is higher than the average adherence of reviews included in the first bin, the same relation exists between their correspondent average scores.

Table 2. Amazon dataset - parameters: Eq. (5), bins = 3.

Category $\mathcal{C}_i$	minRev = 20			minRev = 50			minRev = 100		
	#$\mathcal{I}_j$	#Rev	(%)	#$\mathcal{I}_j$	#Rev	(%)	#$\mathcal{I}_j$	#Rev	(%)
BluetoothHeadsets	817	108693	86	423	96393	93	223	82723	97
BluetoothSpeakers	82	13155	96	54	12278	100	27	10423	100
ScreenProtectors	1741	174320	83	781	144597	90	370	116337	96
UnlockedCellPhones	1116	97049	89	542	78836	94	257	58788	97
Magnifiers	143	8763	72	46	5714	87	18	3694	100
OralIrrigators	48	10301	85	32	9832	91	21	8987	90

For the Booking dataset, we straight consider only hotels with at least 100 reviews. We perform three experiments according to the languages of reviews (English, Italian, and French). For each experiment, $K_{\mathrm{bins}} = 3$ and we report the number of items (#$\mathcal{I}$), the total number of reviews (#Rev) considered and

Table 3. Booking dataset, different languages - parameters: Eq. (5), bins = 3.

Category C_i	English			Italian			French		
	$\#\mathcal{I}_j$	$\#Rev$	(%)	$\#\mathcal{I}_j$	$\#Rev$	(%)	$\#\mathcal{I}_j$	$\#Rev$	(%)
London	356	467863	97	76	11952	96	123	20507	94
LosAngeles	56	46700	93	-	-	-	7	993	100
NewYork	163	182438	95	27	6518	93	60	10753	90
Paris	211	93164	96	6	806	100	72	12623	90
Rome	144	68543	97	64	11040	94	28	4197	93
Sydney	74	126744	100	-	-	-	4	553	100

the percentage of $\mathcal{I}_j \in C_i$ for which (5) is true (%). The results are in Table 3. The percentage of items for which the average score of the last bin is higher than the average score of the first bin is above 90% in all the cases.

Given a set of reviews on, e.g., hotels, or restaurants, in any language, we can identify a group of reviews that, on average, express better opinions than another group of reviews. Noticeably, this analysis works even if the associated score is not available, i.e., it can be applied to general comments about items.

We consider now if also relation (6) is verified for each bin $i = 1, \ldots, K_{\text{bins}}$, i.e., if the function between the ordered sets of average adherence values $\text{Avg}_{\text{adh},i}$ and average score values $\text{Avg}_{score,i}$ is a monotonic function. By plotting the average score *vs* the average adherence, for some items, we found out a general upward trend. Nevertheless, there were many spikes that prevent the function from being monotonic. Then, we tried to smooth down the curves by applying a moving average with $window = 2$ and we then computed the percentage of $\mathcal{I}_j \in C_i$ for which (6) was verified. For the Amazon dataset, we performed three experiments, modifying the minimum number of reviews required ($minRev$) for each item, in order to discard items with few reviews. Results are in Table 4.

Such results are worse with respect to the ones in Table 2. Nevertheless, in all cases (but *Oral Irrigators*), the percentage values increase when $minRev$ increase (for *Magnifiers*, it remains the same with $minRev = 50, 100$). When $minRev = 100$, the percentage of $\mathcal{I}_j \in C_i$ for which (6) is true is above 72%. For the Booking dataset, due to the high number of available reviews, we also varied the number of bins from 3 to 5. We only considered reviews in English and computed the percentage of items for which the Eq. (6) is true. Table 4 shows a clear degradation of performances when the number of bin increases.

So far, the results indicate a relation between the increasing adherence values and the increasing score values. However, we cannot prove a strong correlation between adherence and score, either considering a single review or groups of reviews. Therefore, we followed a different approach, by computing, for each item $\mathcal{I}_j \in C_i$, the *difference* between the average values of the first and last bin, both for the adherence and the score:

$$\Delta_{\text{adh}}(j) = \text{Avg}_{\text{adh}, K_{\text{bins}}} - \text{Avg}_{\text{adh},1}$$

Table 4. Amazon and Booking datasets, respectively - parameters: Eq. (6).

Amazon	minRev (%)			Booking	bins (%)		
Category $\mathcal{C}_i$	20	50	100	Category $\mathcal{C}_i$	3	4	5
BluetoothHeadsets	69	76	82	London	95	83	67
BluetoothSpeakers	88	93	96	LosAngeles	88	75	61
UnlockedCellPhones	69	72	77	New York	88	66	47
Magnifiers	58	72	72	Paris	87	65	40
OralIrrigators	77	81	76	Rome	94	82	67
ScreenProtectors	58	66	73	Sydney	95	92	85

$$\Delta_{\mathrm{score}}(j) = \mathrm{Avg}_{\mathrm{score}, K_{\mathrm{bins}}} - \mathrm{Avg}_{\mathrm{score}, 1}$$

If we average such differences for all the items $\mathcal{I}_j \in \mathcal{C}_i$, both for adherence and score, we obtain an average value for each category $\mathcal{C}_i$:

$$\mathrm{AvgD}_{\mathrm{adh}} = \frac{1}{J} \sum_{j=1}^{J} \Delta_{\mathrm{adh}}(j) \tag{7}$$

$$\mathrm{AvgD}_{\mathrm{score}} = \frac{1}{J} \sum_{j=1}^{J} \Delta_{\mathrm{score}}(j) \tag{8}$$

where J is the total number of items $j \in \mathcal{I}_j$. For page limits, we report two examples in Fig. 5 (detailed results are available online [22]). The x-axis reports the number of bins, whereas the y-axis represents the average differences values. A solid red line and a dashed blue line identify the average differences for the adherence and for the score, respectively. When the number of bin increases, the first and last bin include reviews which describe the product in a considerably different way, in term of positiveness. Thus, given a product category, it

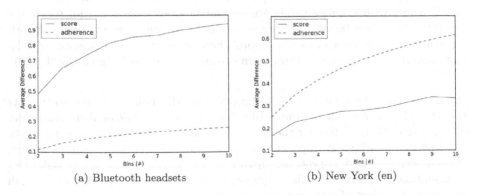

(a) Bluetooth headsets   (b) New York (en)

Fig. 5. Average differences for Amazon and Booking example categories.

is possible to discriminate among groups of related reviews, in such a way that each group expresses an opinion different from the others, ordered from the most negative to the most positive ones (or vice-versa).

4.5 Representative Terms in First and Last Bins

Given an item, we consider the terms included in the positive set and in the negative set (last and first bins, with $K_{bins} = 10$) that can be also found in the extracted terminology. For each term, we compute the term frequency-inverse document frequency (tf-idf) value (tf is the term frequency inside the bin, that is the number of reviews that include such term), we sort the terms accordingly and we select the first 20 ones for both the sets. We then remove the terms in common, in order to identify the most discriminating ones. Table 5 shows an example of the terms extracted for a Mini Speaker. For the reader's convenience, the web page in [22] reports the most relevant positive and negative terms for the single Amazon product categories and for the single Booking hotel categories, for English, Italian and French.

Table 5. The most relevant terms for an example Amazon product

Score	Relevant terms for B005XA0DNQ
2.9	refund, packaging, casing, disconnected, gift, battery, packaged, addition, hooked, plugging, shipping, hook, speaker, purpose, sounds, kitchen
4.3	compact, sound, great, retractable, portable, very, price, unbelievable, satisfied, product, easy, recommend, small, perfect, little, handy, size

5   Related Work

Terminology extraction. Automatic terminology extraction aims at automatically identifying relevant concepts (or terms) from a given domain-specific corpus. Within a collection of candidate domain-relevant terms, actual terms are separated from non-terms by using statistical and machine learning methods [19]. Here, we rely on contrastive approaches, where the identification of relevant candidates is performed through inter-domain contrastive analysis [1,6,20].

Opinion Mining. Opinion mining techniques identify polarities and sentiments in texts [11], by, e.g., extracting subjective expressions, personal opinions and speculations [25] or detecting the polarity acquired by a word contextually to the sentence in which it appears, see, e.g., [16,26,27]. Often, opinion mining rely on lexicon-based approaches, involving the extraction of term polarities from sentiment lexicons and the aggregation of such scores to predict the overall sentiment of a piece of text, see, e.g., [3–5,8].

Clustering Opinions. There exist few research efforts to detect the reviews polarity with standard clustering techniques, like [9,12,17]. Here, we still aggregate reviews based on their polarity, without relying on traditional clustering algorithms nor on linguistic resources. We base our approach on automatic terminology extraction, in a domain and language agnostic fashion.

6 Final Remarks

We presented a novel approach for aggregating reviews, based on their polarity. The methodology did not require pre-labeled reviews and the knowledge of the reviews' domain and language. We introduced the adherence metric and we demonstrated its correlation with the review score. Lastly, we relied on adherence to successfully aggregate reviews, according to the opinions they express.

Acknowledgments. Research partly supported by MSCA-ITN-2015-ETN grant agreement #675320 (*European Network of Excellence in Cybersecurity*) and by Fondazione Cassa di Risparmio di Lucca, financing the project *Reviewland*.

References

1. Basili, R., et al.: A contrastive approach to term extraction. In: Terminologie et intelligence artificielle, pp. 119–128. Rencontres (2001)
2. Bonin, F., et al.: A contrastive approach to multi-word extraction from domain-specific corpora. In: Language Resources and Evaluation, ELRA (2010)
3. Bravo-Marquez, F., et al.: Building a twitter opinion lexicon from automatically-annotated tweets. Knowl.-Based Syst. **108**, 65–78 (2016)
4. Cambria, E., Hussain, A.: Sentic Computing: A Common-Sense-Based Framework for Concept-Level Sentiment Analysis. Springer, Heidelberg (2015)
5. Cambria, E., et al.: SenticNet 3: a common and common-sense knowledge base for cognition-driven sentiment analysis. In: 28th AAAI, pp. 1515–1521 (2014)
6. Chung, T.M., Nation, P.: Identifying technical vocabulary. System **32**(2), 251–263 (2004)
7. Del Vigna, F., Petrocchi, M., Tommasi, A., Zavattari, C., Tesconi, M.: Semi-supervised knowledge extraction for detection of drugs and their effects. In: Social Informatics I (2016)
8. Esuli, A., Sebastiani, F.: SENTIWORDNET: a publicly available lexical resource for opinion mining. In: Language Resources and Evaluation, pp. 417–422 (2006)
9. Li, G., Liu, F.: Application of a clustering method on sentiment analysis. J. Inf. Sci. **38**(2), 127–139 (2012)
10. Ling Lo, S., et al.: A multilingual semi-supervised approach in deriving Singlish sentic patterns for polarity detection. Knowl.-Based Syst. **105**, 236–247 (2016)
11. Liu, B.: Sentiment Analysis and Opinion Mining. Morgan & Claypool, San Rafael (2012)
12. Ma, B., Yuan, H., Wu, Y.: Exploring performance of clustering methods on document sentiment analysis. Inf. Sci. **43**, 54–74 (2015)
13. McAuley, J., Pandey, R., Leskovec, J.: Inferring networks of substitutable and complementary products. In: 21th KDD, pp. 785–794. ACM (2015)

14. McAuley, J., et al.: Image-based recommendations on styles and substitutes. In: 38th Research and Development in Information Retrieval, pp. 43–52. ACM (2015)

15. Mellinas, J.P., María-Dolores, S.M.M., García, J.J.B.: Booking.com: the unexpected scoring system. Tourism Manage. **49**, 72–74 (2015)

16. Muhammad, A., Wiratunga, N., Lothian, R.: Contextual sentiment analysis for social media genres. Knowl.-Based Syst. **108**, 92–101 (2016)

17. Nagamma, P., et al.: An improved sentiment analysis of online movie reviews based on clustering for box-office prediction. In: Computing, Communication and Automation, pp. 933–937 (2015)

18. Pang, B., Lee, L.: Opinion mining and sentiment analysis. Found. Trends Inf. Retr. **2**(1–2), 1–135 (2008)

19. Pazienza, M.T., Zanzotto, F.M.: Terminology extraction: an analysis of linguistic and statistical approaches. In: Sirmakessis, S. (ed.) Knowledge Mining. Studies in Fuzziness and Soft Computing, vol. 185, pp. 255–279. Springer, Heidelberg (2005). doi:10.1007/3-540-32394-5_20

20. Peñas, A., Verdejo, F., Gonzalo, J.: Corpus-based terminology extraction applied to information access. Corpus Linguist. **13**, 458–465 (2001)

21. Ren, Y., Zhang, Y., Zhang, M., Ji, D.: Context-sensitive Twitter sentiment classification using neural network. In: Artificial Intelligence, pp. 215–221. AAAI (2016)

22. Reviewland Project: Additional material associated to ICWE 2017 submission (2017). http://reviewland.projects.iit.cnr.it/publications/icwe2017/icwe2017.html. Accessed 15 Mar 2017

23. Salton, G., Buckley, C.: Term-weighting approaches in automatic text retrieval. Inf. Process. Manage. **24**(5), 513–523 (1988)

24. Turney, P.D.: Thumbs up or thumbs down?: semantic orientation applied to unsupervised classification of reviews. In: Computational Linguistics Meeting, pp. 417–424. ACL (2002)

25. Wilson, T., et al.: OpinionFinder: a system for subjectivity analysis. In: HLT/EMNLP on Interactive Demonstrations, pp. 34–35. ACL (2005)

26. Wilson, T., et al.: Recognizing contextual polarity in phrase-level sentiment analysis. In: HLT/EMNLP, pp. 347–354. ACL (2005)

27. Wilson, T., et al.: Recognizing contextual polarity: an exploration of features for phrase-level sentiment analysis. Comput. Linguist. **35**(3), 399–433 (2009)

XOOM: An End-User Development Tool for Web-Based Wearable Immersive Virtual Tours

Franca Garzotto, Mirko Gelsomini[✉], Vito Matarazzo,
Nicolò Messina, and Daniele Occhiuto

Innovative Interactive Interfaces Lab - Department of Electronics,
Information and Bioengineering, Politecnico di Milano, 20133 Milan, Italy
{franca.garzotto,mirko.gelsomini,
daniele.occhiuto}@polimi.it,
{vito.matarazzo,nicolo.messina}@mail.polimi.it

Abstract. XOOM is a novel interactive tool that allows non ICT-specialists to create web-based applications of Wearable Immersive Virtual Reality (WIVR) technology that use 360° realistic videos as interactive virtual tours. These applications are interesting for various domains that range from gaming, entertainment, cultural heritage, and tourism to education, professional training, therapy and rehabilitation. 360° interactive videos are displayed on smartphones placed on head-mounted VR viewers. Users explore the virtual environment and interact with active elements through head direction and movements. The virtual scenarios can be seen also on external displays (e.g., TV monitors or projections) to enable other users to participate in the experience, and to control the VR space if needed, e.g., for education, training or therapy purposes. XOOM provides the functionality to create applications of this kind, import 360° videos, concatenate them, and superimpose active elements on the virtual scenes, so that the resulting environment is more interactive and is customized to the requirement of a specific domain and user target. XOOM also supports automatic data gathering and visualizations (e.g., through heat-maps) of the users' experience, which can be inspected for analytics purposes, as well as for user evaluation (e.g., in education, training, or therapy contexts). The paper describes the design and implementation of XOOM, and reports a case study in the therapeutic context.

Keywords: Wearable Immersive Virtual Reality · Head-Mounted Display · 360° videos · Virtual tour – End user development – Therapy – Neurodevelopmental disorder · Children

1 Background and Introduction

Wearable Immersive Virtual Reality (WIVR) offers interactive virtual experiences that simulate realistic or imaginary environments on Head-Mounted Displays (HMDs). These are worn on the head as part of helmets and have small display optic devices in front of user's eyes that make users "feel inside" the virtual space. In the last years, the

© Springer International Publishing AG 2017
J. Cabot et al. (Eds.): ICWE 2017, LNCS 10360, pp. 507–519, 2017.
DOI: 10.1007/978-3-319-60131-1_36

evolution of WIVR technology has made HMDs cost-affordable and much more comfortable than previous generation devices. In addition, 360° cameras have enabled the spread of realistic 360° videos that can be played on WIVR viewers. As a result, we witness an increasing popularity of WIVR technology and applications in various domains that range from gaming [24], entertainment [16], cultural heritage [14], and tourism [22] to education, professional training [12], therapy and rehabilitation [13].

An example of WIVR applications used for educational purposes, which is also one of the inspirations for our work, is Google Expeditions [17]. This is a mobile app, that allows groups of students to explore 360° video-based virtual trips around the world under the teacher's control. The teacher can select one of the available tours that is automatically visualized on all the students' HMDs and can put a marker on a specific point of the 360° environment to move their attention on that.

Currently, there are two types of HMDs. *Embedded solutions* provide a complete VR experience without needing any integration with external devices. Examples of these technologies are Oculus Rift [18] and HTC Vive [21]. Embedded solutions usually ensure a more accurate user interaction and a better graphical quality than *modular solutions*, but at a much higher cost. *Modular solutions* exploit external devices, mostly smartphones, as enabling technology for displaying the simulated world, and are much cheaper. Examples are Google Cardboard [20] (starting from 5$) and Samsung Gear VR [25] (starting from 79$). Viewers are composed of two biconvex lenses mounted on a plastic or cardboard structure available in different colors and shapes. The smartphone set inside the visor displays the visual contents as two near-identical bi-dimensional images: the illusion of 3D depth and immersion is the result of the human-brain interpretation of the stereoscopic effect generated by the viewer lenses (Fig. 1).

Fig. 1. On the left, example of Google Cardboard with a smartphone inserted. On the right, a stereoscopic view of a 360° video.

This paper presents the design and implementation of XOOM, a novel tool for the development *web-based WIVR applications*. Specifically, XOOM enables *non ICT experts* to develop wearable immersive *virtual tours based on 360° realistic videos* played on any smart-phone placed on Google Cardboard. Video materials can be any MP4 file downloaded from a video platform or recorded using a 360° camera available on the market. In XOOM applications, the user can change her/his visual perspective on the virtual space, move forward/backward across it, and interact with active areas and active elements that generate visual or audio effects. XOOM provides the functionality to import 360° videos, concatenate them, and superimpose active elements on the virtual scenes, so that the resulting videos and its interaction affordances are customized to the requirement of a specific user group. The VR user's view can be shown on an external display (e.g., TV monitor or projection) through standard casting protocols over Wi-Fi. This enables other users to observe and participate in the WIVR experience, seeing what the VR user sees in the HMD. In addition, XOOM applications are integrated with an *external web-based application* that enable an external supervisor to (i) control the VR user experience, e.g., pausing, restarting, stopping and changing the video while it is playing in the HMD; (ii) add active elements at run-time. (iii) have a real-time perception of the user's behaviors, e.g., to understand what the user wearing the HMD is focusing on the most. Finally, XOOM supports *automatic data gathering and visualizations* (e.g., tracking the user's gaze direction) of relevant data about the users' experience in the immersive VR space that can be inspected for analytics purposes, as well as for user evaluation (e.g., in education, training, or therapy contexts).

The remaining of the paper is organized as follows. Section 2 presents the design of tool in terms of the functionality available to designers. Section 3 describes the implementation approach and the software architecture. Section 4 reports a case study in which XOOM has been used in a therapeutic contexts to create WIVR applications for children with neurodevelopmental disorders. Section 5 discusses the contribution of our work. Section 6 outlines our future research directions.

2 XOOM: Design

XOOM is a web based software platform that, through various interactive tools, supports the creation, customization and analysis of web based WIVR virtual tours called *"XOOM experiences"* (*X-Experiences*). An X-Experience is defined as a sequence of one or more smoothly integrated 360° videos (hereinafter "scenes") enriched with customized interactive elements. The user proceeds along the flow of scenes by effect of his/her specific interactions, or of the commands activated by another external user (if any) or both. An X-Experience is natively integrated with an external web application, for observation, monitoring, control, and evaluation purposes. XOOM distinguishes 3 types of users:

- *developers*, who create an X-Experience; the creation functionalities are organized as a guided, easy-to-understand flow of actions and available through as simple user-friendly interface; therefore the developer does not need to be an ICT expert to use XOOM;
- external *supervisors* (developer, evaluator, or caregiver), who monitor and control the end-user's running X-Experience;
- *end-users,* who experience the immersive VR environment on the HMD.

XOOM is made of *Experience Manager* and *Experience Viewer*. The Experience Manager is used by the developer and the supervisor to personalize the X-Experience. The Experience Viewer is used by the end-user and manages his/her interaction with the virtual environment.

Fig. 2. Creating an X-Experience in a museum using the Creator component of XOOM. On the background there is the original video, while the customization element (in yellow) is superimposed by the developer. Next to it there is the menu to set the element properties. (Color figure online)

2.1 Experience Manager

The *Experience Manager* integrates three main tools called *Creator*, *Runtime Controller*, and *Analyzer* that are used respectively before, during, and after the end-user's experience.

The *Creator* (Fig. 2) allows the developer to create or modify a WIVR tour by:

- Selecting 360° video fragments previously stored in the XOOM repository, and inserting them in the X-Experience under development. They are visible on the bottom part of the window and, by clicking on them, the developer can switch to personalize one or the other.
- Reordering the video fragments, to create a flow of different "scenes" to be shown in sequence.
- Setting specific frames at which the scene must pause at runtime, by clicking the Pause button in correspondence of that instant of the timeline. During the running experience, the video will automatically stop at that frame, allowing for example the end-user to focus on a point of interest before going on.
- Personalize the video and its interactive experience by adding different types of items like arrows, masks, geometrical forms, textual popups, images. These can be: *Static* items that are overlaid on the video with no dynamic effects; *Dynamic* items, which trigger dynamic effects when activated (e.g., showing animations, zooming in-out, rotating); *Control* items that, when selected by the user, control the video execution, e.g., playing a paused video or moving from a scene to the next one. For each of these elements the developer can set position, dimensions, duration from a specific starting frame and its properties. For example, with the *Highlight* function, the developer can select an area of the current frame in a way that, at runtime, it will be lighted up, obscuring the rest with a mask. This functionality is useful to drive the user's attention on a desired point, area, or object of the virtual environment (in the Creator this area is visualized as a yellow sphere).
- Saving the created experience in the XOOM repository.

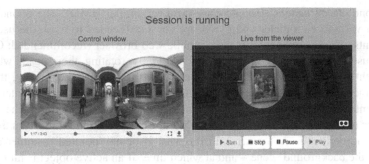

Fig. 3. View of Runtime Controller of the running experience (under creation in Fig. 2), showing the effect of the *Highlight* function at runtime.

The *Runtime Controller* (Fig. 3) enables the supervisor to monitor and control the end-user's running experience. He/she can see what is displayed to the user in the HMD using two windows on the screen. The control window (Fig. 3 – left) shows the complete 360° scene "flattened" so that the external observer can move freely in the timeline of the experience, without affecting the viewer. The monitor window (Fig. 3 – right) shows what the end user is currently seeing through the HMD. Note: this is note what is displayed on the HMD, but what the user actually sees. We inisit on this aspect since we provide the controller with the user's field of view of the 360° video after the transformation performed by the human brain from the 2 almost identical images on the smart phone screen to a single perceived image.

The *Analyzer* visualizes a 360° heatmap of each session and summarizes with different colors the points and areas of the scenes on which the patient focused his attention the most and the least (Fig. 4).

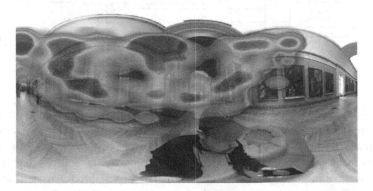

Fig. 4. Heatmap of an X-Experience; red areas are those where the user focused during the interaction (Color figure online)

2.2 Experience Viewer

The Experience Viewer manages the end-user experience: it allows end-users to select an X-Experience and interact with it. The Experience Viewer can run on any end-user smartphone placed inside a Google Cardboard (or similar HMD). The phone must be equipped with accelerometer and gyroscope sensors (needed to track the user's head movements), Wi-Fi connectivity, and a mobile web browser (recommended: Chrome).

End-users can navigate in the virtual world by rotating their head, which will consequently rotate the virtual scene projected in the screen. Technically, the phone sensors track the motion and the orientation of the user's head, which are interpreted as gaze orientation and focus. The head orientation defines the direction of the gaze and the center of the screen is the gaze focus. XOOM casts a ray that intersects the visible "scene" and derives from the ray collision what is being "pointed" – either an active object or the background scene – and at which time. If an active object is hit by the ray for 3 consecutive seconds, the system registers the willingness to interact with it and triggers the reactive behavior defined for that object.

3   XOOM: Implementation

From the software perspective, XOOM is a client-server web application (Fig. 5), in which two different classes of clients (the *Viewer* and the *Manager*) access the two previously specified modules on a web server which hosts our application and is connected to a cloud storage that stores created experiences and the related data.

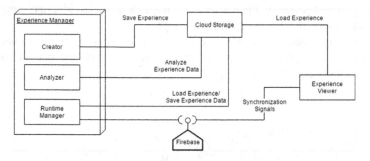

Fig. 5. Component view of XOOM

The videos used for the creation are selected from the ones available on YouTube [26] and must be equirectangular[1]. In case users want to employ their own videos, they are automatically uploaded to the video platform before being inserted into an experience. This means that XOOM privileges videos from external sources (e.g. YouTube) to avoid overloading our system storage, if this is not possible for users (for example with no YouTube account) we provide them the sorage possibility after a registration to our system. The experience is saved as a JSON file with the timeline of each video in the sequence and all the customization elements added with their properties.

---

[1] Equirectangular is the standard projection used in 360° VR content, which applies a certain distortion to a panorama so that it is shown correctly when projected on the inside of a sphere.

In order to start a session, the *Manager* (in particular the *Runtime Controller* component) and the *Viewer* must be univocally associated through a unique code. The connection between them is achieved using Firebase [19], a platform developed by Google that offers several independent features to develop mobile and Internet connected applications. In particular, we use the Real-time Database component, a cloud-hosted NoSQL database to synchronize in both directions the two modules of *XOOM*: the *Controller* sends the commands issued by a supervising user, while the *Viewer* sends synch-signals (time and orientation of the video) every 100 ms to enable the synchronization of the monitor window of the supervising user's screen. The experience is not streamed from the *Viewer* to the *Controller*, but it runs in parallel in the two devices, showing the same images at any time thanks to the specified synchronization information. We chose this solution because it would have been complex to support a real-time streaming, especially in case of slow Internet connection. In a slow network scenario caching strategies may be implemented to temper the perceived slowliness of the bandwidth. The Real-time Database, instead, allows to obtain a fluid experience reproduction, due to the small quantity of exchanged data and the fast response time of Firebase. Direct communication between the two modules would have required to create an ad-hoc tunnel and so a more complex network configuration.

Moreover, the time and orientation values sent from the *Viewer* are saved by the application and will be used by the *Analyzer* to build the final heatmap of the current experience. The usage of external platforms, in particular YouTube, Firebase and a cloud storage improves the scalability of our system, allowing the management of a large quantity of data and users. The management of the virtual environment, both at creation time and at runtime, is supported by A-Frame, an open-source web framework for creating 3D and virtual reality applications with JavaScript and HTML [15].

4 Case Study

We have applied XOOM in our research on WIVR technology for persons with Neurodevelopmental Disorders (NDDs). NDD is an umbrella term for a group of disabilities that appear during the developmental period and are characterized by deficits and limitations in the cognitive, emotional, motor and intellectual spheres [2]. The use of WIVR as a therapeutic or education tool for people with NDD was explored in the past and abandoned because of the drawbacks of first generation viewers [13] (e.g., high cost, weight, motion sickness effects). With the advent of cheaper, more comfortable, and technically more accurate hardware and software solutions, WIVR has raised a growing interest in the NDD [6, 11]. Our research in this area have so far considered virtual spaces based on the contents, characters and environments of fantasy tales used with NDD children at the therapeutic centers that collaborate in our research. Storytelling plays an important role in educational practices, in particular for children with NDD, to promote from high-level to more basic skills, such as the ability to focus on what is most interesting at a given moment, generalization[2] skills and development

---

[2] A generalization is a concept in the inductive sense of that word, or an extension of the concept to less-specific criteria; i.e. to abstract an action from the story to the real world.

of appropriate elementary behaviour. A number of empirical studies at therapeutic centers, involving overall 8 children with NDD and 4 therapists, has proved the effectiveness of our applications to promote attentional skills and cause-understanding capability. For example, we have witnessed that the highlight effect helps the users to understand which are the important details of the scene. In fact, while in the first therapy sessions this customization effect has been necessary to guide children's attention, in the following ones with the same X-Experience they have progressively learnt to focus on the same points without the help of the H*ighlight.*

Three therapists at the centers we are collaborating with in a number of national and international projects have used XOOM to create a new class of applications for this target groups, called *WIVR social stories.* The term "social story" is used in NDD therapy to denote visual materials (paper- or video-based) that describe everyday life situations as simple short narratives (Fig. 6 - left). These tools are used in the treatment of persons with NDD, particularly ASD (Autism Spectrum Disorder), to help them develop social and practical skills and to learn appropriate behaviour and norms [1, 3, 11].

XOOM has enabled the therapists to mix the power of traditional social stories with that of WIVR. Using XOOM, they have created a set of WIVR-based social stories on everyday situations (e.g., going to school, shopping at a grocery store, visiting a museum). An example is visible in Fig. 6: therapists have recorded a 360° video in a supermarket and have built an X-Experience from it to teach patients to search and buy specific products. As shown in the mentioned figure, a PCS (Picture Communication Symbols)[3] has been inserted over the video, at creation time, to suggest the end-user to look for some fruit.

Fig. 6. An example of paper-based social story (left), a WIVR-based social story (right).

Transforming social stories into interactive immersive virtual narratives can increase the benefits of the traditional social story approach. WIVR-based social stories allow persons with NDD to train in everyday life's activities in a secure environment, while VR headsets promote attention and engagement as they remove the distractions

---

[3] Picture Communication symbols are a set of color and black & white drawings originally developed by Mayer-Johnson, LLC [23] for use in augmentative and alternative communication (AAC) systems. Several studies report children with cognitive disabilities learn PCS easily.

caused by external visual stimuli [5] and help users focus on therapy or learning tasks. The increasingly low-cost of WIVR technology paves the ground towards large scale adoption of this class of assistive applications, at therapeutic centers and in other contexts of life (e.g. at home and at school). The customization features offered by XOOM empower caregivers and give them a control of the patient's experience with WIVR technology at a degree which is not allowed by any other existing tool. Finally, the nature of the contents exploited in WIVR social stories offers novel opportunities to give an active role to family caregivers in the therapeutic process. Patients' parents can be involved to record videos in the real contexts of their children's life to feed the XOOM video repository.

5   Contribution and Discussion

Our work provides several contributions to the current state of the art in WIVR.

The *first contribution* concerns the *nature* of the applications supported by our tool. The applications developed using XOOM are *interaction-rich*. The gamut of interaction affordances go beyond the typical virtual tour "navigation" of most 360° videos for HMDs, which support only changing of the visual perspective of the virtual space, or moving forward/backward across it. In XOOM applications the user can interact also with a countless number of "active elements" that are rendered as active areas or graphic elements superimposed on the video content and generate engaging visual or audio effects. In addition, while most existing interaction-rich WIVR applications are delivered as native apps, XOOM applications are *web-based* and embrace a VRaaS (VR as a Service) paradigm. Few years ago there has been a migration from desktop native apps to cloud served solution (e.g. SaaS), and now we witness a similar trend on mobile platforms. WIVR end users can derive a set of benefits from a web-based migration of this class of applications. They are not bothered with installing apps that they may visit only once. Web guarantees easy access to contents, skipping the installation process overhead (download, install, grant permissions, open) with respect to just clicking a link to open the VR environment in a browser. Web apps are intrinsically cross-platform, as they require a browser alone, and are agnostic of the underlying operating system, thus they are easier to distribute and maintain. Their distribution mechanism does not need a dedicated market place and the associated approval period to be published: they are accessible "as-is". Nonetheless, web application must still face some challenges. They strongly depend on connectivity (even if service-workers in progressive web applications are remarkably addressing this issue). They have access to a limited set of hardware features from the web interface (anew this limitation is being tackled by the physical web open approach). Being so easily served, they suffer from a broader attack surface than mobile apps and may be more subjected to security issues.

As *second contribution* of XOOM concerns the original functionality of the applications created with the tool. XOOM application are natively integrated with features that supports *participation to, control and analysis* of the VR experience by *external* users. A control panel enables them to see what the VR user wearing the HMD sees and to *customize the experience content and the video behavior at run time*;

existing WIVR applications support the run-time presentation of the contents of the HMD on an external display only through screen casting. In addition, XOOM applications are integrated with an *interaction analysis tool*. Both these features are particularly useful for caregivers in education or therapeutic contexts. Educators or therapists need to know what the learner or the patient is currently watching at within the HMD, so to be able to intervene at the right moment of the experience if needed. In addition, these stakeholders enormously benefit from automatic data gathering and visualization. Properly aggregated and presented interaction data offer useful valuable information that would otherwise been collected manually, through textual reporting or video analysis. Heat-maps for example offer insights to evaluate the user experience and the user's progressions, enabling to identify where and when during the experience the user's attention and concentration were lower or higher.

As a *third contribution*, XOOM supports an *end-user development process* (EUD) [9]. The EUD approach goes beyond conventional methodologies for the design of interactive systems since its goal is to shift control of application design, development, and evolution from skilled ICT professionals to people who are the key owners of problems in a domain, creating opportunities for extensions and modifications that are appropriate for those who need to make changes and are not technology experts. EUD tools provide non ICT professional with tools to customize applications or create brand new ones to support personal, situational needs in order to address the requirements of a specific domain and intended users. An EUD approach is particularly relevant in domains such as training, education, and therapy, where the capability of autonomous customization by therapists or educators is fundamental, as the value of any interactive technology in these contexts is directly related to its ability to meet to the specific characteristics and the needs of the target user group(s). There is empirical evidence for example that digital therapeutic tools for persons with disability that cannot be easily customized, or can be customized by ICT professionals only, are more often abandoned [10]. To our knowledge, the EUD approach has never been applied in the WIVR domain at the degree achieved in XOOM. The only similar idea comes from the previously cited Google Expedition, which is a WIVR application easily usable by non ICT experts. However, the EUD features of this platform are currently limited. In fact, as already said, teachers cannot create their own experiences but can only select one of the available "explorations", and the only possible customization consists in the insertion of passive "markers" in specific areas of interest of the virtual environment. Starting from this concept, we have enriched our tool with a lot more customization possibilities from the insertion of images in the 360° environment to the ability to control the video execution through control items.

The *fourth contribution* of our work is from a software engineering perspective. XOOM integrates several software frameworks adopting a modular approach and organizing each module by feature. This modularity opens to a top-down problem decomposition. For example, concerning the graphical modelling, XOOM is powered by A-Frame framework, establishing a high level abstraction to address graphical virtual elements with as less complexity as possible; if A-Frame capabilities are not sufficient our system deals with the problem by leveraging Three.js underneath layer, and eventually, it can interrogate WebGL library for virtual content manipulation. This vertical problem resolution is accompanied with a horizontal procedure to interconnect modules in a seamless way (e.g. real-time adaptation of the virtual world). This is done by leveraging fast online synchronization. XOOM benefits from low coupling in the sense that each module is not strongly dependent on others: for example if the Runtime Controller fails, this failure will not affect the experience viewer module. Moreover, XOOM was designed with high cohesion in mind in fact we organized components by features (e.g. geometries folder to group shapes that can be added, shaders folder to implement the highlight effect, …). Following this approach XOOM results in a flexible, robust and easy to maintain application.

6 Future Work

So far XOOM has been used to develop WIVR applications in therapeutic contexts, as the ones described in the previous section. According to the feedbacks received by the specialists that have used XOOM and its applications, or tried them in demo events, XOOM may pave the ground towards large scale adoption of WIVR assistive technology at therapeutic centers and in other contexts of a patient's life (e.g. at home and at school).

Still, the spectrum of XOOM application domains is much wider. People can use this simple tool to create videogame-like WIVR experiences, gaming gadgets in brick-and-mortar shops, educational contents for professional training, or virtual experiences in cultural heritage or tourism contexts. For example, visitors of a cultural heritage site can use on the HMD display to zoom-in and explore details of the physical space they are visiting and get insights on its artifacts. Customers at a travel agency office can experience a possible destination virtually through an HMD instead of just being presented with the photos of a location on a generic catalogue. Each situation in each specific domain requires different videos and interaction elements, but the generality and power of XOOM enables non-ICT-expert domain operators to create VR experiences that can be quickly adapted to every different goal and target.

Our research agenda in the short term envisions an improvement of the user interface, the full-size engineering of XOOM functionalities, and its extension. We will simplify the UX features of some existing functions (e.g., move, rotate and scale commands, which created some initial difficulty to the users in our case study). We will enrich the available customization features and their interaction properties. Finally, we will improve the management system that manages accounts, authentications and end-user profile. In addition, we have designed two empirical studies that will start in Spring 2017. The first study will explore the adoptability of XOOM in therapeutic contexts. We will systematically evaluate with 10 NDD specialists the usability of XOOM and will elicit the socio-organizational requirements for its deployment and adoption as commercial tool. The second study (controlled study) will involve 20 children with NDD aged 6–12 and will focus on the therapeutic benefits of WIVR-based social stories, also compared to traditional social stories approaches.

Acknowledgements. This work is partially supported by TIM/Telecom Italia, under the PhD Grant "Internet of Products".

References

1. Adams, L., Gouvousis, A., VanLue, M., Waldron, C.: Social story intervention: improving communication skills in a child with an autism spectrum disorder. Focus Autism Other Dev. Disabil. **19**(2), 87–94 (2004)
2. American Psychiatric Association. Diagnostic and statistical manual of mental disorders (DSM-5®). American Psychiatric Pub. (2013)
3. Andersson, U., Josefsson, P., Pareto, L.: Challenges in designing virtual environments: training social skills for children with autism. Int. J. Disabil. Hum. Dev. (2006)
4. Cruz-Neira, C., Sandin, D.J., DeFanti, T.A., Kenyon, R.V., Hart, J.C.: The CAVE: audio visual experience automatic virtual environment. Commun. ACM **35**, 64–72 (1992)
5. Ehrlich, J., Munger, J.: Utilizing head mounted displays as a learning tool for children with autism. In: Proceedings of 4th European Immersive Education Summit, Vienna, Austria (2012)
6. Finkelstein, S., Barnes, T., Wartell, Z., Suma, E.A.: Evaluation of the exertion and motivation factors of a virtual reality exercise game for children with autism. In: Proceedings of Virtual and Augmented Assistive Technology (VAAT) (2013)
7. Gelsomini, M.: An affordable virtual reality learning framework for children with neuro-developmental disorder. In: Proceedings of the 18th International ACM SIGACCESS Conference on Computers and Accessibility (ASSETS 2016), pp. 343–344. ACM (2016). Best student research paper award
8. Lányi, C.S., Tilinger, Á.: Multimedia and virtual reality in the rehabilitation of autistic children. In: Miesenberger, K., Klaus, J., Zagler, W.L., Burger, D. (eds.) ICCHP 2004. LNCS, vol. 3118, pp. 22–28. Springer, Heidelberg (2004). doi:10.1007/978-3-540-27817-7_4
9. Lieberman, H., Paternò, F., Wulf, V. (eds.): End User Development. Springer, Netherlands (2006)
10. Phillips, B., Zhao, H.: Predictors of assistive technology abandonment. Assistive Technol. **5**(1), 36–45 (1993). Taylor & Francis
11. Quirmbach, L.M., Lincoln, A.J., Feinberg-Gizzo, M.J., Ingersoll, B.R., Andrews, S.M.: Social stories: mechanisms of effectiveness in increasing game play skills in children diagnosed with autism spectrum disorder using a pretest posttest repeated measures randomized control group design (2008)
12. Seymour, N.E., Gallagher, A.G., Roman, S.A., O'Brien, M.K., Bansal, V.K., Andersen, D.K., Satava, R.M.: Virtual reality training improves operating room performance: results of a randomized, double-blinded study. Ann. Surg. **236**(4), 458–464 (2002)
13. Strickland, D.C., Marcus, L.M., Mesibov, G.B., Hogan, K.: Brief report: two case studies using virtual reality as a learning tool for autistic children. J. Autism Dev. Disorders **26**, 651–659 (1996)
14. Wojciechowski, R., Walczak, K., White, M., Cellary, W.: Building virtual and augmented reality museum exhibitions. In: Proceedings of the Ninth International Conference on 3D Web Technology (Web3D 2004), pp. 135–144. ACM, New York (2004)
15. A-Frame. https://aframe.io/
16. Disney Movies VR. https://www.disneymoviesvr.com/
17. Google Expeditions. https://www.google.com/edu/expeditions/
18. Facebook Oculus Rift. https://www.oculus.com
19. Firebase. https://firebase.google.com/
20. Google Cardboard. https://www.google.com/intl/en/get/cardboard/
21. HTC Vive. https://www.vive.com/eu/

22. MSC Crociere - Virtual Reality Brochure. https://www.msccrociere.it/it-it/Offerte-Crociera/Tariffe-MSC/Virtual-Reality-Brochure-MSC-Crociere.aspx
23. PCS Collection, Mayer Johnson. http://www.mayer-johnson.com/pcs-collections/
24. Playstation VR. https://www.playstation.com/en-us/explore/playstation-vr/
25. Samsung Gear VR. http://www.samsung.com/us/explore/gear-vr/
26. Youtube. https://www.youtube.com/

Public Debates on the Web

Fabian Gilson[1]([⊠]), André Bittar[2], and Pierre-Yves Schobbens[1]

[1] PReCISE Research Center, University of Namur, Namur, Belgium
{fabian.gilson,pierre-yves.schobbens}@unamur.be
[2] Centre for Natural Language Processing, Université catholique de Louvain,
Louvain-la-neuve, Belgium
andre.bittar@uclouvain.be

Abstract. With the advent of social media, any piece of information may be spread all over the world in no time. Furthermore, the vast number of available communication channels makes it difficult to cross-check information that has been (re-)published on different media in real time. In this context where people may express their positions on many subjects, as well as launching new open initiatives, the public needs a mean to gather and compare ideas and opinions in a structured manner. The present paper presents the WebDEB project, which aims to develop a collaborative platform where opinions, namely arguments, are gathered, analyzed and linked to one another via explicit relations. WebDEB relies on various Natural Language Processing modules to semi-automatically extract information from the web and propose meaningful visualizations to the platform's contributors. Furthermore, public actors may be identified and attached to the ideas they publish to create a structured knowledge base where annotated texts, extracted positions and alliances may be identified.

Keywords: Social web · Public debate · Argumentation · Speech acts · Natural language processing

1 Introduction

Today, social media has become a key platform for public debates. Presidential elections, financial scandals, document leaks and so forth are among the many subjects that are attracting public persons and individuals to express their opinions in real time. As more people jump into debate arenas, more opinions, facts or statistics are published by many sources. Due to this complexity, any attempt to sort through or structure this mass of information requires a significant summarization effort [2].

Many initiatives following *Wikipedia*[1]'s example have appeared in recent years allowing any person to create and record information, making it openly available across the web. Furthermore, one of the main focuses in the field of

---

[1] https://wikipedia.org.

© Springer International Publishing AG 2017
J. Cabot et al. (Eds.): ICWE 2017, LNCS 10360, pp. 520–532, 2017.
DOI: 10.1007/978-3-319-60131-1_37

Natural Language Processing (NLP) is the extraction and structuring of textual data, with an increasing focus on the social web [5,8].

Public debates are a gold mine of ideas, facts and opinions, but extracting the main structures from such material is tough work for the human mind [2]. Furthermore, such a flow of information is difficult to channel and post-process by the public. To this end, we introduce WebDEB (WD), an openly accessible platform dedicated to structure and analyze arguments expressed publicly or retrieved from press articles. Centered on graphical representations of arguments, involved actors and the relations between them, WD is designed to help citizens in today's understanding of public debates. As an initial goal, the WebDEB platform is intended to offer an online facility to concentrate arguments gathered from many sources, introduced by contributors as well as automatically fed from various sources. As a second target audience, WebDEB is built as a pedagogical support in language teaching in the field of, among others, discourse analysis and reasoning for secondary and higher education students [7].

In the present paper, we first describe the current design of the platform, with a series of captures of its implementation in Sect. 2. We then discuss the state of validation and list current limitations as observed by the present community of users in Sect. 3. In Sect. 4, we look over related work in discourse analysis, argumentation theories and debate representations. We finally sum up the contribution of this paper in Sect. 5 and review future work on the conception and validation of the platform.

2   The WebDEB Platform

WebDEB is initially designed as a pedagogical platform for the teaching of sociological issues. It is also intended as an open platform to gather any form of debates. With industrial partners, WD has also been developed to offer support to journalists in their information cross-checking and archiving tasks. We first conceptually depict the structuring of arguments in WD and describe the NLP features we put in place to support users in their encoding tasks and to build rich visualizations. We also provide some captures of the current implementation.

2.1   Conceptual Description

The two main concepts in WebDEB are Contributors and Contributions, as shown in Fig. 1. A Contributor is any registered user that contributes to the WD database. A clear hierarchy of contributors has been defined for pedagogical purposes on the one hand, and for content monitoring on the other hand. A standard separation has been specified to distinguish Groups and Permissions in order to make user management flexible and extensible. Also, a particular attention has been paid to data access, integrity and visibility since WD is meant to be openly accessible, while holding sensitive data.

Contributions are any data inserted (and validated when necessary) by Contributors. Possible types of Contributions are Actor, Text, Argument and Argument-Link. Actors may be affiliated to other Actors with the aim of tracing associations of Persons to Organizations, and also partnerships between Organizations. Actors may be involved in many ways in TextualContributions, *e.g.* as authors or publishers. TextualContributions have Topics associated to them for language processing, search and filtering purposes.

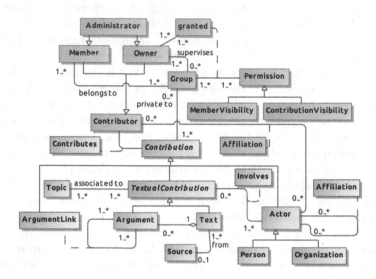

Fig. 1. WebDEB domain model

Arguments are extracted from Texts either manually or semi-automatically. Arguments are a rewriting of a text excerpt expressing a single idea in as simple form as possible. They also must be self-contained and unambiguous. Following the theory of *speech of acts* [1,10], Arguments are classified into appreciative expressions of an emotion or a feeling, performative expressions of an action that changes the described situation, prescriptive expression of the necessity of taking an action or finally constative being any other type of expressions, like simple ascertainment. Temporal attributes (TimingType) and degree of certainty (ShadeType) are used to refine ArgumentTypes, adding more precision for NLP, searching and filtering purposes. Depending on the actual argument type, this degree is either specified on a three or five scale of belief in predefined textual form such as *"I believe"* or *"It is likely that"*.

Arguments may be interconnected to each other by justification links where one argument *supports*, *qualifies* (not really justifies but neither refutes) or *refutes* another one, or by similarity links where two arguments are *similar* (express the same point of view on a given subject), *nuanced* (somehow similar or dissimilar) or *dissimilar* (express opposing points of view on the same subject).

From ArgumentLinks between Arguments, accompanied by their authorships and affiliations of Actors, we are able to create valuable visualizations and aggregations that represent different viewpoints regarding the four aforementioned types of Contributions, as we will detail in Sect. 2.4.

2.2 Natural Language Processing and Data Retrieval Tools

A set of aids have been developed to assist contributors in encoding data or make suggestions based on semantic analysis of existing arguments.

Argument classification. Based on the aforementioned four types of *speech of act* [1,10], a classifier described in a separated work is invoked to automatically suggest a classification when adding new arguments [3]. This classes are meant to describe the linguistic mechanisms used in the discourse and is used to discover argumentation links automatically. A second classifier is also invoked to retrieve the *discourse connectives*, *e.g.* comparison, condition or correlation.

Automated data retrieval. When adding press articles to the platform, web content may be imported with other properties such as the authors, simplifying the encoding of new texts that are saved either in the private library of a contributor, or when permitted by copyright, in the publicly visible library. Also, from partnerships with press editors, an *Rich Site Summary* (RSS) feeder regularly injects new content into the public database on which topic extraction processes are run. Another service listens to *Twitter* feeds and process them to create potential arguments that will be further validated by contributors. A final data retrieval service extracts actors' details, such as affiliations, from the *wikidata.org* open database based on a user-provided name or *Wikipedia* url.

Similarity detection. Last, when new arguments are inserted into the database, they are compared to all current ones in order to identify potentially similar arguments and to enrich the various visualizations presented in Sect. 2.4. Potential pairs of semantically similar arguments are then subject to manual validation by contributors before being effectively added into the WD platform.

2.3 Typical Usage

The core business of the WebDeb platform resides in arguments. In order to import them into the database, contributors either validate automatically extracted ones in a dedicated *yes/no* webpage where a list of suggestions is available, or they may start by importing a text like a press article. From any text, contributors may start to annotate them using the dedicated screen shown in Fig. 2.

Named entities, *e.g.* persons, organizations, dates or professions are highlighted from the text displayed in the top area with paging facility. For all arguments already extracted from the current text, their corresponding excerpts are highlighted too, giving the possibility to see the corresponding transformed arguments as well as going to their own visualizations.

Contributors are then able to select part of the text (with smart paging in case of an excerpt would be split onto multiple pages) to display an *add-new* pop-up window and fill in the details regarding that argument.

At the bottom of this screen, the list of already extracted arguments, the text structure in terms of these arguments, the properties and all involved actors are also viewable. The discourse structure is editable with a *drag and drop* feature, as visible in Fig. 3.

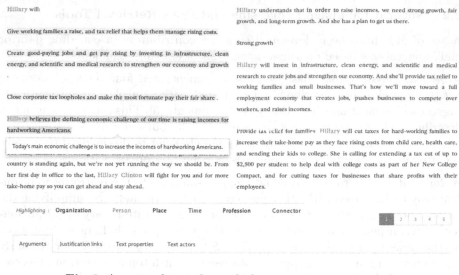

Fig. 2. Annotated text, from which arguments are extracted

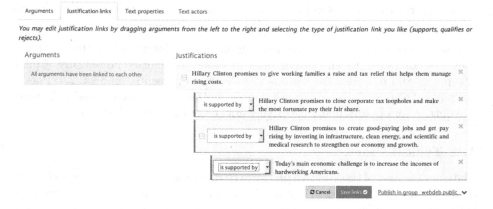

Fig. 3. Defining the argumentation structure

In order to add arguments, like other types of contributions, contributors are invited to fill in forms as presented in Fig. 4. The type of argument is proposed by the aforementioned annotator and the argument shades are proposed accordingly to the selected type. An automated topic extraction module has also been developed to help contributors when filling meta-data and suggestions of argument standardizations are currently under development.

When submitted, the argument will be highlighted in the text and compared to the existing database to find similarity matches to be validated later on.

Similar forms are available to encode the other types of contributions into the platform, but they are not detailed here for space reason.

Please fill in the properties regarding your argument. Fields marked with a * are mandatory

" Hillary believes the defining economic challenge of our time is raising incomes for hardworking Americans.
"

Who enunciates the argument? * ⓘ

| Person or organization | Function (optional) | Affiliation (optional) |
| Hillary Clinton | Function or mandate | Affiliation | + − |

Who reports the argument? ⓘ

| Name | Function (optional) | Affiliation (optional) |
| Person or organization | Function or mandate | Affiliation | + − |

Type of argument * ⓘ

☑ Constative It is a true fact, It is not true that,...

☐ Prescriptive Something/someone should/must,...

☐ Opinion I love, I hate, I feel,...

☐ Performative I promise, I decide, I commit myself, I demand,...

Select one type from the available propositions

📢 Our annotation service suggests the following types: Constative (100%)

Ⓧ Cancel   Next ❯

Fig. 4. First screen of the two step process to add a new argument

2.4   Rich Visualizations

For all types of Contributions, a set of visualizations have been created to trace arguments and explore related authors, texts and sources. Other visualizations have also been created to check opponents of Actors or summarize any argument made by an Actor that either has been encoded by users or automatically imported from external sources. Finally, a representation has been developed that presents the hierarchy of arguments inside a particular text, depicting its logical structure. On top of all these depictions, all Contributions, apart from the ArgumentLink which has no meaning *per se*, have a summary page where all their inherent properties are displayed[2].

For Actors, a first view traces all affiliations and, for organizations, their affiliated Actors, as shown in Fig. 5. As in any other visualization, users are also invited to add more information. Furthermore, as for this view, graphs are exportable in many formats, *e.g.* PDF or PNG.

Fig. 5. Person's affiliations (historic view)

---

[2] These summary pages are not shown here due to space limitations.

All **Arguments** expressed by a given **Actor** may be displayed with an *"agreement graph"* depicting how many other **Actors** agree or not for each **Argument** this **Actor** has produced. This view may be sorted either by date, according to the amount of similar **Arguments** this particular Actor made, *i.e.* number of times this **Actor** said the same thing, or by the degree of agreement of other **Actors**, as visible in Fig. 6.

Fig. 6. All talks of a given **Actor** (sorted according to the degree of agreement)

For **Actors**, all their talks may be compared to the talks of all other **Actors** to build aggregated views of allies and opponents that may be sorted on an individual basis, or grouped by ages, functions, countries or affiliation (organization), as shown in Fig. 7.

Fig. 7. All allies and opponents for a given **Actor** (grouped by ages)

The most significant visualizations regarding public debates are the similarity and justification maps of **Arguments**. First, the justification map, as presented in Fig. 8, articulates around a chosen **Argument** the other **Arguments** that support or refute it. We use transitivity rules over similarity relations to enhance justification maps. In the following rules, let:

- A, B, C be **Arguments**
- ⌢ and ⧸ respectively denote *similarity* and *dissimilarity* between **Arguments**

– ,  and  respectively denote *justification*, *qualification* and *refutation* between **Arguments**.

We then apply the following transitivity rules (note that we do not use the *nuance* relationship to explore the similarity and they are only presented at the first *degree* of the justification map, as visible in Fig. 8):

$$(A \simeq B) \land (B \to C) \quad \lor \quad (A \lesseqgtr B) \land (B \twoheadrightarrow C) \quad \Rightarrow \quad A \to C$$
$$(A \simeq B) \land (B \rightsquigarrow C) \quad \lor \quad (A \lesseqgtr B) \land (B \rightsquigarrow C) \quad \Rightarrow \quad A \rightsquigarrow C$$
$$(A \simeq B) \land (B \twoheadrightarrow C) \quad \lor \quad (A \lesseqgtr B) \land (B \to C) \quad \Rightarrow \quad A \twoheadrightarrow C$$

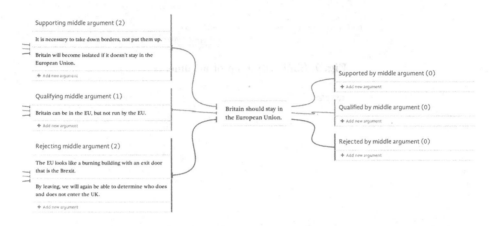

Fig. 8. Justification map of arguments

Similarity maps are displayed as sortable **Argument** lists where all **Actors** that have taken sides for or against a chosen **Argument** are shown, as presented in Fig. 9. Some statistics are also displayed to give a quick overview of the amount of **Arguments** having a similarity relationship with the **Argument** under investigation and the amount of **Actors** having partaken on the same subject. As for many other views, various grouping possibilities are available to the user.

2.5   Smart Search with Filters

Another reason of enforcing such structured details regarding contributions in general is to empower the ability to query the database and filter results in a user-friendly way. Since we are targeting a wide range of user profiles, from non-experts to journalists or sociologists, we made a point of providing an effective way of searching through the contributions. To this end, coupled to a common search bar, we added a filtering feature *à la* amazon, as shown in Fig. 10.

Fig. 9. Similarity map of arguments

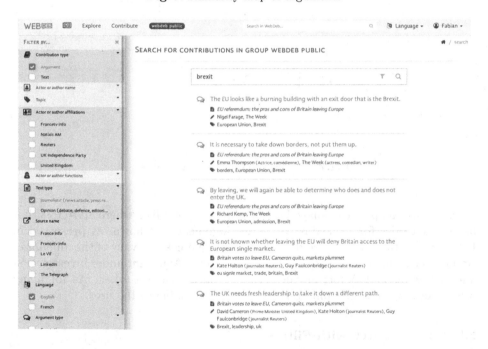

Fig. 10. Search and filter contributions

Filtering values on the left side are dynamically calculated based on the request made by the user. In the example shown, we may refine between all sources from which arguments have been extracted, the functions (professions) of the involved actors, their affiliations or their names, the topics, and so forth.

2.6 Group Management

As a pedagogic platform, WD also provides the possibility to create *closed* environments, named groups, where teachers may work with their students either collaboratively, or individually. Specific features have been put in place to this end such as the ability to activate or not the various NLP helps, the possibility to invite users to join groups, validate and mark students' contributions or to push validated contributions into the public database.

3 Discussion and Limitations

WebDEB provides dedicated visualizations to explore arguments expressed in the public arena. It is meant to provide graphical overviews of public actors with respect to particular topics or other actors' positions. Enhanced by NLP tools meant to facilitate the contributors' job while adding new data into the database, the data insertion burden that could lead to undesirable and discouraging effects, is minimized as much as possible.

We decided to work from a structured representation of arguments, texts and actors in order to ease the work of NLP-based discovery of relations between arguments as well as effectively tracing actor's positions. The argument's degree of certainty is meant to avoid formally linking arguments that have inappropriate levels of confidence. Thought, this *rigourous* structuring requires a couple of hours of familiarization to contributors before making things straightforward.

Content monitoring is also a rather harsh task, even if content may be deleted easily by (group) administrators. Improvements must be made regarding offensive words and fake news, at least by warning content administrators when detecting such contributions. However, since the platform is not intended to record individual's own opinions but dedicated to gather politicians or scientists' statements, such a risk is minimized, but must be still taken into account.

We conducted preliminary qualitative evaluations of the platform between May and November 2016. Researchers from the *Université Catholique de Louvain* and students of a secondary school in Namur were asked to evaluate the ergonomic and aesthetic aspects as well as the platform's effectiveness as a search engine and as a societal need. From these experiments, a list of limitations were identified from which a series of visualizations are still under development, *e.g. coalition of arguments* where all actors' positions are aggregated from similar arguments regarding a particular argument, or an aggregated view of texts that relate to each other through their respective linked arguments. Some NLP tools are also under development, especially as aids in encoding tasks regarding argument extraction and standardization, as well as automatic suggestions from press articles themselves.

4 Related Work

A series of approaches and applications have been proposed in the field of mind-mapping, public debates and discourse analysis. Thought, most of these approaches concentrate on argument diagrams, mainly following Toulmin's [11] or Walton's [13] methods [9]. Among such approaches, *Compendium*[3], *ThruthMapping*[4] and *Rationale*[5] focus on argumentation graphs, sometimes collaboratively. WebDeb, on the other hand, makes use of the theory of speech acts [1,10] and focuses on the collection and aggregation of statements from various sources in order to build summary visualizations.

Some recent research focus on the detection of speech act types on web forums and emails [6] and in *Twitter* feeds [12,14]. Contrary to this work, we use *Twitter* as a source from which to extract and transform (as well as classify) tweets to populate our database with arguments.

In the context of (public) debates, many didactic initiatives have emerged in recent years, such as the *International Debate Education Association* (idea)[6], *ArgueHow*[7] or *CreateDebate*[8], where students learn to argue about specific topics. Although we target a pedagogical purpose close to *idea*'s view, we also aim to concentrate actual public debates into a centralized database thus empowering individuals with an open social network where they are able to browse and visualize substantial data. *Debategraph*[9] is a very close system to ours where people may explore argumentation maps. However, on top of the representation of argumentation graphs, our purpose is to relate arguments to their authors and to build richer aggregations on arguments and actors. Also, we provide a search engine that is more user-friendly thanks to our filtering capabilities.

5   Conclusion and Future Work

We have presented WebDeb, a collaborative and open platform dedicated to structuring and tracing public debates as well as building valuable visualizations for end users. This project has a democratic objective of allowing people to search for public actors and review their positions over topics or regarding other actors. By automatically integrating sources from partner media, Wd centralizes opinions and builds summarized representations of public debates, as well as gathering in one place as much information as possible regarding public statements and controversy. WebDeb also follows other existing platforms dedicated to teaching argumentation theory, but provides richer data visualization and makes use of Natural Language Processing aids.

At present time, a series of limitations have been pointed out by early testers, and some functionalities are still missing in our approach. In future, we plan to build more visualizations, especially regarding public actors and more exportation facilities are targeted to generate CSV files, aggregated PDF documents or

---

[3] http://compendiuminstitute.net/.
[4] http://www.thruthmapping.com.
[5] http://www.reasoninglab.com.
[6] http://idebate.org and http://debatepedia.idebate.org.
[7] http://arguehow.com.
[8] http://www.createdebate.com.
[9] http://debategraph.org/.

even *Argument Interchange Format* (AIF) [4]. We are also gathering more feedback from post-graduate students currently using the platform for educational purposes in a linguistic course. Last, we will investigate the possibility to analyze comments on *Facebook* the same way we are doing for *Twitter*. As an extend to our current approach, we plan to use replies and *retweets* to discover similar or opposed opinions of persons, and this way, enrich our visualizations.

Acknowledgment. This work is funded by the Walloon Region of Belgium, under convention no. 1318202 (*Programme Germaine Tillion*). The authors also thank, in alphabetic order, L.-A. Cougnon, B. Delvaux, C. Fairon, P. Francq, S. Roekhaut and D. Uygur for their various contributions to this project.
Website and sources. A running instance of the platform can be found at https://webdeb.be. The sources are distributed under the LGPL license and available at https://bitbucket.org/fabgilson/webdeb-sources.

References

1. Austin, J.L.: How to Do Things with Words. Harvard University Press, Cambridge (1962). 2nd edn. (2005)
2. Bex, F., Lawrence, J., Snaith, M., Reed, C.: Implementing the argument web. Commun. ACM **56**(10), 66–73 (2013). http://doi.acm.org/10.1145/2500891
3. Bittar, A., Cougnon, A., Delvaux, B., Gilson, F., Fairon, C., Schobbens, P.Y.: Webdeb: learning to analyse public debate with the help of NLP. Submitted to Traitement Automatique des Langues on 28 October 2016. https://staff.info.unamur.be/fgi/webdeb/TAL_en_ex.pdf
4. Chesñevar, C., Modgil, S., Rahwan, I., Reed, C., Simari, G., South, M., Vreeswijk, G., Willmott, S.: Towards an argument interchange format. Knowl. Eng. Rev. **21**(4), 293–316 (2006)
5. Erhmann, M.: Les entités nommées, de la linguistique au TAL: statut théorique et méthodes de désambiguïsation. Ph.D. thesis, Université Paris 7, France (2008)
6. Jeong, M., Lin, C.Y., Lee, G.G.: Semi-supervised speech act recognition in emails and forums. In: Proceedings of the 2009 Conference on Empirical Methods in Natural Language Processing, EMNLP 2009, vol. 3, pp. 1250–1259. Association for Computational Linguistics, Stroudsburg (2009). http://dl.acm.org/citation.cfm?id=1699648.1699671
7. Litman, D.: Natural language processing for enhancing teaching and learning. In: Proceedings 30th Conference on Artificial Intelligence (AAAI), pp. 4170–4176. Phoenix (AZ), USA (2016)
8. Nadeau, D., Sekine, S.: A survey of named entity recognition and classification. In: Lingvisticæ Investigationes, vol. 30, pp. 3–26. John Benjamins Publishing Company (2007)
9. Schneider, J., Groza, T., Passant, A.: A review of argumentation for the social semantic web. Semant. web **4**(2), 159–218 (2013). http://dl.acm.org/citation.cfm?id=2590215.2590218
10. Searle, J.R.: Speech Acts: An Essay in the Philosophy of Language. Cambridge University Press, Cambridge (1969)
11. Toulmin, S., Rieke, R., Janik, A.: An Introduction to Reasoning, 2nd edn. Macmillian Publishing Co., Inc., New York (1984)

12. Vosoughi, S., Roy, D.: Tweet acts: a speech act classifier for twitter. In: Proceedings of the 10th International Conference on Web and Social Media, pp. 711–715 (2016). http://www.aaai.org/ocs/index.php/ICWSM/ICWSM16/paper/view/13171
13. Walton, D.: Fundamentals of Critical Argumentation. Critical Reasoning and Argumentation. Cambridge University Press, Cambridge (2005)
14. Zhang, R., Gao, D., Li, W.: What are tweeters doing: recognizing speech acts in twitter. In: Proceedings of the 5th AAAI Conference on Analyzing Microtext, AAAIWS 2011-05, pp. 86–91. AAAI Press (2011). http://dl.acm.org/citation.cfm?id=2908630.2908645

Demonstration Papers

A Web Tool for Type Checking and Testing of SPARQL Queries

Jesús M. Almendros-Jiménez(✉) and Antonio Becerra-Terón

Department of Informatics, University of Almería, 04120 Almería, Spain
{jalmen,abecerra}@ual.es

Abstract. In this paper a property-based testing tool for SPARQL is described. The tool randomly generates test cases in the form of instances of an ontology. The tool checks the well typed-ness of the SPARQL query as well as the consistency of the test cases with the ontology axioms. Test cases are after used to execute queries. The output of the queries is tested with a Boolean property which is defined in terms of membership of ontology individuals to classes. The testing tool reports counterexamples when the Boolean property is not satisfied.

1 Introduction

Property-based testing (PBT) is a well-known technique of program testing [9] involving the specification of *properties/assertions* on the output of a program to be tested. Properties on the output describe the required relationships between output data, which should be ensured by *free bug programs*. Normally, *test cases* are generated as input of the program and the properties/assertions are checked on the output of the test cases. When a counterexample is found, that is, an input test case in which the property is not satisfied by the result, the program has a bug. PBT can use either *black-box* techniques (i.e., *randomly generated test cases*) or *white-box* techniques (i.e., *test cases generated from code analysis*) for each kind of program. Among others, PBT has been studied in Java [8], functional languages [5], XQuery [2], model transformation languages [1] and relational databases [4].

In this paper a black-box tool for property-based testing has been designed in which input and output data are modeled in *RDF* and *OWL*, the program is a SPARQL query, and properties have the form of *membership of individuals to classes*. A RDF/OWL ontology to XML Schema *automatic mapping* is carried out by the tool in which classes and properties of the ontology **TBox** (i.e., ontology axioms) are mapped into XML Schema labels and attributes. The XML Schema is used to generate test cases. However, in order to generate useful test cases two additional checks are carried out. Firstly, the tool checks whether the SPARQL query is *well-typed*. A method for *type checking* is used *based on ontology consistency*. Wrongly typed queries *prevent testing*. Secondly, the tool

This work was supported by the EU (FEDER) and the Spanish MINECO Ministry (*Ministerio de Economía y Competitividad*) under grant TIN2013-44742-C4-4-R.

J. Cabot et al. (Eds.): ICWE 2017, LNCS 10360, pp. 535–538, 2017.
DOI: 10.1007/978-3-319-60131-1_38

```
      SELECT ?event ?user2
      WHERE {
(1)       ?event sn:created_by ?user1 . ?event  sn:likes ?user2 .
          ?user2 sn:invited_to ?event
          }
```

```
      SELECT ?msg1
(2)   WHERE {
          ?msg1 sn:sent_by ?user . ?msg1 sn:replied_by ?msg1
          }
```

```
      SELECT ?user1 ?event ?user2
      WHERE {
(3)       ?event sn:added_by ?user1 . ?event sn:date ?date .
          ?user1 sn:friend_of ?user2 . ?user2 sn:age ?age .
          FILTER (?age >= 40 && ?date < '2017-01-01T00:00:00Z'^^xsd:dateTime)
          }
```

Fig. 1. Examples of buggy SPARQL queries

checks the *consistency* of test cases with ontology axioms. Inconsistent test cases lead to *wrong testing results*.

While query testing has been studied in other contexts (e.g., SQL [6]), and programming bugs are well stablished for SQL (e.g., [3]), as far as we know SPARQL testing has not been studied yet. Additionally, type checking is not currently supported by SPARQL implementations. This has as consequence empty/wrong/missing answers for queries.

2 Type Checking and Testing of SPARQL Queries

In order to illustrate the work, an example of ontology defining a *social network* is considered. Testing is a mechanism to detect bugs. In our case, testing is used for the detection of bugs in SPARQL queries. The question now is what does a bug mean in a SPARQL query? Fig. 1 shows some examples of SPARQL queries having a bug. Due to the lack of space, only three cases are shown, but more examples are available at the Web site of the tool (http://minerva.ual.es:8080/SPARQL/). Each case represents a different kind of bug in our approach.

Case (1). In this case the bug is due to *typing*. More concretely it is due to the domain and range of properties. The triple pattern *?event sn:likes ?user2* is wrong, because users like activities (in particular, events), and thus the order on the triple is wrong. From user's point of view, this query returns an empty answer when the bug is present, but some hint could be given. The testing tool uses HermiT ontology reasoner to check types, reporting the following diagnosis:

```
Test cases cannot be generated:
DisjointClasses(#Activity #User)
ClassAssertion(#Activity #event)
ClassAssertion(#User #event)
```

Case (2). In this case, the bug is due to *constraints* on properties (i.e., ontology axioms for properties). The triple pattern *?message1 sn:replied_by ?message1* is wrong, because *replied_by* is an *irreflexive* property: a message cannot be answered by itself. Thus, again the *answer will be empty*. Consistency of test cases (i.e., agreement with constraints on classes and properties) is checked by the tool through HermiT ontology reasoner:

```
Unable to test the property.
It was not possible to find consistent tests.
```

Case (3). In this case, the query is well-typed and it does not contradict constraints. However, the user intention has to be taken into account. The intended meaning of the query is *"Retrieve events before this year added by users older than 40, and friends of these users"*. Here, the answer could be not empty but the programmer can find *wrong answers*: events added by users younger than 40. Here, there is a *mistake using variables*. The triple pattern *?user2 sn:age ?age* is wrong because *?user1* should be older than 40, instead of *?user2*. Here, a case in which testing of output properties is useful arises. Let us suppose that the testing tool is called with the following property: *Mature(?user1)* where *Mature ≡ age some integer*[>= 40]. The testing tool reports (using values: tennis for events, 30 and 50 for ages, jesus and antonio for users, and 2016-01-01, 2018-01-01 for dates[1]) the following message:

```
Output Property Falsifiable after 256 tests.
Counterexample:

<rdf:RDF >
  <sn:Event   rdf:about="#tennis">
    <sn:date rdf:datatype="#dateTime">2016-01-01T00:00:00Z</sn:date>
    <sn:added_by rdf:resource="#jesus"/>
  </sn:Event>
  <sn:User   rdf:about="#antonio">
    <sn:age rdf:datatype="#integer">50</sn:age>
    <sn:friend_of rdf:resource="#jesus"/>
  </sn:User>
  <sn:User   rdf:about="#jesus">
    <sn:age rdf:datatype="#integer">30</sn:age>
    <sn:friend_of rdf:resource="#antonio"/>
  </sn:User>
</rdf:RDF>
```

It means that after 256 test cases, the testing tool has found a counterexample for *Mature(?user1)*. The counterexample shows an event *tennis* which has been added by *jesus*, and *antonio* is a friend of *jesus* who is 50 years old, and *jesus* is 30 year old. Thus, *?user1* which is bound to *jesus* is not *Mature*. This counterexample serves as witness of the bug, which can be found in *?user2 sn:age ?age*. Replacing this triple pattern by *?user1 sn:age ?age*, the testing tool answers as follows:

```
Ok: passed 256 tests.
```

3   Web Tool

A *Web tool* available at http://minerva.ual.es:8080/SPARQL/ has been developed enabling the transformation of ontologies into XML Schemas facilitating

---

[1] Values are added to the XML Schema in our approach.

the customization of XML Schemas, for test case generation, by automatically pruning the XML Schemas. The tool has been implemented under the BaseX XQuery interpreter mainly responsible of test case generation. SPARQL (*Apache Jena ARQ engine*) has been embedded into XQuery thanks to *Java binding* capabilities of BaseX. Also the *ontology reasoner HermiT* [7] has been embedded into XQuery in order to check well typed-ness of SPARQL queries and consistency of test cases.

4   Conclusions and Future Work

A limitation of our approach is that properties can only be specified for individuals with membership to classes, and not for classes and properties. This a limitation of the current implementation but we plan to extend property definition to classes (for instance, disjointness of output classes) and properties (for instance, sub-property relationships). With regard to SPARQL coverage, the testing tool is able to test any SELECT query. ASK, DESCRIBE and CONSTRUCT queries are out of the scope of the testing tool. Finally, our testing tool generates test cases without taking into account the code, and the human tester intervention is required. We plan to extend our work to white box testing which means to automatically generate/prune the XML Schema from the code. The testing tool will become completely automatic without human tester intervention.

References

1. Almendros-Jiménez, J.M., Becerra-Terón, A.: Automatic generation of ecore models for testing ATL transformations. In: Bellatreche, L., Pastor, Ó., Almendros Jiménez, J.M., Aït-Ameur, Y. (eds.) MEDI 2016. LNCS, vol. 9893, pp. 16–30. Springer, Cham (2016). doi:10.1007/978-3-319-45547-1_2
2. Almendros-Jiménez, J.M., Becerra-Terón, A.: Automatic property-based testing and path validation of XQuery programs. Softw. Test. Verif. Reliab. **27**(1–2), 1–29 (2017)
3. Brass, S., Goldberg, C.: Semantic errors in SQL queries: a quite complete list. J. Syst. Softw. **79**(5), 630–644 (2006)
4. Chays, D., Deng, Y., Frankl, P.G., Dan, S., Vokolos, F.I., Weyuker, E.J.: An AGENDA for testing relational database applications. Softw. Test. Verif. Reliab. **14**(1), 17–44 (2004)
5. Claessen, K., Hughes, J.: QuickCheck: a lightweight tool for random testing of Haskell programs. ACM SIGPLAN Not. **46**(4), 53–64 (2011)
6. De La Riva, C., Suárez-Cabal, M.J., Tuya, J.: Constraint-based test database generation for SQL queries. In: Proceedings of the 5th Workshop on Automation of Software Test, pp. 67–74. ACM (2010)
7. Glimm, B., Horrocks, I., Motik, B., Stoilos, G., Wang, Z.: HermiT: an OWL 2 reasoner. J. Autom. Reasoning **53**(3), 245–269 (2014)
8. Khurshid, S., Marinov, D.: TestEra: specification-based testing of Java programs using SAT. Autom. Softw. Eng. **11**(4), 403–434 (2004)
9. Utting, M., Pretschner, A., Legeard, B.: A taxonomy of model-based testing approaches. Softw. Test. Verif. Reliab. **22**(5), 297–312 (2012)

Supporting Mobile Web Augmentation by End Users

Gabriela Bosetti[1]([⊠]) [iD], Sergio Firmenich[1,2] [iD], Gustavo Rossi[1,2] [iD],
and Marco Winckler[3] [iD]

[1] LIFIA, Facultad de Informática, Universidad Nacional de La Plata,
La Plata, Argentina
{gabriela.bosetti,sergio.firmenich,
gustavo}@lifia.info.unlp.edu.ar
[2] CONICET, La Plata, Argentina
[3] ICS-IRIT, University of Toulouse 3, Toulouse, France
winckler@irit.fr

Abstract. This article presents MoWA Authoring, an End User Development platform supporting the improvement of existing –usually third party– Web applications with mobile features. This enhancement is carried out by the addition of specific behaviours, mostly dependent on context values. The tool assists the user in the construction of applications by easily selecting the components that fit his needs. A series of forms allows selecting sensors, context values of interest and digital counterparts to define an augmentation layer. The latter is composed of augmentation units, called augmenters, which are configurable through widgets that can be placed over the presentation layer of any Web application.

Keywords: Web Augmentation · Mobile web · End-User Development

1 Introduction

A single Web application not always fits a particular user's need; lacking some required information or features. Web Augmentation (WA) [1] allows users to manipulate the Web according to their own requirements. In turn, mobile devices bring a new challenge: creating more comprehensive experiences by taking advantage of the devices' capability of sensing the context. Many sites already take advantage of such features or provide a native mobile counterpart, but we can still find many of them with a poor –or no– mobile Web counterpart. Mobile WA can help to add such features when required and End User Development (EUD) [2] techniques may help to add them on-the-fly by the users themselves. Although some approaches exist for augmenting Web applications [3] and even with mobile features [4, 5], these are aimed and limited to developers or users with programming skills. Similarly, in [6], the system supports the automatic triggering of adaptations when a concrete context of use is perceived, but it does not contemplate EUD techniques either. A vast number of authoring tools allow the creation of mobile applications from desktop [7], native mobile [8] or Mobile Web

© Springer International Publishing AG 2017
J. Cabot et al. (Eds.): ICWE 2017, LNCS 10360, pp. 539–543, 2017.
DOI: 10.1007/978-3-319-60131-1_39

environments [9], but none of them generates pure Web applications. All of them rely on some native component for their execution instead of reusing a standard and popular Web browser to which the user is possibly used. With this aim, we present Mobile Web Augmentation (MoWA) authoring [10], an EUD approach empowering end users to create MoWA applications through a form-based process combined with composable visual widgets and live programming capabilities. In [10] we demonstrated its feasibility by conducting an experiment with 22 end users; results showed that the approach was not only feasible but promising, since users were able to complete, in average, 84% of the requirements of the experiment.

2 MoWA Authoring: The Support Tool

In our approach, augmentations are executed by MoWA applications, which are loaded and run by a Web browser weaver, deployed as a Firefox for Android extension. Such applications are notified by sensors when the context changes, and are capable of manipulating the Document Object Model (DOM) of any Web page through specialised augmenters that know how to interpret the perceived context values. At execution time, the sensors observe changes in the user's context. The common way is through the device's physical sensors, as shown in Fig. 1. This way, when a change in the context is perceived, the sensed value is notified to the MoWA sensor, which in turn notifies to all its subscribed applications. Applications can be instantiated with different execution strategies that execute the augmenters under different criteria.

Broadly speaking, the process of creating an application consists of four steps, shown at the left of Fig. 1. First, the user selects the sensors that the application will be subscribed to. He must also define which context values are of interest for the application, as a particular 3-axes value for the orientation. Then, he should define which will be the digital counterpart(s) to associate with the defined context values of interest, e.g. all the articles in an online newspaper. Finally, he must select and configure the augmenters of his preference, in charge of manipulating the DOM of the underlying application in consequence of its purposes and the sensed values.

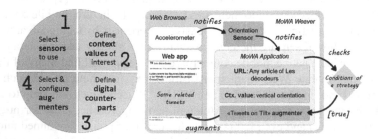

Fig. 1. Main stages of the authoring process and the expected application execution

3   A Motivational Scenario

Wayra is a regular reader of *Les Decodeurs*, a section of *Le Monde* specialised in fact-checking and contextualization of popular assertions. She reads it daily on her phone, and sometimes she wants to know what people say about a concrete topic. In Fig. 2, Wayra is reading about "CrossCheck" (a), but the article has just two comments at the bottom of the page (b). She wants to read more opinions so, in this kind of situation, she used to copy the title, open a new tab, visit Twitter and perform a search using the copied text as keywords. However, this process became tedious for her and she eventually stopped doing it. She would like to retrieve tweets in the same context of the article easily. Besides, the extra interactions make her feel uncomfortable to perform it with a single hand, and she would like to use tilting gestures to show and hide an extra section with related tweets at the top of the article, as in steps "c" and "d".

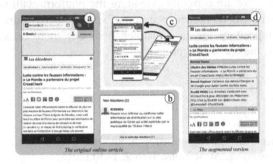

Fig. 2. The original mobile Web site and the augmented version

MoWA authoring empowers Wayra to create her own solution; Fig. 3 shows the screenshots of the main stages of the process. It starts when the user opens the tool in the browser (0) and the full-screen mode is enabled. He chooses to create a new application and sets up some base data. Then, he selects his required sensors (1), and manages the context values of interest according to the selected sensor, as an orientation in (2). In such a concrete case, values for the axes are automatically updated; the user should move the device to the desired position and save it. Then, he defines at least one digital counterpart in (3) and eventually sets up the augmentation layer (4).

Fig. 3. Main stages of the authoring process through the tool

He can use as many augmenters as required. The last screenshot shows that an instance of the *TweetsOnTilt* augmenter has been created. This augmenter is capable of showing related Tweets through a sliding down effect that –as a physical metaphor of an object sliding on the surface– makes the interaction memorable for the user. The augmenter just requires the user to pick a DOM element for generating an XPath, so each time the document is loaded, the DOM element can be retrieved and its *textContent* attribute can be used as the keywords to search on Twitter.

4   Conclusions and Further Work

This paper presents MoWA Authoring, a platform supporting an approach for enhancing Web applications with mobile features. The tool currently supports a set of extensible and specialised sensors, context types and augmenters for covering scenarios in different domains. We are currently extending the platform with new augmenters to provide the end users with a broader spectrum of specific behaviours and making them available from a repository for easy maintenance. For more information about the project, please visit the site of the project[1].

References

1. Díaz, O., Arellano, C.: The augmented web: rationales, opportunities, and challenges on browser-side transcoding. ACM Trans. Web **9**, 1–30 (2015)
2. Lieberman, H., Paternò, F., Klann, M., Wulf, V.: End-user development: an emerging paradigm. End User Dev, pp. 1–8. Springer, Netherlands (2006)
3. Díaz, O., Arellano, C., Azanza, M.: A language for end-user web augmentation. ACM Trans. Web **7**, 1–51 (2013)
4. Carlson, D., Ruge, L.: Ambient amp: an open framework for dynamically augmenting legacy websites with context-awareness. In: 2014 IEEE Ninth International Conference on Intelligent Sensors, Sensor Networks and Information Processing, pp. 1–6. IEEE (2014)
5. Van Woensel, W., Casteleyn, S., De Troyer, O.: A generic approach for on-the-fly adding of context-aware features to existing websites. In: Proceedings of the 22nd ACM Conference on Hypertext Hypermedia - HT 2011, p. 143. ACM Press, New York (2011)
6. Ceri, S., Daniel, F., Facca, F.M., Matera, M.: Model-driven engineering of active context-awareness. World Wide Web **10**, 387–413 (2007)
7. Corvetta, F., Matera, M., Medana, R., Quintarelli, E., Rizzo, V., Tanca, L.: Designing and developing context-aware mobile mashups: the CAMUS approach. In: Cimiano, P., Frasincar, F., Houben, G.-J., Schwabe, D. (eds.) ICWE 2015. LNCS, vol. 9114, pp. 651–654. Springer, Cham (2015). doi:10.1007/978-3-319-19890-3_49
8. Pittarello, F., Bertani, L.: CASTOR: learning to create context-sensitive and emotionally engaging narrations in-situ. In: Proceedings of the 11th International Conference on Interaction Design and Children - IDC 2012, p. 1. ACM Press, New York (2012)

---

[1] http://www.goo.gl/kQyKe8.

9. Cappiello, C., Matera, M., Picozzi, M.: A UI-centric approach for the end-user development of multidevice mashups. ACM Trans. Web **9**, 1–40 (2015)
10. Bosetti, G., Firmenich, S., Gordillo, S.E., Rossi, G., Winckler, M.: An end user development approach for mobile web augmentation. Mob. Inf. Syst. **2017**, 1–28 (2017)

Rapid Engineering of QA Systems Using the Light-Weight Qanary Architecture

Andreas Both[1(✉)], Kuldeep Singh[3], Dennis Diefenbach[2], and Ioanna Lytra[3,4]

[1] DATEV eG, Nuremberg, Germany
contact@andreasboth.de
[2] Laboratoire Hubert Curien, Saint Etienne, France
[3] Fraunhofer IAIS, Sankt Augustin, Germany
[4] Enterprise Information Systems, University of Bonn, Bonn, Germany

Abstract. Establishing a Question Answering (QA) system is time consuming. One main reason is the involved fields, as solving a Question Answering task, i.e., answering a user's question with the correct fact(s), might require functionalities from different fields like information retrieval, natural language processing, and linked data. The architecture used for Qanary supports the derived need for easy collaboration on the level of QA processes. The focus of the design of Qanary was to enable rapid engineering of QA systems as same as a high flexibility of the component functionality. In this paper, we will present the engineering approach leading to re-usable components, high flexibility, and easy-to-compose QA systems.

Keywords: Software reusability · Question Answering · Light-weight web architectures · Service composition · Semantic search · Ontologies · Annotation model

1 Introduction

The Web of Data is growing permanently as well as the industrial data sets. Induced by this movement the challenge for retrieving knowledge from such data sets has gained much importance in research and industry. Question Answering (QA) is tackling this challenge by providing an easy-to-use natural language interface for retrieving knowledge from large data sets. However, as QA is a challenge requiring to solve research questions from many different fields, a QA system is mostly consisting of many different components (from different research fields). Hence, enabling easy collaboration between researchers is an important engineering path while aiming at supporting the research community. Additionally, a reasonable engineering approach is required to enable a loose cooperation of different researchers.

Earlier, we established a component-oriented approach named Qanary [1] on top of a RDF vocabulary qa [6]. This approach provides a methodology for

© Springer International Publishing AG 2017
J. Cabot et al. (Eds.): ICWE 2017, LNCS 10360, pp. 544–548, 2017.
DOI: 10.1007/978-3-319-60131-1_40

creating QA processes using a central knowledge base (KB) to store all available QA process data. Here, we will focus on the component model and service composition following the Qanary methodology. In the demonstration, we will use the Qanary reference implementation to show the achievement w.r.t. to the rapid engineering process that was established. We will show the engineering process for creating a Qanary Web service as well as a complete Qanary-based QA system.

2   Related Work

In the context of QA, a large number of systems and frameworks have been developed in the last years. For example, more than 20 QA systems (in the last 5 years) were evaluated against the QALD benchmark (cf., http://qald.sebastianwalter. org). These reasons led to the idea of developing component-based frameworks that make parts of QA systems reusable. We are aware of three frameworks that attempt to provide a reusable architecture for QA systems. QALL-ME [4] provides a reusable architecture skeleton for building multilingual QA systems. openQA [5] provides a mechanism to combine different QA systems and evaluate their performance using the QALD-3 benchmark. The Open Knowledge Base and Question Answering (OKBQA) challenge (cf., http://www.okbqa.org/) is a community effort to develop a QA system that defines rigid JSON interfaces between the components. In contrast, Qanary [1] does not propose a rigid skeleton for QA pipelines, instead we allow multiple levels of granularity, enable the community to develop new types of QA systems (not only pipelines), and focus on the research tasks.

3   The Qanary Component Engineering Process

Requirements. The core requirements of the Qanary architecture are: R1 programming language independent approach, R2 combining of components to different QA processes as easy as possible (no predefinition of specific pattern, e.g., QA pipeline), and R3 enabling the researches from different communities to follow their own research tasks with as few as possible restrictions.

The Qanary Component Model. Each Qanary component is an independent Web service implementing the tiny RESTful interface: process(M). Via the synchronous interface the component is triggered to process the current user question. The question (and any process data) is not contained in the message M, instead it was stored in an RDF KB. Consequently, $M = (T, G_i, G_o)$ contains the endpoint URI of the KB T, and the graph G_i in T containing the inbound information as well as the graph G_o in T that should be used to store the computed information (i.e., outbound data flow) for further use in the QA process (by other components). Finally, the component is returning the focus to the QA process where other QA components might be called which can use the generated data. Hence, after being notified by the QA process a component will fetch the information required for its task from G_i (in T) and perform its task using this information, cf., Fig. 1. To enable an easy data exchange on common ground, the RDF vocabulary qa [6] was established (built on top of the W3C WADM, cf., w3.org/TR/annotation-model) holding the computed information as annotations of the question. Therefore, within the process the computed data can be interpreted by any Qanary component. Note that all information stored in G_i is retrievable by each Qanary component. Hence, no restrictions w.r.t. the accessible data are imposed (cf., [R3]).

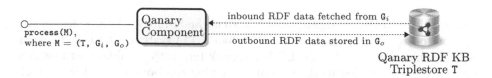

Fig. 1. Qanary component model (note that G_i might be equal to G_o).

Service Composition. All Qanary components implement the same lightweight interface and retrieve/store the data using the qa vocabulary. Hence, the Qanary services can be integrated by combining these components, analogously to the Pipes and Filters [2] architecture pattern. The Qanary reference implementation (cf., github.com/WDAqua/Qanary→qanary_pipeline-template) takes advantage of the characteristics of Qanary components. It contains a service registry (AdminServer) which is called automatically during the start-up phase by all components. Hence, all Qanary components are known and can be easily composed (cf., [R2]), e.g., the following simple user interfaces are provided to create a QA pipeline using a textual or audio question. There components can easily be activated and combined by drag and drop (define order in QA pipeline):

http://www.wdaqua.eu/qanary/startquestionansweringwithtextquestion,
http://www.wdaqua.eu/qanary/startquestionansweringwithaudioquestion

Service Implementation. The implementation of a Qanary component is supported using a Maven archetype (cf., github.com/WDAqua/Qanary→qanary-component-archetype). It already contains the registration to the AdminServer and several other functionalities for rapid engineering. Note: There are no restrictions on the functionality nor the programming language (cf., [R1]); however, the reference implementation is in Java.

Demonstration. As an example, we show how to create a QA pipeline providing the functionality focusing on the engineering tasks. The pipeline is aiming at answering the question "What is the real name of Batman?"[1] (cf., QALD

---

[1] Full description at https://github.com/WDAqua/Qanary/wiki/ICWE-2017-demo.

question no. 92). It will use a component that already exists in the Qanary ecosystem providing functionality for Named Entity Recognition and Disambiguation (NER/NED), e.g., the Qanary DBpedia Spotlight component (cf., [3]). It will interlink the sub-string "Batman" to the DBpedia resource `dbr:Batman`. However, additional semantics is required to map the textual question to an interpretable representation. Therefore, we will interactively implement a new component C (using Qanary's Maven archetype) which adds new annotations to the Qanary KB T while analyzing the user's question. C will serve only the purpose to identify the relation `dbp:alterEgo` (i.e., a DBpedia property) while searching for the sub-string "real name" in the question.

The demonstration will finish while creating and executing the QA pipeline using the service composition and showing the result of the question.

Discussion. Here, we have demonstrated the main advantage a developers receives while integrating a component in the Qanary ecosystem. A rapid engineering process is provided and a created component can easily be interweaved with the already existing ones. A basic installation of a QA pipeline provided with a user interface called Trill can be found at http://www.wdaqua.eu/qa.

4   Conclusion

In this paper we presented the component model of the reference implementation of the Qanary framework. Qanary components are easy to implement as it was shown in the paper. However, one of the core features is the option to (re)combine components to QA systems without adopting the component's source code, while still having the full freedom of dedicating a (new) component to a completely new functionality. This new functionality might use data from the Qanary triplestore never used before in this particular combination. Hence, as all features are data-driven, allowing to add new functionality to the whole QA system from a local component independently. Additionally, the component model is language-independent and driven by the power of linked data which enables additional features like polymorph data types included in the inbound data. Our main contribution is a component-based architecture enabling developers to create or re-combine components following a plug-and-play approach. While aiming at an optimal system w.r.t. a given use case (scientific) developers are enabled to rapidly create new/adapted QA systems from the set of Qanary components available. Hence, we are handing the scientific QA community an easy-to-use approach reducing the investments for engineering tasks during typical tasks.

Acknowledgments. This project has received funding from the European Union's Horizon 2020 research and innovation program under the Marie Sklodowska-Curie grant agreement No 642795.

References

1. Both, A., Diefenbach, D., Singh, K., Shekarpour, S., Cherix, D., Lange, C.: Qanary – a methodology for vocabulary-driven open Question Answering systems. In: Sack, H., Blomqvist, E., d'Aquin, M., Ghidini, C., Ponzetto, S.P., Lange, C. (eds.) ESWC 2016. LNCS, vol. 9678, pp. 625–641. Springer, Cham (2016). doi:10.1007/978-3-319-34129-3_38
2. Buschmann, F., Meunier, R., Rohnert, H., Sommerlad, P., Stal, M.: Pattern-Oriented Software Architecture - Volume 1. A System of Patterns. Wiley, New York (1996)
3. Diefenbach, D., Singh, K., Both, A., Cherix, D., Lange, C., Auer, S.: The Qanary ecosystem: getting new insights by composing Question Answering pipelines. In: International Conference on Web Engineering, ICWE. Springer (2017)
4. Ferrández, Ó., Spurk, Ch., Kouylekov, M., Dornescu, I., Ferrández, S., Negri, M., Izquierdo, R., Tomás, D., Orasan, C., Neumann, G., Magnini, B., González, J.L.V.: The QALL-ME framework: a specifiable-domain multilingual Question Answering architecture. J. Web Seman. Sci. Serv. Agents WWW **9**, 137–145 (2011)
5. Marx, E., Usbeck, R., Ngonga Ngomo, A., Höffner, K., Lehmann, J., Auer, S.: Towards an open Question Answering architecture. In: SEMANTiCS (2014)
6. Singh, K., Both, A., Diefenbach, D., Shekarpour, S.: Towards a message-driven vocabulary for promoting the interoperability of Question Answering systems. In: IEEE International Conference on Semantic Computing, ICSC (2016)

Improving GISBuilder
with Runtime Product Preview

Alejandro Cortiñas[1]($\boxtimes$), Carlo Bernaschina[2],
Miguel R. Luaces[1], and Piero Fraternali[2]

[1] Databases Laboratory, Universidade da Coruña, A Coruña, Spain
{alejandro.cortinas,luacos}@udc.es
[2] DEIB, Politecnico di Milano, Milan, Italy
{carlo.bernaschina,piero.fraternali}@polimi.it

Abstract. Software product lines allow users with little development experience to configure and generate applications. On the web this approach is becoming more and more popular due to the low time required to bring a new release to the final users. The architecture of web applications though require complex development environments in order to allow users to test and evaluate a new configuration. In this work we present a novel approach, based on in-browser generation and emulation techniques, which can be applied to real-world state of the art software product lines, reducing test deployment complexity and enabling an agile development cycle.

Keywords: Software product lines · Agile software development · Rapid prototyping · Geographic information systems

1 Introduction

Software Product Lines Engineering is a discipline that tries to apply industrialisation techniques, such as mass-customisation and reusing strategies, to software development with the focus on improving quality of products and, at the same time, decreasing costs and time-to-market of new products [1]. Developing costs of a SPL are compensated from the third product built [1], so their application is very convenient for software development companies that usually create similar systems, i.e., a product family. This is the case of Enxenio[1], a Spanish SME (small and medium-sized enterprise) with a certain grade of expertise in GIS [2-4]. In a previous work [5], we have approached the design of a SPL for the automatic generation of web-based GIS applications for its usage at Enxenio as an internal tool. GISBuilder provides a web interface where an analyst can design and generate the source-code of web-based GIS. The generated applications must be deployed afterwards within a Java web server. From the point of view of *agile* development, this is far from optimum since it requires a full redeployment for any small change that the analysts want to try in a product. [6]

---

[1] http://www.enxenio.es.

© Springer International Publishing AG 2017
J. Cabot et al. (Eds.): ICWE 2017, LNCS 10360, pp. 549–553, 2017.
DOI: 10.1007/978-3-319-60131-1_41

describes a totally different approach, illustrated with a web tool able to design, generate and run web/mobile applications, doing it all within a web browser.

Following the mentioned approach, we have adapted GISBuilder web specification interface to provide the analyst a run-time preview of the product he or she is designing. In the demo, we will use the adapted version of GISBuilder to design, preview and generate some simple web-based GIS applications.

2   GISBuilder

GISBuilder architecture is shown in Fig. 1a. We describe it thoroughly in [5], but we can summarize its workflow as follows. When a new application has to be created, an analyst interacts with the *specification interface*, a Node.js Web Application (more specifically, an Express app). In this interface he or she can decide: (1) Which features the application provides. Examples of features are *csv importer* or *user management*. (2) The data model for the application: entities, properties and relationships. The analyst can define lists, forms and maps from this data model that can be then linked through menu items. (3) Several aspects of the graphical user interface, such as the menu configuration, the static pages or the theme.

GISBuilder is not a traditional SPL because its capabilities are enhanced through the usage of a *scaffolding-based derivation engine*, able not only to assemble static software assets common to every product (the *features*) but also to generate product-specific code (lists related to data model or specific menu configuration for an application). This **derivation engine** is invoked with the product specification, which is nothing but a JSON document that complies with a JSON Schema, when the analyst finishes the configuration of the product. The product specification is also stored in the **project repository**, a MongoDB instance. The **derivation engine** takes the product specification and assembles/generates the source code of the final products getting the required components/templates from the **component repository**. Then, the product source code is generated and the analyst gets it as a downloadable zip file.

In order to deploy and try the generated product, the analyst needs a computer with some previously installed software: Node.js, npm or yarn, Java 8 and PostgreSQL with PostGIS. Then, he or she needs to run some processes to download all the dependencies and compile the product, and also to modify a text file to set the database connection configuration. These steps are very easy to follow for a developer, but (i) it is still somehow slow to do that for every little change on the product configuration and (ii) one of the goals of GISBuilder is that the analyst does not need to have any advanced knowledge on IT.

3   GISBuilder with Runtime Live Preview

In [6] we have presented a methodology for Rapid Prototyping in a Model-Driven Development context. IFMLEdit.org, a prototype created to validate this approach, is a web tool that allows the user to design web applications

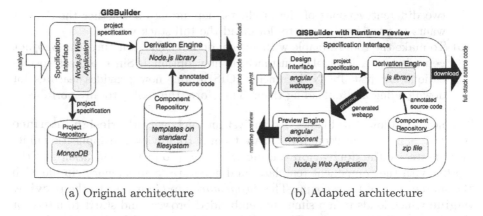

(a) Original architecture   (b) Adapted architecture

Fig. 1. Architecture changes in GISBuilder

using IFML [7], and to try them directly in the web browser before downloading them. If the user makes a change in the design, he or she can relaunch the application within the web browser and examine the change at runtime.

We have applied a similar approach on GISBuilder. This is, we have made GISBuilder able to generate and show a preview of the designed products at runtime, directly on the browser, without the need of any server-side structure. To achieve that, we have made several changes on the architecture of the SPL, as we can see in Fig. 1b. The biggest difference with respect to IFMLEdit.org is that its web applications are just prototypes, and the previewing engines are insufficient to deal with the complex single web page applications created by GISBuilder. The main changes made are:

(1) We have implemented a new version of the **derivation engine** able to run entirely on the web browser. Instead of getting the component templates from the file system, they are loaded from a zip file. Similarly, it can generate the products in memory and provide a zip file with them. The **derivation engine** is the component analogous to IFMLEdit.org's *transformation engine*.

(2) The **derivation engine** is integrated within the **specification interface**, as well as the **component repository**.

(3) In IFMLEdit.org there are two components to preview applications: one for thin-client web-applications and another for web based thick-client mobile applications. To address for the advanced architecture of GISBuilder we have fused and extended these components, introducing full support for AJAX requests on which advanced frameworks, like Angular, are based. The component is able to load a dynamic webapp in an iframe, which intercepts XHR requests and returns mock responses to each specific REST petition.

(4) GISBuilder produces full-stack web applications with Spring in the server side and Angular in client side. In the adapted version, GISBuilder creates

two different versions of the products, depending on whether the analyst wants to preview them or to download the full-stack version.

(5) To make the tool runnable without any deployment structure, the **project repository**, a MongoDB instance, is now optional. Since project specifications are just JSON documents, GISBuilder now provides features for downloading and uploading them. Therefore, the tool is totally portable.

The new workflow is as follows. An analyst interacts with the **design interface** to specify a new application. If the analyst wants to preview the application in its current status, the **derivation engine** takes the product specification and the zip file with the component templates and generates a *in-memory* zip file with the source code of the webapp. This *in-memory* zip file is sent to the **preview engine** which loads it in a simulated embedded browser and starts to intercept the Angular XHR requests. The analyst can then return back to the **design interface** and modify the configuration of the product. When the product is finished, the analyst can download the full-stack version of the application.

4   Conclusions and Future Work

In this paper we present a new version of GISBuilder, a SPL for web-based GIS, that provides *runtime live previewing* capabilities. This is, an analyst can see the product to build *on the fly*. We have shown how the approach in [6] can be applied in a different context to provide live previewing to a different automatic software generation tool. As future work, we want to apply Continuous Integration techniques to the GISBuilder, allowing it not only to facilitate the preview to the analyst but also the final deployment of the full-stack built products.

Acknowledgments. The work of the authors from UDC has been funded by MINECO (PGE & FEDER) [TIN2016-78011-C4-1-R, TIN2016-77158-C4-3-R, TIN2013-46238-C4-3-R, TIN2013-46801-C4-3-R]; CDTI and MINECO [Ref. IDI-20141259, Ref. ITC-20151305, Ref. ITC-20151247]; Xunta de Galicia (FEDER) [Ref. ED431G/01].

References

1. Pohl, K., Böckle, G., Van Der Linden, F.: Software Product Line Engineering Foundations, Principles and Techniques. Springer, New York (2005)
2. Luaces, M.R., Pérez, D.T., Fonte, J.I.L., Cerdeira-Pena, A.: An urban planning web viewer based on AJAX. In: Vossen, G., Long, D.D.E., Yu, J.X. (eds.) WISE 2009. LNCS, vol. 5802, pp. 443–453. Springer, Heidelberg (2009). doi:10.1007/978-3-642-04409-0_43
3. Brisaboa, N.R., Cotelo, J.A., Fariña, A., Luaces, M.R., Paramá, J.R., Viqueira, J.R.: Collecting and Publishing large multiscale geographic datasets. Softw. Pract. Experience **37**, 1319–1348 (2007)
4. Places, A.S., Brisaboa, N.R., Fariña, A., Luaces, M.R., Paramá, J.R., Penabad, M.R.: The galician virtual library. Online Inf. Rev. **31**, 333–352 (2007)

5. Brisaboa, N.R., Cortiñas, A., Luaces, M.R., Pedreira, O.: GISBuilder: a framework for the semi-automatic generation of web-based geographic information systems. In: PACIS 2016 (2016)
6. Bernaschina, C., Comai, S., Fraternali, P.: Online model editing, simulation and code generation for web and mobile applications. In: MiSE 2017 (2017)
7. Brambilla, M., Fraternali, P.: Interaction Flow Modeling Language: Model-Driven UI Engineering of Web and Mobile Apps with IFML. Morgan Kaufmann (2014)

ALMOsT-Trace: A Web Based Embeddable Tracing Tool for ALMOsT.js

Rocio Nahime Torres and Carlo Bernaschina[✉]

Dipartimento di Elettronica, Informazione e Bioingegneria,
Politecnico di Milano, Piazza Leonardo da Vinci, 32, 20133 Milano, Italy
{rocionahime.torres,carlo.bernaschina}@polimi.it
http://www.deib.polimi.it

Abstract. Model Driven Development (MDD) requires model-to-model and/or model-to-text transformations to produce application code from high level descriptions. Debugging and evaluating such transformations is in itself a complex task; complexity which can be mitigated through the usage of advanced developer tools. We demonstrate ALMOsT-Trace, a plug-in for ALMOsT.js, which allows developers to debug and analyze their model transformations from within their applications. In the demo, attendees will be able to experiment with ALMOsT-Trace by evaluating it in IFMLEdit.org, an online tool for the rapid prototyping of web and mobile applications, and by means of examples that can be customized by the attendees themself.

Keywords: Agile development · Model-driven development · Computer aided software engineering

1 Introduction

Model Driven Development (MDD) is the branch of software engineering that advocates the use of *models*, i.e., abstract representations of a system, and of *model transformations* as key ingredients of software development [9]. With MDD, developers use a general purpose (e.g. UML [1]) or domain specific (e.g., IFML [10]) modeling language to portrait the essential aspects of a system, under one or more perspectives, and use (or build) suitable chains of *transformations* to progressively refine the models into executable code.

Online platforms like [4] can help reduce the complexity of models and model transformations management, execution and validation.

Traceability of artifacts during the evolution of a software project is a key aspect for regression testing and monitoring. In model transformations the ability to trace models between transformation steps [2,3] enables advanced analysis and rapid error detection.

In a previous work [5] we have presented ALMOsT.js [1] (AgiLe MOdel Transformations for JavaScript), an agile, in-browser framework for the rapid prototyping of MDD transformations, which lowers the technical skills required for

---

[1] www.npmjs.com/package/almost/.

© Springer International Publishing AG 2017
J. Cabot et al. (Eds.): ICWE 2017, LNCS 10360, pp. 554–558, 2017.
DOI: 10.1007/978-3-319-60131-1_42

Web and Mobile developers to start be proficient with modeling and code generation. The design philosophy of ALMOsT.js stemmed from 6 requirements (#1 *No installation*, #2 *No new language*, #3 *Fast start-up*, #4 *Parallel development*, #5 *Customized output*, #6 *Customized generation*), which could be summarized in one: keep it simple. It allows the developer to rapidly bootstrap and evolve an MDD project, by storing models in a flexible JSON based format and by defining model transformation rules by means of simple JavaScript functions. The following code shows a template example to create a rule.

```
// Create a rule
createRule(
    // Condition function
    function (element, model) { return /*condition*/;},
    // Action function
    function (element, model) { return /*result*/; }
);

// Create a transformer
var transform = createTransformer(rules, 'm2m');

// Execute transformer;
var output_model = transform(input_model);
```

We present ALMOsT-Trace, a plug-in for ALMOsT.js, able to trace elements and relations during the execution of model transformations. The novelty of ALMOsT-Trace is the ability to trace model during transformations in-browser, by easily integrating the tracing tool inside the final MDD environment.

2 Framework

Following the *keep it simple* and plug-in based nature of ALMOsT.js the introduction of ALMOsT-Trace in a project is completely optional and as for other plug-ins [5] does not require any changes to the existing system.

ALMOsT-Trace is based on two components:

1. A drop-in replacement for the **createTransformer** creator function which enhances the created transformer with tracing abilities, as shown below.

```
// Traced transformer
var transformer = createTracedTransformer(rules, 'm2m');

// Events
transformer.on('begin', function (model) { ... });
transformer.on('skipped', function (rule, input) { ... });
transformer.on('executed', function (rule, input, output) { ... });
transformer.on('end', function (result) { ... });
```

Tracing aware rules can add custom tracing data to the final report, via a **trace** function that is injected as last parameter in each one of them.

2. A **dashboard** which listens for events on a transformer and allows the developer to analyze the traces recorded during every execution. It can be easily integrated inside any web based tool already using ALMOsT.js and enabled on demand.

```
// Create the dashboard
var dashboard = createDashboard(transformer);
// Show the dashboard
dashboard.show();
```

At the end of the execution of each rule the tuple $< rule, input, output >$ is stored. This lookup table allows the dashboard to lookup all the tuples related to each *rule*, *input* element and relation or *output* object components.

While forward tuples lookups (from *rules* and *input* to the *output*) are naïve to implement, backward lookups (from *output* to *rules* and *input*) are not. If during the aggregation phase two or more objects are merged together it is not possible to identify them from their root reference, it is instead required to search for outputs in the tuples collection which contain a particular subpart that survived unaltered till the end of the aggregation.

While reference-based types (*Object, Array, ...*) are easy to identify, it is enough to compare the reference, primitive types (*number, string, ...*) cannot be identified by means of a simple comparison, it is not possible to distinguish a number from another which contains the same value.

Enhanced rules map all the primitive values to their *Object* wrapper counterpart (*Number, String, ...*) the wrappers will generate the same result during aggregation, but will be uniquely identifiable. The enhanced transformer is responsible to restore each one of the wrapper objects to their original form before returning the result to the caller, preserving though the drop-in nature of ALMOsT-Trace.

By analyzing the stored tuples in the lookup table the dashboard of ALMOsT-Trace generates four different reports:

1. **Rule execution statistics**
   All the defined rules are listed and for each one of them it is possible to know: (a) the number of evaluations (b) the number of executions (c) the enabling elements or relations (d) the output related to the execution against each enabling element.
2. **Input Model statistics**
   All the elements and relations defined in the input model are listed and for each of them it is possible to know: (a) the number of enabled rules (b) the output related to the execution of the enabled rules against the element or relation itself.
3. **Output Model statistics**
   For *model to model* and *model to text* transformations all the elements, relations, folders or files defined in the output are listed and for each one of

them it is possible to know: (a) the rules which generated them (b) the input against which the rules have been executed to generate the element, relation, folder or file itself.

4. **Output Object statistics**

For *custom* transformations (transformations with an output format defined by the developer) the final output JSON Object is presented in an expandable tree-view. For each *Object* attribute or *Array* item it is possible to know if they were generated directly from a rule or if they are the result of an aggregation step. For each attribute or item that can be identified in one of the rules outputs the corresponding $< rule, input >$ pair is shown.

3   Conclusions

This demo presents ALMOsT-Trace, a transformations tracing plug-in for ALMOsT.js [5]. We show how it can be integrated inside an existing project and how it can be used to analyze the model transformations execution. Attendees will be able to experiment with ALMOsT-Trace by seeing its impact inside IFMLEdit.org [6–8], a web based model driven tool, and by means of examples that the can be customized by the attendees themself.

The future work will focus on the experimentation and further assessment of ALMOsT-Trace in order to validate its impact on the development cycle, in both industry and academia.

References

1. UML unified modeling language. www.uml.org/. Accessed 17 Mar 2017
2. MeTAGeM-trace: improving trace generation in model transformation by leveraging the role of transformation models. Sci. Comput. Program. **98**, Part 1, 3–27 (2015). Fifth Issue of Experimental Software and Toolkits (EST): A Special Issue on Academics Modelling with Eclipse (ACME 2012)
3. Aranega, V., Mottu, J.M., Etien, A., Dekeyser, J.L.: Using Trace to Situate Errors in Model Transformations
4. Basciani, F., Di Rocco, J., Di Ruscio, D., Di Salle, A., Iovino, L., Pierantonio, A.: MDEForge: an extensible web-based modeling platform. In: CloudMDE@ MoDELS
5. Bernaschina, C.: ALMOsT.js: an agile model to model and model to text transformation framework. In: International Conference on Web Engineering (ICWE) 2017
6. Bernaschina, C., Brambilla, M., Koga, T., Mauri, A., Umuhoza, E.: Integrating modeling languages and web logs for enhanced user behavior analytics. In: International Conference on World Wide Web (WWW) 2017
7. Bernaschina, C., Brambilla, M., Mauri, A., Umuhoza, E.: A big data analysis framework for model-based web user behavior analytics. In: International Conference on Web Engineering (ICWE) 2017

8. Bernaschina, C., Comai, S., Fraternali, P.: IFMLEdit.org: model driven rapid prototyping of mobile apps. In: International Conference on Mobile Software Engineering and Systems (MOBILESoft) 2017

9. Kleppe, A., Warmer, J., Bast, W.: MDA Explained - The Model Driven Architecture: Practice and Promise. Addison Wesley object technology series. Addison-Wesley, Reading (2003)

10. OMG: Interaction flow modeling language (ifml), version 1.0. (2015). http://www.omg.org/spec/IFML/1.0/

TweetCric: A Twitter-Based Accountability Mechanism for Cricket

Arjumand Younus[1]($\boxtimes$), M. Atif Qureshi[1], Naif R. Aljohani[2], Derek Greene[1], and Michael P. O'Mahony[1]

[1] Insight Centre for Data Analytics, University College Dublin, Dublin, Ireland
{arjumand.younus,muhammad.qureshi,derek.greene,michael.omahony}@ucd.ie
[2] Faculty of Computing and Information Technology, King Abdul Aziz University, Jeddah, Saudi Arabia
nraljohani@kau.edu.sa

Abstract. This paper demonstrates a Web service called *TweetCric* to uncover cricket insights from Twitter with the aim of facilitating sports analysts and journalists. It essentially arranges crowdsourced Twitter data about a team in comprehensive visualizations by incorporating domain-specific approaches to sentiment analysis.

1 Introduction

Cricket is an international sport with a massive number of fans from around the world. Regions within South Asia in particular have a massive fan base, making the game analogous to a religion [1]. Over the years there have been cases of corruption within the game [3], and in many instances selection of cricketers for future matches is driven by politics [6]. Towards the aim of facilitating greater scrutiny of cricket events, this demonstration paper proposes a Twitter data aggregator designed to help sports analysts and journalists in deriving deeper insights into a game which in turn can lead towards better decision-making processes within cricket. Similar to some of the techniques deployed by the World Bank [2], *TweetCric* attempts to utilise big social media data for innovative analytics in cricket such as highlighting certain cricketers associated with negative sentiment on a frequent basis.

TweetCric essentially mines the crowdsourced Twitter data posted by cricket fans and presents it in an exploratory search interface. The interface provides interactive visualizations which supports the user by (1) extracting overall tweet volume activity for a team, (2) highlighting cricketers who receive more attention at different time intervals and (3) displaying sentiment expressed for a cricketer at different time intervals of the match.

In line with *TwitInfo* [5], our system allows a user to monitor activity peaks around an event (which in this case is a cricket game) along with associated sentiment information. However, *TweetCric* also provides entity-specific sentiments to evaluate performance of various entities within the event[1].

---

[1] In the context of *TweetCric* an entity represents a cricketer.

© Springer International Publishing AG 2017
J. Cabot et al. (Eds.): ICWE 2017, LNCS 10360, pp. 559–563, 2017.
DOI: 10.1007/978-3-319-60131-1_43

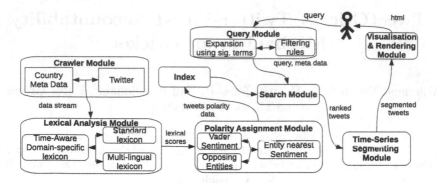

Fig. 1. Architecture of the *TweetCric* system.

2   System Architecture

Figure 1 shows an overview of the *TweetCric* architecture. The *crawler module* is responsible for back-end data acquisition, continuously collecting tweets from the live stream using the Twitter Streaming API[2]. The *crawler module* also gathers metadata (e.g., player names and their Twitter accounts, teams' management, match venues etc.) relating to countries that play cricket.

When the user issues a query, it is processed by the *query module* to produce a ranked list of relevant tweets. Note that within *TweetCric*, a query represents a certain player of interest (which we refer to as an entity) in the game for which the user of our system wishes to conduct a performance analysis. The query is automatically expanded through the use of significant terms appearing in the tweets, where significance is calculated by the chi-square test of independence. The chi-square test is able to detect term dependencies, and is hence useful in identifying useful terms to include together with an entity-based query. For example, the query "Shahid Afridi" is expanded with the term "Lala" which is a popular nickname of cricketer "Shahid Afridi". The query module also applies a set of filtering rules to eliminate noisy tweets[3]. The *time-series segmenting module* partitions the ranked list of tweets into segments which represent equal-sized time intervals between the timestamps of the first and last tweets among those retrieved for the query. Note that the distribution into equal-sized time intervals is a design choice motivated by the nature of cricket wherein a single time interval influences the entire match, and activity peaks are not as pronounced as in other sports. Further, this module also determines informative terms for each segment by calculating their importance using standard tf-idf weights. In the *polarity assignment module*, each tweet is assigned a polarity (sentiment) score as explained in the next section. Finally, the *visualisation and rendering module*

---

[2] https://dev.twitter.com.

[3] Tweets directed to political accounts, off-topic tweets such as related to showbiz, and tweets from known bots.

generates the final HTML output in the form of visualizations and a ranked list of retrieved tweets.

3   Domain-Specific Sentiment Analysis

TweetCric incorporates a domain-specific approach to sentiment analysis whereby the *lexical analysis module* and *polarity assignment module* (see Fig. 1) work in combination with each other to incorporate algorithmic changes during the calculation of sentiment scores. Firstly, multi-lingual terms are added to a standard lexicon (the VADER lexicon [4]) to produce an extended lexicon denoted by *ExtLex*. This is done after a manual inspection of partially non-English tweets (i.e., tweets posted in Indic languages but written in the Roman script) to introduce non-standard sentiment terms (65 in total which comprise 28 positive terms and 37 negative terms). Additionally, the *lexical analysis module* combines some cricket terminology with context-specific information to compute a final sentiment score. In this particular case, time is the significant contextual feature and we compute a sentiment score for cricket terms while taking into account timestamps. For example, terms such as "boundary" and "six", which denote the scoring of runs, are considered positive for the *batting* team but negative for the *fielding* team; similarly, terms such as "wicket" and "dismissal" are considered positive for the *fielding* team but negative for the *batting* team. Hence, the sentiment score of terms in the context-aware, domain-specific lexicon are sensitive to time implying positive scores for terms denoting *good batting* when our team of interest is the *batting* side, and negative scores for terms denoting *good batting* when our team of interest is the *fielding* side.

TweetCric performs sentiment aggregation at the following granularity levels: (1) player, (2) team, (3) game, and (4) time segment within a game. Once the set of tweets related to each of the above levels are identified, aggregation of individual tweet-level sentiments is then performed to produce an overall sentiment score.

In the case of entities (players), the following refinements are also considered. The *entity-nearest sentiment* scores a particular entity through a decay factor based on word distance (d) from the entity. Specifically, the scoring function is $Entity_{sentiment} = \sum_{w \in ExtLex} \frac{1}{d} w$ where d (set to 5 by experiment) measures the word distance between our entity of interest and a sentiment word w.

In addition, certain tweets mention two entities; as an example consider the tweet "Brilliant innings @imVkohli on a tough surface... Well done!!! Expected more from @SAfridiOfficial with the ball". Here, assume "SAfridiOfficial" is our entity of interest and "imVkohli" is a player from the opposing team. In these cases, we utilise information about such *opposing entities* (from within teams that are opposed to our team of interest[4]) to produce a sentiment score for

---

[4] The metadata about entities pertaining to different countries is obtained by the *crawling module* (see Fig. 1).

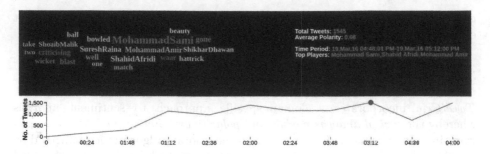

Fig. 2. *TweetCric* interface showing results for team "Pakistan" in the "India vs. Pakistan" World T20 game

our entity of interest. Here, the sentiment score of an opposing entity is simply subtracted to produce a sentiment score for our entity of interest.

Finally, we note that the opposing entities approach to sentiment can be further refined and that the approach has wider application – e.g. in the politics and technology domains, where opposing politicians and competing electronic products can be identified. These matters are left to in future work.

4   Demonstration Plan

TweetCric encompasses a query-based exploratory UI; a full walkthrough of the system can be accessed at http://mlg.ucd.ie/twitcric/. As mentioned in Sect. 2, the interface supports various modalities showing summary information for an entire game and a certain cricketer together with retrieved tweets. Furthermore, users are provided with navigation capabilities for significant entities (i.e., cricketers) at various points of the game. As an example, Fig. 2 shows results for a crucial game between "India" and "Pakistan" with the query on cricketer "Muhammad Sami"; various points in the game can be clicked and a word cloud summary for each point is presented.

References

1. Livin' on a prayer - espncricinfo. http://www.espncricinfo.com/blogs/content/story/996281.html
2. Calderon, N.A., Fisher, B., Hemsley, J., Ceskavich, B., Jansen, G., Marciano, R., Lemieux, V.L.: Mixed-initiative social media analytics at the world bank: observations of citizen sentiment in twitter data to explore "trust" of political actors and state institutions and its relationship to social protest. In: Big Data 2015 (2015)
3. Davies, C.: Match and spot-fixing: the challenges for the international cricket council. Sports Law ejournal **2015**, 1–9 (2015)

4. Hutto, C.J., Gilbert, E.: Vader: a parsimonious rule-based model for sentiment analysis of social media text. In: ICWSM (2014)
5. Marcus, A., Bernstein, M.S., Badar, O., Karger, D.R., Madden, S., Miller, R.C.: Twitinfo: aggregating and visualizing microblogs for event exploration. In: SIGCHI 2011 (2011)
6. Radford, B., Hair, D.: Caught Out-Shocking Revelations of Corruption in International Cricket. John Blake Publishing, London (2012)

PhD Symposium Papers

Extending the Genomic Data Model and the Genometric Query Language with Domain Taxonomies

Eleonora Cappelli[1](✉) and Emanuel Weitschek[2]

[1] Department of Engineering, Roma Tre University,
Via della Vasca Navale 79, 00146 Rome, Italy
eleonora.cappelli@uniroma3.it
[2] Department of Engineering, Uninettuno International University,
Corso Vittorio Emanuele II 39, 00186 Rome, Italy
emanuel@iasi.cnr.it

Abstract. In bioinformatics and biology researchers annotate experimental data in many different ways. When other researchers need to query these data, they are typically unaware of the specificity of the annotations; often they encounter possible mismatches between the granularity of the query and the granularity of the annotations. In this work, we propose an extension of the Genomic Data Model and the GenoMetric Query Language (a well established framework for biomedical data), able to search, integrate, and extend genomic data. The extension is going to be performed through domain taxonomies and by considering many external ontologies and databases. An ad-hoc software system and query language will be implemented for the storage, management, search, retrieval, and integration of biomedical data.

Keywords: Bioinformatics · Domain taxonomies · Semantic web · Big data · Data extension · Data integration

1 Introduction and Motivation

In recent years, biology and computer science came together with the aim of addressing the advancements of biomedicine. These advancements are producing huge amounts of data and their management becomes crucial in the field of healthcare. Therefore an innovative branch of science emerges, Big Data Bioinformatics, which is closely connected to the contemporary biomedical developments, i.e., high throughput genome sequencing. Big Data Bioinformatics deals with the manipulation of biological information through computer science methods that are specially engineered to manage large amounts of data generated by biomedical devices and to extract knowledge from them. In particular, the sequencing of the human genome is a research field that has evolved greatly with the spread of Next Generation Sequencing (NGS), i.e., high-throughput sequencing technology [1]. The data science in bioinformatics is mainly related

© Springer International Publishing AG 2017
J. Cabot et al. (Eds.): ICWE 2017, LNCS 10360, pp. 567–574, 2017.
DOI: 10.1007/978-3-319-60131-1_44

to two aspects, the management of the enormous amount of data produced by biomedical devices and the analysis of them. On one hand, the management should support the ability to organize data and to carry out analyses, whose aim is to achieve knowledge of the roots that can determine a disease. On the other hand, the heterogeneity of the sources does not simplify the extraction and the search for such information. Because of the advances in biomedicine and of the large amount of data generated by biomedical devices, it becomes necessary to define an infrastructure that enables the data management through Big Data technologies, which allow to search, to query, to extract, to integrate, to analyze, and to visualize biomedical data from heterogeneous sources [2]. Actually, the Genomic Data Model (GDM) [3] defines a gold standard for storing and dealing with big biomedical data, i.e., genomic and clinical data. Additionally, it provides the GenoMetric Query Language (GMQL) [4] for effectively and efficiently searching them. The Genomic Data Model allows the integration of multiple heterogeneous data sets from different sources and GMQL computes massive operations on genomic data, which take into account regions, relative positions, and distances.

2   Research Issues, Objectives, and Questions (and Hypotheses)

The PhD research project proposes to extend the Genomic Data Model and the GenoMetric Query Language through the use of domain taxonomies [5].

In bioinformatics and biology, the annotations [6] of experiments are fundamental. Researchers annotate experimental data that will be collected in large databases for being queried. When these data are interrogated by external users, who do not know the specific nature of these records, the granularity of the query may not always correspond to what is present in the database. Thanks to the application of taxonomies, we can figure out what kind of relationship exists between the information sought by a given user through a query and the data that are actually present in the records, which may not be exactly the requested ones. Therefore, the bioinformatics community published several taxonomies and ontologies to solve these problems. The vastness of ontologies and taxonomies mostly consist of relationships and links between terms, and thus conveys significant additional taxonomic information that can be integrated in state of the art biomedical databases. The ontology is a fundamental concept of the semantic web, which becomes a key resource to describe information domains of any types around the web; the information coming from ontologies may be used to do an ontology-based integration of heterogeneous sources to enable researchers to relax their queries and allow the results to match the requested records despite syntactic mismatches. Many other works address the extension and integration of classical databases with ontology like interfaces (e.g., [7,8]), conversely in our proposal we focus on genomic databases and ontologies. Additionally, we consider a novel ad-hoc defined data model and query language.

During the PhD project a new query language that integrates taxonomies will be designed and developed. It will be capable to manage huge amounts of biomedical data through big data technologies and will permit to perform efficiently search, retrieval, extraction, management, and integration of genomic and clinical data. It will rely on the extension of the Genomic Data Model and of the GenoMetric Query Language and will be implemented in a novel software system. This system will permit to obtain additional information through the integration of several open databases and domain taxonomies, thus extending temporarily (or permanently) the initial data set. The system will be also able to perform integrated analyses through the query language. The final objective is to support the study of biomedical experts for advancing research against diseases such as cancer, and to facilitate the search for precious information in genomic and clinical data.

3   Research Methodology and Research Design

The research proposal will be based on the Genomic Data Model and the Geno-Metric Query Language. The Genomic Data Model (GDM) represents each genomic sample through two fundamental concepts, the genomic regions and their metadata. The genomic regions are described by coordinates and some high-level properties. Conversely, the metadata describes the biological and the clinical properties associated with each sample, not related specifically to the genomic regions. The metadata associated with the sample are extremely heterogeneous and they are represented as attribute-value pairs. Therefore, in GDM each sample is described through two different data structures: the first one encodes the characteristics of the genomic regions; the second one represents the metadata. Each data structure has its own standardized format, with a defined data schema that defines a precise structure of the genomic regions. The records in the genomic region data structure have values that are compatible with the schema. The GenoMetric Query Language (GMQL), is a query language for genomic data, based on the GDM data model described above. GMQL is inspired by Pig Latin, so it is a language that combines the high-level declarative style (SQL-like language) with the low level procedural style of map-reduce. GMQL is structurally consistent with the traditional database management techniques, but it also aims improving the interaction between biologists and biomedical data. GMQL algebraic operations are designed for the management of genomic data, specifically for the bioinformatics domain. GMQL operators admit parameters based on metadata or on the attributes of the schema of the genomic regions. In particular this last operations are related to the distances between the genomic regions. The methodology for extending the GDM and the GMQL is based on the integration of external sources (i.e., taxonomies) to provide an extended data schema and to allow improved query capabilities. Indeed, taxonomies deal with the classifications of terms in a hierarchical structure. These classifications provide information about a topic in different levels of granularity. Taxonomies are a special form of ontologies, i.e., formal representations of the knowledge, where a particular domain is schematically represented in a data structure containing all the relevant entities and their relationships. The most famous ontology in bioinformatics is Gene Ontology [9]. The extension of the data and of the GDM will take place, as described in [5]. We will apply two methods, which differ in the granularity of data. In the following, we define formally the concepts of *upward* and *downward extension*. Consider a taxonomy T, a set of levels $L = l_1, ..., l_n$ of T, a set of data S and an attribute a of S. If a is provided at a level l_i, you can use T for extending the data set to a level l_j with $l_j > l_i$. To move to a higher level means getting a coarser granularity of the data, when the attribute a is initially at a fine granularity. The *upward extension* has the objective of storing new information, but at a higher level in the taxonomy, i.e., at a less fine granularity than the one that is available.

Consider a taxonomy T, a set of levels $L = l_1, ..., l_n$ of T, a set of data S and an attribute a of S. If a is provided at a level l_i, you can use T for extending the data set to a level l_j with $l_j < l_i$. In this case, we are considering the *downward extension*, because the taxonomy is traversed from top to bottom to achieve a lower level that means obtain a finer granularity of the data. The a attribute is initially at a coarse granularity and the objective is to apply the downward extension to store new information at a more detailed level, i.e., at a finer grain than that available.

4   Preliminary Key Results or Contributions

Part of the preliminary study that was carried out for this work concerns the individuation of the external data sources that can be integrated in our system to allow the extensions. Specifically, the ontologies and the databases of particular interest for our work and that cover our domain of interest are reported in the following. The first one is the National Center for Biotechnology Information (NCBI) [10], which hosts a number of relevant biomedical databases and important resources for bioinformatics. The main databases include GenBank [11] for DNA sequences and PubMed [12], a bibliographic database for biomedical literature. All these databases are available online through the Entrez search engine [13]. Additionally, we will take into account Gene Ontology [9], a bioinformatics database that aims to maintain and to develop a description of genes and gene products for every living organism, to annotate genes and gene products and to disseminate such data, providing tools for easy access to them.

The upward and downward extension operations are made possible by the use of the above mentioned sources that allow to increase the expressive power of the available data. The data that we considered in our preliminary work are related to Next Generation Sequencing experiments of various forms of cancer. We extracted these data from The Cancer Genome Atlas (TCGA) [14], a comprehensive open-access biomedical database containing more than 1.5 TB of cancer data, and we implemented already extensions to them. For example we applied the downward extension method to sequencing data of breast cancer and of kidney cancer. This example of extension included the addition of a new attribute to the schema of the genomic regions, which represents the possible

Chrom	Start	End	St	Symbol	Gene_id
chr8	144801161	144801161	+	MAPK15	225689
chr7	20824445	20824445	+	SP8	221833
chr9	32988092	32988092	+	APTX	54840

Chrom	Start	End	St	Symbol	Symbol_syn	Gene_id
chr8	144801161	144801161	+	MAPK15	ERK7	225689
chr8	144801161	144801161	+	MAPK15	ERK8	225689
chr7	20824445	20824445	+	SP8	BTD	221833
chr9	32988092	32988092	+	APTX	AOA	54840
chr9	32988092	32988092	+	APTX	AOA1	54840
chr9	32988092	32988092	+	APTX	AXA1	54840

Fig. 1. The genomic regions before and after the downward extension.

synonyms of a gene. In Fig. 1 we show genomic regions before and after the downward extension. As result, for each genomic region the records are temporarily duplicated as many times as there exists synonyms for the gene. In this way the queries will match also if the user adopts a synonym of the interested gene.

5 Work Plan and Conclusions

The work plan will focus on the two considered cases concerning the extension of the metadata and of the schema of genomic regions. Different extensions for these two cases will be implemented. For helping the reader in understanding our methodology, we will describe a metadata extension use case and two schema extension use cases, considering data of TCGA related to sequencing experiments for different forms of cancer. During the PhD project several other extension use cases are going to be addressed.

The metadata of the samples in TCGA have an attribute that represents the processing center, which has executed the sequencing of a given tissue. In particular, there are at most two attributes of this type: "center", which as the name of the processing center; "center_id", with the identifier value of the center. We want to add the State, where the sequencing center is located. In order to get the State from the name of the sequencing center, we can consider the geographical taxonomy of the sequencing centers which provides information about name, city and state of the centers. We will traverse the levels of the taxonomy starting from bottom (center name) to top (state of the center) performing an upward extension, for upgrading the information about the sequencing centers to a coarser grain. In order to get this extension, it is necessary to retrieve and store the cities of each center, and then query geographical ontologies such as GeoNames [15], to get the state from the city. GeoNames provides access to geographic information (features), represented by the Resource Description Framework (RDF) data model which can be queried using SPARQL, the query language for RDF data.

For the extension of the schema of genomic regions we discuss two examples. This extension starts from the gene identifiers of all genomic regions of the samples of TCGA. The schema defines for each genomic region the *gene_symbol* and the *entrez_gene_id* of the corresponding gene:

$< field\ type = "STRING" > gene\_symbol < /field >$
$< field\ type = "STRING" > entrez\_gene\_id < /field >.$

In the first example, we already implemented a solution for extending the schema of the genomic regions with a field that corresponds to a possible synonym for the gene related to a given genomic region. Consider a taxonomy of the genes and starting from the *entrez_gene_id* of a gene, we retrieve synonyms corresponding to it. The taxonomy is traversed from top to bottom in order to increase the level of detail and to consider the characteristics of the gene to a finer grain. The schema of each experiment is subjected to downward extensions by adding a new field corresponding to the synonym for the gene (example of line in the schema: $<$ *field type* $=$ "*STRING*" $>$ *gene_symbol_synonyms* $<$ /*field* $>$), while for each line of each sample the id of the corresponding gene has been selected in order to obtain all the possible synonyms. The data source used for the recovery of synonyms is NCBI, and through the Entrez search engine, it is possible to access the Gene database in NCBI, which collects information on individual genes; this database has been used to obtain synonyms of genes available in the various samples. Entrez uses the API Entrez uses Entrez Programming Utilities (E-utilities), to access different databases. The E-utilities use a fixed URL-based syntax to recover the requested data. In this case the Java language has been used to send the URL to NCBI with following parameters: the Gene database, to which we request the information, the id of the gene, and the desired output format (xml). From the received xml file it has been possible to get all the synonyms of the input gene, thus obtaining for each sample, a corresponding list with all the gene ids and its synonyms. Then these synonyms are integrated in each sample file of genomic regions.

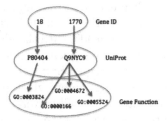

Fig. 2. Gene taxonomy (for gene functions).

In the second example, we report a downward extension of the schema of the genomic regions with a field related to the function of a gene. Consider a taxonomy of genes (Fig. 2) that provides the information about the functions of a particular gene, having his *entrez_gene_id* as input. The goal is to be able to extend the genomic regions with the information on elementary activities of the gene product at molecular level. For this information we can use the GeneOntology, a bioinformatics project that unifies all the descriptions of the characteristics of the products of the genes in all species. The databases that mainly populate the information exposed by GeneOntology are the ones of the UniProt project [16], the largest bioinformatics database for protein sequences of all living organisms. In this example, it is necessary to access to UniProt identifiers of each gene and then take advantage of the GeneOntology web services to get the names and the identification of the molecular functions of the gene.

Finally, all the extensions, which enable a user to be able to obtain information that are not previously found in the original dataset (e.g., synonyms of a gene, gene functions, the state in which a sample is sequenced, etc.) will be integrated in the GDM and the GMQL. The user will be able to perform a query to a database that can automatically activate the extension process. The aim is to apply the defined extensions to operations such as selection or join as explained in [5]. The relaxation of the query can be provided by introducing the derived operators in taxonomy-based relaxation modalities, i.e., by considering the upward and downward extension of the data. The extension is based on the field requested by the user, using a specific operator in the relaxed query: the user is guided in the selection of the field that is most suitable to his needs, then an appropriate taxonomy is queried to proceed with the data extension. The extensions will be implemented in a software system, whose architectural diagram is shown in Fig. 3. The flow process begins with the relaxation of the original GMQL query. The next step provides the extension of the GMQL operators (the metadata-based and region field-based operators). Afterwards the taxonomy engine is called for retrieving the appropriate taxonomy from external data sources. Different parsers for connecting the taxonomy engine to external sources are applied. After that the GDM extension is applied with the new features to the original dataset by using the upward or downward extension.

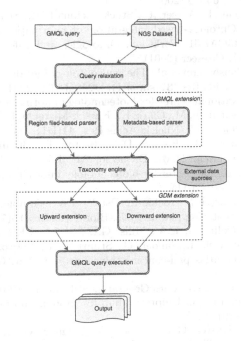

Fig. 3. Architectural diagram of the proposed system.

In conclusion, the extended GMQL query is executed to provide the requested output. In future, we intend to integrate the system in a framework for biomedical data management (e.g., Galaxy [17]) in order to provide enhanced search, access, extraction, and analysis of biomedical information.

References

1. Weitschek, E., Cumbo, F., Cappelli, E., Felici, G.: Genomic data integration: a case study on next generation sequencing of cancer. In: 2016 27th International Workshop on Database and Expert Systems Applications (DEXA), pp. 49–53. IEEE (2016)
2. Cumbo, F., Fiscon, G., Ceri, S., Masseroli, M., Weitschek, E.: Tcga2bed: extracting, extending, integrating, and querying the cancer genome atlas. BMC Bioinform. **18**(1), 6 (2017)
3. Ceri, S., Kaitoua, A., Masseroli, M., Pinoli, P., Venco, F.: Data management for next generation genomic computing. In: EDBT, pp. 485–490 (2016)
4. Masseroli, M., Pinoli, P., Venco, F., Kaitoua, A., Jalili, V., Palluzzi, F., Muller, H., Ceri, S.: Genometric query language: a novel approach to large-scale genomic data management. Bioinformatics **31**(12), 1881–1888 (2015)
5. Martinenghi, D., Torlone, R.: Taxonomy-based relaxation of query answering in relational databases. VLDB J. **23**(5), 747–769 (2014)
6. Stein, L.: Genome annotation: from sequence to biology. Nat. Rev. Genet. **2**(7), 493–503 (2001)
7. Noy, N.F.: Semantic integration: a survey of ontology-based approaches. ACM Sigmod Rec. **33**(4), 65–70 (2004)
8. Wache, H., Voegele, T., Visser, U., Stuckenschmidt, H., Schuster, G., Neumann, H., Hübner, S.: Ontology-based integration of information-a survey of existing approaches. In: IJCAI-01 Workshop: Ontologies and Information Sharing, vol. 2001, pp. 108–117. Citeseer (2001)
9. Gene Ontology Consortium et al.: The gene ontology (go) database and informatics resource. Nucleic Acids Res. **32**(suppl 1), D258–D261 (2004)
10. Cates, S.: Ncbi: National center for biotechnology information (2006)
11. Benson, D.A., Cavanaugh, M., Clark, K., Karsch-Mizrachi, I., Lipman, D.J., Ostell, J., Sayers, E.W.: Genbank. Nucleic Acids Res. **41**(D1), D36–D42 (2013)
12. Sayers, E.W., Barrett, T., Benson, D.A., Bolton, E., Bryant, S.H., Canese, K., Chetvernin, V., Church, D.M., DiCuccio, M., Federhen, S., et al.: Database resources of the national center for biotechnology information. Nucleic Acids Res. **39**(suppl 1), D38–D51 (2011)
13. Tatusova, T.A., Karsch-Mizrachi, I., Ostell, J.A.: Complete genomes in www entrez: data representation and analysis. Bioinformatics **15**(7), 536–543 (1999)
14. Weinstein, J.N., Collisson, E.A., Mills, G.B., Shaw, K.R.M., Ozenberger, B.A., Ellrott, K., Shmulevich, I., Sander, C., Stuart, J.M., Network, C.G.A.R., et al.: The cancer genome atlas pan-cancer analysis project. Nat. Genet. **45**(10), 1113–1120 (2013)
15. Wick, M.: Geonames. GeoNames Geographical Database (2011)
16. Uniprot Consortium et al.: Uniprot: a hub for protein information. Nucleic Acids Res. **43**, gku989 (2014)
17. Blankenberg, D., Kuster, G.V., Coraor, N., Ananda, G., Lazarus, R., Mangan, M., Nekrutenko, A., Taylor, J.: Galaxy: a web-based genome analysis tool for experimentalists. Curr. Protoc. Mol. Biol. **10**, 1–21 (2010)

A Semantic Integration Approach for Building Knowledge Graphs On-Demand

Diego Collarana[1,2][✉]

[1] Enterprise Information Systems (EIS), University of Bonn, Bonn, Germany
collaran@cs.uni-bonn.de
[2] Fraunhofer Institute for Intelligent Analysis and Information Systems (IAIS),
Sankt Augustin, Germany

Abstract. Information about the same entity may be spread across several Web data sources, e.g., people on the social networks (Social Web), product descriptions on e-commerce sites (Deep Web) or in public Knowledge Graphs (Web of data). The problem of integrating entities from heterogeneous Web data sources on-demand is still a challenge. Existing approaches propose expensive Extraction Transformation Loading (ETL) processes and rely on syntactic comparison of entity properties, leaving aside the semantics encoded in the data. We devise *FuhSen*, an integration approach that exploits search capabilities of Web data sources and semantics encoded in the data. *FuhSen* generates Knowledge Graphs in response to keyword-based queries. Resulting Knowledge Graphs describe the semantics of the integrated entities, as well as the relationships among these entities. *FuhSen* approach utilizes an ontology to describe the Web data sources in terms of content and search capabilities, and exploits this knowledge to select the sources relevant for answering a keyword-based query on-demand. The results of various empirical studies of the effectiveness of *FuhSen* suggest that the proposed integration technique is able to accurately integrate data from heterogeneous Web data sources into a Knowledge Graph.

1 Problem Statement

The strong support that Web based technologies have received from researchers, developers, and practitioners has resulted in the publication of data from almost any domain on the Web. Additionally, standards and technologies have been defined to query, search, and manage Web accessible data sources. For example, Web access interfaces or APIs allow for querying and searching sources like Twitter, Google+, or the DBpedia, Wikidata. Web data sources make overlapping as well as complementary data available about entities, e.g., people or products. Although these entities may be described in terms of different vocabularies by these Web data sources, the data correspond to the same real-world entities. Thus, the distributed data needs to be integrated in order to have a more complete description of these entities (Fig. 1).

As example, consider a *distributed* and *heterogeneous* search scenario in the context of crime investigation. During a crime investigation process, collecting,

© Springer International Publishing AG 2017
J. Cabot et al. (Eds.): ICWE 2017, LNCS 10360, pp. 575–583, 2017.
DOI: 10.1007/978-3-319-60131-1_45

and analyzing information from different sources is a key step performed by investigators. Although scene analysis is always required, a crime investigation process greatly benefits from searching information about people, products, and organisations on the Web. Typically, data collected from the following data sources is utilised for enhancing crime analysis processes: (1) The *Social Web* encompasses user generated content and personal profiles. (2) The *Deep Web* advertises products and services offered by organisations, e.g., the eBay e-commerce platform. (3) The *Web of Data* includes billions of machine-comprehensible facts, which can serve as background knowledge for collecting information about different types of entities. (4) The *Dark Web* refers to sites accessible only with specific software, and restricted trading of goods that can be accessed through the so-called dark-net markets.

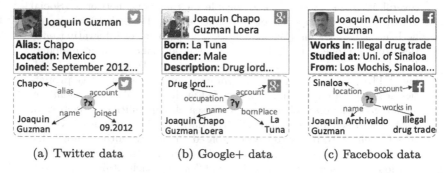

(a) Twitter data   (b) Google+ data   (c) Facebook data

Fig. 1. Motivating Example. Pieces of data (RDF molecules) about *Joaquin Chapo Guzman* collected from different Web social networks.

To solve this data integration scenario, in this doctoral work we propose *FuhSen*, a semantic integration approach that exploits Web APIs (e.g., REST APIs) provided by Web data sources to collect, and integrate molecules of data, to then enrich and summarize information about an entity (e.g., a suspect). Using Linked Data as the core technology, the objectives of *FuhSen* approach is to provide a novel integration technique able to: (1) integrate heterogeneous data extracted from APIs into a unified data schema on-demand; (2) create a Knowledge Graph on-demand with the data extracted from the different data sources; and (3) enrich this Knowledge Graph using semantic algorithms e.g., entity disambiguation, typing and entity summarization, and ranking.

2   Research Objectives

This doctoral work attempts to answer the following research questions:

> **RQ1:** Can Knowledge Graphs be populated with data collected from heterogeneous Web data sources on-demand?

To answer **RQ1**, we plan to explore and evaluate the use of RDF vocabularies to facilitate source selection and data fusion tasks.

> **RQ2**: Can semantics encoded in RDF graphs be exploited to integrate data collected from heterogeneous data sources?

To answer **RQ2**, we will analyse how to use both the explicit semantics, e.g., Properties, Relations, and Hierarchy of classes, and also the implicit semantics.

> **RQ3**: Can semantic similarity measures be able to enhance accuracy of the integration of data collected from heterogeneous data sources?

To answer **RQ3**, we will evaluate semantic similarity approaches that can be used in the context of data integration. Propose a new semantic similarity metric with the goal of data integration is also an option to answer RQ3.

3   State-of-the-art

Traditional approaches toward constructing **Knowledge Graphs (KG)**, e.g., NOUS [3], Knowledge Vault [7], or DeepDive [12], imply materialization of the designed graph built from (un-, semi-)structured sources. Therefore, a heavy Extraction Transformation Loading (ETL) process needs to be executed to integrate the data. In comparison, the novelty of *FuhSen* approach resides in a non-materialized Knowledge Graph and usage of semantics encoded in the data. Non-materialization supports efficient knowledge delivery on-demand. Further, *FuhSen* creates RDF molecules that unify and encode hybrid knowledge from heterogeneous sources in an abstract entity. Moreover, the problem of integrating RDF graphs is in the research focus for many years. Knoblock et al. [10] propose KARMA, a framework for integrating structured data sources. Schultz et al. [15] describe a Linked Data Integration Framework (LDIF) that provides a set of independent tools to support the process of interlinking RDF datasets. For instance, SILK [9] identifies `owl:sameAs` links among entities of two datasets and Sieve [11] performs data fusion. Although the aforementioned approaches are fast and effective, they require domain knowledge and significant manual effort while configuring the pipeline. In contrast, *FuhSen* is a universal **black box** technique that requires only a small number of high-level parameters, while enables users to adjust the system according to the application domain.

4   Proposed Approach

Given a keyword query, *FuhSen* executes the query over the relevant sources, and utilizes semantic similarity measures to determine the relatedness among the entities to be integrated. *FuhSen* creates a Knowledge Graph with the integrated entities at query time. A Knowledge Graph is composed of a set of entities, their properties, and relations among these entities. The Semantic Web technology stack provides the pieces required to define and build a Knowledge Graph. To properly understand these concepts, we follow the notation proposed by Arenas et al. [1], Piro et al. [13], and Fernandez et al. [8] to define RDF triples, Knowledge Graphs, and RDF molecules, respectively.

Definition 1 (RDF triple [1]). *Let* $\mathbf{I}$, $\mathbf{B}$, $\mathbf{L}$ *be disjoint infinite sets of URIs, blank nodes, and literals, respectively. A tuple* $(s, p, o) \in (\mathbf{I} \cup \mathbf{B}) \times \mathbf{I} \times (\mathbf{I} \cup \mathbf{B} \cup \mathbf{L})$ *is denominated an RDF triple, where* s *is called the subject,* p *the predicate, and* o *the object.*

Definition 2 (Knowledge Graph [13]). *Given a set* T *of RDF triples, a Knowledge Graph is a pair* $G = (V, E)$, *where* $V = \{s \mid (s, p, o) \in T\} \cup \{o \mid (s, p, o) \in T\}$ *and* $E = \{(s, p, o) \in T\}$.

Definition 3 (RDF Subject Molecule [8]). *Given an RDF graph* G, *an RDF subject-molecule* $M \subseteq G$ *is a set of triples* $t_1, t_2, \ldots, t_n$ *in which* $subject(t_1) = subject(t_2) = \cdots = subject(t_n)$.

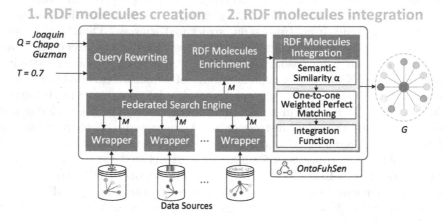

Fig. 2. The *FuhSen* Architecture. *FuhSen receives a keyword query* Q *and a threshold* T, *and produces a Knowledge Graph* G *populated with the entities associated with the keywords*

FuhSen is a two-fold approach, the architecture is shown in Fig. 2. The first step is the creation of RDF molecules of data from the heterogeneous Web data sources. Web services facilitate accessibility of data on-demand e.g., using REST services. In this step, RDF molecules from the same entities have to be recognized and integrated in order to build complete Knowledge Graphs

4.1   Creation of RDF Molecules

As an input, *FuhSen* receives a keyword query Q, e.g., *Joaquin Chapo Guzman*, and a similarity threshold value T, e.g., 0.7. The input values are processed by the *Query Rewriting* module, which formulates a correct query to be sent to the *Search Engine* module. The *Search Engine* explores several wrappers and transforms the output into RDF molecules. Intermediate results are enriched with additional knowledge in the *RDF Molecules Enrichment* module.

4.2   Integration of RDF Molecules

This module constructs a Knowledge Graph out of the enriched molecules. The input is a set of RDF molecules, and the output is an integrated RDF graph. The module consists of three sub-modules:

- *Computing Similarity of RDF Molecules.* Similar RDF molecules should be integrated in order to create a fused, universal representation of a certain entity. In contrast with triple-based linking engines like Silk [15], we employ a RDF molecule-based approach increasing the complexity level and considering the semantics of molecules. That is, we do not work with independent triples, but rather with a set of triples belonging to a certain subject. The RDF molecule-based approach allows for natural clustering of a Knowledge Graph, reducing the complexity of the linking algorithm.
- *1-1 Weighted Perfect Matching.* Given a weighted bipartite graph BG of RDF molecules, where weights correspond to values of semantic similarity between the RDF molecules in BG, a matching of BG corresponds to a set of edges that do not share an RDF molecule, and where each RDF molecule of BG is incident to exactly one edge of the matching. The problem of the 1-1 weighted perfect matching of BG corresponds to a matching where the sum of the values of edge weights in the matching have a maximal value.
- *Integration functions.* When similar molecules are identified under the desired conditions, the last step of the pipeline is to integrate them into an RDF Knowledge Graph. The result Knowledge Graph contains all the unique facts of the analyzed set of RDF molecules. The implementation of the integration function in *FuhSen* is the union, i.e., the logical disjunction, of the molecules identified as similar during the previous steps.

5   Research Methodology and Research Design

The research methodology of this doctoral work includes the following steps:

1. Review the literature to evaluate the state-of-the-art approaches relevant to the problem of integrating heterogeneous Web data sources on-demand.
2. Formalise an on-demand semantic integration approach named *FuhSen*.
3. Empirically evaluate different properties of the approach, e.g., effectiveness and performance. Evaluate different components of the architecture and propose new algorithms and operators to realize the vision of this work.

6   Results and Contributions

So far, we have evaluated the architecture and the effectiveness of *FuhSen*. In [4,6], we proposed and implemented an RDF vocabulary mediator-wrapper architecture and proposed an evaluation study to answer **RQ1**.

Lessons Learned: The proposed architecture implemented in *FuhSen* is able to query heterogeneous Web data sources and create RDF molecules of data in a federated manner. The results of our evaluations shows that the vocabulary approach defined in *FuhSen* architecture allows for handling heterogeneity of data in an effective way. At the same time more Web data sources can be plug in/out in an easy manner, as consequence, the integration process is reduced. However, the experiments show problems in terms of scalability, the more Web data sources the slower the integration process becomes. Thus, a better resource selection approach should be investigated to answer research question **RQ1**.

In [5], we propose a two-fold approach to integrate RDF molecules from different data sources, with this approach and its evaluation we answer research questions **RQ2** and **RQ3**. We evaluate the effectiveness of integration on different datasets. We also experiment with two similarity metrics: Jaccard and GADES [14]. Our goal was to determine the impact of similarity function on the integration approach. Therefore, a triplet-based similarity metric Jaccard is compared against a semantic similarity function GADES [14].

We created a **Gold Standard (GS)** of the type Person extracted from DBpedia, which results in 829,184 triples. Two **Test data Sets (TS)** were created from the Gold Standard with their properties and values randomly split among two test datasets. Each triple is randomly assigned to one or several test datasets. We measure the behavior of our integration approach *FuhSen* [5] in terms of the following metrics: *Precision*, *Recall*, and *F-measure* during the experiments. Precision is the fraction of RDF molecules that has been identified and integrated by the approach (M) that intersects with the Gold Standard (GS), i.e., $Precision = \frac{|M \cap GS|}{|M|}$. Recall corresponds to the fraction of the identified similar molecules in the Gold Standard, i.e., $Recall = \frac{|M \cap GS|}{|GS|}$.

Lessons Learned: Table 1 shows the effectiveness of *FuhSen* on the integration task over 20,000 molecules. Jaccard demonstrates lower performance on the data set as its algorithm just relies on the particular properties of the RDF molecule. Jaccard does not utilize semantics encoded in the Knowledge Graph. On the other hand, GADES exhibit a good performance and it might be used as a **black box** in *FuhSen* approach.

Performance of the integration depends on the threshold parameter. As a simple set-based approach, the performance (precision, recall, and F-measure) of the Jaccard similarity quickly decreases with higher thresholds, while GADES remains stable. These insights suggest a positive answer to research questions **RQ2** and **RQ3**. However, GADES performance is impacted by the quality of the schema e.g., a good design of the hierarchy of classes, properties, and relations. Thus, enrichment of the molecules and tuning process is a pre-requisite in GADES. This impacts on an automatic nature of the problem we are trying to solve. Therefore, a pre-trained and automatic similarity function to compare RDF Molecules is required to answer research questions **RQ2** and **RQ3**.

Table 1. Effectiveness of *FuhSen* **on 20,000 RDF molecules.** Jaccard vs GADES approach using different thresholds (T). Highest values of Recall and F-measure are highlighted in bold.

	T0.0	T0.1	T0.2	T0.3	T0.4	T0.5	T0.6	T0.7	T0.8	T0.9
Precision										
Jaccard	0.72	0.77	0.44	0.34	0.37	0.36	0.27	0.21	0.21	0.21
GADES	0.76	**0.80**	**0.80**	0.79	0.79	0.79	0.79	0.76	0.70	0.65
	T0.0	T0.1	T0.2	T0.3	T0.4	T0.5	T0.6	T0.7	T0.8	T0.9
Recall										
Jaccard	0.72	0.42	0.09	0.05	0.02	0.02	0.01	0.01	0.01	0.01
GADES	**0.76**	**0.76**	**0.76**	**0.76**	**0.76**	**0.76**	0.68	0.46	0.22	0.06
	T0.0	T0.1	T0.2	T0.3	T0.4	T0.5	T0.6	T0.7	T0.8	T0.9
F-Measure										
Jaccard	0.72	0.54	0.15	0.08	0.04	0.04	0.02	0.02	0.02	0.02
GADES	0.76	**0.78**	**0.78**	0.77	0.77	0.77	0.73	0.57	0.33	0.11

Although significant progress has been done in the context of this doctoral work, more empirically results are needed to fully answer research questions **RQ1**, **RQ2**, and **RQ3**. Next section describes the plan for the next year.

7 Work Plan

This doctoral work is entering in its final stage (3rd year). To completely answer the defined research questions the following research tasks remain:

1. Propose a novel RDF fusion operator, the operator should be able to determine the relatedness between two RDF molecules and integrating them. This task is related to **RQ2**, and the target publication is a research paper.
2. Present a semantic similarity measure based on TransE [2] which utilizes the gradient descent optimization method to learn the features representation of RDF entities automatically. This task is related to **RQ3**, and the target publication is a research paper.
3. Present a scalable and efficient source selection based on the semantic description of the Web sources and keyword query. This task is related to **RQ3**, and we plan to publish a research paper.

8   Conclusions

In this doctoral work, we address the problem of data integration about the same entity that is spread in different Web data sources. We propose *FuhSen*, an *on-demand semantic integration approach* that creates Knowledge Graphs on-demand by integrating data collected from a federation of heterogeneous data sources using an RDF molecule integration approach. We have explained the creation of RDF molecules by using Linked Data wrappers; we have also presented how semantic similarity measures can be used to determine the relatedness of two resources in terms of the relatedness of their RDF molecules. Results of the empirical evaluation suggest that *FuhSen* is able to effectively integrate pieces of information spread across different data sources. The experiments suggest that the molecule based integration technique implemented in *FuhSen* integrates data in a Knowledge Graph more accurately than existing integration techniques.

Acknowledgments. I would like to thank Maria-Esther Vidal, Christoph Lange, Sören Auer, and the German Ministry of Education and Research with grant no. 13N13627 project LiDaKrA for supporting this work.

References

1. Arenas, M., Gutierrez, C., Pérez, J.: Foundations of RDF databases. In: Tessaris, S., Franconi, E., Eiter, T., Gutierrez, C., Handschuh, S., Rousset, M.-C., Schmidt, R.A. (eds.) Reasoning Web 2009. LNCS, vol. 5689, pp. 158–204. Springer, Heidelberg (2009). doi:10.1007/978-3-642-03754-2_4
2. Bordes, A., Usunier, N., Garcia-Duran, A., Weston, J., Yakhnenko, O.: Translating embeddings for modeling multi-relational data. In: Advances in Neural Information Processing Systems 26. Curran Associates Inc. (2013)
3. Choudhury, S., Agarwal, K., Purohit, S., Zhang, B., Pirrung, M., Smith, W., Thomas, M.: Nous: construction and querying of dynamic knowledge graphs. arXiv preprint arXiv:1606.02314 (2016)
4. Collarana, D., Galkin, M., Lange, C., Grangel-González, I., Vidal, M., Auer, S.: Fuhsen: a federated hybrid search engine for building a knowledge graph on-demand (short paper). In: OTM Conferences - ODBASE (2016)
5. Collarana, D., Galkin, M., Traverso-Ribón, I., Lange, C., Vidal, M.-E., Auer, S.: Semantic data integration for knowledge graph construction at query time. In: ICSC (2017)
6. Collarana, D., Lange, C., Auer, S.: Fuhsen: a platform for federated, RDF-based hybrid search. In: WWW Companion Volume (2016)
7. Dong, X., Gabrilovich, E., Heitz, G., Horn, W., Lao, N., Murphy, K., Strohmann, T., Sun, S., Zhang, W.: Knowledge vault: a web-scale approach to probabilistic knowledge fusion. In: SIGKDD. ACM (2014)
8. Fernández, J.D., Llaves, A., Corcho, O.: Efficient RDF Interchange (ERI) format for RDF data streams. In: Mika, P., Tudorache, T., Bernstein, A., Welty, C., Knoblock, C., Vrandečić, D., Groth, P., Noy, N., Janowicz, K., Goble, C. (eds.) ISWC 2014. LNCS, vol. 8797, pp. 244–259. Springer, Cham (2014). doi:10.1007/978-3-319-11915-1_16

9. Isele, R., Bizer, C.: Active learning of expressive linkage rules using genetic programming. J. Web Seman. **23**, 2–15 (2013)
10. Knoblock, C.A., et al.: Semi-automatically mapping structured sources into the semantic web. In: Simperl, E., Cimiano, P., Polleres, A., Corcho, O., Presutti, V. (eds.) ESWC 2012. LNCS, vol. 7295, pp. 375–390. Springer, Heidelberg (2012). doi:10.1007/978-3-642-30284-8_32
11. Mendes, P.N., Mühleisen, H., Bizer, C.: Sieve: linked data quality assessment and fusion. In: Proceedings of the 2012 Joint EDBT/ICDT Workshops, Berlin, Germany, 30 March 2012 (2012)
12. Palomares, T., Ahres, Y., Kangaspunta, J., Ré, C.: Wikipedia knowledge graph with DeepDive. In: 10th AAAI Conference on Web and Social Media (2016)
13. Pirrò, G.: Explaining and suggesting relatedness in knowledge graphs. In: Arenas, M., et al. (eds.) ISWC 2015. LNCS, vol. 9366, pp. 622–639. Springer, Cham (2015). doi:10.1007/978-3-319-25007-6_36
14. Ribón, I.T., Vidal, M., Kämpgen, B., Sure-Vetter, Y.: GADES: a graph-based semantic similarity measure. In: SEMANTiCS (2016)
15. Schultz, A., Matteini, A., Isele, R., Mendes, P.N., Bizer, C., Becker, C.: LDIF - a framework for large-scale linked data integration. In: 21st International World Wide Web Conference, Developers Track, April 2012

A People-Oriented Paradigm for Smart Cities

Alejandro Pérez-Vereda[⊠] and Carlos Canal

Department of Computer Science, University of Malaga, Malaga, Spain
apvereda@uma.es, canal@lcc.uma.es

Abstract. Most works in the literature agree on considering the Internet of Things (IoT) as the base technology to collect information related to smart cities. This information is usually offered as open data for its analysis, and to elaborate statistics or provide services which improve the management of the city, making it more efficient and more comfortable to live in. However, it is not possible to actually improve the quality of life of smart cities' inhabitants if there is no direct information about them and their experiences. To address this problem, we propose using a social and mobile computation model, called the Internet of People (IoP) which empowers smartphones to recollect information about their users, analyze it to obtain knowledge about their habits, and provide this knowledge as a service creating a collaborative information network. Combining IoT and IoP, we allow the smart city to dynamically adapt its services to the needs of its citizens, promoting their welfare as the main objective of the city.

1 Introduction

Recent initiatives are starting to develop software systems based on the Internet of Things (IoT) in order to better managing the resources and services provided by a city. The cities that integrate this kind of solutions are those we call smart cities. Their purpose is to allow public institutions to get a better control of the city, and to improve the quality of life of their inhabitants [1].

Nowadays, many cities are offering publicly on the Internet open data collected from very different kinds of sensors. Hence, we can get valuable and updated information about air pollution, the availability and schedule of public transport, etc. However, this is not enough for making a city smart. Cohen [2] envisions smart cities with a broad, integrated approach, with the aim of improving the efficiency of the services they offer, making local economy to grow, and developing stronger citizen engagement. Indeed, as stated in [3], people are one of the pillars supporting the smart city.

Hence, it is necessary to put the citizens themselves in the smart city picture. For that we need to obtain information about how its inhabitants use the services offered by the city, and what is their user experience. This would help to improve city infrastructures with better plans, and to adapt them to the context of the people using them.

Our proposal consists in creating an ecosystem of shared data where citizens act as sensors providing information to others and to the city itself. The institutions will be able to obtain a rich feedback, to react to the needs of the citizens, and to dynamically adapt to them the services they offer. For that, we propose a distributed architecture

© Springer International Publishing AG 2017
J. Cabot et al. (Eds.): ICWE 2017, LNCS 10360, pp. 584–591, 2017.
DOI: 10.1007/978-3-319-60131-1_46

which uses smartphones for the creation, storage and sharing of sociological profiles with information about their users' experiences and habits.

One of distinctive features of our proposal is that the information about a user never leaves her smartphone, it is stored only in it. In order to access a particular piece of information, we need to request it to the smartphone, which may decide to provide it or not, according to its applicability for this particular user and her privacy preferences. Our proposal is based on the Internet of People (IoP) [4], a social computation model that was thought to improve the integration of people in the IoT, combining the information coming from the IoT with that from the people using it.

The structure of this paper is as follows. In the next section, we present our research objectives, further discussing the problem we are trying to solve, which is illustrated by means of an urban mobility scenario. In Sect. 3, we discuss several works in the literature that are somehow related to our proposal. Then, Sect. 4 presents our research methodology, divided into the different tasks to be addressed. Finally, the last section deal with the preliminary results.

2 Research Objectives

As mentioned above, we detect a deficit of information related to the inhabitants of a smart city and their habits and experiences using its services in their daily life. This way, the analysis of the information provided by the smart city sensors cannot take into account the context of the people, nor it is able to adapt to their needs.

Aiming to fill this gap, we propose a social and collaborative architecture focused on the people. The architecture relies on the use of smartphones, which we always carry on with us, to elaborate sociological profiles with information about our daily activities, and to offer them as a service in a controlled and secure way. Our proposal gives smartphones the role of autonomous devices able to analyze the data they obtain from their users and share the results with other users or third-party institutions. A collaborative network is created between users, offering them the opportunity to benefit from each other. The traditional notion of a centralized server containing all the information in the system disappears, letting every smartphone be a server.

To show how our proposal and the combination between IoP and the open data from IoT works, we now present an illustrating scenario of urban mobility.

Peter is a university student living in a medium size city. Every working day he catches the bus for going to the university. As always, he carries his smartphone with him, letting it to collect data about all his movements. Analyzing the information recollected about Peter's locations and movements, it is easy to determine the location of Peter's home, the places he usually visits, like the university, as well as the routes he follows to go to these places, the transportation means used, and his daily schedule.

By combining this information, stored in Peter's mobile phone as part of his sociological profile, with some of the open data sets provided by the city, we would be able to suggest Peter alternative means of transportation, such as a public bicycle, if there are public bike racks both near his home and the university, or if the city traffic control center has detected a jam in the road by which Peter's bus passes. This way, we promote the reduction of pollutants to the atmosphere, one of the goals of city representatives (Fig. 1).

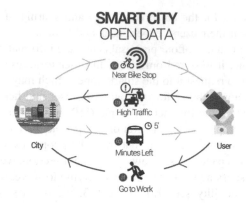

Fig. 1. Urban mobility scenario

However, Peter has not got his bike helmet with him, so he discards the recommendation. At the same time, John, one of Peter's neighbours, has decided to go to the university with his car. In order to share expenses, John's smartphone gets in contact with Peter's, offering him a ride by an alternative route that avoids the traffic jam.

Accordingly to the scenario just described, the goal of this PhD. project is to prove whether (a) smartphones can be used to infer information about the daily activities of their users, in particular their routines of movement in the city, (b) this information can be combined with open data publicly available on the web in order to build higher level knowledge of how the user is actually moving, (c) all the inferred knowledge can be represented as a sociological profile of the user, stored in the smartphones themselves, instead of in a centralized server, and (d) these profiles can be efficiently offered as a service to third parties, also from the smartphones.

In order to accomplish this goal, it will be necessary to tackle a number of different problems and open issues: open data integration, machine learning techniques for inferring the profiles, complex queries resolution, etc. Therefore, this not a PhD. that studies in depth a particular problem to find out a specific solution to it, but an in-width approach, proposing an alternative framework to how a kind of problems is addressed, combining and improving when necessary techniques coming from the fields of smart cities, collaborative architectures, and social and mobile computing.

3   Related Work

The literature provides a body of research works related to the goals of this PhD. project. If we focus on the data offered by smart cities, Jin et al. [5] present a framework for the realization of smart cities through the IoT. Their framework supports from the sensors installed in the city, to the management of the data and the cloud-based integration of the respective services. However, we will show that there still is a big issue to address: the standardization of the data the IoT offers. To this extent, the reference initiative we are considering to apply is the Web of Things (WoT) [6].

If we focus on collaborative scenarios, there are also precedents for smart cities. Cardone et al. [7] talk about the power the collectives could give to the smart cities and propose a crowdsensing platform for this purpose. In fact, many recent research works agree on solving the problems of IoT and the smart cities by adding a paradigm focused on the people [8, 9]. The social architecture that inspires ours is called People as a Service (PeaaS) [10]. It is a mobile computation architecture based on the IoP model. PeaaS promotes exploiting the smartphones' potential for inferring, storing and sharing a user profile with personal information about their owner.

It is also worth to pay a look to industrial approaches such as Google's or Apple's assistants, which retrieve lots of personal data from their users. Considering the scenario above, they are able to get enough information about their users so that they can provide notifications about traffic jams or bus schedules before leaving. However, the drawback here comes from the fact that they only consider the interests of part of the stakeholders. Users can get information from other people's experiences, that's true, but all the information is stored and analyzed in the central servers of the corresponding companies, which exploit it mainly for their own benefit. Contrarily, ours is a decentralized model that allows establishing synergies between the different actors, being them people or institutions like the municipality, that would be able to improve their infrastructures by having more precise information about their use.

4   Methodology

To carry out with this work we have designed a research methodology consisting on different steps that are built one on another. At each step, we will need to define the problem being addressed, review the literature for current approaches to it, decide the technologies to be applied, devise solutions one step ahead the current state of the art, and define validation methods and processes according to the characteristics of the problem and the solution provided. Each one of the steps builds on the previous one to obtain more complex and higher-level resulting information.

Data acquisition. The first step is data recollection. For our motivating scenario, we just need to monitor the GPS sensor of the smartphone, but other sources of information can also be considered. Locations are timestamped and stored in the phone, but we need to be aware of battery and memory consumption. Hence, the location is collected only when the user is moving. To deal with indoor GPS signal problems, alternative means like wifi triangulation and beacons have to be considered.

Event detection. The next thing to do is to analyze these raw data for extracting more relevant information. We follow a model based on the DIKW pyramid (Fig. 2) [11] to reach a higher level of complexity with each step. The timestamps with GPS positioning correspond to the data level of the pyramid. From that we can infer events of interest, such as the places visited (those in which the user stayed for a certain time), or sharp changes of direction or speed when moving between them (which may correspond to a change in the transportation mean). These events constitute the information level of the DIKW pyramid. As we will explain in Sect. 5, we have already developed an ad hoc algorithm for detecting this kind of events in our mobility scenario.

However, for more general situations we need to consider alternatives like Complex Event Processing [12], an emerging technology for defining events and that allows processing, analyzing and correlating big amounts of data to detect these relevant situations.

Fig. 2. DIKW pyramid

Routine inference. At this point, we need to recognize event patterns in order to reach the knowledge level. The idea is to detect the places visited and the routes and speed of movement between them. The final goal is to be able to predict at least with a certain degree of confidence, daily movement routines and other relevant knowledge, for instance which of the places visited corresponds to the user's home, or to his workplace, and when is he expected to be there, or even his habits (e.g. going to the gym every Tuesday afternoon). Several approaches can be considered to address this problem, such as machine learning techniques like Bayesian networks [13]. However, for a first approach, we are planning to use using Social Workflow techniques [14], which are in fact focused on people's procedures for completing tasks, and that could help us representing and analyzing the information we have learnt up to this point. It is important to remember that all the information recollected and analyzed for a given user is stored only in her smartphone.

At this step, a first validation process must be made. We will install the system in the smartphones of a test population with different transportations habits. Then this people will be interviewed to validate the correctness of the knowledge inferred.

Integration with the IoT. Now that we have knowledge about people, the next thing to do is to integrate this knowledge with the IoT information coming as open data from the smart city. Combining these two data sources, it would be possible both to get a more elaborate knowledge about the users (for instance, that Peter takes bus no. 11 for going to the university), and to adapt the services of the city to the needs and context of its citizens. One of the problems to be solved here is open data standardization. We know the data we want to use from the city. However, the description, schema and format of these data are not the same for all smart cities. We need some kind of semantic description or ontology for smart cities [15] in order to access and use IoT data independently of the city we are considering.

This is one of the issues addressed by SMART-FI (http://www.smart-fi.eu), a EU-funded research project in which we are currently involved. It is based on the FIWARE (http://www.fiware.org) platform and aims to create business opportunities for applications about smart cities, and to improve the citizens' quality of life by using open data. Although the SMART-FI project is more oriented to development than to research, one of its case studies is actually an urban mobility scenario similar to that presented above.

Collaborative IoP. Once people's and IoT information is integrated, we would be able to reach the wisdom level of the DIKW pyramid, and to extract statistical information about the use of city resources, or even to elaborate recommendations, as for instance to extend a given bus line with an additional stop due to a high percentage of users of this line walking for a while in the same direction after leaving the bus.

However, as mentioned before, all the information is stored in the smartphone, so we will need a tool for querying it, and reaching for instance the users of a given bus line. Hence, we have to develop a high-level language that allows making abstract queries to the whole collaborative network of citizens: a language to interact with the IoP.

At this point, additional validations can be performed, using the smartphone itself to obtain feedback and determining whether the recommendations are being followed by the user. If this was not the case, interviews can again be used to find the reasons why.

5 Preliminary Results

Although this PhD. project is in an initial state, part of the work has already been done. GPS data recollection and a first attempt of its analysis have been designed and implemented in order to obtain some preliminary validation results. Considering several existing approaches of the smartphone use to monitor people with cognitive impairment [16], we have applied the concepts of the IoP model, and the reference PeaaS architecture that inspire ours in order to develop SafeWalks, a mobile app that determines the movement routines of people attained with initial stages of Alzheimer [17], sending alerts to caregivers in case of deviations from routine. This app has been developed as an initial proof of concept of our proposal, but certainly a more general solution to the problem, involving more sophisticated techniques needs to be devised.

The SafeWalks app monitors the GPS sensor of the smartphone and stores timestamps with locations every time the user moves. We established a configurable 35 m threshold, so every time the user moves this limit away, we collect the GPS position and also speed measurements, as shown in Fig. 3.

The next step is obtaining the places where the user has been, which correspond with periods of time when we do not collect data. We have developed an ad hoc analysis that detects a visited place when we find two or more continuous timestamps that are separated by at least five minutes (again, configurable), but not by more than 35 m, with a five-meter margin for errors. Then we store the mean point among these points as an instance of a Place.

Here it is important to pay attention in filtering possible GPS errors. Due to different reasons, it is not unlikely that the GPS sensor reports anomalous positioning. A given

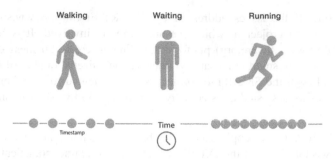

Fig. 3. Timestamps recollection over time

position is compared with previous and subsequent lectures and discarded if considered wrong. Also, anomalous speeds are discarded: when the user changes her speed, she needs to maintain it at least for the following timestamp with a ten percent error margin.

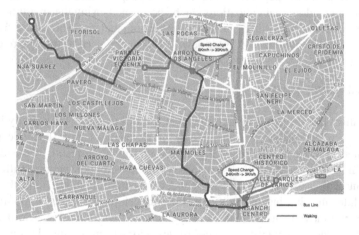

Fig. 4. Superposition of a user's movement with a public bus line (in blue) (Color figure online)

Once we have started to obtain instances of places, it is easy to determine the Routes between them. We only have to pair the instances of Places following a chronological order. The route is represented by the positions, speed and timestamps between them. We then compute points of interest in the route, corresponding to sharp changes in direction or speed. We need these points because sometimes the speed is not enough for determining the mean of transportation, for instance to distinguish between a car and a bus. We use the points of interest to compare the route with the ones followed by the buses of the city and their stops as shown in Fig. 4. Finally, when a Place or a Route is repeated a given number of times, it becomes a Routine or habit, and the app starts computing its frequency and schedule.

Acknowledgments. This work has been partially supported by the Spanish Ministry of Science, under project TIN2015-67083-R.

References

1. Nam, T., Pardo, T.A.: Conceptualizing smart city with dimensions of technology, people and institutions. In: Proceedings of the 12th Annual International Digital Government Research Conference: Digital Government Innovation in Challenging Times, Maryland, USA, pp. 282–291 (2011)
2. Cohen, B.: What exactly is a smart city. Co. Exist 19 (2012)
3. Manville, C., Cochrane, G., Cave, J., Millard, J., Pederson, J.K., Thaarup, R.K., Liebe, A., Wissner, M., Massnink, R., Kotterink, B.: Mapping Smart Cities in the EU (2014)
4. Miranda, J., Mäkitalo, N., Garcia-Alonso, J., Berrocal, J., Mikkonen, T., Canal, C., Murillo, J.M.: From the internet of things to the internet of people. IEEE Internet Comput. Mag. **19** (2), 40–47 (2015)
5. Jin, J., Gubbi, J., Marusic, S., Palaniswami, M.: An information framework for creating a smart city through internet of things. IEEE Internet Things J. **1**(2), 112–121 (2014)
6. Guinard, D., Trifa, V., Mattern, F., Wilde, E.: From the Internet of Things to the Web of Things: Resource-Oriented Architecture and Best Practices. Architecting the Internet of Things, pp. 97–129. Springer, Heidelberg (2011)
7. Cardone, G., Foschini, L., Bellavista, P., Corradi, A., Borcea, C., Talasila, M., Curtmola, R.: Fostering participation in smart cities: a geo-social crowdsensing platform. IEEE Commun. Mag. **51**(6), 112–119 (2013)
8. Sheth, A.: Computing for human experience: semantics-empowered sensors, services, and social computing on the ubiquitous web. IEEE Internet Comput. **14**(1), 88–91 (2010)
9. Wang, F.Y., Carley, K.M., Zeng, D., Mao, W.: Social computing: from social informatics to social intelligence. IEEE Intell. Syst. **22**(2), 79–83 (2007)
10. Guillen, J., Miranda, J., Berrocal, J., Garcia-Alonso, J., Murillo, J.M., Canal, C.: People as a service: a mobile-centric model for providing collective sociological profiles. IEEE Softw. **31**(2), 48–59 (2014)
11. Davenport, T.H., Prusak, L.: Working Knowledge: How Organizations Manage What They Know. Harvard Business School Press, Boston (1998)
12. Luckham, D.: Event Processing for Business: Organizing the Real-Time Enterprise. John Wiley & Sons, Hoboken (2011)
13. Tang, L., Chen, H., Ku, W.S., Sun, M.T.: Exploiting location-aware social networks for efficient spatial query processing. GeoInformatica **21**(1), 33–55 (2017)
14. Vázquez-Barreiros, B., Mucientes, M., Lama, M.: ProDiGen: mining complete, precise and minimal structure process models with a genetic algorithm. Inf. Sci. **294**, 315–333 (2015)
15. Schleicher, J.M., Vögler, M., Dustdar, S., Inzinger, C.: Enabling a smart city application ecosystem: requirements and architectural aspects. IEEE Internet Comput. **20**(2), 58–65 (2016)
16. Lin, Q., Zhang, D., Connelly, K., Ni, H., Yu, Z., Zhou, X.: Disorientation detection by mining GPS trajectories for cognitively-impaired elders. Pervasive Mob. Comput. **19**, 71–85 (2015)
17. Pérez-Lozano, P., Pérez-Vereda, A., Murillo, J.M., Canal, C.: SafeWalks: aplicación móvil de supervisión de pacientes de Alzheimer. XI Jornadas de Ciencia e Ingeniería de Servicios, Santander (2015)

CSQuaRE: Approach for Quality Control in Crowdsourcing

Lalit Mohan Sanagavarapu[✉]

Software Engineering Research Center,
International Institute of Information Technology,
Gachibowli, Hyderabad 500036, India
lalit.mohan@research.iiit.ac.in

Abstract. Quality control of responses enhances sustainability and adoption of crowdsourcing. Expert and peer reviews, majority voting, machine learning, game theory, etc. are some of the practices for quality control in crowdsourcing. However, quality of crowdsourced responses is still a concern. We propose a quality control approach drawing inspiration from Requirements Engineering quality attributes - Completeness, Consistency and Correctness (3Cs). The 3Cs of a response are assessed and displayed as a CSQuaRE score based on coverage with reference to knowledge base (ontology), cohesiveness and contributor credibility in the domain. The knowledge base would evolve with continued extraction of instances of information and thus responses would be re-calibrated for relevance. The suggested approach would be demonstrated for Information Security related Question and Answers on a crowdsourcing platform. The evaluation of the approach would be based on comparison with existing quality control techniques and feedback from security experts.

Keywords: Quality control · 3Cs · Crowdsourcing · Q&A

1 Motivation and Background

Improving penetration of internet[1] and digital literacy is leading to the growth of crowdsourcing (image recognition, language translation, responses to questions and other micro and macro tasks). For quick money or due to lack of complete and correct knowledge, responses suffer in quality. The current quality control processes use majority voting, peer-reviews, data mining, fault-tolerant sub-tasks, game theory and other hybrid modes [6]. Crowdsourcing Q&A platforms such as StackExchange and Quora use majority voting for assessing the quality of responses, this would mean manual intervention and latency till viewers rate/vote. We extracted 135 responses of a sample 50 questions[2] on 'Phishing' from Quora. More than 77% of the responses were not related answers. Among the related and relevant responses, greater than 90% responses had less/no votes

---

[1] http://www.internetlivestats.com/.
[2] http://goo.gl/ocEg4G.

© Springer International Publishing AG 2017
J. Cabot et al. (Eds.): ICWE 2017, LNCS 10360, pp. 592–599, 2017.
DOI: 10.1007/978-3-319-60131-1_47

from crowd though they were semantically same from a relevant and higher rated answer. Lack of recognition (viewer rating) could de-motivate intrinsic contributors. Also, we conducted an online survey[3] in Sep'2016 approach to understand the quality concerns in crowdsourcing. Social media sites including CrowdsourcingWeek[4] Linkedin group were used for survey participation. The majority of the survey respondents were IT savvy working professionals from Asian countries. 76% of these 212 respondents stated that crowdsourced responses have poor quality. The quality gaps in crowdsourcing of Q&A is the motivation for this research.

We propose Completeness, Consistency and Correctness (3Cs) approach adopted from Software Requirements Engineering (RE) for quality control of crowdsourced Question and Answers. Like software products that are built based on stakeholder(s) requirements (understanding and knowledge level), the responses in crowdsourcing are also based on workers' knowledge level. Hence, we hypothesize that the rigor of 3Cs will differentiate good from bad responses, thus leads to quality control. Though there are many other RE quality characteristics such as traceability, modifiability, unambiguity, etc., the importance of 3Cs is unequivocally stated in research publications [1,10,14], ISO/IEC 25010:2011[5] standards and Gartner[6] market research.

An example of completeness of a response for a question on 'what are the key characteristics of Information Security' is 'Confidentially, Integrity and Availability'. Completeness has puritan view with many forms such as Functional, Syntactic, Semantic, etc. Taking direction from Gabriel's comments in 'The rise of worse is better', we measure Completeness as the degree of coverage of real world situations in the response(s) ensuring unnecessary or irrelevant features are not captured. Obtaining complete information for a domain is never ending problem [3]. Hence, our completeness measure for a response is with reference to extracted knowledge base (KB), termed as Adequate Completeness ACP.

Consistency is the measure of conflict free sentences of the response with respect to the objective (question). An example of consistency in a response as extracted from a crowdsourcing platform for a question on 'What are the security features of Amex credit card' is 'Amex credit card has 2 levels of security: they have the normal CVV (Card Verification Value) and the 3 digits are a CID (Customer Card Identity). CVV is a calculated highly secure 4 digit code based on your card number that is not contained in the card magnetic strip'. Based on evolving Ontology[7] with increasing instances of KB, our consistency ACN is measure of conflict free tuples (Concept + Relationship + Concept) in the response. This would also mean that the response is not just a bag of words but sentences that are cohesive and are conflict free. The history of past contributions (credibility) of workers' in the topic (Question-Answer) is also a factor in our

---

[3] https://goo.gl/forms/AhEA8fu8pSAOwxjm1.

[4] http://tinyurl.com/crowdsourcingtrends.

[5] https://www.iso.org/standard/35733.html.

[6] http://tinyurl.com/Gartner3Cs.

[7] https://www.w3.org/OWL/.

consistency measure. We relate credibility to consistency rather correctness as they are measures of trust rather rightness.

Correctness is the degree to which a response contains conditions and limitations for the desired capability (question). Hence, a response correctness is not necessarily a binary (Yes/No or True/False) but a degree of match/similarity. An example of correctness of a response is *'Authentication is used for providing an access entry into the system'*. Like completeness, our correctness [3] is a measurement with respect to the extracted KB. The adequate correctness ACR of a crowdsourced response is based on the occurrences of semantically similar content in the extracted KB and the relation to question type (What, Why, When, Where, Who and How - 5W & 1H).

We propose $CSQuaRE$ score based on ACP, ACN and ACR for assessing *C*rowd*S*ourced responses *Qua*lity using *R*equirements *E*ngineering approach of information security related questions. Our proposed approach would be demonstrated for 'Information Security' crowdsourced responses. Our past experiences[8] and existence of security related information exchange platforms such as StackExchange, AlienVault[9], etc. provide confidence that individuals are comfortable in seeking and responding to security related questions on public platforms. Crowdsourcing Week, a leading website on crowdsourcing has identified security information exchange as one of the top emerging trends.

2   Research Questions

Addressing the following research questions would provide a quantitative measure $CSQuaRE$, for assessing 3Cs in a response.

- (Q1) What are the dimensions of completeness, correctness and consistency of a response that can be measured automatically?
  Increasing KB for completeness leads to inconsistency. Completeness and consistency of a response enhances correctness. The interplay among these 3Cs has to be identified to avoid double-counting or negation in $CSQuaRE$ calculation, this includes the degree of relationship - linear/polynomial.
- (Q2) What is the credibility of a worker in the past while responding to questions in a specific topic/domain?
  Most of the existing crowdsourcing platforms limit credibility assessment at platform and/or tasks level. The crowd worker may not be active on the crowdsourcing platform but may have deep knowledge in the question domain and could be prolific contributor on other internet sites. This research includes identification of the person and his/her credibility in the specific question domain from the obtained KB.
- (Q3) What is the temporal effect on the response of a question with respect to completeness, consistency and correctness?
  As KB increases with time, the response that had a certain $CSQuaRE$ may

---

[8] http://idrbt.ac.in/ib-cart.html.
[9] https://otx.alienvault.com/.

change over a period. As an example, strength of cryptography algorithms has improved from SHA1 to SHA2 and so on. Hence, responses' *CSQuaRE* requires re-calibration to maintain 3Cs.

Part of our research, we also plan to crawl internet to extract security related information, conduct a study on importance of text cohesion in crowdsourced responses and develop an evolving ontology (KB) based on newer instances of extracted information.

3   Related Work

The related work describes quality control in crowdsourcing, 3Cs and its attributes for quality control, credibility assessment and Q&A platforms.

3.1   Quality in Crowdsourcing

Afra et al. [8] used credibility based on past contributions and contributors mobility pattern for quality control. Aroyo et al. [11] performed quality assessments of Q&A postings on disagreement-based metrics to harness human interpretation. In a recent study, Bernstein et al. [12] discusses on reputation of crowd workers and importance of peer-reviews. The existing quality control mechanisms use hard wired mechanism and are not multi-dimensional model. Some of the other related literature discusses usage of game theory such as multi-armed bandit, better task clarity, effect of cascade model, Groundtruth, experience and language nativity for evaluating quality of workers/tasks. We plan to use reputation/credibility of crowd workers based on past contributions and Ground truth in the form of KB in our quality control approach.

3.2   Completeness, Consistency and Correctness

Siegemund et al. [9] uses ontology model for identification of consistency and completeness of evolving requirements. Lami et al. [7] presented a methodology and tool for evaluation of natural language based requirements for consistency and completeness. The work of McCalls quality model and Zowghi et al. [15] identified the interplay among 3Cs that we plan to extend in our *CSQuaRE* measurement. The behavioural aspects of the worker such as credibility based on past contributions and profile attribute to consistency in the quality. We plan to extend the work of Kumaraguru et al. [4] on Twitter tweets for credibility score of responses specific to the question domain.

3.3   Question and Answers

Question Answering systems transformed much in the last four decades on par with natural language processing (NLP) techniques. In 1978, the first classic Q&A book published based on Lehnert's thesis provided fundamental basis for research. Availability of TREC corpus, research in Biomedical domain gave impetus to Q&A platforms. Hirschman et al. [5] factored importance of completeness and correctness in Q&A platforms. The articles and publications of AnswerBus [13], START from MIT, etc. state the usage of advancements in NLP and AI for providing the services. Publicly available literature on IBM Watson, Apple Siri, etc. states the Q&A usefulness and the importance of continuous evolution/training based on KB. While there is no involvement of crowd workers in any of these platforms for quality control, their success on relying evolving KB and ensuring cohesion in responses aligns with our approach.

As evident from the reviewed literature, there is no comprehensive approach for quality control of crowdsourced responses using credibility of worker in the internet, temporal affect on the quality of response, domain knowledge for completeness, consistency and correctness measurement.

4   Proposed Approach

To demonstrate 3Cs approach for quality control, we plan to build a crowdsourcing platform prototype using available open source Q&A software after technical and functional evaluation[10]. The following sections describe the progress of work in terms of 'In-progress' and 'Yet to begin' for addressing the research questions. The schematic Fig. 1 depicts the approach for implementing quality control of crowdsourced responses.

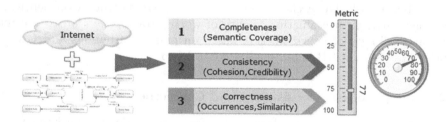

Fig. 1. Approach for quality control in crowdsourcing

4.1   Building Domain Repository: In-Progress

The basis for measurement of $CSQuaRE$ is based on the extracted domain content. 934,000+ security related URLs[11] are obtained from Wikipedia and Twitter. These URLs are categorized into 14 groups and 114 controls as available in ISO/IEC 27001:2013[12] to ensure representation across sub-domains. The crawled content based on seed URLs are cleansed (stop word removal and stemming)

---

[10] http://tinyurl.com/hcuvhnd.
[11] https://goo.gl/DjR4bA.
[12] http://iso27001security.com/html/27002.html.

and classified into security sub-domains. As there is no prevalent security search engine, we plan to provide an interface (Search Engine) to the extracted domain content by June 2017. This would also provide a user base and an opportunity to seek feedback on data relevance by security experts associated with Banking Community and DSCI[13], a voluntary organization with security experts as its members.

4.2   Ontology Evolution: In-Progress

The extracted domain content would be represented as an ordered pair of TBox (Concepts and Relationships) and ABox (Assertions) using existing Security Ontology [2] and Word2Vec. We use Word2Vec for similarity mapping between Ontology terms and extracted internet content, Word2Vec is trained on 100 Billion words of GoogleNews Archives and provides. This ordered pair (TBox and ABox) of ontology would be considered as knowledge base (KB). This KB would be used for evaluating the ACP, ACN and ACR of crowdsourced responses. However, the KB needs update with change in time as the concepts and relationships of a domain evolve or change and thus re-calibration of the past responses. Automatically updating ontology based on increased domain content may not be acceptable to ontologists. An observable pattern of KB would be identified for ontologists to update the security ontology. The observable pattern of extracted domain content would be assessed on text cohesion, relevance of the text to 114 controls of security and credibility of the information source. The text of the responses will also be used for ontology evolution as the responses may contain information that is not available in the crawled content and this text could be used for quality control of related future questions.

4.3   Credibility: Yet to Begin

As stated earlier, credibility in a question domain is part of our consistency measure. We plan to have login to our crowdsourcing platform based on user's Twitter ID. The crowd worker credibility would be based on past contributions on crowdsourcing platform and the credibility score on Twitter in the question-answer related topic. Also, we are in the process of evaluating credibility[14] of website containing the information security content, to ensure not every available content is being used for ontology evolution.

$$Credibility = \{TwitterCred, Site, Contributions...\}$$

4.4   Assignment of $CSQuaRE$: Yet to Begin

A question posted on the crowdsourcing platform may have one or more responses. Every response would be assigned a $CSQuaRE$ score based on ACP,

---

[13] https://www.dsci.in/.

[14] https://serc.iiit.ac.in/WEBCred/.

ACN and ACR. Information Retrieval evaluation metrics such as Recall, Latent Semantic Index, etc. are being explored for measuring Completeness (ACP). NLP techniques such as cohesiveness (part of discourse analysis to ensure responses are not just bag of words but sentences that are related) and individual credibility in the domain based on past contributions would be part of Consistency (ACN) measure. The users in our proposed crowdsourcing platform will also have voting option for the responses, this voting would act as feedback loop for improving the credibility of the crowd worker. For the Correctness (ACR) measure, we are exploring Machine Learning (Decision Tree and SVM) approaches for matching Question Type vis-a-vis response and FrameNet[15] to obtain semantic similarity of response with reference to the KB.

The initial weights for each of the components (ACP, ACN and ACR) of $CSQuaRE$ score would be equal, scaled to 10 (0 being unrelated and 10 being highest) and will refine based on the feedback loop. Some of the factors in calculating score are

$$ACP = \{TermCoverage, OntologyDepth, ...\}$$

$$ACN = \{DLMatch, IndividualCredibility, ...\}$$

$$ACR = \{QuestionType, ResponseSimilarity, ...\}$$

$$CSQuaRE = \{ACP, ACN, ACR\}$$

5   Evaluation Plan

An empirical approach would be used for validating the solutioning of research problems. We plan to extract questions and responses from StackExchange that are related to 114 control groups of ISO 27001, with more than 3 respondents and are rated by viewers. We provide these responses to Security Experts and ask them to evaluate the relevance of response on a scale 0–10, 0 - being unrelated and 10 - being highest. We assess the $CSQuaRE$ of these responses on our crowdsourcing platform making credibility score a constant. We hypothesize that $CSQuaRE$ score should be similar to the score assessed by security experts. The viewer rating of StackExchange responses may also be high for responses that are assessed high by our approach. We also plan to perform a control data experiment to assess $CSQuaRE$ applicability. As part of the evaluation, each of the measures such as ACP, ACN and ACR would be made constant to measure their effectiveness for quality control.

Also, a survey would be conducted to get the feedback on $CSQuaRE$ from security experts. This survey would guide us in identifying gaps and scope for further refinement of the approach.

---

[15] https://framenet.icsi.berkeley.edu/fndrupal/.

References

1. Boehm, B.W., Brown, J.R., Lipow, M.: Quantitative evaluation of software quality. In: Proceedings of the 2nd International Conference on Software Engineering, pp. 592–605. IEEE Computer Society Press (1976)
2. Ekelhart, A., Fenz, S., Klemen, M.D., Weippl, E.R.: Security ontology: simulating threats to corporate assets. In: Bagchi, A., Atluri, V. (eds.) ICISS 2006. LNCS, vol. 4332, pp. 249–259. Springer, Heidelberg (2006). doi:10.1007/11961635_17
3. Gardner, M., Talukdar, P.P., Kisiel, B., Mitchell, T.: Improving learning and inference in a large knowledge-base using latent syntactic cues (2013)
4. Gupta, A., Kumaraguru, P., Castillo, C., Meier, P.: TweetCred: real-time credibility assessment of content on Twitter. In: Aiello, L.M., McFarland, D. (eds.) SocInfo 2014. LNCS, vol. 8851, pp. 228–243. Springer, Cham (2014). doi:10.1007/978-3-319-13734-6_16
5. Hirschman, L., Gaizauskas, R.: Natural language question answering: the view from here. Nat. Lang. Eng. 7(04), 275–300 (2001)
6. Ipeirotis, P.G., Provost, F., Wang, J.: Quality management on Amazon mechanical Turk. In: Proceedings of the ACM SIGKDD Workshop on Human Computation, pp. 64–67. ACM (2010)
7. Lami, G., Gnesi, S., Fabbrini, F., Fusani, M., Trentanni, G.: An automatic tool for the analysis of natural language requirements. Informe técnico, CNR Information Science and Technology Institute, Pisa, Italia, Setiembre (2004)
8. Mashhadi, A.J., Capra, L.: Quality control for real-time ubiquitous crowdsourcing. In: Proceedings of the 2nd International Workshop on Ubiquitous Crowdsouring, pp. 5–8. ACM (2011)
9. Siegemund, K., Thomas, E.J., Zhao, Y., Pan, J., Assmann, U.: Towards ontology-driven requirements engineering. In: Workshop Semantic Web Enabled Software Engineering at 10th International Semantic Web Conference (ISWC), Bonn (2011)
10. Tamai, T., Kamata, M.I.: Impact of requirements quality on project success or failure. In: Lyytinen, K., Loucopoulos, P., Mylopoulos, J., Robinson, B. (eds.) Design Requirements Engineering: A Ten-Year Perspective. LNBIP, vol. 14, pp. 258–275. Springer, Heidelberg (2009). doi:10.1007/978-3-540-92966-6_15
11. Timmermans, B., Aroyo, L., Welty, C.: Crowdsourcing ground truth for question answering using crowdtruth. In: Proceedings of the ACM Web Science Conference, p. 61. ACM (2015)
12. Whiting, M.E., Gamage, D., Gaikwad, S., Gilbee, A., Goyal, S., Ballav, A., Majeti, D., Chhibber, N., Richmond-Fuller, A., Vargus, F., et al.: Crowd guilds: worker-led reputation and feedback on crowdsourcing platforms. arXiv preprint arXiv:1611.01572 (2016)
13. Zheng, Z.: Answerbus question answering system. In: Proceedings of the Second International Conference on Human Language Technology Research, pp. 399–404. Morgan Kaufmann Publishers Inc. (2002)
14. Zowghi, D., Gervasi, V.: The three Cs of requirements: consistency, completeness, and correctness. In: International Workshop on Requirements Engineering: Foundations for Software Quality, pp. 155–164. Essener Informatik Beitiage, Essen (2002)
15. Zowghi, D., Gervasi, V.: On the interplay between consistency, completeness, and correctness in requirements evolution. Inf. Softw. Technol. 45(14), 993–1009 (2003)

Design of a Small and Medium Enterprise Growth Prediction Model Based on Web Mining

Yiea-Funk Te[(⊠)] and Irena Pletikosa Cvijikj

ETH Zurich, Weinberstrasse 56/58, 8092 Zurich, Switzerland
{fte,ipletikosa}@ethz.ch

Abstract. Small and medium enterprises (SMEs) play an important role in the economy of many countries. Still, due to the highly turbulent business environment, SMEs experience more severe challenges in maintaining and expanding their business. To support SMEs at improving their competitiveness, researchers recently turned their focus on applying web mining (WM) to build growth prediction models. WM enables automatic and large-scale collection and analysis of potentially valuable data from various online platforms, thus bearing a great potential for extracting SME growth factors, and enhancing existing SME growth prediction models. This study aims at developing an automated system to collect business-relevant data from the Web and predict future growth trends of SMEs by means of WM and machine learning (ML) techniques.

Keywords: Data mining · Web mining · SME growth prediction

1 Research Problem

Small and medium enterprises (SMEs) play an important role in the world economy. SMEs represent 95% of all businesses, accounting for 66% of the total employment and 55% of the total production [11]. However, existing studies show that the current business environment is highly turbulent and influenced by modern information and communication technologies, globalization and employee mobility [1, 14]. Additionally, the growing number of SMEs caused competition to become increasingly intensive, forcing SMEs to experience more severe challenges in maintaining their existence and expanding their business. Thus, understanding the SMEs success factors is of great importance.

In order to support SMEs at improving their competitiveness, scholars and practitioners have been analyzing growth and success factors of SMEs for decades. With the emergence of big data, the focus was placed on applying data mining techniques to build novel risk and growth prediction models. However, existing models only include few data types, e.g. financial or operational data, and thus cannot fully explain the complex context of SME growth [13]. Moreover, data is mostly obtained via questionnaires, which is a time-consuming process, or is provided by financial institutes, thus not publicly available and sensitive to privacy issues. In addition, existing models mostly focus on risk evaluation and bankruptcy prediction. Although numerous studies on SME growth prediction exist, studies applying data mining techniques are scarce.

© Springer International Publishing AG 2017
J. Cabot et al. (Eds.): ICWE 2017, LNCS 10360, pp. 600–607, 2017.
DOI: 10.1007/978-3-319-60131-1_48

Recently, web mining (WM) has emerged as a new approach towards obtaining valuable business insights. WM enables an automated and large scale collection and analysis of potentially valuable data from the web such as the national commercial register and company websites. While WM methods have been frequently studied to anticipate growth of sales volume for e-commerce platforms, their application for assessment of SME growth factors is still scarce. Considering the large and increasing amount of data freely available online, WM bears a great potential in revealing valuable information hidden in web, which can be further used to build a SME growth prediction model.

2 Research Objectives

I aim at developing a growth prediction model for SMEs based on publicly available data. Therefore, I explore the potential of using WM to access a wide range of factors influencing growth from various web data sources. Finally, I aim at developing an automated system to collect business-relevant data from the web and predict future growth trends of SMEs by means of WM and data mining techniques. The envisioned system should serve as an "early recognition system" for future growth opportunities.

Business growth can be measured as (1) non-financial, i.e., growth of employment, customer satisfaction and loyalty, or (2) financial, such as growth of revenues, profits and assets. The focus of my work is on the revenue growth, due to its importance to the economy. To address the mentioned issues, I state the following research questions:

- RQ1. How can web data be used to leverage SME growth modelling? - The proposed research project explores the potential of WM to access growth factors from the web in order to build a SME growth prediction model. In order to answer this generally formulated question, three sub-questions are stated as follows.
- RQ2. Which growth factors can be extracted from the web? - Various web data sources have been identified which will be further analyzed with regard to the usability and accuracy to build a SME growth prediction model. Given a large set of well-studied growth factors, the goal is to identify suitable web data sources for feature extraction through WM.
- RQ3. From all the potential growth factors, which one prove to be discriminative? - Since a large variety of features are included in the growth prediction model, it is crucial to understand which factors are essential. To answer this question, different machine learning (ML) algorithms capable of feature selection or feature ranking will be applied [19]. The goal is to (1) confirm existing SME growth studies in a data-driven and model-based manner, (2) identify new growth correlates to extend our understanding of SME growth mechanisms, and finally (3) remove redundant features, thus reduce overfitting and improve the generalization of the models.
- RQ4. Which ML techniques can improve current state-of-the-art SME growth models? - Different ML algorithms will be tested with the goal to optimize the performance in predicting the future growth development of SMEs. Furthermore, in order to evaluate the additional value of our web data based growth model, the accuracy of our final model will be compared to a baseline growth model built on conventional, i.e. not publicly accessible data provided by a large insurance company.

3   State of the Art

3.1   SME Growth Research

The literature on business growth models dates back to 1967 and has proliferated since then into different streams addressing specific industries and business sizes. Lippitt and Schmidt [8] developed a general growth model for all sizes of businesses by examining how personality development theories influences the creation, growth and maturation of a business. Steinmetz [17] qualitatively analyzed the growth of small enterprises by partitioning the growth curve of small enterprises into different stages and assessing the characteristic attributes of each stage. A qualitative study conducted by Scott and Bruce [16] suggested a model for small business growth supporting managers to plan for future growth. The proposed model isolates five growth stages characterized by an unique combination of firm attributes. As a small company goes through different growth stages, attributes such as the management style, organizational structure and the use of technology changes. Although stage models are widely accepted among researchers and practitioners, stage models are criticized on some counts [12]. In particular, empirical research are only conducted on small sample sizes and specific types of businesses via questionnaires studies and thus, stage models are not generalizable.

3.2   Data Mining in SME Risk and Growth Research

Data mining techniques such as artificial neural network and decision tree have been extensively studied with a strong focus on SME risk evaluation and bankruptcy prediction rather than SME growth modelling. An early study indicated that backpropagation neural network (BPNN) were the most popular ML techniques among researchers in the finance and business domain during the 1990's [20]. For instance, Zhang et al. [22] provide a comprehensive review of Artificial Neural Network (ANN) applications for bankruptcy prediction. Their findings indicated that ANN perform significantly better than logistic regression models. West [19] investigated the credit scoring accuracy of various ANN models (e.g. multilayer perceptron, mixture-of-experts, radial basis function) in comparison with traditional methods such as logistic regression and discriminant analysis model. Consistent with the findings of Zhang et al. [22], ANN models perform slightly better than traditional methods.

Other techniques widely applied in the domain of risk evaluation and bankruptcy prediction includes decision trees (DT) and their ensemble variations such as random forests (RF). For instance, Fantazzini and Figini [4] developed a model based on RF for SME credit risk measurement and compared its performance with the traditional logistic regression approach. They came to the conclusion that both models provided similar results in terms of performance, highlighting the potential of RF for credit risk modelling. Another recent and more application-oriented study conducted by Ozgulbas and Koyuncugil [13] proposed an early warning system based on DT-algorithms for SMEs to detect risk profiles. The proposed system uses financial data to identify risk indicators and early warning signs, and create risk profiles for the classification of SMEs into different risk levels.

In summary, data mining techniques such as ANN and DT are extensively used for SME risk evaluation and bankruptcy prediction. However, research reporting data mining based SME growth prediction cannot be identified. Furthermore, current prediction models usually include one type of data sources and thus cannot explain the whole and complex context of SME growth [13].

3.3   Web Mining in E-Commerce and SME Growth Research

The web is a popular and interactive medium with intense amount of freely available data. It is a collection of documents, audios, videos and other multimedia data [9]. With billions of web pages available in the web, it is a rapidly growing key source of information, presenting an opportunity for businesses and researchers to derive useful knowledge out of it. Therefore, WM research for knowledge discovery has emerged.

WM has been proved very useful in the business world, especially in e-commerce [6, 7, 10, 15, 18]. For instance, Lin and Ho [7] proposed a system, which extracts content information from a set of web pages with a goal of extracting the informative content. Morinaga et al. [10] presented a system for finding the reputation of products from the internet to support marketing and customer relationship management in order to increase the sales volume. Finally, Thorleuchter et al. [18] analyzed the impact of textual information from e-commerce companies' websites on their commercial success by extracting web content data from the most successful top 500 worldwide companies.

While WM methods has been well researched and used in the field of e-commerce research to increase the sales volume, it has barely been applied for SME growth research. The study conducted by Antlová, Popelínsky and Tandler [2] demonstrated the potential of WM for SME growth prediction. In their paper, they examined the relationship between long-term growth of SMEs and a web presentation. Li et al. [6] recently explored micro-level characteristics and impacts of external relationships such as government or university relations on the growth of SMEs. However, these studies only focus on the information available in company websites and thus, restricting the amount and spectrum of growth factors. Hence, further research exploiting the full potential of web data for SME growth prediction is required.

4   Methodology

4.1   Research Design

In order to answer the stated research questions, the research is structured as follows. In the first step, a large set of well-studied growth factors is identified from an extensive literature review, which will serve as a groundwork for feature generation.

In the preliminary study, a SME growth model based on conventional and mostly well-structured data will be developed, serving as a baseline model. Therefore, data provided by a large Swiss insurance company will be used. The data consist of a wide range of firm attributes. Furthermore, the data contain information about the annual revenues, which will serve as "growth label" for supervised ML (Fig. 1: left).

In the main study, WM methods for automatic data collection and pre-processing will be applied. Various web sources, such as company websites, commercial register websites and social media will be studied. The retrieved web data will be used to develop a web data based growth prediction model. Finally, the accuracy of the final model will be compared to that of the baseline model (Fig. 1: right).

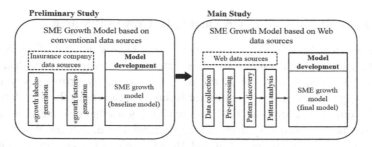

Fig. 1. Research design.

4.2   Identification of Growth Factors

The current business environment is influenced by a variety of firm internal and external factors [21], as illustrated in Fig. 2. Firm-internal factors can be divided into: (1) the characteristics of the firm, such as firm attributes (age, size, location) and firm strategies (marketing, training strategies), and (2) the characteristics of the entrepreneur, such as socio-demographic characteristics (age, gender, family background) and the personality of the entrepreneur (need for achievement, risk-taking propensity). Firm-external factors can be divided into factors reflecting (1) the immediate and (2) the contextual environment. The immediate environment includes supplier and customer relationship, competition, labor market and resource market. In contrast, the contextual environment comprises macro-environmental factors, such as economic, political, socio-cultural, technological and legal influences on the growth of businesses, which can emanate from local and national sources, but also from international developments [21].

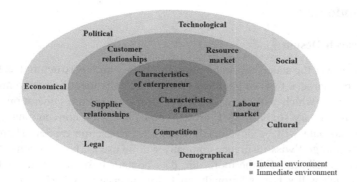

Fig. 2. Factors influencing growth of SMEs.

4.3   Data Collection

Two different types of data will be used: (1) not publicly available data provided by a large Swiss insurer, and (2) publicly available web data.

The data provided by the insurance company cover a 7-year period from 2010–2016 and contain detailed information about a large set of Swiss SMEs including their annual revenue. This data source will be used to: (1) build the baseline growth model, and (2) serve as ground truth data (i.e. growth-labeled data) to train the final model.

The web data related to a large set of Swiss SMEs with known revenues (i.e. ground truth) are collected. Therefore, the usability of various web data sources will be evaluated with respect to the identified growth factors. The data sources include governmental websites such as the Swiss Federal Statistical Office, commercial websites such as the Central Business Names Index, social media like Twitter and LinkedIn, and company websites. First, I plan to manually inspect the data sources in order to assess their usability. The next step is to develop a python-based program which allows an automated extraction of previously identified growth factors for a large set of companies. Table 1 summarizes the data sources which will be examined.

Table 1. Web data sources to be examined.

Web sources	Description
Central Business Names Index	Federal registry about firm location, founding year, board members, capital structure, legal authority
Swiss Federal Statistical Office	Granular information about the Swiss socio-demographics, economics, culture, education and political orientation
Federal Institute of Intellectual Property	Intellectual protected property
Job postings	Indicator for demand, skill level and sourcing
Twitter, LinkedIn, Xing	Information about company owners, company postings and mentions
Company websites	Information about the firm such as location, team, products, partners and company news

5   Expected Outcome

In my PhD thesis, I aim to enable the prediction of SME growth based on publicly available web data, currently not a common approach in the domain of SME growth prediction. Therefore, various web data sources will be analyzed with respect to their usability for SME growth prediction. WM will applied to collect web data and feature generation, whereas different ML algorithms will be studied to build a SME growth model with high accuracy. Furthermore, my research aims to expand the existing knowledge of several closely-related domains, including business intelligence, business information systems, web mining and SME growth research.

Furthermore, the research carried out in the scope of this thesis will have important practical implications. For the Swiss SME organizations, the insights generated in my thesis may support the Swiss SME organizations at understanding the success and failure of SMEs, thus strengthen their supportive role for SMEs. For the investment companies, the built system can be used to monitor the development of SMEs by mining the changes in the website of firms, serving as an "early recognition system" for future opportunities for growth. Finally, for SMEs, the system can be used to evaluate the characteristics of firms based on the information given in the web. The absence of important key success factors can be pointed out to firms, thus serving as a consulting program.

6 Stage of the Research

At the time of writing, the factors influencing growth of SMEs have been determined through an extensive literature review. The potential web sources for feature generation are identified and will be further investigated with regard to the usability to build a SME growth model. In particular, the data from the Central Business Names Index and Swiss Federal Statistical Office are completely collected, which gives us an overall picture of the SME landscape and the general socio-economic situation in Switzerland. Furthermore, the data from company websites of our ground truth (approximately 9000 SMEs) are collected by a web crawler and stored in the HTML format. In the next step, the focus lies in the extraction of firm characteristics such as age of firm, size, financial resources and human capital by means of text mining. A first SME growth model will be built based on the collected data. Therefore, different ML algorithms will be used und compared with regard to the accuracy. In an iterative and incremental approach, features reflecting the characteristics of the entrepreneur and environmental factors will be extracted from other web sources and added to the growth model.

7 Advice Sought

I hope to get valuable feedback from senior WM researchers on potential techniques suitable for extraction of growth factors from the web data sources. In particular, handling company websites is very challenging because the information available on company websites is not standardized, and varies according to the company and how it wishes to present itself. In addition, I would like to discuss the prediction ML algorithms I plan to use, as well as and further useful datasets and potential collaborations.

References

1. Antlová, K.: Motivation and barriers of ICT adoption in small and medium-sized enterprises. E+M Ekonomie Manage. **22**(2), 140–154 (2009)
2. Antlová, K., Popelínsky, L., Tandler, L.: Long term growth of SME from the view of ICT competencies and web presentations. E+M Ekonomie Manage. **14**(4), 125 (2011)
3. Breiman, L.: Decision-tree forests. Mach. Learn. **45**(1), 5–32 (2001)
4. Fantazzini, D., Figini, S.: Random survival forests models for SME credit risk measurement. Methodol. Comput. Appl. Probab. **11**(1), 29–45 (2009)

5. Hastie, T., Tibshirani, R., Friedman, J., Franklin, J.: The elements of statistical learning: data mining, inference and prediction. Math. Intell. **27**(2), 83–85 (2005)
6. Li, Y., Arora, S., Youtie, J., Shapira, P.: Using web mining to explore Triple Helix influences on growth in small and mid-size firms. Technovation (2016)
7. Lin, S.H., Ho, J.M.: Discovering informative content blocks from Web documents. In: Proceedings of the Eighth ACM SIGKDD International Conference on Knowledge Discovery and Data Mining, pp. 588–593. ACM (2002)
8. Lippitt, G.L., Schmidt, W.H.: Crises in a developing organization. Harvard Bus. Rev. **45**, 102–112 (1967)
9. Malarvizhi, R., Saraswathi, K.: Web content mining techniques tools & algorithms-a comprehensive study. Int. J. Comput. Trends Technol. (IJCTT) **4**(8), 2940–2945 (2013)
10. Morinaga, S., Yamanishi, K., Tateishi, K., Fukushima, T.: Mining product reputations on the web. In: Proceedings of the Eighth ACM SIGKDD International Conference on Knowledge Discovery and Data Mining, pp. 341–349. ACM (2002)
11. OECD: Small and medium-sized enterprises. In: Turkey Issues and Policies. Organization for Economic Cooperation and Development. OECD Press (2004)
12. O'Farrell, P.N., Hitchens, D.M.: Alternative theories of small-firm growth: a critical review. Environ. Plann. A **20**(10), 1365–1383 (1988)
13. Ozgulbas, N., Koyuncugil, A.S.: Risk Classification of SMEs by Early Warning Model Based on Data Mining. World Acad. Sci., Eng. Technol., Int. J. Soc., Behav., Educ., Econ., Bus. Ind. Eng. **6**(10), 2649–2660 (2012)
14. Post, H.A.: Building a Strategy on Competencies. Long Range Plan. **30**(5), 733–740 (1997)
15. Saini, S., Pandey, H.M.: Review on web content mining techniques. Int. J. Comput. Appl. **118**(18), 33 (2015)
16. Scott, M., Bruce, R.: Five stages of growth in small business. Long Range Plan. **20**, 45–52 (1987)
17. Steinmetz, L.L.: Critical stages of small business growth: When they occur and how to survive them. Bus. Horiz. **12**(1), 29–36 (1969)
18. Thorleuchter, D., Van Den Poel, D.: Predicting e-commerce company success by mining the text of its publicly-accessible website. Expert Syst. Appl. **39**(17), 13026–13034 (2012)
19. West, D.: Neural network credit scoring models. Comput. Oper. Res. **27**(11), 1131–1152 (2000)
20. Wong, B.K., Lai, V.S., Lam, J.: A bibliography of neural network business applications research: 1994–1998. Comput. Oper. Res. **27**, 1045–1076 (2000)
21. Worthington, C., Britton, I.: The Business Environment. Financial Times, Harlow (2006)
22. Zhang, G., Hu, M.Y., Patuwo, B.E., Indro, D.C.: Artificial neural networks in bankruptcy prediction: General framework and cross-validation analysis. Eur. J. Oper. Res. **116**(1), 16–32 (1999)

Intelligent End User Development Platform Towards Enhanced Decision-Making

Bahareh Zarei[(⊠)] and Martin Gaedke[(⊠)]

Technische Universität Chemnitz, 09111 Chemnitz, Germany
{bahareh.zarei,martin.gaedke}@informatik.tu-chemnitz.de

Abstract. From a decision-making perspective, the web is an emerging information domain. It makes a large amount of data available for a large group of users in different domains. In recent years, the dramatic growth of data accessible through the web and the development of large-scale distributed web services have presented new challenges for users. Web services generate data in an ad hoc manner; hence, the systematic management of data has become an obstacle for efficient decision making. On the other hand, the emergence of IoT devices exaggerates the data production rate. Therefore, due to the overwhelming amount of data online, it is essential to support end users, who have limited knowledge of programming, in accessing the relevant information. During the last few decades, many approaches have been proposed to collect and process heterogeneous data from distributed sources in a more uniform way. However, existing solutions have failed to provide the flexibility required for data integration and management. The goal of this project is providing a systematic approach for end users to access and analyze exact data at the right time. The goal of this PhD project is to support knowledge workers in decision making with an end user development approach.

Keywords: End User Development · Decision-making · Data integration · Knowledge extraction

1 Introduction and Motivation

The ubiquity of the World Wide Web and the development of large-scale distributed web services promote a dramatic growth in the amount of accessible data through the web and introduces new challenges for end users. Due to the large amount of online data sources, end users are now – in principle – be able to access and integrate the relevant exact information. On the other hand, web services generate data in an ad hoc manner; hence, the systematic management of data has become an obstacle for efficient knowledge extraction and decision making.

For non-expert users, who do not have programming skills, information management and data integration is very complicated, and hinders them to make use of the full potential of the provided data and services for decision making. End User Development (EUD) is a research field with focus on techniques and methods to support the development needs of non-expert users. In the domain of Web Engineering EUD concentrates on empowering non-expert web users to create web applications, which

J. Cabot et al. (Eds.): ICWE 2017, LNCS 10360, pp. 608–615, 2017.
DOI: 10.1007/978-3-319-60131-1_49

satisfy their need [1]. Expertness in here refers to web technology knowledge in other words non-expert users can be domain or technical experts without web engineering knowledge.

The problem that we aim to tackle in this PhD research is the lack of suitable tool support to enable end users in systematically achieving their data-driven decision-making goals. Integrating and analyzing the huge amount of data from different sources to create new functionalities and values demand sophisticated skills, which non-expert users lack [2].

One of the promising approaches in EUD domain is Mashup technology [3]. Mashup tools have been introduced as a solution for achieving new functionalities by combining various data resources and services across the web [4]. The emergence of mashup technics empower non-expert end users to develop their own application [5]. However, even in semi-automatic mashup tools, users still need background knowledge regarding web technologies and programming. In the semi-automatic mashup approach, end users manually select data sources and services. This selection does not only require knowledge about data source locations, but also about the semantics and structures of data. Moreover, to fulfill the reliability and security requirements end users should be aware of all kinds of related data source aspects; otherwise, due to lack of sufficient information, decision quality might be affected negatively [6]. Another limitation of this approach is a lack of adequate flexibility for more dynamic scenarios where the data providers change their structure or behavior. Therefore, end users need a more flexible and systematic platform, which enables them to create their own tools using the big data produced by the Internet of Services (IoS) and Internet of Things (IoT).

A potential platform to remedy the situation should be able to integrate data from various sources and to provide users with a unified view of data. The platform should relieve users from navigating one source to another to access the data. The data is retrieved from different sources with different structures, therefore preprocessing is required. The next step is knowledge extraction from existing sources and visualizing the result.

A potential platform should fulfill the following requirements. These requirements are divided into two groups, namely framework requirement and End user requirement.

Framework requirements:

- Extracting, integrating and analyzing relevant data
- Flexible enough to handle data from various domains of knowledge to provide domain specific solutions
- Flexible enough to handle data from different sources such as social streams, open APIs and IoT streams
- Ability to work in heterogeneous, distributed and dynamic environment
- Ability to represent information in visual component

The following requirements are extracted from [7] as some general user interface design principles:

End user requirements:

- Easy to learn the workflow
- Using familiar terms
- Consistent user interface
- Minimizing the element of surprise for users

The proposed platform is a web-based application, mainly implemented by Python programming language and Flask framework[1]. For the visualization part, the d3.js[2] library was used. A Deep Learning algorithm provided by genism[3] python library was used as well to increase the intelligence and efficiency of the solution.

2 Research Issues, Objectives and Questions

In this section an overview of research issues and objectives will be given. The main objective of this project is "to support knowledge workers in decision making with a systematic end user development approach". Using the envisioned platform enables non-expert users to make informed decisions by spending less effort on data definition, integration and knowledge extraction. Using knowledge extraction and Artificial Intelligence techniques, increases decision precision in the overwhelming data environment.

Some of the issues, which need to be addressed in our solution are compatibility with different domains of knowledge and capability of analyzing the different forms of data to produce knowledge. Other issues we faced while solving the problem are reusability, sustainability and security issues.

This research aims to prove the following statistical hypothesis:

- H_1: The average user's ratings are higher for the envisioned platform compared to other existing mashup platforms.
- H_2: the average time required to access the exact information and make a decision is lower compared to the average time needed for the same task without using the platform.

3 Research Methodology and Research Design

In this section, we discuss the used methods and techniques in this research project. One of the used methodologies is Design Science. Based on the definition in [8], Design Science is a research paradigm in which human problems are addressed via innovative artifacts, thereby contributing new knowledge to the body of scientific evidence. In Design Science research the key focus is contribution of innovative and sound knowledge [9].

---

[1] http://flask.pocoo.org/.

[2] https://d3js.org/.

[3] https://github.com/RaRe-Technologies/gensim.

Another tool to guarantee a clear and organized project management procedure is Logical Framework Approach. This approach is one of the European Commission's suggestions for project management. LFA is an analytical process which aids us during planning and management phase and involves stakeholder and problem analysis, objective setting and strategy selection. This approach provides a structured way to organize the information so that the weaknesses can be identified and important questions can be asked [10].

Throughout the rest of this section we discuss the necessary steps to accomplish the project. The first step was performing a literature review to identify the existing solutions and requirements. For the next step, problems and objectives were identified and formulated in the form of problem and objective trees. The Logical Framework Matrix was developed to capture the main focus points of research. After identifying the stakeholders and their needs the platform requirements can be derived respectively. To prove the feasibility of the solution and perform the first evaluation, a prototype of the platform will be implemented. The prototype should fulfill the main functional and non-functional requirements. A user journey map is another tool that will be used to gain a better understanding of the interaction between users and the platform. This tool illustrates user's needs, requirements and expectations. Therefore, it can be used to gain useful information regarding the success of the solution.

To evaluate different platform's functionalities, we plan to conduct a series of validation phases (levels). For the first level of validation, prototyping will be used. Prototyping will help us to evaluate the platform in the terms of feasibility and functionality. In the next level after developing the first version, a series of user studies will be conducted by different groups of users. At the end of each series the platform will be improved and modified according to the feedbacks and newly found functionalities. The important criteria are usability, flexibility and efficiency. New criteria will be determined throughout the process. The evaluation attempts to answer the following questions:

- **Q1:** How usable is the platform from an end user's perspective with different levels of programming skills?
- **Q2:** How much time spent for decision making will the use of the platform reduce and how much effort is required to make a decision?
- **Q3:** How precise and intelligent are the decisions?

To evaluate the usability, users should rate the platform on a 1 to 5 Likert scale regarding the complexity and ease of use. Then, the average ratings will be used to assess the usability. In terms of efficiency, the required time and effort to find the information and make the decision will be considered. The effort can be expressed in the number of times the user switches from one data source to other.

4   Preliminary Key Results or Contributions

To address the mentioned problems, we propose a platform for End User Development. The IKEV (Intelligent Knowledge Extraction and Visualization) platform should fulfill the end user's requirements as well as resolving the drawbacks of previous works.

For the clarity's sake, we use a scenario in which the user can employ the IKEV platform to achieve her goals and later in this section the platform architecture will be introduced.

4.1 Use Case Scenario

A research group is trying to find the best fitting scholarship for funding their projects in the field of computer science. To find the best funding opportunity, researchers should go through all calls for proposals. This procedure requires time and effort to gather information about the funding agencies, compatibility to the group's field of research and knowledge regarding each member's interests and skills. The IKEV platform will assist researchers to create their own personalized profile based on their publications and projects. Required information for creating the profile will be extracted from agreed-on data sources such as, ResearchGate, Google Scholar etc. Each profile contains information regarding the member's research interests and other personalized information. By using this information and data regarding the call for proposals, the platform can choose the best fitting call. This scenario demonstrates the platform's ability to provide a simple, customized and intelligent solution which saves the user huge amount of time and effort. The involved actors of this scenario are academics who are experienced in terms of searching and using the internet. Due to their experience in academic research they possess higher skills in using web sources compared to normal users. This target group can be of any research discipline other than computer science or web engineering, therefore they might not have skills in using web technologies to create their own customized intelligent solution. Our solution is also applicable for normal users with limited "internet using skills", i.e. users who are able to conduct normal use of the web, but not programming like tasks. The critical difference between normal and experienced users in this scenario is their ability to identify and access the relevant web sources - but in the terms of web technology knowledge they both are at the same low level. The following Table 1 differentiates the different end users:

Table 1. End users definition

Platform user	Technical web knowledge	Internet using skills
Normal users	NO	LIMITED
Experienced users	NO	YES

4.2 IKEV Platform Architecture

The IKEV platform architecture is presented in Fig. 1. This architecture should incorporate the essential components to fulfill the mentioned requirements and support the end users in our scenario.

The main sources of information for the platform are web sources, which consist of the social stream such as crowd data, open APIs and IoT streams such the data provided by sensors. The end users use client devices like mobile and web clients to interact with the platform. The main blocks of the IKEV platform are *Authentication/Authorization Manager*, *Planning Assistant Module*, *Source/Service Editor* and *Information Store*.

- *Authentication/Authorization Manager* is responsible for granting access to the users based on their credential and information stored in *User's Profile* repository.
- If access is granted to the user, he can use the *User Dashboard* to make a new planning, edit existing one or manage the data sources.
- *Source/Service Editor*'s main functionality is identifying and providing access to relevant data sources and analyzing the data. The *Source Discovery* and *Data Analysis* components are respectively responsible for these functions. *Source Compatibility Checker* checks whether the data source is reasonable to be used for decision making. This component is important for the normal users with limited experience in using the web sources to help them identify the relevant information. Finally, the *Visualization Component* provides visual results by using the visualization template stored in *Visualization Repository*.

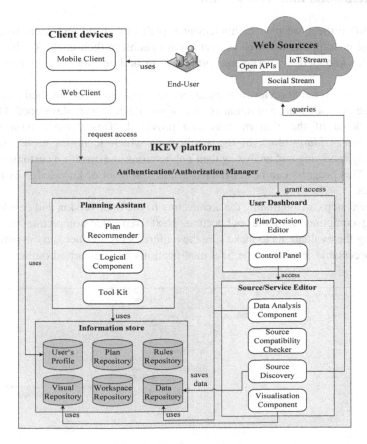

Fig. 1. Platform architecture

- *Planning Assistant* helps end users to make a sound decision based on the provided data. The more optimized the plan becomes, the higher the user's satisfaction that can be achieved. The *Logical Component* uses the *Rules Repository* to support end users in making logical decisions. Another component in this block is *Tool Kit* which enables end users to benefit from several project management tools such as various diagrams, problem and objective tree, user journey map etc. New plans (strategic decisions) can be suggested to the users based on the previous decisions and results from *Logical Component*.
- *Information Store* contains all the required repositories for the platform.

The main expected contributions of the project in the first place, is enhancing usability and efficiency of End User Development platforms for making better decisions in less time with less effort. Additionally, increasing the intelligence and also introducing User Journey-based solution are other contributions of this work.

5   Conclusions and Work Plan

In this PhD project we plan to implement a platform to aid end users with limited knowledge in computer and web programming to achieve their goals. We introduced a scenario for knowledge workers and presented an architecture for a platform to help them.

So far, the literature review and management phase has been finished and the LFA matrix, user journey map, and initial version of architecture were developed. Moreover, the first demo of the platform was also provided. This demo demonstrates the *Source/Service Editor* block of the architecture which identifies each researcher's specialty by accessing and analyzing their publications and creates a customized profile for each. This progress can be considered as the first level of knowledge extraction in the project.

The next steps are upgrading the architecture to the final version and developing the remaining components of the architecture. Next step after completing the demo is conducting the evaluations to make sure the requirements are met and the performance is at an acceptable level then the final modifications can be carried out accordingly.

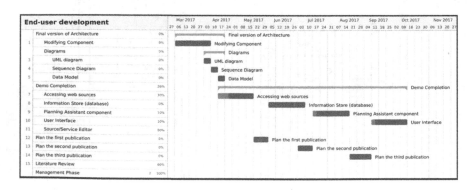

Fig. 2. Gantt chart

The next prospective paper is associated with the initial version of the demo and the achieved results after the evaluation. Since the blocks of the architecture are distinct and each has a different responsibility, after completing and enhancing each block a new paper can be proposed to show and discuss the result. To illustrate the detailed plan following Gantt chart is provided (Fig. 2):

Acknowledgement. This work has been supported by the ESF and the Free State of Saxony.

References

1. Rode, J., Rosson, M.M.B.: End user development of web applications. In: Lieberman, H., Paternò, F., Wulf, V. (eds.) End User Development. Human-Computer Interaction Series, vol. 9, pp. 161–182. Springer, Netherlands (2006)
2. Hummer, W., Leitner, P., Dustdar, S.: WS-aggregation: distributed aggregation of web services data. In: Proceedings of the 2011 ACM Symposium on Applied Computing (SAC 2011), pp. 1590–1597 (2011)
3. Daniel, F., Matera, M.: Mashups: concepts, models and architectures
4. Anjomshoaa, H.A., Tjoa, A.M.: Towards semantic mashup tools for big data analysis. In: Linawati, M.M.S., Neuhold, E.J., Tjoa, A.M., You, I. (eds.) ICT-EurAsia 2014. LNCS, vol. 8407, pp. 129–138. Springer, Heidelberg (2014). doi:10.1007/978-3-642-55032-4_13
5. Imran, M.: DISI - University of Trento An effective end-user development approach through domain-specific mashups for Research Impact Evaluation, March 2013
6. Fischer, T., Bakalov, F., Nauerz, A.: An overview of current approaches to mashup generation. In: Proceedings of the Fifth Conference Professional Knowledge Management: Experiences and Visions, pp. 254–259 (2009)
7. Bourque, P., Fairley, R.E.: Guide to the Software Engineering - Body of Knowledge (2014)
8. Hevner, A., Chatterjee, S.: Design Research in Information Systems: Theory and Practice. Integrated Series in Information Systems, vol. 22. Springer, US (2010). Google Books
9. Vaishnavi, V., Kuechler, B.: Design Science Research in Information Systems Overview of Design Science Research, p. 45. AIS (2004)
10. E. Commission, Supporting effective implementation of EC external assistance European commission, vol. 1, p. 149 (2004)

Author Index

Printed in the United States
By Bookmasters